Disease-Modifying Targets in Neurodegenerative Disorders

Disease-Modifying Targets in Neurodegenerative Disorders

Paving the Way for Disease-Modifying Therapies

Edited by

Veerle Baekelandt

Evy Lobbestael

ACADEMIC PRESS
An imprint of Elsevier

Academic Press is an imprint of Elsevier
125 London Wall, London EC2Y 5AS, United Kingdom
525 B Street, Suite 1800, San Diego, CA 92101-4495, United States
50 Hampshire Street, 5th Floor, Cambridge, MA 02139, United States
The Boulevard, Langford Lane, Kidlington, Oxford OX5 1GB, UK

Library of Congress Cataloging-in-Publication Data
A catalog record for this book is available from the Library of Congress

British Library Cataloguing-in-Publication Data
A catalogue record for this book is available from the British Library

ISBN: 978-0-12-805120-7

For information on all Academic Press publications visit our website at
https://www.elsevier.com/books-and-journals

 Working together
to grow libraries in
developing countries

www.elsevier.com • www.bookaid.org

Publisher: Mara Conner
Acquisition Editor: Natalie Farra
Editorial Project Manager: Kathy Padilla
Production Project Manager: Karen East and Kirsty Halterman
Designer: Mark Rogers

Typeset by Thomson Digital

Contents

List of Contributors...xi

Editor Biographies xiii

Foreword...xiv

Preface...xvi

CHAPTER 1   The Multitude of Therapeutic Targets in Neurodegenerative Proteinopathies...1

R. Melki

Protein Misfolding and Aggregation...1

Mechanism of Assembly..2

Prion-Like Propagation of Protein Assemblies..4

Propagation Routes of Pathogenic Protein Aggregates....................................5

Limiting Steps in the Propagation of Pathogenic Protein Assemblies.......................6

Therapeutic Strategies Targeting the Physiological Levels of Aggregation-Prone Proteins Involved in Neurodegeneration 7

Therapeutic Strategies Targeting the Misfolding and Aggregation of Proteins Involved in Neurodegeneration 7

Therapeutic Strategies Targeting the Accumulation of Misfolded Pathologic Protein Aggregates..8

Therapeutic Strategies Targeting the Cell-to-Cell Propagation of Pathogenic Protein Aggregates 9

Therapeutic Strategies Aimed at Restoring the Damage Pathogenic Protein Aggregates Induce 9

Therapeutic Strategies Targeting Misfolded Pathologic Protein Aggregate–Mediated Neuroinflammation...10

Limitations of the Different Therapeutic Strategies..11

References 12

CHAPTER 2   Synuclein Misfolding as a Therapeutic Target 21

W. Peelaerts, T.F. Outeiro

Introduction 21

The Conformational Landscape of α-Synuclein: the Native State 22

Cytosolic Folding States 24

Membrane-Bound Folding States 24

The Conformational Landscape of α-Synuclein: Misfolded Variants....................25

Folding Intermediates..25

Oligomeric Variants..27

Fibrillar Variants...29

　　　　Misfolding in Complex Environments..30
　　　　Synuclein Strains..32
　　　　Glycation of α-SYN—an Age-Associated Posttranslational Modification 32
　　Misfolding of α-Synuclein as a Therapeutic Target 33
　　　　Preserving the Native State..33
　　　　Preventing the Formation of Folding Intermediates.............................33
　　　　Targeting Aggregation 35
　　　　Stabilizing Membrane Interactions..38
　　Conclusions 38
　　References 39

CHAPTER 3   Neuroinflammation as a Therapeutic Target in Neurodegenerative Diseases 49
　　R. Gordon, T.M. Woodruff
　　Introduction 50
　　　　The Global Burden of Neurodegenerative Diseases and the
　　　　Challenge of Developing Effective Therapeutics.................................50
　　Glial Cells in CNS Development, Homeostasis, and Pathology..............51
　　　　Microglia...51
　　　　Astrocytes...53
　　　　Oligodendrocytes and Ependymal Cells..54
　　Chronic Neuroinflammation as a Common Pathophysiological
　　Mediator in Progressive Neurodegenerative Diseases 54
　　　　Mechanisms of Neuroinflammation-Mediated Neurodegeneration 56
　　　　Alzheimer's Disease...58
　　　　Parkinson's Disease 60
　　　　Amyotrophic Lateral Sclerosis...64
　　　　Huntington's Disease 65
　　　　Frontotemporal Dementia and Lewy Body Dementia...........................66
　　Therapeutic Strategies Targeting Neuroinflammation in Progressive
　　Neurodegenerative Diseases 66
　　Future Perspectives 69
　　Conclusions 70
　　References 70

CHAPTER 4A Stem Cells in Neurodegeneration: Mind the Gap 81
　　C. Claes, J. Terryn, C.M. Verfaillie
　　Introduction 81
　　Part I: Stem Cells..82
　　　　Genome Editing of Stem Cells 83
　　　　The Generation of Specific Cell Types 84

Part II: Stem Cells as Regenerative Therapy 85
   Clinical Translation and Immunology 88
   Keeping Track of Cell Transplants..90
Part III: Stem Cells to Model Neurodegenerative Diseases....................91
   iPSC Characteristics: The Role of Development in
   Neurodegenerative Diseases 91
   iPSC Technology: Disease Modeling and Drug Screening.............92
   Future Prospects 96
References 96

**CHAPTER 4B The Potential of Stem Cells in Tackling
Neurodegenerative Diseases 101**
S. Libbrecht
Endogenous Stem Cells as a Therapeutic Target...................................101
   What are Endogenous Stem Cells?...101
   Adult Neurogenesis in a Neurodegenerative Environment 104
   Adult Brain Regeneration to Tackle Neurodegeneration 108
References 110

**CHAPTER 5 Preclinical Models of Alzheimer's Disease for Identification and
Preclinical Validation of Therapeutic Targets: From Fine-Tuning
Strategies for Validated Targets to New Venues for Therapy.............115**
B. Vasconcelos, M. Bird, I.-C. Stancu, D. Terwel, I. Dewachter
Introduction 116
   Alzheimer's Disease: Clinical Features, Pathology, and Genetics
   Leading to the Formulation of the Amyloid Cascade Hypothesis 116
   Transgenic Models Recapitulating Amyloid Pathology as
   Preclinical Models to Identify Targets for Anti-Aβ Therapies 118
   Transgenic Models Recapitulating Tau-Pathology as Preclinical
   Models to Identify Targets for Anti-Tau-Directed Therapies 124
   Transgenic Mouse Models Recapitulating Prion-Like Spreading and
   Propagation of Tau-Pathology: New Venues for Therapeutic Strategies 128
   Transgenic Mouse Models Recapitulating Aβ-to-Tau Axis: New Venues to
   Identify Novel Therapeutic Targets Aiming at the Molecular and Cellular
   Mechanisms Linking Aβ and Tau 132
Conclusions 137
References 138

CHAPTER 6 Parkinson's Disease 157
M.R. Cookson
Introduction 157
Current Treatment Approaches in Parkinson's Disease..........................158

Pharmacological Approaches: Replacing Dopamine 158
Nonpharmacological Approaches: Deep Brain Stimulation and
Cell Replacement 160
Drugs That are Being Evaluated Clinically for Disease Modification in PD 160
Novel Targets for Disease-Modifying Therapies in PD 161
α-Synuclein 161
LRRK2 165
Recessive Genes 166
Conclusions 167
References 168

CHAPTER 7   Lewy Body Dementia **175**
M. Delenclos, S. Moussaud, P.J. McLean
Lewy Body Dementia 175
Concept of Lewy Body Dementia 175
Clinical Symptoms 177
Genetic Association 177
Pathophysiology 178
Management of LBD 182
Current Symptomatic Treatment 182
Disease-Modifying Therapy 183
Concluding Remarks 188
References 188

CHAPTER 8   Frontotemporal Dementia **199**
E. Wauters, K. Sleegers, M. Cruts, C. Van Broeckhoven
Overview of Frontotemporal Dementia 199
Clinical Presentations 200
Pathological Heterogeneity 201
Genetics of FTD 202
FTD: A Heterogeneous Disorder 205
Disease Management 206
Pharmacologic Symptomatic Treatment 206
Nonpharmacologic Management 209
Novel Possibilities in Disease-Modifying Drug Development 209
Tau 209
Granulin 214
C9orf72 216
TDP-43 219
Cellular Waste Clearance Systems 220
Heterogeneity = Opportunity 222

Discussion: Essentials for the Development of a Disease-Modifying Therapy 224
    Gaps in our Understanding of FTD 224
    Toward Early and Targeted Intervention................................226
Conclusions 227
References 227

CHAPTER 9 **From Huntingtin Gene to Huntington's Disease-Altering Strategies** **251**
N. Déglon
Huntington's Disease 251
Huntingtin Gene and Transcripts 253
Huntingtin Protein 257
HD Pathogenic Mechanisms 258
Molecular Strategies for HD 259
ASO 260
    RNA Interference 261
Genome Editing 262
Conclusions and Perspectives 263
References 264

CHAPTER 10 Amyotrophic Lateral Sclerosis: Mechanisms and Therapeutic Strategies **277**
L. Van Den Bosch
Introduction 277
Excitotoxicity 280
Hyperexcitability 281
Pathogenic Role of Non-neuronal Cells 281
    Astrocyte Dysfunction 281
    Oligodendrocyte Dysfunction 282
    Microglial Dysfunction and T Cells 282
Shortage of Neurotrophic Factors 283
Mitochondrial Dysfunction 284
Axonal Defects 284
Altered Proteostasis and Autophagy 285
Altered RNA Metabolism and Stress Granule Formation 286
Hexanucleotide Repeats in C9ORF72 and Disturbances in Nucleocytoplasmic Transport 286
Conclusions 287
References 288

Index 297

List of Contributors

Matthew Bird
Institute of Neuroscience, Catholic University of Louvain, Brussels, Belgium

Christel Claes
KU Leuven Stem Cell Institute; VIB Center for the Biology of Disease, Leuven, Belgium

Mark R. Cookson
National Institute on Aging, National Institutes of Health, Bethesda, MD, United States

Marc Cruts
VIB Center of Molecular Neurology; Institute Born–Bunge, University of Antwerp, Antwerp, Belgium

Nicole Déglon
Neuroscience Research Center (CRN); Lausanne University Hospital (CHUV), Lausanne, Switzerland

Marion Delenclos
Jacksonville, FL, United States

Ilse Dewachter
Research Group Physiology, University of Hasselt, Belgium

Richard Gordon
The University of Queensland, St. Lucia, QLD, Australia

Sarah Libbrecht
KU Leuven, Leuven, Belgium

Pamela J. McLean
Jacksonville, FL, United States

Ronald Melki
Paris–Saclay Institute of Neuroscience, French National Center for Scientific Research, University of Paris–Saclay, Gif-sur-Yvette, France

Simon Moussaud
Jacksonville, FL, United States

Tiago F. Outeiro
Institute of Molecular Medicine, University of Lisbon, Lisboa, Portugal; Center for Nanoscale Microscopy and Molecular Physiology of the Brain, University Medical Center Goettingen, Goettingen, Germany

Wouter Peelaerts
KU Leuven, Leuven, Belgium

Kristel Sleegers
VIB Center of Molecular Neurology; Institute Born–Bunge, University of Antwerp, Antwerp, Belgium

Ilie-Cosmin Stancu
Institute of Neuroscience, Catholic University of Louvain, Brussels, Belgium

Joke Terryn
KU Leuven Stem Cell Institute; VIB Vesalius Research Center; University Hospitals Leuven, Leuven, Belgium

Dick Terwel
reMYND NV, Leuven, Belgium

Christine Van Broeckhoven
VIB Center of Molecular Neurology; Institute Born–Bunge, University of Antwerp, Antwerp, Belgium

Ludo Van Den Bosch
Leuven Research Institute for Neuroscience and Disease (LIND), KU Leuven; VIB, Center for Brain & Disease Research, Leuven, Belgium

Bruno Vasconcelos
Institute of Neuroscience, Catholic University of Louvain, Brussels, Belgium

Catherine M. Verfaillie
KU Leuven Stem Cell Institute, Leuven, Belgium

Eline Wauters
VIB Center of Molecular Neurology; Institute Born–Bunge, University of Antwerp, Antwerp, Belgium

Trent M. Woodruff
The University of Queensland, St. Lucia, QLD, Australia

Editor Biographies

Veerle Baekelandt, PhD, holds a Master of Romance Languages, a Master of Biology, and a PhD in Biology degrees at the Katholieke Universiteit Leuven. In 1992 she received a Frank Boas Fulbright scholarship for graduate study at Harvard University and became a research fellow in the Laboratory for Neuroscience Research headed by Dr. Larry Benowitz, Children's Hospital, Harvard Medical School, Boston (1992–1993). In 1999 she joined as a postdoctoral fellow in a new gene therapy project for neurodegenerative diseases, which was the start of her own research group. In 2003 she was appointed as assistant professor at the KU Leuven and in 2007 as full-time research professor (BOF-ZAP). She is now head of the Laboratory for Neurobiology and Gene Therapy. Her research focuses on disease modeling and therapy for Parkinson's disease using viral vectors in cell culture and in vivo. The underlying rationale is that the generation of more relevant models in cells and in rodent brain will lead to a better insight into the molecular pathogenesis of PD and to the development of new therapeutic strategies and drugs.

Evy Lobbestael, PhD, obtained her Master's degree in biomedical sciences in the faculty of medicine at the KU Leuven (Belgium) in 2007. She received a PhD scholarship from the Research Foundation Flanders and did her PhD training in the laboratory for neurobiology and gene therapy at the KU Leuven under the supervision of Professor Baekelandt (2013). Her doctoral research focused on the function and dysfunction of the Parkinson's disease linked gene *LRRK2*. Besides the development of multiple tools to study *LRRK2*, which are currently used worldwide, she identified protein phosphatase 1 as a physiological regulator of cellular LRRK2 phosphorylation. Currently, she heads the LRRK2 group in the lab of Professor Baekelandt with a focus on LRRK2 signaling.

Foreword

The increasing prevalence of neurodegenerative disorders worldwide is closely linked with the demographical changes related to aging populations and represents a major socioeconomic challenge. Currently untreatable beyond late-stage symptomatic therapy, the invariably progressive neurodegenerative disorders are presently affecting 45 million people worldwide and expected to reach 76 million in 2030. This relevant burden with increasing morbidity and mortality for the respective healthcare systems provides an increasing demand for novel treatment strategies including disease-modifying approaches to slow down or even halt disease progression in neurodegenerative disorders as a major unmet medical need.

Recent progress in genetics and molecular biology has greatly enhanced our understanding of many neurodegenerative disorders. The spectrum of the hereditary component ranges from clear monogenic forms of neurodegenerative diseases (e.g., Huntington's disease; HD) to highly heterogeneous disorders, such as Alzheimer's disease (AD) or Parkinson's disease (PD) with the majority of sporadic cases without evidence for heritability of the disease trait. Despite this remarkable progress, translation into clinical application is still rare. It is therefore necessary to strengthen translational research into the molecular basis of neuronal dysfunction to identify targets for causative treatment options and disease modification.

Interestingly even for highly complex and heterogenous neurodegenerative diseases, such as Parkinson's disease, where monogenic forms contribute to only 5%–10% of all cases, the genetics still seem to be an important entry point for our understanding the underlying mechanisms. Thus from more than 30 identified genes or genetic risk factors a number of different molecular pathways were identified, that may serve as prototypes and help to stratify subgroups within the heterogenous group of sporadic PD patients. For instance, rare mutations in the alpha-synuclein gene that cause an earlier onset and more rapidly progressive form of PD related to an increased aggregation-forming capacity lead to the identification of pathological alpha-synuclein aggregation in the brains of all PD patients. This supported findings in other neurodegenerative diseases, such as HD, AD, or Prion disease, that accumulation and aggregation of misfolded proteins in affected brain regions is a major hallmark of the neurodegenerative process. Based on findings in prion disease, it even was postulated that small proteinaceous particles could be the infectious agents responsible for the transmission of not only spongiform encephalopathies, but also more common neurodegenerative diseases, such as AD and PD, that due to "prion-like" behavior of tau- or alpha-synuclein contribute to the formation and pathological spreading of these misfolded protein aggregates. This opens new avenues for understanding the fundamental mechanisms governing seeding, spreading, and aggregation of pathological proteins in different neurodegenerative disorders and therefore to identify viable targets and assess their druggability for future therapeutic interventions, for example, via immunotherapy directed against aggregating proteins.

Previous efforts to develop novel treatments were hampered by the lack of appropriate in vitro and in vivo models and the increasingly recognized pathomechanistic heterogeneity of the respective neurodegenerative disorders and therefore failed to develop strategies to intervene with the chronic progressive neurodegenerative process. A major methodological advancement came from the application of new technologies to reprogram fibroblasts into induced pluripotent stem cells (iPSC). Disease-specific neuronal populations can be differentiated from iPSC ex vivo, giving insight into the affected cell-type in monogenic forms of PD possible for the first time. Recently first proof-of-principle evidence was provided, that neurons with the genome of a sporadic AD or sporadic PD patient may share phenotypes also seen in iPSC-derived neurons from patients with monogenic forms of these diseases. This strongly supported the concept that sporadic and familial PD share multiple genetic risks of variable effect strengths and that monogenic forms with defined disease-causing mutations can serve as prototypes for the underlying mechanisms. Here different pathways were identified, implicating mitochondrial dysfunction, impaired calcium homeostasis, pathological protein aggregation, endosomal–lysosomal dysfunction, or neuroimmunological dysregulation leading to neurodegeneration.

These emerging concepts of the complex genetic and molecular architecture of neurodegenerative diseases are represented in the contributions by the different experts to this book and underscore the imperative for more individualized therapeutic strategies according to different pathophysiological subtypes. Therefore, current strategies in drug target discovery focus on the implementation of patient-based cellular models as "closer-to-human-disease" models to identify molecules able to revert disease-associated cellular phenotypes. Indeed, first proof-of-concept studies in iPSC-based models for familial dysautonomia (FD), a rare monogenic neurodegenerative disorder, showed that cell-autonomous disease phenotypes in neurons derived from iPSC can be reverted by pharmacological intervention ex vivo and were translated into clinical trials. Based on the above-mentioned pathomechanistic overlaps between different neurodegenerative disorders, for example, aggregation-prone proteins in AD, PD, and HD, it can be speculated that identification of novel targets and disease modifiers may be effective not only in one disease.

Thus, the time is ripe for *Disease-Modifying Targets in Neurodegenerative Disorders*, in which experts gathered by the editor summarize the state-of-the-art of disease models and strategies to define novel targets that provide a successful synthesis of fundamental and translational research on future neuroprotective treatments for these still incurable diseases.

Prof. Dr. Rejko Krüger
Head of the Clinical & Experimental Neuroscience Group
University of Luxembourg
Neurologist at Centre Hospitalier du Luxembourg

Preface

Neurodegenerative disorders such as Alzheimer's disease and Parkinson's disease represent a heavy personal, societal, and economic burden worldwide (current estimates indicate Alzheimer's disease as the third leading cause of death in the United states, just behind heart disease and cancer). Given the high prevalence in elderly people and the increasing life expectancy in developed countries, the impact of neurodegenerative diseases will only further increase in the coming years.

For most of these diseases, current therapy is at best restricted to relieving symptoms. It does not interfere with the disease progress; hence, the pathological processes cannot be reversed, stopped, or slowed down. This shortcoming has stimulated both academia and industry to unravel the molecular events underlying the pathology, in the hope to be able to interfere with these processes and ultimately cure patients.

Besides providing a concise summary of the current disease management for the most common neurodegenerative disorders, this book intends to give an overview of our current knowledge on underlying disease mechanisms and how this insight can lead to new therapeutic strategies, which will hopefully be translated in a disease-modifying therapy.

The first four chapters are dedicated to therapeutic strategies that might be beneficial for several neurodegenerative disorders. Each of the subsequent six chapters is devoted to one specific disorder, written and peer reviewed by experts in the respective fields.

With this book, we want to offer a glance at future disease management and we hope that this book will be appreciated by (neuro-)scientists, medical doctors, caregivers as well as patients and their relatives.

Finally, we would like to express our sincere appreciation to the many authors and reviewers that made this book a state-of-the-art piece of work, as well as to the efforts of the Elsevier staff.

Veerle Baekelandt
Evy Lobbestael

THE MULTITUDE OF THERAPEUTIC TARGETS IN NEURODEGENERATIVE PROTEINOPATHIES

Ronald Melki

Paris–Saclay Institute of Neuroscience, French National Center for Scientific Research (CNRS), University of Paris–Saclay, Gif-sur-Yvette, France

CHAPTER OUTLINE

Protein Misfolding and Aggregation 1
Mechanism of Assembly... 2
Prion-Like Propagation of Protein Assemblies 4
Propagation Routes of Pathogenic Protein Aggregates 5
Limiting Steps in the Propagation of Pathogenic Protein Assemblies.. 6
Therapeutic Strategies Targeting the Physiological Levels of Aggregation-Prone
Proteins Involved in Neurodegeneration 7
Therapeutic Strategies Targeting the Misfolding and Aggregation of Proteins
Involved in Neurodegeneration 7
Therapeutic Strategies Targeting the Accumulation of Misfolded Pathologic Protein Aggregates 8
Therapeutic Strategies Targeting the Cell-to-Cell Propagation of Pathogenic Protein Aggregates 9
Therapeutic Strategies Aimed at Restoring the Damage Pathogenic Protein Aggregates Induce...................... 9
Therapeutic Strategies Targeting Misfolded Pathologic Protein
Aggregate–Mediated Neuroinflammation 10
Limitations of the Different Therapeutic Strategies... 11
Acknowledgments.. 12
References 12

PROTEIN MISFOLDING AND AGGREGATION

Within our cells, proteins are folded to be functional. Following or concomitant to synthesis, with or without the assistance of molecular chaperones, newly synthesized proteins acquire native folds [1]. During this process, polypeptides populate billions of folding intermediates, a fraction of which cannot convert into functional proteins [2,3]. In addition, the folded state of functional proteins is not static. Indeed, proteins oscillate between different shapes depending on, for example, their enzymatic activity or interaction with partner proteins and populate nonnative conformations that are trapped thermodynamically in a misfolded state [4]. Some misfolded proteins or their degradation products have the ability to interact and coalesce into intra- or extracellular protein clumps, a fraction of which resembles orderly elongated aggregates that are named amyloid fibrils [5]. These fibrillar protein clumps are the common hallmark of Parkinson's disease (PD) and other synucleinopathies (which are delineated in Chapters 2, 6, and 7), Alzheimer's disease (which is introduced in Chapter 5), amyotrophic lateral sclerosis (Chapter 10), Creutzfeldt–Jakob, Huntington's disease (Chapter 9), and other dementias [6,7] (Chapter 8).

The diseases caused by misfolded proteins aggregates are either the consequence of the loss of function of proteins following trapping into irreversible aggregates or of a gain of pathologic function of the macromolecular assembly made of misfolded proteins [8–10]. Misfolded proteins aggregates can indeed: (1) permeabilize the plasma membrane and/or intracellular membranous compartments, (2) perturb membrane protein dynamics and distribution, (3) constitute novel pathogenic signaling platforms, and (4) trap significant amounts of molecular chaperones and other partner proteins. In addition, as misfolded protein aggregates recruit the soluble form of their constituting proteins, they grow in size and amplify when they break into pieces.

MECHANISM OF ASSEMBLY

Under physiological pH, salt, and temperature conditions, a variety of proteins, or domains within proteins, whose intrinsic structures are ill defined (i.e, they adopt many conformations in equilibrium) self-associate into assemblies that can elongate by incorporation of additional monomers. Among these proteins are the prion protein (PrP); the amyloid β (Aβ) peptide; the proteins Tau and α-synuclein (α-syn), whose aggregation process is detailed in Chapter 2; huntingtin exon 1 (HTTExon1) with a pathologic polyQ stretch; mutant superoxide dismutase (SOD1); and TAR DNA-binding protein 43 (TDP43). These assemblies grow in an unlimited manner, as the incorporation of a monomer creates a new binding site for another monomer [11]. The early assemblies, considered as the precursors of fibrils, are unstable, as they result from the transient longitudinal or lateral interactions between protein molecules in the same conformation. When longitudinal and lateral interactions between polypeptides in the same conformation are established, the intermolecular interactions outweigh the entropic energy in solution under physiological conditions, yielding nearly irreversible seeds unless a destabilizing irreversible reaction, such as nucleotide hydrolysis, occurs within the assemblies [11]. The latter can grow indefinitely by the incorporation of protein molecules at their ends, as the binding of an additional nonnative monomer to the seed ends generates an incorporation site for another subunit. The rate limiting step in a nonnative polypeptide aggregation is, therefore, the formation of stable seeds. Brownian movement [12] and severing and/or disaggregating factors [13–15] generate increased numbers of ends and increase the likelihood of new subunits being incorporated. As the numerous folding intermediates are in equilibrium, a given conformer recruitment by seed or fibril ends displaces the equilibrium toward repopulating this precise conformer. This leads ultimately to a near complete aggregation of the protein. This cooperative aggregation process is schematized in Fig. 1.1.

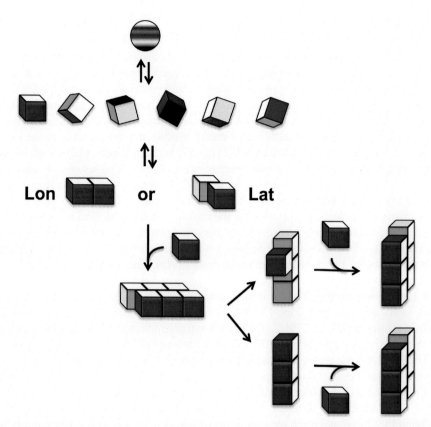

FIGURE 1.1 Model for Protein Misfolding and Aggregation into Fibrillar Assemblies that can Grow Indefinitely

Native or natively unfolded polypeptides *(sphere)* undergo conformational changes that lead to an abnormal assembly-competent conformation *(cube)*. The latter form is in equilibrium with the native or natively unfolded polypeptide. The different surfaces of the abnormal, assembly-competent form are represented (in six different colors). The abnormal form has the ability to interact transiently either longitudinally *(Lon)* or laterally *(Lat)* with another molecule in the same conformation. Here the longitudinal interaction occurs between the surfaces colored *yellow* and *blue* and the lateral interactions occur between two surfaces colored *green*. As long as the interactions are only longitudinal or lateral, the oligomers remain unstable and dissociate because the intermolecular interactions do not outweigh the entropic cost of binding. Once longitudinal and lateral interactions have been established between abnormal assembly-competent protein molecules, stable seeds are formed. Seed growth is unlimited by the incorporation of molecules in the same conformation, yielding fibrils. The fibrils can break into smaller fragments because, among other things, of Brownian movement. The rate of breakage depends on the number of bonds established between the molecules. Each resulting fragment acts as a seed. The fibrils represented here have distinct surfaces: *red*, *white*, and *black*. They interact through these defined surfaces with protein partners or lipids.

PRION-LIKE PROPAGATION OF PROTEIN ASSEMBLIES

Until recently, the spread and transmission of disease via misfolded protein aggregates was thought to be restricted to PrP [16]. A wide range of recent experimental evidence suggest that other protein aggregates that are the hallmarks of major neurodegenerative diseases spread in a prion-like manner [7,17–19]. The first evidence for the transmission of misfolded protein aggregates not involving PrP in man came from the observation that the injection of brain homogenates containing amyloid plaques, whose main constituent is the aggregated form of the Aβ peptide, into the brains of model mice induced lesions characteristic of Alzheimer's disease [20,21], while no such lesions where observed when control brain extracts lacking protein aggregates were injected. The induced lesions, initially confined to the injected brain region, propagated to neighboring and/or axonally connected areas over time. The intraperitoneal injection of brain homogenates containing amyloid plaques also induced cerebral lesions characteristic of Aβ deposits [22]. Finally, the insertion of stainless steel needles preincubated in brain homogenates containing aggregated Aβ also triggered plaque formation in recipient animals [23] in a manner similar to what was observed decades before when stainless steel needles contaminated by the infectious form of the PrP were inserted in the brain of recipient animals [24].

The first evidence supporting the propagation of aggregated α-syn from cell to cell came from the finding that Lewy bodies from PD brains invade grafts of fetal dopaminergic neurons transplanted into these brains, years after transplantation [25,26]. Intercellular transfer of α-syn has subsequently been demonstrated in cultured cells [27–30]. Aggregated α-syn assemblies were shown to be taken up by neurons and transported anterogradely and retrogradely in vitro and in vivo [31–33]. Injections of recombinant α-syn and brain homogenates of transgenic mice or patients in transgenic and wild type mice, rats, and nonhuman primates clearly showed spreading and acceleration of α-syn pathology, with significant differences in the nature of the pathologic α-syn species, the time frame, and robustness of the seeding/amplification reported between different studies [30,34–39]. α-Syn assemblies amplify when injected within the brain [35,37,39] and reach the central nervous system (CNS) where they amplify by seeding the aggregation of soluble α-syn when injected in the muscle [38] or the blood stream [40].

Recent in vitro and in vivo studies have shown that Tau aggregates seed–soluble Tau aggregation within neurons [41–44]. Tau assemblies spread to neighboring cells and functionally connected brain regions [41,45]. Tau transsynaptic transfer has been shown, for instance, to allow the spread of neurofibrillary degeneration in a hierarchical pattern along existing neural networks [46–48]. Tau fibrils made in vitro were shown to trigger Tau aggregation in cultured cells [49] and in Tau transgenic mice [50], and the intracerebral injections of brain extracts from various human tauopathies have shown that Tau lesions in mice can be induced to resemble those in the corresponding human diseases [51].

Similarly, huntingtin with expanded polyQ stretches aggregates have been shown to transfer to neighboring cells and seed misfolding of the wild type protein in vitro [52–54] and in vivo [55–57], TDP43 pathologic aggregates from patient brains revealed capable seeding the aggregation of soluble TDP43 in vitro [58]. Finally, mutant SOD1 aggregation was demonstrated to be self-perpetuating, and the aggregate shown to propagate from cell to cell [59].

These and other data support the idea that neurodegenerative diseases associated with protein misfolding and aggregation could propagate in a prion-like manner. This process is schematized in Fig. 1.2. Furthermore, considering that these diseases follow anatomical pathways for their propagation in the brain, it is tempting to speculate on the existence of common spreading mechanisms of different proteinaceous aggregates that might contribute to the progression of neurodegeneration.

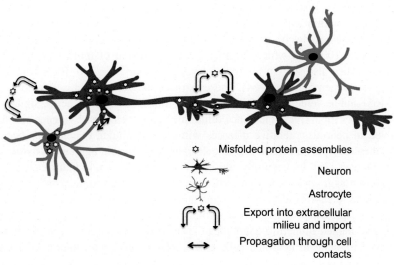

FIGURE 1.2 Schematic Representation of the Intercellular Propagation of Protein Assemblies

Misfolded pathologic protein assemblies *(stars)* can be released from neurons *(colored red)* after anterograde or retrograde active transport or from astrocytes *(colored blue)* in the medium and taken up by naïve neurons or astrocytes. They can also move from neuron to neuron or neuron to astrocytes, and vice versa, through cell-to-cell contacts. While propagating, misfolded pathologic protein assemblies grow and multiply by recruitment of misfolded protein monomers.

PROPAGATION ROUTES OF PATHOGENIC PROTEIN AGGREGATES

For extracellular assemblies of insoluble nonnative proteins to come in contact with the cytoplasm, where they can amplify, they must cross a lipid bilayer. Biological membranes are selectively permeable to ions and small solutes, such as sugars and amino acids, but are generally impermeable to large polar macromolecules. The only known nonvesicular mechanisms for protein translocation across electrically tight bilayers are either ABC transporters [60] or translocons [61] that only can accommodate completely unfolded monomeric polypeptides. Given the high-thermodynamic stability of the protein assemblies involved in disease, the possibility that aggregates disassemble outside the cell, cross the membrane as monomers or oligomers, and reassemble inside the cell is excluded. Thus, one needs to consider that pathologic protein assemblies enter the cells either by nonconventional cellular mechanisms, such as cell membrane puncturing or hijacking of existing cellular processes.

   The observation of "unroofed" cells that had been exposed briefly to fibrillar polyQ assemblies by deep-etch transmission electron microscopy revealed the existence of fibrils, neither surrounded by endomembranous structures nor by clathrin, that were attached to the inner surface of the plasma membrane [52]. This strongly suggests that amyloid fibrils can reach the cytosol after the physical breaching of the plasma membrane. This, together with the passive release of cytosolic aggregates upon cell death within the extracellular milieu, may contribute to the cell-to-cell propagation of protein aggregates associated with neurodegenerative diseases [17].

Proteins and larger macromolecular assemblies, including viruses, enter and leave eukaryotic cells through encapsulation within and fusion of vesicles, for example via endocytosis and phagocytosis. Cells secrete membrane vesicles in the extracellular space [62–64]. These extracellular vesicles comprise exosomes (~30–100 nm), which originate from the fusion of multivesicular bodies with the plasma membrane. They also secrete microvesicles, or ectosomes (~100–1000 nm), which form directly from the plasma membrane [62–64]. These extracellular vesicles act as a interneuronal communication route [65] and as vehicles for the intercellular transfer of nucleic acids, signaling molecules, and pathogenic factors [66,67] and clearance/dilution of these vehicles. Evidence for the presence of proteins involved in neurodegenerative diseases within extracellular vesicles has been documented [68–70]. Thus, they may play a role in cell-to-cell transfer of protein aggregates associated with neurodegenerative diseases [71]. Interestingly and in agreement with this, in many cases, the propagation of pathologic protein aggregates from one cell to another does not require contact between cells but includes extracellular release by affected cells and take-up by naïve cells [30,31,59].

Evidence for the presence of pathologic protein aggregates within dynamic filamentous actin-containing bridges, named tunneling nanotubes (TNTs) [72] that directly connect the cytoplasm of distant cells, from immune to neuronal cells exists. These long-range intercellular communication routes selectively transfer a wide variety of cellular materials, such as cytoplasmic macromolecules, plasma membrane components, vesicles, and organelles [73]. TNTs were shown to allow the passage of infectious prions particles and that of HTTExon1, α-syn and Tau aggregates from affected donor to naïve recipient cells [74–77], and may contribute to the spread of pathogenic protein assemblies involved in neurodegenerative diseases. Other yet to be described cellular processes may also contribute to the uptake and propagation of protein aggregates involved in neurodegenerative diseases.

LIMITING STEPS IN THE PROPAGATION OF PATHOGENIC PROTEIN ASSEMBLIES

The binding and docking of naked or membrane-encapsulated misfolded pathogenic protein assemblies to the plasma membrane are not only rate limiting steps in their prion-like propagation process but also key steps that define their tropism for specific cell types within the CNS. Their penetration into neurons controls the amplification of aggregates by recruitment of the endogenous soluble form of their constituting proteins and disease progression.

A number of proteins that are prone to aggregate into pathogenic assemblies have been shown to sense membrane curvature [78] and alter the distribution of lipid molecules within the membrane to the extent where it is remodeled and/or disrupted [79]. Membrane-bound naked misfolded pathogenic protein assemblies impinge on the physiologic distribution of membrane proteins and lipids fluidity, thus inducing the reorganization of membrane components and properties. Upon encountering/binding the plasma membrane of neurons, astrocytes, oligodendrocytes, or glial cells, the diffusion of naked misfolded pathogenic protein assemblies is "restricted" to a two-dimensional space. Bound pathogenic protein assemblies diffuse in the plane of the plasma membrane. Their lateral diffusion favors the formation of clusters that affects the overall distribution of extracellularly exposed membranous proteins through a selective recruitment process. This may not only lead to membrane lipid bilayer curving and/or rupture following molecular crowding, but also to the formation of platforms with undesirable properties, including calcium signaling [80,81]. Indeed, the increased confinement of pathogenic aggregates and membrane proteins and their lateral diffusion favor molecular interactions [82]. The recruitment of specific phospholipids and neuronal membrane proteins, such as receptors, channels, adhesion molecules, etc. by bound misfolded pathogenic proteins aggregates affects the lipid bilayer geometry and the normal dynamics, distribution, and function of the recruited molecules. Such redistribution will facilitate the passage of bound pathogenic protein assemblies into the cytosol where they can amplify.

Naked misfolded pathogenic protein assemblies take-up through endocytic events and amplification by recruitment of the soluble form of their constituting proteins are limited by their ability to bind partner proteins at the cell surface and the rates of (1) endocytosis and (2) disruption and/or crossing of the endocytic vesicle membranes.

Misfolded pathogenic protein assemblies may traffic between donor and naïve cells within extracellular membrane vesicles or through TNTs. In the latter cases, the propagation rate depends on the physiology of the cells that regulates membrane fusion events and the physical constraints imposed by TNTs, in particular those pertaining to the size of the protein assemblies.

THERAPEUTIC STRATEGIES TARGETING THE PHYSIOLOGICAL LEVELS OF AGGREGATION-PRONE PROTEINS INVOLVED IN NEURODEGENERATION

The finding that elevated expression levels of a number of proteins whose misfolding is associated to early onset forms of neurodegenerative diseases [83–88] suggests that the expression of aggregation-prone proteins involved in neurodegeneration may represent a therapeutic target. The proof of concept of strategies aimed at interfering with RNA transcription, processing, or translation was demonstrated in several diseases. In the case of the autosomal dominant Huntington's disease, the silencing of the mutant allele that causes disease by antisense RNAs and/or short-hairpin complementary DNA of microRNAs expressed in adeno-associated virus decreased mutant huntingtin synthesis [89,90]. In familial forms of PD due to *SNCA* triplication and overexpression, gene silencing therapies based on small interfering RNA or short-hairpin RNA were also demonstrated to have disease-modifying effects [91–93]. Further details and limitations are given in Chapter 2. In the case of amyotrophic lateral sclerosis, completed phase I clinical human trials demonstrated the feasibility of neuroprotective strategies based on the use of antisense oligonucleotides [94]. For tauopathies, the feasibility of strategies based on the use of oligonucleotides that mask the 5′-splice sites of exons 1 and 5 were demonstrated [95]. A reduction of *MAPT* mRNA and Tau level was observed following an oligonucleotide-mediated frame shift leading to a downstream premature stop codon, resulting in nonsense-mediated decay. Similarly, when the activities of enzymes involved in APP processing and cleavage were affected, Aβ concentration decreased [96,97]. It is worth noting that the practical success of these therapies in humans has been limited thus far.

THERAPEUTIC STRATEGIES TARGETING THE MISFOLDING AND AGGREGATION OF PROTEINS INVOLVED IN NEURODEGENERATION

Stabilizing proteins in an aggregation-incompetent state and/or activating the clearance of misfolded proteins within cells represent therapeutic strategies targeting disease-associated protein assemblies. Indeed, as stated earlier, given that the numerous folding intermediates of aggregation-prone proteins are in equilibrium, the binding of ligands to a conformer can displace the equilibrium between the different conformers toward this precise conformation that may be assembly incompetent either because they are unable to establish interactions with like conformers and/or are unsuited for recruitment by preformed seeds. Small molecules of synthetic or natural origin, such as the polyphenols rottlerin, curcumin, and epigallocatechin-3-gallate, have been shown to interfere with the aggregation of disease-associated polypeptides [98–102]. The mechanisms through which polyphenol and other molecules, identified through compounds screening, interfere with α-syn aggregation are detailed in Chapter 2. An alternative strategy consists of utilizing molecular chaperones to inhibit the misfolding of aggregation-prone proteins. Indeed, in our cells, misfolded polypeptides are given additional chances to reach their native, functional state by a variety of molecular chaperones that possess either unfolding activities, named unfoldases, or that sequester misfolded proteins, are called holdases. Some of these proteins act as triage platforms, directing misfolded proteins to degradation by the ubiquitin proteasome system. Molecular chaperones form functional networks that can not only prevent the aggregation of one specific client protein and facilitate the aggregation of other polypeptides, but can also do the opposite. Therefore, researchers have been seeking for molecules that restore global proteostasis. A novel class of small molecules has been identified that restores proteome balance and repairs proteostasis networks specifically through HSF1, FOXO, and Nrf2 [103]. These molecules have a promising therapeutic potential in a variety of protein conformational diseases, as they enhance cell stress pathways and chaperone activity [104]. They, therefore, protect the cells in a generic manner against misfolding and/or aggregation and the resulting pleiotropic damage. Alternatively, the expression or delivery of peptides derived from molecular chaperones could disfavor misfolding and the aggregation of proteins into pathogenic assemblies [105–107].

THERAPEUTIC STRATEGIES TARGETING THE ACCUMULATION OF MISFOLDED PATHOLOGIC PROTEIN AGGREGATES

The accumulation of misfolded protein aggregates in the intercellular milieu correlates most often with disease severity. Targeting and clearing these aggregates may be an effective way to reduce cytotoxicity. Clearing protein aggregates while they form in affected cells may, therefore, not only be an effective way to reduce their toxicity, but also a mean to abolish their ability to propagate to naïve cells. Depending on the nature of the protein and the aggregation stage, clearance is managed by cellular processes, such as autophagy and the proteasome or cytosolic proteases, such as calpains [108–112]. Solid evidence for molecular chaperone–mediated disaggregation of toxic fibrils have been very recently highlighted [13]. Thus, the pharmacological stimulation of misfolded protein seeds, disaggregating activity of specific molecular chaperones, and/or restoration of normal clearance should have an impact on the cytotoxicity associated with their accumulation within the cells and their cell-to-cell propagation. Additional and mechanistic details illustrating the case of α-syn are given in Chapter 2.

Clearing protein aggregates that have been released in the extracellular milieu is also an effective mechanism to reduce their pathogenic potential. Plasmin is an extracellular serine protease derived from its inactive form, plasminogen. Plasmin not only cleaves a variety of extracellular substrates, such as fibrin, fibronectin, laminin, and matrix metalloproteinases [113], but also aggregated Aβ [114] and α-syn [115]. Neurosin (kallikrein 6) is another protease that possesses extracellular protein aggregate–clearing activity [116,118]. Thus, strategies aimed at using this protease to clear extracellular toxic aggregates may have therapeutic potential. In the case of Aβ aggregates, three strategies based on reducing the amount of aggregates have progressed to phase II clinical trials [118].

THERAPEUTIC STRATEGIES TARGETING THE CELL-TO-CELL PROPAGATION OF PATHOGENIC PROTEIN AGGREGATES

The propagation of pathogenic protein aggregates from affected cells to naïve neurons contributes to the amplification of aggregates and, as a consequence, to disease progression. Blocking the propagation of protein assemblies involved in neurodegeneration and their subsequent amplification represent, therefore, a possible therapeutic avenue. The strategies that need to be implemented depend on whether the propagating misfolded protein assemblies are naked or comprised within membrane structures (e.g., within extracellular vesicles or TNTs), as the mechanism of propagation may include both aggregates that are naked or under the protection of membranes.

Several strategies may be considered for naked pathogenic protein aggregates. The first consists of identifying the receptors/protein partners of pathogenic protein assemblies so that they are made unavailable for protein aggregate anchoring to the cell membrane. A second strategy may consist of affecting yet unidentified uptake routes. In both cases, pleiotropic effects are expected. A third strategy consists of blocking the pathogenic protein aggregates outside naïve cells following their release from donor cells. This can be achieved by changing their surface properties in such a way that they are no more capable of binding to and being internalized by neuronal cells. This strategy has the advantage of targeting the toxic protein aggregates specifically. One possible approach is to block pathogenic protein aggregates outside the cells using specific antibodies [43,119] or following immunization [120–125]. The permeability of the blood–brain barrier to antibodies is, however, limited and constitutes a serious drawback. Changing the surface properties of pathologic protein aggregates using relatively small macromolecules that bind to their surfaces and that can be functionalized represents an alternative strategy. Molecular chaperones have been shown to bind tightly fibrillar protein assemblies and to diminish their toxicity [126]. As such proteins may not reach the target tissue, an alternative consists of using smaller polypeptides derived from macromolecules that bind and change the surface properties of toxic protein aggregates. Identification of the surface interfaces involved in ligand-toxic protein aggregates is a prerequisite for the design of such molecules. Such approaches have been initiated and may yield promising molecules [127–129].

When pathogenic aggregates are under the protection of membranes, they are no longer targeted by the strategies listed earlier. Their propagation through exosomes and TNTs may, nonetheless, be modulated by inhibitors of exosome release (e.g., spireopoxide or imipramine) or by targeting the ESCRT complexes or the actin cytoskeleton [130]. The pleiotropic effects of such therapeutic strategies may, nonetheless, outweigh the expected benefits.

THERAPEUTIC STRATEGIES AIMED AT RESTORING THE DAMAGE PATHOGENIC PROTEIN AGGREGATES INDUCE

Upon encountering/binding the plasma membrane of neurons, astrocytes, oligodendrocytes, or glial cells, the diffusion of misfolded protein assemblies is "restricted" to a two-dimensional space. The increased confinement of pathogenic and membrane proteins and their lateral diffusion favor molecular interactions. Neuronal membrane protein recruitment by bound pathogenic proteins affects their normal dynamics and/or distribution, compromises their function, or leads to novel signaling platforms. Moreover, the binding and docking of misfolded protein assemblies should theoretically affect the overall normal distribution of membrane proteins through a passive exclusion mechanism that may lead to a pathologic distribution. Aβ, α-syn, and SOD high-molecular weight assemblies were shown to interact, on either sides of the plasma membrane, with a number of neuronal membrane proteins, one of which is common, the sodium potassium ATPase with functional consequences, such as abnormal distribution and pumping activity [80,81,131]. A clustering of this essential protein was observed. Inhibiting the interaction between misfolded protein assemblies and membrane proteins and/or restoring the normal distribution of the latter within the membrane would certainly prevent the damage pathologic protein aggregates cause. Besides interacting with sodium potassium ATPase, Aβ assemblies have been shown to interact with membrane proteins, playing critical roles in signaling pathways, such as PrP [132] and mGluR5 [133,134].

Interfering with pathogenic protein assembly penetration into neuronal cells and restoring the damage they cause represent a rationale and a promising therapeutic avenue in neurodegenerative diseases. It has been demonstrated that pathogenic protein assemblies are taken up by endocytosis. However, interfering with endocytosis may have pleiotropic effects, as this process is essential in the cell. Interestingly, several studies report that cells respond to extracellular protein aggregates by changing their proteomes [135]. Thus, the variations in proteins expression following cell exposure to pathogenic protein aggregates may represent therapeutic targets. Indeed, the overexpression of proteins with protective activity and the downregulation of proteins that trigger, for example, apoptosis may have additional therapeutic potential.

THERAPEUTIC STRATEGIES TARGETING MISFOLDED PATHOLOGIC PROTEIN AGGREGATE–MEDIATED NEUROINFLAMMATION

Neuroinflammation characterizes a number of neurodegenerative diseases [136–139]. Indeed, numerous studies established that neurodegenerative diseases are characterized by an immune response orchestrated by microglial cells, activated astrocytes, and infiltrating T lymphocytes [140–142]. The secretion of proinflammatory factors, such as NO, TNFα, or IL1β have been reported [143–145]. Inhibiting neuroinflammation represents, therefore, a therapeutic avenue. The inflammatory process within the CNS is controlled by a signaling pathway leading to the activation of the transcription factor, NFκB or AP1 that stimulate the expression of genes involved in inflammation. However, to avoid cell damage, inflammation is tightly regulated. Glucocorticoids are the most potent antiinflammatory molecules. They bind to their nuclear receptors and adapt the transcription and expression of various genes involved in neuroinflammation. Given the potency of glucocorticoids, it may be difficult to use them in a generic manner for therapeutic purpose. Other nuclear membrane receptors and transcription factors modulate the neuroinflammatory response some of which are NFκB or AP1 interactors [146,147]. Identifying and targeting those proteins may lead to tailored therapeutic strategies. Gordon and Woodruff further detail, in the chapter they coauthor, the strategies aimed at targeting neuroninflammation in neurodegenerative diseases specifically (Chapter 3).

LIMITATIONS OF THE DIFFERENT THERAPEUTIC STRATEGIES

The different processes that can be theoretically targeted by therapeutic strategies are summarized in Fig. 1.3. One approach that is not considered in this figure is a promising repair strategy based on the use of stem cells, whose potential is detailed in the Chapter 4. This is because repairing the lesions proteins misfolding, aggregation, and propagation and the associated inflammation can only take place once these pathogenic processes are halted. Otherwise, stem cells or derived lineages would be targeted by the misfolded proteins that have accumulated within the CNS in a manner similar to what was reported upon grafting of embryonic cells within the CNS [25,26,56].

When must a therapy be implemented? As long as we are unable to restore the damage caused by pathologic protein aggregates through the replacement of degenerating neurons and the reestablishment of relevant and functional synaptic networks, it is critical to implement therapeutic strategies while neuronal loss is not yet massive, for example, at a prodromal stage [148] preceding the appearance of distinctive pathologic phenotypes, such as movement disorders. This relies on highly specific protein aggregate detection and/or the identification of very early markers of disease. Such tools are lacking so far and a particular effort is required to meet this need. When very early diagnostic tools will be available, it is very likely that complex therapeutic strategies aimed at (1) decreasing the concentration of aggregation-prone proteins and (2) sequestering and/or clearing their misfolded states before highly stable and resistant assemblies are generated in a stochastic manner, will need to be implemented.

Target	Action	Tool
Expression level of aggregation-prone proteins with prion-like properties	↘	siRNA, shRNA, antisense RNA
Stability of aggregation-prone proteins with prion-like properties	↗	Stabilizing antiaggregation agents, molecular chaperones
Clearance of aggregation-prone proteins with prion-like properties		
• Prior to aggregation	↗	Molecular chaperones–UPS system
• Following aggregation	↗	autophagy inducers
Cell-to-cell propagation		
• Following export into the extracellular environment	↘	Immmunotherapy/ligand-degrading enzymes
• Through cell-to-cell contacts	↘	Modulators of transport from cell to cell
Damages pathogenic protein assemblies cause	↘	Restoration of damages when available
Neuroinflammation	↘	Antiinflammatories

FIGURE 1.3 List of the Steps that are Potential Targets of Therapeutic Strategies in Prion-Like Propagation of Misfolded Protein Assemblies

The actions, increase *(in green)* or decrease *(in red)* are indicated as are the tools *(in blue)*.

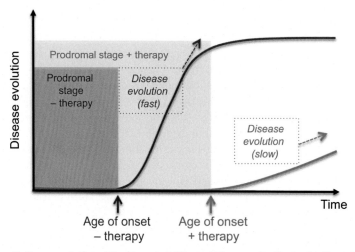

FIGURE 1.4 Schematic Representation of the Expected Effect of Therapeutic Strategies Targeting the Prion-Like Propagation of Misfolded Protein Assemblies

Therapeutic strategies targeting the misfolding, aggregation, and amplification of misfolded protein aggregates in a prion-like manner should delay the onset of disease and extend the prodromal state significantly. They should also slow down disease evolution.

These strategies have an impact on the intrinsic toxicity of misfolded protein aggregates and their ability to propagate from cell to cell. In addition, strategies targeting the propagation of misfolded protein assemblies from affected to naïve cells and their amplification in a catalytic manner in the latter cells should be implemented, as they slow down disease progression. Finally, therapeutic tools aimed at counteracting and/or restoring the damage misfolded protein aggregates cause at the surface and within the cells should also slow down disease progression and the associated neuroinflammation. The effect of the therapeutic strategies on disease onset and progression is schematized in Fig. 1.4.

ACKNOWLEDGMENTS

The research performed in R.M. laboratory is supported by the Centre National de la Recherche Scientifique, the Agence Nationale de la Recherche, the Fondation de France, France-Parkinson, The Fondation Recherche Médicale, the Fondation Simone et Cino Del Duca of the Institut de France, and the Fondation Bettencourt-Schueller.

REFERENCES

[1] F.U. Hartl, A. Bracher, M. Hayer-Hartl, Molecular chaperones in protein folding and proteostasis, Nature 475 (2011) 324–332.
[2] A.R. Fersht, V. Daggett, Protein folding and unfolding at atomic resolution, Cell 108 (2002) 573–582.
[3] J.N. Onuchic, N.D. Socci, Z. Luthey-Schulten, P.G. Wolynes, Protein folding funnels: the nature of the transition state ensemble, Fold. Des. 1 (1996) 441–450.

[4] A.R. Fersht, Structure and Mechanism in Protein Science: A Guide to Enzyme Catalysis and Protein Folding, W.H. Freeman, New York, NY, (1999).

[5] T.P. Knowles, M. Vendruscolo, C.M. Dobson, The amyloid state and its association with protein misfolding diseases, Nat. Rev. Mol. Cell. Biol. 15 (2014) 384–396.

[6] D. Eisenberg, M. Jucker, The amyloid state of proteins in human diseases, Cell 148 (2012) 1188–1203.

[7] M. Goedert, Alzheimer's and Parkinson's diseases: the prion concept in relation to assembled Aβ, tau, and α-synuclein, Science 349 (2015) 1255555.

[8] C.A. Ross, M.A. Poirier, Protein aggregation and neurodegenerative disease, Nat. Med. 10 (2014) S10–S17.

[9] S.H. Park, Y. Kukushkin, R. Gupta, T. Chen, A. Konagai, M.S. Hipp, M. Hayer-Hartl, F.U. Hartl, PolyQ proteins interfere with nuclear degradation of cytosolic proteins by sequestering the Sis1p chaperone, Cell 154 (2013) 134–145.

[10] A. Yu, Y. Shibata, B. Shah, B. Calamini, D.C. Lo, R.I. Morimoto, Protein aggregation can inhibit clathrin-mediated endocytosis by chaperone competition, Proc. Natl. Acad. Sci. USA 111 (2014) E1481–E1490.

[11] F. Oosawa, S. Asakura, in: B. Horecker, N.O. Kaplan, J. Matmur, H.A. Scheraga (Eds.), Thermodynamics of the Polymerization of Protein, Academic Press, London, 1975.

[12] R. Brown, A brief account of microscopical observations made in the month of June, July and August, 1827, on the particles contained in the pollen of plants; and on the general existence of active molecules in organic and inorganic bodies, Phil. Mag. 4 (1828) 161–173.

[13] X. Gao, M. Carroni, C. Nussbaum-Krammer, A. Mogk, N.B. Nillegoda, A. Szlachcic, D.L. Guilbride, H.R. Saibil, M.P. Mayer, B. Bukau, Human Hsp70 disaggregase reverses Parkinson's-linked α-synuclein amyloid fibrils, Mol. Cell. 59 (2015) 781–793.

[14] N.B. Nillegoda, J. Kirstein, A. Szlachcic, M. Berynskyy, A. Stank, F. Stengel, K. Arnsburg, X. Gao, A. Scior, R. Aebersold, D.L. Guilbride, R.C. Wade, R.I. Morimoto, M.P. Mayer, B. Bukau, Crucial HSP70 co-chaperone complex unlocks metazoan protein disaggregation, Nature 524 (2015) 247–251.

[15] M.E. Jackrel, J. Shorter, Engineering enhanced protein disaggregases for neurodegenerative disease, Prion 9 (2015) 90–109.

[16] S.B. Prusiner, Cell biology. A unifying role for prions in neurodegenerative diseases, Science 336 (2012) 1511–1513.

[17] P. Brundin, R. Melki, R. Kopito, Prion-like transmission of protein aggregates in neurodegenerative diseases, Nat. Rev. Mol. Cell. Biol. 11 (2010) 301–307.

[18] M. Jucker, L.C. Walker, Self-propagation of pathogenic protein aggregates in neurodegenerative diseases, Nature 501 (2013) 45–51.

[19] S.J. Lee, P. Desplats, C. Sigurdson, I. Tsigelny, E. Masliah, Cell-to-cell transmission of non-prion protein aggregates, Nat. Rev. Neurol. 6 (2010) 702–706.

[20] M.D. Kane, W.J. Lipinski, M.J. Callahan, F. Bian, R.A. Durham, R.D. Schwarz, A.E. Roher, L.C. Walker, Evidence for seeding of beta -amyloid by intracerebral infusion of Alzheimer brain extracts in beta -amyloid precursor protein-transgenic mice, J. Neurosci. 20 (2000) 3606–3611.

[21] M. Meyer-Luehmann, J. Coomaraswamy, T. Bolmont, S. Kaeser, C. Schaefer, E. Kilger, A. Neuenschwander, D. Abramowski, P. Frey, A.L. Jaton, J.M. Vigouret, P. Paganetti, D.M. Walsh, P.M. Mathews, J. Ghiso, M. Staufenbiel, L.C. Walker, M. Jucker, Exogenous induction of cerebral beta-amyloidogenesis is governed by agent and host, Science 313 (2006) 1781–1784.

[22] Y.S. Eisele, U. Obermüller, G. Heilbronner, F. Baumann, S.A. Kaeser, H. Wolburg, L.C. Walker, M. Staufenbiel, M. Heikenwalder, M. Jucker, Peripherally applied Abeta-containing inoculates induce cerebral beta-amyloidosis, Science 330 (2010) 980–982.

[23] Y.S. Eisele, T. Bolmont, M. Heikenwalder, F. Langer, L.H. Jacobson, Z.X. Yan, K. Roth, A. Aguzzi, M. Staufenbiel, L.C. Walker, M. Jucker, Induction of cerebral beta-amyloidosis: intracerebral versus systemic Abeta inoculation, Proc. Natl. Acad. Sci. USA 106 (2009) 12926–12931.

[24] E. Zobeley, E. Flechsig, A. Cozzio, M. Enari, C. Weissmann, Infectivity of scrapie prions bound to a stainless steel surface, Mol. Med. 5 (1999) 240–243.

[25] J.H. Kordower, Y. Chu, R.A. Hauser, T.B. Freeman, C.W. Olanow, Lewy body-like pathology in long-term embryonic nigral transplants in Parkinson's disease, Nat. Med. 14 (2008) 504–506.

[26] J.Y. Li, E. Englund, J.L. Holton, D. Soulet, P. Hagell, A.J. Lees, T. Lashley, N.P. Quinn, S. Rehncrona, A. Björklund, H. Widner, T. Revesz, O. Lindvall, P. Brundin, Lewy bodies in grafted neurons in subjects with Parkinson's disease suggest host-to-graft disease propagation, Nat. Med. 14 (2008) 501–503.

[27] K.M. Danzer, S.K. Krebs, M. Wolff, G. Birk, B. Hengerer, Seeding induced by alpha-synuclein oligomers provides evidence for spreading of alpha-synuclein pathology, J. Neurochem. 111 (2009) 192–203.

[28] P. Desplats, H.-J. Lee, E. Bae, C. Patrick, E. Rockenstein, L. Crews, B. Spencer, E. Masliah, S.J. Lee, Inclusion formation and neuronal cell death through neuron-to-neuron transmission of -synuclein, Proc. Natl. Acad. Sci. USA 106 (2009) 13010–13015.

[29] E. Emmanouilidou, K. Melachroinou, T. Roumeliotis, S.D. Garbis, M. Ntzouni, L.H. Margaritis, L. Stefanis, K. Vekrellis, Cell produced alpha-synuclein is secreted in a calcium-dependent manner by exosomes and impacts neuronal survival, J. Neurosci. 30 (2010) 6838–6851.

[30] C. Hansen, E. Angot, A.L. Bergström, J.A. Steiner, L. Pieri, G. Paul, T.F. Outeiro, R. Melki, P. Kallunki, K. Fog, J.Y. Li, P. Brundin, α-Synuclein propagates from mouse brain to grafted dopaminergic neurons and seeds aggregation in cultured human cells, J. Clin. Invest. 121 (2011) 715–725.

[31] E.C. Freundt, N. Maynard, E.K. Clancy, S. Roy, L. Bousset, Y. Sourigues, M. Covert, R. Melki, K. Kirkegaard, M. Brahic, Neuron-to-neuron transmission of alpha-synuclein fibrils through axonal transport, Ann. Neurol. 72 (2012) 517–524.

[32] S. Holmqvist, O. Chutna, L. Bousset, P. Aldrin-Kirk, W. Li, T. Björklund, Z.Y. Wang, L. Roybon, R. Melki, J.Y. Li, Direct evidence of Parkinson pathology spread from the gastrointestinal tract to the brain in rats, Acta Neuropathol. 128 (2014) 805–820.

[33] M. Brahic, L. Bousset, G. Bieri, R. Melki, A.D. Gitler, Axonal transport and secretion of fibrillar forms of α-synuclein, Aβ42 peptide and HTTExon 1, Acta Neuropathol. 131 (2016) 539–548.

[34] K.C. Luk, V. Kehm, J. Carroll, B. Zhang, P. O'Brien, J.Q. Trojanowski, V.M. Lee, Pathological α-synuclein transmission initiates Parkinson-like neurodegeneration in nontransgenic mice, Science 338 (2012) 949–953.

[35] K.C. Luk, V.M. Kehm, B. Zhang, P. O'Brien, J.Q. Trojanowski, V.M. Lee, Intracerebral inoculation of pathological alpha-synuclein initiates a rapidly progressive neurodegenerative alphasynucleinopathy in mice, J. Exp. Med. 209 (2012) 975–986.

[36] M. Masuda-Suzukake, T. Nonaka, M. Hosokawa, T. Oikawa, T. Arai, H. Akiyama, D.M. Mann, M. Hasegawa, Prion-like spreading of pathological α-synuclein in brain, Brain 136 (2013) 1128–1138.

[37] A.L. Mougenot, S. Nicot, A. Bencsik, E. Morignat, J. Verchère, L. Lakhdar, S. Legastelois, T. Baron, Prion-like acceleration of a synucleinopathy in a transgenic mouse model, Neurobiol. Aging 33 (2012) 2225–2228.

[38] A.N. Sacino, M. Brooks, M.A. Thomas, A.B. McKinney, S. Lee, R.W. Regenhardt, N.H. McGarvey, J.I. Ayers, L. Notterpek, D.R. Borchelt, T.E. Golde, B.I. Giasson, Intramuscular injection of a-synuclein induces CNS a-synuclein pathology and a rapid-onset motor phenotype in transgenic mice, Proc. Natl. Acad. Sci. USA 111 (2014) 10732–10737.

[39] A. Recasens, B. Dehay, J. Bové, I. Carballo-Carbajal, S. Dovero, A. Pérez-Villalba, P.O. Fernagut, J. Blesa, A. Parent, C. Perier, I. Fariñas, J.A. Obeso, E. Bezard, M. Vila, Lewy body extracts from Parkinson disease brains trigger α-synuclein pathology and neurodegeneration in mice and monkeys, Ann. Neurol. 75 (2014) 351–362.

[40] W. Peelaerts, L. Bousset, A. Van der Perren, A. Moskalyuk, R. Pulizzi, G. Giugliano, C. Van den Haute, R. Melki, V. Baekelandt, Alpha-synuclein strains cause distinct synucleinopathies after local and systemic administration, Nature 522 (2015) 340–344.

[41] F. Clavaguera, T. Bolmont, R.A. Crowther, D. Abramowski, S. Frank, A. Probst, G. Fraser, A.K. Stalder, M. Beibel, M. Staufenbiel, M. Jucker, M. Goedert, M. Tolnay, Transmission and spreading of tauopathy in transgenic mouse brain, Nat. Cell. Biol. 11 (2009) 909–913.

[42] B. Frost, R.L. Jacks, M.I. Diamond, Propagation of tau misfolding from the outside to the inside of a cell, J. Biol. Chem. 284 (2009) 12845–12852.

[43] N. Kfoury, B.B. Holmes, H. Jiang, D.M. Holtzman, M.I. Diamond, Trans-cellular propagation of Tau aggregation by fibrillar species, J. Biol. Chem. 287 (2012) 19440–19451.

[44] J.W. Wu, M. Herman, L. Liu, S. Simoes, C.M. Acker, H. Figueroa, J.I. Steinberg, M. Margittai, R. Kayed, C. Zurzolo, G. Di Paolo, K.E. Duff, Small misfolded Tau species are internalized via bulk endocytosis and anterogradely and retrogradely transported in neurons, J. Biol. Chem. 288 (2013) 1856–1870.

[45] Z. Ahmed, J. Cooper, T.K. Murray, K. Garn, E. McNaughton, H. Clarke, S. Parhizkar, M.A. Ward, A. Cavallini, S. Jackson, S. Bose, F. Clavaguera, M. Tolnay, I. Lavenir, M. Goedert, M.L. Hutton, M.J. O'Neill, A novel in vivo model of tau propagation with rapid and progressive neurofibrillary tangle pathology: the pattern of spread is determined by connectivity, not proximity, Acta Neuropathol. 127 (2014) 667–683.

[46] A. de Calignon, M. Polydoro, M. Suarez-Calvet, C. William, D.H. Adamowicz, K.J. Kopeikina, R. Pitstick, N. Sahara, K.H. Ashe, G.A. Carlson, T.L. Spires-Jones, B.T. Hyman, Propagation of tau pathology in a model of early Alzheimer's disease, Neuron 73 (2012) 685–697.

[47] L. Liu, V. Drouet, J.W. Wu, M.P. Witter, S.A. Small, C. Clelland, K. Duff, Trans-synaptic spread of tau pathology in vivo, PLoS One 7 (2012) e31302.

[48] S. Dujardin, S. Bégard, R. Caillierez, C. Lachaud, L. Delattre, S. Carrier, A. Loyens, M.C. Galas, L. Bousset, R. Melki, G. Aurégan, P. Hantraye, E. Brouillet, L. Buée, M. Colin, Ectosomes: a new mechanism for non-exosomal secretion of Tau protein, PLoS One 9 (6) (2014) e100760.

[49] J.L. Guo, V.M. Lee, Neurofibrillary tangle-like tau pathology induced by synthetic tau fibrils in primary neurons over-expressing mutant tau, FEBS Lett. 587 (2013) 717–723.

[50] M. Iba, J.L. Guo, J.D. McBride, B. Zhang, J.Q. Trojanowski, V.M. Lee, Synthetic tau fibrils mediate transmission of neurofibrillary tangles in a transgenic mouse model of Alzheimer's-like tauopathy, J. Neurosci. 33 (2013) 1024–1037.

[51] F. Clavaguera, H. Akatsu, G. Fraser, R.A. Crowther, S. Frank, J. Hench, A. Probst, D.T. Winkler, J. Reichwald, M. Staufenbiel, B. Ghetti, M. Goedert, M. Tolnay, Brain homogenates from human tauopathies induce tau inclusions in mouse brain, Proc. Natl. Acad. Sci. USA 110 (2013) 9535–9540.

[52] P. Ren, J.E. Lauckner, I. Kachirskaia, J.E. Heuser, R. Melki, R.R. Kopito, Cytoplasmic penetration and persistent infection of mammalian cells by polyglutamine aggregates, Nat. Cell. Biol. 11 (2009) 219–225.

[53] E. Pecho-Vrieseling, C. Rieker, S. Fuchs, D. Bleckmann, M.S. Esposito, P. Botta, C. Goldstein, M. Bernhard, I. Galimberti, M. Müller, A. Lüthi, S. Arber, T. Bouwmeester, H. van der Putten, F.P. Di Giorgio, Trans-neuronal propagation of mutant huntingtin contributes to non-cell autonomous pathology in neurons, Nat. Neurosci. 17 (2014) 1064–1072.

[54] F. Herrera, S. Tenreiro, L. Miller-Fleming, T.F. Outeiro, Visualization of cell-to-cell transmission of mutant huntingtin oligomers, PLoS Curr. 3 (2011) RRN1210.

[55] F. Cicchetti, S. Saporta, R.A. Hauser, M. Parent, M. Saint-Pierre, P.R. Sanberg, X.J. Li, J.R. Parker, Y. Chu, E.J. Mufson, J.H. Kordower, T.B. Freeman, Neural transplants in patients with Huntington's disease undergo disease-like neuronal degeneration, Proc. Natl. Acad. Sci. USA 106 (2009) 12483–12488.

[56] F. Cicchetti, S. Lacroix, G. Cisbani, N. Vallières, M. Saint-Pierre, I. St-Amour, R. Tolouei, J.N. Skepper, R.A. Hauser, D. Mantovani, R.A. Barker, T.B. Freeman, Mutant huntingtin is present in neuronal grafts in Huntington disease patients, Ann. Neurol. 76 (2014) 31–42.

[57] M.M. Pearce, E.J. Spartz, W. Hong, L. Luo, R.R. Kopito, Prion-like transmission of neuronal huntingtin aggregates to phagocytic glia in the *Drosophila* brain, Nat. Commun. 6 (2015) 6768.

[58] T. Nonaka, M. Masuda-Suzukake, T. Arai, Y. Hasegawa, H. Akatsu, T. Obi, M. Yoshida, S. Murayama, D.M. Mann, H. Akiyama, M. Hasegawa, Prion-like properties of pathological TDP-43 aggregates from diseased brains, Cell Rep. 4 (2013) 124–134.

[59] C. Münch, J. O'Brien, A. Bertolotti, Prion-like propagation of mutant superoxide dismutase-1 misfolding in neuronal cells, Proc. Natl. Acad. Sci. USA 108 (2011) 3548–3553.

[60] K. Kuchler, J. Thorner, Secretion of peptides and proteins lacking hydrophobic signal sequences: the role of adenosine triphosphate-driven membrane translocators, Endocr. Rev. 13 (1992) 499–514.

[61] S.H. White, G. von Heijne, The machinery of membrane protein assembly, Curr. Opin. Struct. Biol. 14 (2004) 397–404.

[62] M. Colombo, G. Raposo, C. Thery, Biogenesis, secretion, and intercellular interactions of exosomes and other extracellular vesicles, Annu. Rev. Cell Dev. Biol. 30 (2014) 255–289.

[63] B.L. Deatherage, B.T. Cookson, Membrane vesicle release in bacteria, eukaryotes, and archaea: a conserved yet underappreciated aspect of microbial life, Infect. Immun. 80 (2012) 1948–1957.

[64] J.M. Wolf, A. Casadevall, Challenges posed by extracellular vesicles from eukaryotic microbes, Curr. Opin. Microbiol. 22 (2014) 73–78.

[65] M. Chivet, C. Javalet, F. Hemming, K. Pernet-Gallay, K. Laulagnier, S. Fraboulet, R. Sadoul, Exosomes as a novel way of interneuronal communication, Biochem. Soc. Trans. 41 (2013) 241–244.

[66] J.S. Schorey, Y. Cheng, P.P. Singh, V.L. Smith, Exosomes and other extracellular vesicles in host-pathogen interactions, EMBO Rep. 16 (2015) 24–43.

[67] M. Simons, G. Raposo, Exosomes—vesicular carriers for intercellular communication, Curr. Opin. Cell Biol. 21 (2009) 575–581.

[68] D. Simon, E. Garcia-Garcia, F. Royo, J.M. Falcon-Perez, J. Avila, Proteostasis of tau. Tau overexpression results in its secretion via membrane vesicles, FEBS Lett. 586 (2012) 47–54.

[69] S. Saman, W. Kim, M. Raya, Y. Visnick, S. Miro, S. Saman, B. Jackson, A.C. McKee, V.E. Alvarez, N.C. Lee, G.F. Hall, Exosome-associated tau is secreted in tauopathy models and is selectively phosphorylated in cerebrospinal fluid in early Alzheimer disease, J. Biol. Chem. 287 (2012) 3842–3849.

[70] S. Dujardin, K. Lecolle, R. Caillierez, S. Begard, N. Zommer, C. Lachaud, S. Carrier, N. Dufour, G. Aurégan, J. Winderickx, P. Hantraye, N. Déglon, M. Colin, L. Buée, Neuron-to-neuron wild-type Tau protein transfer through a trans-synaptic mechanism: relevance to sporadic tauopathies, Acta Neuropathol. Commun. 2 (2014) 14.

[71] L. Rajendran, J. Bali, M.M. Barr, F.A. Court, E.M. Kramer-Albers, F. Picou, G. Raposo, K.E. van der Vos, G. van Niel, J. Wang, X.O. Breakefield, Emerging roles of extracellular vesicles in the nervous system, J. Neurosci. 34 (2014) 15482–15489.

[72] A. Rustom, R. Saffrich, I. Markovic, P. Walther, H.H. Gerdes, Nanotubular highways for intercellular organelle transport, Science 303 (2004) 1007–1010.

[73] S. Abounit, C. Zurzolo, Wiring through tunneling nanotubes—from electrical signals to organelle transfer, J. Cell Sci. 125 (2012) 1089–1098.

[74] K. Gousset, E. Schiff, C. Langevin, Z. Marijanovic, A. Caputo, D.T. Browman, N. Chenouard, F. de Chaumont, A. Martino, J. Enninga, J.C. Olivo-Marin, D. Männel, C. Zurzolo, Prions hijack tunnelling nanotubes for intercellular spread, Nat. Cell. Biol. 11 (2009) 328–336.

[75] M. Costanzo, S. Abounit, L. Marzo, A. Danckaert, Z. Chamoun, P. Roux, C. Zurzolo, Transfer of polyglutamine aggregates in neuronal cells occurs in tunneling nanotubes, J. Cell. Sci. 126 (2013) 3678–3685.

[76] S. Abounit, L. Bousset, F. Loria, S. Zhu, F. de Chaumont, L. Pieri, J.C. Olivo-Marin, R. Melki, C. Zurzolo, Tunneling nanotubes spread fibrillar α-synuclein by intercellular trafficking of lysosomes, EMBO J. 35 (2016) 2120–2138.

[77] M. Tardivel, S. Bégard, L. Bousset, S. Dujardin, A. Coens, R. Melki, L. Buée, M. Colin, Tunneling nanotube (TNT)-mediated neuron-to neuron transfer of pathological Tau protein assemblies, Acta. Neuropathol. Commun. 4 (2016) 117.

[78] B. Antonny, Mechanisms of membrane curvature sensing, Annu. Rev. Biochem. 80 (2011) 101–123.

[79] J. Varkey, J.M. Isas, N. Mizuno, M.B. Jensen, V.K. Bhatia, C.C. Jao, J. Petrlova, J.C. Voss, D.G. Stamou, A.C. Steven, R. Langen, Membrane curvature induction and tubulation are common features of synucleins and apolipoproteins, J. Biol. Chem. 285 (2010) 32486–32493.

[80] T. Ohnishi, M. Yanazawa, T. Sasahara, Y. Kitamura, H. Hiroaki, Y. Fukazawa, I. Kii, T. Nishiyama, A. Kakita, H. Takeda, A. Takeuchi, Y. Arai, A. Ito, H. Komura, H. Hirao, K. Satomura, M. Inoue, S. Muramatsu, K. Matsui, M. Tada, M. Sato, E. Saijo, Y. Shigemitsu, S. Sakai, Y. Umetsu, N. Goda, N. Takino, H. Takahashi, M. Hagiwara, T. Sawasaki, G. Iwasaki, Y. Nakamura, Y. Nabeshima, D.B. Teplow, M. Hoshi, Na,K-ATPase α3 is a death target of Alzheimer patient amyloid-β assembly, Proc. Natl. Acad. Sci. USA 112 (2015) E4465–E4474.

[81] A.N. Shrivastava, V. Redeker, N. Fritz, L. Pieri, L.G. Almeida, M. Spolidoro, T. Liebmann, L. Bousset, M. Renner, C. Léna, A. Aperia, R. Melki, A. Triller, α-synuclein assemblies sequester neuronal α3-Na$^+$/K$^+$-ATPase and impair Na$^+$ gradient, EMBO J. 34 (2015) 2408–2423.

[82] M. Bokvist, G. Gröbner, Misfolding of amyloidogenic proteins at membrane surfaces: the impact of macromolecular crowding, J. Am. Chem. Soc. 129 (2007) 14848–14849.

[83] A.B. Singleton, M. Farrer, J. Johnson, A. Singleton, S. Hague, J. Kachergus, M. Hulihan, T. Peuralinna, A. Dutra, R. Nussbaum, S. Lincoln, A. Crawley, M. Hanson, D. Maraganore, C. Adler, M.R. Cookson, M. Muenter, M. Baptista, D. Miller, J. Blancato, J. Hardy, K. Gwinn-Hardy, Alpha-synuclein locus triplication causes Parkinson's disease, Science 302 (2003) 841.

[84] M.C. Chartier-Harlin, J. Kachergus, C. Roumier, V. Mouroux, X. Douay, S. Lincoln, C. Levecque, L. Larvor, J. Andrieux, M. Hulihan, N. Waucquier, L. Defebvre, P. Amouyel, M. Farrer, A. Destée, Alpha-synuclein locus duplication as a cause of familial Parkinson's disease, Lancet 364 (2004) 1167–1169.

[85] O. Chiba-Falek, G.J. Lopez, R.L. Nussbaum, Levels of alpha-synuclein mRNA in sporadic Parkinson disease patients, Mov. Disord. 21 (2006) 1703–1708.

[86] P. Ibáñez, A.M. Bonnet, B. Débarges, E. Lohmann, F. Tison, P. Pollak, Y. Agid, A. Dürr, A. Brice, Causal relation between alpha-synuclein gene duplication and familial Parkinson's disease, Lancet 364 (2004) 1169–1171.

[87] A. Rovelet-Lecrux, D. Hannequin, G. Raux, N. Le Meur, A. Laquerrière, A. Vital, C. Dumanchin, S. Feuillette, A. Brice, M. Vercelletto, F. Dubas, T. Frebourg, D. Campion, APP locus duplication causes autosomal dominant early-onset Alzheimer disease with cerebral amyloid angiopathy, Nat. Genet. 38 (2006) 24–26.

[88] M.S. Cheon, M. Dierssen, S.H. Kim, G. Lubec, Protein expression of BACE1, BACE2 and APP in Down syndrome brains, Amino Acids 35 (2008) 339–343.

[89] H.B. Kordasiewicz, L.M. Stanek, E.V. Wancewicz, C. Mazur, M.M. McAlonis, K.A. Pytel, J.W. Artates, A. Weiss, S.H. Cheng, L.S. Shihabuddin, G. Hung, C.F. Bennett, D.W. Cleveland, Sustained therapeutic reversal of Huntington's disease by transient repression of huntingtin synthesis, Neuron 74 (2012) 1031–1044.

[90] D. Yu, H. Pendergraff, J. Liu, H.B. Kordasiewicz, D.W. Cleveland, E.E. Swayze, W.F. Lima, S.T. Crooke, T.P. Prakash, D.R. Corey, Single-stranded RNAs use RNAi to potently and allele-selectively inhibit mutant huntingtin expression, Cell 150 (2012) 895–908.

[91] M. Takahashi, M. Suzuki, M. Fukuoka, N. Fujikake, S. Watanabe, M. Murata, K. Wada, Y. Nagai, H. Hohjoh, Normalization of overexpressed alpha-synuclein causing Parkinson's disease by a moderate gene silencing with RNA interference, Mol. Ther. Nucleic Acids 4 (2015) e241.

[92] A.D. Zharikov, J.R. Cannon, V. Tapias, Q. Bai, M.P. Horowitz, V. Shah, A. El Ayadi, T.G. Hastings, J.T. Greenamyre, E.A. Burton, shRNA targeting alpha-synuclein prevents neurodegeneration in a Parkinson's disease model, J. Clin. Invest. 125 (2015) 2721–2735.

[93] D.M. Maraganore, Rationale for therapeutic silencing of alpha-synuclein in Parkinson's disease, J. Mov. Disord. 4 (2011) 1–7.

[94] R.A. Smith, T.M. Miller, K. Yamanaka, B.P. Monia, T.P. Condon, G. Hung, C.S. Lobsiger, C.M. Ward, M. McAlonis-Downes, H. Wei, E.V. Wancewicz, C.F. Bennett, D.W. Cleveland, Antisense oligonucleotide therapy for neurodegenerative disease, J. Clin. Invest. 116 (2006) 2290–2296.

[95] R. Sud, E.T. Geller, G.D. Schellenberg, Antisense-mediated exon skipping decreases tau protein expression: a potential therapy for tauopathies, Mol. Ther. Nucleic Acids 3 (2014) e180.

[96] C. Haass, C. Kaether, G. Thinakaran, S. Sisodia, Trafficking and proteolytic processing of APP, Cold Spring Harb. Perspect. Med. 2 (2012) a006270.

[97] R. Vassar, P.H. Kuhn, C. Haass, M.E. Kennedy, L. Rajendran, P.C. Wong, S.F. Lichtenthaler, Function, therapeutic potential and cell biology of BACE proteases: current status and future prospects, J. Neurochem. 130 (2014) 4–28.

[98] A. Francioso, P. Punzi, A. Boffi, C. Lori, S. Martire, C. Giordano, M. D'Erme, L. Mosca, β-sheet interfering molecules acting against β-amyloid aggregation and fibrillogenesis, Bioorg. Med. Chem. 23 (2015) 1671–1683.

[99] B. Bulic, M. Pickhardt, E. Mandelkow, Progress and developments in tau aggregation inhibitors for Alzheimer disease, J. Med. Chem. 56 (2013) 4135–4155.

[100] A. Mähler, S. Mandel, M. Lorenz, U. Ruegg, E.E. Wanker, M. Boschmann, F. Paul, Epigallocatechin-3-gallate: a useful, effective and safe clinical approach for targeted prevention and individualised treatment of neurological diseases? EPMA J. 4 (2013) 5.

[101] M. Masuda, N. Suzuki, S. Taniguchi, T. Oikawa, T. Nonaka, T. Iwatsubo, S. Hisanaga, M. Goedert, M. Hasegawa, Small molecule inhibitors of alpha-synuclein filament assembly, Biochemistry 45 (2006) 6085–6094.

[102] E. Maioli, C. Torricelli, G. Valacchi, Rottlerin and curcumin: a comparative analysis, Ann. NY Acad. Sci. 1259 (2012) 65–76.

[103] B. Calamini, M.C. Silva, F. Madoux, D.M. Hutt, S. Khanna, M.A. Chalfant, S.A. Saldanha, P. Hodder, B.D. Tait, D. Garza, W.E. Balch, R.I. Morimoto, Small-molecule proteostasis regulators for protein conformational diseases, Nat. Chem. Biol. 8 (2011) 185–196.

[104] B. Calamini, R.I. Morimoto, Protein homeostasis as a therapeutic target for diseases of protein conformation, Curr. Top. Med. Chem. 12 (2012) 2623–2640.

[105] J. Chatellier, F. Hill, P.A. Lund, A.R. Fersht, In vivo activities of GroEL minichaperones, Proc. Natl. Acad. Sci. USA 95 (1998) 9861–9866.

[106] R.B. Nahomi, B. Wang, C.T. Raghavan, O. Voss, A.I. Doseff, P. anthoshkumar, R.H. Nagaraj, Chaperone peptides of α-crystallin inhibit epithelial cell apoptosis, protein insolubilization, and opacification in experimental cataracts, J. Biol. Chem. 288 (2013) 13022–13035.

[107] M. Raju, P. Santhoshkumar, K. Krishna Sharma, Alpha-crystallin-derived peptides as therapeutic chaperones, Biochim. Biophys. Acta. 1860 (1 Pt. B) (2016) 246–251.

[108] R.I. Morimoto, A.M. Cuervo, Proteostasis and the aging proteome in health and disease, J. Gerontol. A Biol. Sci. Med. Sci. 69 (Suppl. 1) (2014) S33–S38.

[109] Y. Kiriyama, H. Nochi, The function of autophagy in neurodegenerative diseases, Int. J. Mol. Sci. 16 (2015) 26797–26812.

[110] J.L. Schneider, A.M. Cuervo, Autophagy and human disease: emerging themes, Curr. Opin. Genet. Dev. 26 (2014) 16–23.

[111] C.J. Cortes, A.R. La Spada, Autophagy in polyglutamine disease: imposing order on disorder or contributing to the chaos? Mol Cell. Neurosci. 66 (Pt. A) (2015) 53–61.

[112] A. Ciechanover, Y.T. Kwon, Degradation of misfolded proteins in neurodegenerative diseases: therapeutic targets and strategies, Exp. Mol. Med. 47 (2015) e147.

[113] J.J. Sheehan, S.E. Tsirka, Fibrin-modifying serine proteases thrombin, tPA, and plasmin in ischemic stroke: a review, Glia 50 (2005) 340–350.

[114] H.M. Tucker, M. Kihiko, J.N. Caldwell, S. Wright, T. Kawarabayashi, D. Price, D. Walker, S. Scheff, J.P. McGillis, R.E. Rydel, S. Estus, The plasmin system is induced by and degrades amyloid-beta aggregates, J. Neurosci. 20 (2000) 3937–3946.

[115] K.S. Kim, Y.R. Choi, J.Y. Park, J.H. Lee, D.K. Kim, S.J. Lee, S.R. Paik, I. Jou, S.M. Park, Proteolytic cleavage of extracellular α-synuclein by plasmin: implications for Parkinson disease, J. Biol. Chem. 287 (2012) 24862–24872.

[116] S.M. Park, K.S. Kim, Proteolytic clearance of extracellular α-synuclein as a new therapeutic approach against Parkinson disease, Prion 7 (2013) 121–126.

[117] B. Spencer, E. Valera, E. Rockenstein, M. Trejo-Morales, A. Adame, E. Masliah, A brain-targeted, modified neurosin (kallikrein-6) reduces alpha-synuclein accumulation in a mouse model of multiple system atrophy, Mol. Neurodegener. 10 (2015) 48.

[118] P.T. Lansbury, H.A. Lashuel, A century-old debate on protein aggregation and neurodegeneration enters the clinic, Nature 443 (2006) 774–779.

[119] K. Yanamandra, N. Kfoury, H. Jiang, T.E. Mahan, S. Ma, S.E. Maloney, D.F. Wozniak, M.I. Diamond, D.M. Holtzman, Anti-tau antibodies that block tau aggregate seeding in vitro markedly decrease pathology and improve cognition in vivo, Neuron 80 (2013) 402–414.

[120] M. Mandler, E. Valera, E. Rockenstein, H. Weninger, C. Patrick, A. Adame, R. Santic, S. Meindl, B. Vigl, O. Smrzka, A. Schneeberger, F. Mattner, E. Masliah, Next-generation active immunization approach for synucleinopathies: implications for Parkinson's disease clinical trials, Acta Neuropathol. 127 (2014) 861–879.

[121] M. Mandler, E. Valera, E. Rockenstein, M. Mante, H. Weninger, C. Patrick, A. Adame, S. Schmidhuber, R. Santic, A. Schneeberger, W. Schmidt, F. Mattner, E. Masliah, Active immunization against alpha-synuclein ameliorates the degenerative pathology and prevents demyelination in a model of multiple system atrophy, Mol. Neurodegener. 10 (2015) 10.

[122] M.A. Busche, C. Grienberger, A.D. Keskin, B. Song, U. Neumann, M. Staufenbiel, H. Förstl, A. Konnerth, Decreased amyloid-β and increased neuronal hyperactivity by immunotherapy in Alzheimer's models, Nat. Neurosci. 18 (2015) 1725–1727.

[123] K.R. Bales, S.M. O'Neill, N. Pozdnyakov, F. Pan, D. Caouette, Y. Pi, K.M. Wood, D. Volfson, J.R. Cirrito, B.H. Han, A.W. Johnson, G.J. Zipfel, T.A. Samad, Passive immunotherapy targeting amyloid-β reduces cerebral amyloid angiopathy and improves vascular reactivity, Brain 139 (2016) 563–577.

[124] J.T. Pedersen, E.M. Sigurdsson, Tau immunotherapy for Alzheimer's disease, Trends Mol. Med. 21 (2015) 394–402.

[125] D.L. Castillo-Carranza, M.J. Guerrero-Muñoz, U. Sengupta, C. Hernandez, A.D. Barrett, K. Dineley, R. Kayed, Tau immunotherapy modulates both pathological tau and upstream amyloid pathology in an Alzheimer's disease mouse model, J. Neurosci. 35 (2015) 4857–4868.

[126] S. Pemberton, K. Madiona, L. Pieri, M. Kabani, L. Bousset, R. Melki, Hsc70 protein interaction with soluble and fibrillar alpha-synuclein, J. Biol. Chem. 286 (2011) 34690–34699.

[127] V. Redeker, S. Pemberton, W. Bienvenut, L. Bousset, R. Melki, Identification of protein interfaces between α-synuclein, the principal component of Lewy bodies in Parkinson disease, and the molecular chaperones human Hsc70 and the yeast Ssa1p, J. Biol. Chem. 287 (2012) 32630–32639.

[128] C. Nury, V. Redeker, S. Dautrey, A. Romieu, G. van der Rest, P.Y. Renard, R. Melki, J. Chamot-Rooke, A novel bio-orthogonal cross-linker for improved protein/protein interaction analysis, Anal. Chem. 87 (2015) 1853–1860.

[129] E. Monsellier, V. Redeker, G. Ruiz-Arlandis, L. Bousset, R. Melki, Molecular interaction between the chaperone Hsc70 and the N-terminal flank of huntingtin exon 1 modulates aggregation, J. Biol. Chem. 290 (2015) 2560–2576.

[130] J. Li, K. Liu, Y. Liu, Y. Xu, F. Zhang, H. Yang, J. Liu, T. Pan, J. Chen, M. Wu, X. Zhou, Z. Yuan, Exosomes mediate the cell-to-cell transmission of IFN-a-induced antiviral activity, Nat. Immunol. 14 (2013) 793–803.

[131] C. Ruegsegger, N. Maharjan, A. Goswami, A. Filézac de L'Etang, J. Weis, D. Troost, M. Heller, H. Gut, S. Saxena, Aberrant association of misfolded SOD1 with Na⁺/K⁺ATPaseα-3 impairs its activity and contributes to motor neuron vulnerability in ALS, Acta Neuropathol. 131 (2016) 427–451.

[132] J. Lauren, D.A. Gimbel, H.B. Nygaard, J.W. Gilbert, S.M. Strittmatter, Cellular prion protein mediates impairement of synaptic plasticity by amyloid-β oligomers, Nature 457 (2009) 1128–1134.

[133] M. Renner, P.N. Lacor, P.T. Velasco, J. Xu, A. Contractor, W.L. Klein, A. Triller, Deleterious effects of amyloid beta oligomers acting as an extracellular scaffold for mGluR5, Neuron 66 (2010) 739–754.

[134] A.N. Shrivastava, J.M. Kowalewski, M. Renner, L. Bousset, A. Koulakoff, R. Melki, C. Giaume, A. Triller, β-amyloid and ATP-induced diffusional trapping of astrocyte and neuronal metabotropic glutamate type-5 receptors, Glia 61 (2013) 1673–1686.

[135] L. Pieri, P. Chafey, M. Le Gall, G. Clary, R. Melki, V. Redeker, Cellular response of human neuroblastoma cells to α-synuclein fibrils, the main constituent of Lewy bodies, Biochim. Biophys. Acta. 1860 (2016) 8–19.

[136] F.L. Heppner, R.M. Ransohoff, B. Becher, Immune attack: the role of inflammation in Alzheimer disease, Nat. Rev. Neurosci. 16 (2015) 358–372.

[137] A. Crotti, C.K. Glass, The choreography of neuroinflammation in Huntington's disease, Trends Immunol. 36 (2015) 364–373.

[138] K.G. Hooten, D.R. Beers, W. Zhao, S.H. Appel, Protective and toxic neuroinflammation in amyotrophic lateral sclerosis, Neurotherapeutics 12 (2015) 364–375.

[139] E.C. Hirsch, S. Hunot, Neuroinflammation in Parkinson's disease: a target for neuroprotection? Lancet Neurol. 8 (2009) 382–397.

[140] S. Shavali, C.K. Combs, M. Ebadi, Reactive macrophages increase oxidative stress and alpha-synuclein nitration during death of dopaminergic neuronal cells in co-culture: relevance to Parkinson's disease, Neurochem. Res. 31 (2006) 85–94.

[141] R. Griffin, R. Nally, Y. Nolan, Y. McCartney, J. Linden, M.A. Lynch, The age-related attenuation in long-term potentiation is associated with microglial activation, J. Neurochem. 99 (2006) 1263–1272.

[142] W. Zhang, T. Wang, Z. Pei, D.S. Miller, X. Wu, M.L. Block, B. Wilson, W. Zhang, Y. Zhou, J.S. Hong, J. Zhang, Aggregated alpha-synuclein activates microglia: a process leading to disease progression in Parkinson's disease, FASEB J. 19 (2005) 533–542.

[143] W.Y. Wang, M.S. Tan, J.T. Yu, L. Tan, Role of pro-inflammatory cytokines released from microglia in Alzheimer's disease, Ann. Transl. Med. 3 (2015) 136.

[144] S.V. More, H. Kumar, I.S. Kim, S.Y. Song, D.K. Choi, Cellular and molecular mediators of neuroinflammation in the pathogenesis of Parkinson's disease, Mediators Inflamm. 2013 (2013) 952375.

[145] P. Ghezzi, T. Mennini, Tumor necrosis factor and motoneuronal degeneration: an open problem, Neuroimmunomodulation 9 (2001) 178–182.

[146] M.K. McCoy, M.G. Tansey, TNF signaling inhibition in the CNS: implications for normal brain function and neurodegenerative disease, J. Neuroinflammation 5 (2008) 45.

[147] S. Vyas, L. Maatouk, Contribution of glucocorticoids and glucocorticoid receptors to the regulation of neurodegenerative processes, CNS Neurol. Disord. Drug Targets 12 (2013) 1175–1193.

[148] D. Berg, R.B. Postuma, C.H. Adler, B.R. Bloem, P. Chan, B. Dubois, T. Gasser, C.G. Goetz, G. Halliday, L. Joseph, A.E. Lang, I. Liepelt-Scarfone, I. Litvan, K. Marek, J. Obeso, W. Oertel, C.W. Olanow, W. Poewe, M. Stern, G. Deuschl, MDS research criteria for prodromal Parkinson's disease, Mov. Disord. 30 (2015) 1600–1611.

SYNUCLEIN MISFOLDING AS A THERAPEUTIC TARGET

2

Wouter Peelaerts*, Tiago F. Outeiro**,†

*KU Leuven, Leuven, Belgium
**Institute of Molecular Medicine, University of Lisbon, Lisboa, Portugal
†Center for Nanoscale Microscopy and Molecular Physiology of the Brain,
University Medical Center Goettingen, Goettingen, Germany

CHAPTER OUTLINE

Introduction..21
The Conformational Landscape of α-Synuclein: the Native State...............................22
    Cytosolic Folding States...24
    Membrane-Bound Folding States 24
The Conformational Landscape of α-Synuclein: Misfolded Variants 25
    Folding Intermediates..25
    Oligomeric Variants...27
    Fibrillar Variants...29
    Misfolding in Complex Environments...30
    Synuclein Strains 32
    Glycation of α-SYN—an Age-Associated Posttranslational Modification...................32
Misfolding of α-Synuclein as a Therapeutic Target 33
    Preserving the Native State 33
    Preventing the Formation of Folding Intermediates 33
    Targeting Aggregation 35
    Stabilizing Membrane Interactions..38
Conclusions...38
References...39

INTRODUCTION

Deposits of misfolded proteins are a key hallmark of many neurodegenerative disorders. They form proteinaceous inclusion bodies at specific predilection sites within the central nervous system and are characteristic for neurodegenerative protein misfolding diseases. Already in the beginning of 20th century, Friederich Lewy identified large inclusion bodies in the brain of a patient who was diagnosed with Parkinson's disease (PD) [1]. They were found inside neuronal cell bodies and axons and were termed accordingly, Lewy bodies (LB) and Lewy neurites (LN). Lewy described these inclusions or *"Negrischen Körperchen"* and compared them with lesions that are caused in the central nervous system after exposure to infectious viral pathogens [2]. Although it was clear that these characteristic histopathological hallmarks had an important role in the disease process, the exact anatomical substrate of these large inclusions remained elusive and a search was set out to identify this pathognomonic component.

The search for this enigmatic protein lasted eventually for almost a decade. Starting the early 1990s, two short peptides were isolated from amyloid plaques from the brain of Alzheimer patients, which were termed the nonamyloid component of Aβ plaques (NAC) [3]. These peptides appeared to overlap with residues 61–95 of a larger precursor protein of which the sequence was subsequently cloned and identified as the human ortholog of a synaptic vesicular protein of the pacific electric eel *Torpedo californica* [4]. This human ortholog was found in association with the synapse and the nuclear envelope, and was therefore termed "synuclein" even though the link of α-synuclein (α-SYN) with the nucleus is still a subject of debate. Following the identification of the synuclein protein, as the first member of the synuclein family, different and independent research lines finally imparted α-SYN as the pathogenic agent involved in PD etiopathogenesis: α-SYN was identified as the major constituent of LB and at the same time, a missense mutation in the α-SYN encoding gene (*SNCA*) was found in a patient with a familial form of PD [5,6]. Since then, many studies have been trying to elucidate the role of α-SYN in the disease process and confirmed its role as the causative agent for disease initiation and propagation.

Now it is well established that α-SYN is crucial in the etiopathogenesis of a large group of heterogeneous diseases that are all characterized by the deposition of α-SYN throughout the central and peripheral nervous system. These neurodegenerative disorders include PD, multiple system atrophy, dementia with LB and neurodegeneration with brain iron accumulation and their shared histopathological features classify these heterogeneous diseases as the "synucleinopathies." Many genetic and environmental factors can disturb the native folding of α-SYN and diverging the folding process of α-SYN onto a pathogenic pathway will lead to the formation of heterogeneous toxic assemblies. Understanding how α-SYN propagates through this large conformational landscape is crucial to develop therapeutic strategies that can tackle the formation of toxic aggregates that give rise to synucleinopathies.

THE CONFORMATIONAL LANDSCAPE OF α-SYNUCLEIN: THE NATIVE STATE

α-SYN is abundantly expressed in neuronal cells and accounts for 0.1% of all proteins in the whole brain [7]. Initial ultrastructural immunolabeling studies showed that α-SYN mainly distributes throughout the axonal terminals, within the presynaptic cytoplasmic matrix and adjacent to synaptic vesicles [8]. This was further confirmed by several biochemical and cellular fractionation experiments that examined the specific localization of α-SYN. Here, α-SYN cosedimented with the synaptosomal fraction and within this fraction α-SYN is enriched in the synaptic supernatant [4,7]. Although only low amounts were found directly to bind synaptic vesicles, the localization of α-SYN at different sites within the presynapse is important since α-SYN is thought to function at the synapse, where it exists in equilibrium between cytosolic and membrane-bound states in a process that is tightly regulated [9]. It is able to dynamically bind the lipid membrane of small vesicles and in order to faithfully shuttle between cytosolic and membrane-bound states, the native state of α-SYN needs to partition rapidly and efficiently between different conformational states.

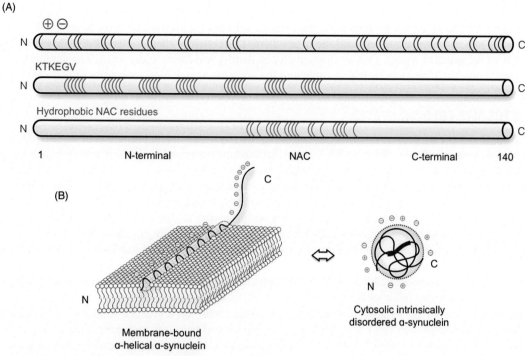

FIGURE 2.1 The Physicochemical Properties of α-Synuclein

α-SYN is a small protein with high conformational flexibility. (A) It comprises three main domains including the N-terminus, the hydrophobic NAC domain, and the negatively charged C-terminus. The high charge density of α-SYN in combination with its high hydrophobicity allows α-SYN to exist as an unfolded protein in the cytosol as depicted in (B). The imperfect KTKEGV consensus motif with alternating positively and negatively charged residues confers the amphipathic character of α-SYN and by adopting a α-helical conformation α-SYN inserts itself in the lipid bilayer with its negative residues exposed to the solvent. Due to its structural flexibility, a dynamic equilibrium exists between the membrane-bound and cytosolic states of α-SYN, the latter being monomeric and unfolded or multimeric and α-helical.

This remarkable conformational flexibility is made possible by the unique physicochemical properties of α-SYN [10] as shown in Fig. 2.1A. α-SYN is a small protein of 140 amino acids and consists out of 3 distinct domains. The central part of the protein consists out of a long hydrophobic stretch and still owes its name to its original identification in amyloid plaques, the nonamyloid component or NAC domain. Over 40% of the NAC domain is made up of alanine and valine residues that significantly contributes to its hydrophobicity. With its hydrophobic and amyloidogenic character and a net charge of zero, isolated fragments of the NAC region can readily form amyloid, which makes the NAC domain a crucial segment in the fibrillation of α-SYN [11,12]. An N-terminal domain precedes the central NAC region and contains several imperfect KTKEGV repeats that are included within a sequence of eleven amino acids long. These repeats alternate seven times within both the N-terminal and NAC domains, and confer the amphipathic character of α-SYN [13]. The alternating positively charged lysine residues interact with the negatively charged phosphate head groups of the lipid membrane and by adopting a twisting helical conformation α-SYN is able to superficially bind the lipid bilayer with its negatively charged residues of the same segments exposed to the solvent [14]. In contrast to the N-terminal region and the NAC domain, the C-terminal region contains 15 negatively charged residues, which gives α-SYN its net negative charge. Due to this high charge density and electrostatic repulsions within the C-terminal domain, it remains largely unstructured lowering the aggregation propensity of α-SYN [15].

CYTOSOLIC FOLDING STATES

Although numerous studies have been investigating the physicochemical properties of α-SYN, its native state remains elusive. When α-SYN is purified and characterized in vitro, it mainly exists as an unfolded monomer with little or no tertiary structure, which is due to its large net charge and hydrophobicity [16]. However, the idea that α-SYN exists as a single monomeric protein was recently challenged when new lines of evidence showed that cytosolic α-SYN can adapt a quaternary multimeric conformation via self-interaction [17]. This discrepancy with previous studies that found α-SYN to exist as an unfolded monomer was hypothesized to be the result of the instability of these α-SYN quaternary complexes that readily dissociate during purification procedures [18]. After stabilizing α-SYN multimers via chemical cross-linking, α-SYN acquires an α-helix-rich conformation that is the result of intermolecular associations of multiple α-SYN molecules that together form the multimeric complex [19]. A crucial role appears to lie within the KTKEGV consensus motif with a certain degree of redundancy between the alternating sequences. Alteration within this conserved KTKEGV sequence abolishes the formation of α-helix-rich α-SYN multimers in the cytosol and increases their propensity to aggregate [20]. A caveat to these findings, however, is that the KTKEGV sequence is crucial for membrane association (discussed later), and disrupting this sequence and therefore the balance between unstructured cytosolic and helical membrane-bound states also leads to aggregation [21–23]. It is thus not clear whether these folding assemblies reflect kinetically trapped folding intermediates or functional relevant species. Although the real stoichiometric contributions of these different species are not yet elucidated, and until these discrepancies are further clarified, α-SYN appears to mainly exist in the cytosol as an unfolded molecule with no apparent conformation [24]. When α-SYN is in its disordered state, it can shield its hydrophobic NAC core from the cytoplasm via transient intramolecular interactions, which prevents aberrant misfolding and aggregation (Fig. 2.1A–B) [24].

MEMBRANE-BOUND FOLDING STATES

α-SYN is loosely associated with membrane vesicles in the presynapse and is able to directly bind small negatively charged vesicles because of its amphipathic features. α-SYN possesses several imperfect KTKEGV repeats with alternating positively charged lysine, negatively charged glutamate and hydrophobic residues. These canonical repeats result in a reciprocal interaction with negatively charged phospholipid head groups, the hydrophobic inner membrane and the cytoplasmic solvent (Fig. 2.1B) [25]. The lysine residues point sideward from the helix, which allow favorable interactions with the charged head groups at the phospholipid interface, a process that is also referred to as "snorkeling" [26]. The hydrophobic threonine or alanine residues point toward the nonpolar face while the positively charged glutamate residues will be positioned opposite toward the polar hydrated cytosolic side. Every repeat in the amphipathic N-terminal region takes up three helical turns with low immersion depths and as a result, α-SYN adopts a helical curvature that binds peripherally on the outside of small vesicles [27] in a bent or extended helical conformation [27–30].

α-SYN has a specific affinity for lipid membrane composition that is regulated by two regions that bind the membrane [10]. An initial sequence that comprises the first 25 amino acids of the N-terminus inserts itself in the outer membrane leaflet and serves as a lipid anchor [10]. The adjacent and more central region then acts as a membrane sensor that determines the strength of the interaction with the membrane in a lipid-dependent manner [10,31,32]. These membrane structures include acidic phospholipids, such as phosphatidic acid (PA) or phosphatidylserines (PS), and the glycosphingolipid GM1 [33,34], which are enriched in secretory and endocytic vesicles, but also cardiolipin that is found at the outer and inner mitochondrial membrane [35,36]. As a consequence, different membrane properties, such as curvature and fatty acid composition will affect the interaction of α-SYN with the membrane [32,37]. Additionally, α-SYN can also form multimers via insertion of its amphipathic α-helix that bends into a broken helix or hairpin structure within the membrane forming multiple α-SYN molecules that associate together on the membrane surface [38].

Opposed to the N-terminal and central regions, the C-terminal region does not participate in membrane interactions. It does not possess any amphipathic features and is rich of negatively charged amino acid residues. The entire C-terminal region therefore remains unstructured and when α-SYN binds to the phospholipid membrane electrostatic repulsion between lipid membrane and the C-terminal sequence will remain in the cytoplasm. The C-terminal region therefore remains freely available to bind and interact with other synaptic proteins, and together they bind and cluster synaptic vesicles at the synaptic membrane where they regulate synaptic vesicle kinetics [30,39–44]. α-SYN can furthermore cluster lipid vesicles through a mechanism in which the entire sequence spans two separate membrane surfaces via its bent helix [45]. This way, α-SYN is proposed to bind and link two separate membrane components where it is involved in regulating the fusion and the stability of docked synaptic vesicles and neurotransmitter release. α-SYN conformation and function are thus intimately linked and recent observations that α-SYN needs to adopt a multimeric conformation to restrict the motility of synaptic vesicles further underscores this view [46,47].

THE CONFORMATIONAL LANDSCAPE OF α-SYNUCLEIN: MISFOLDED VARIANTS

Inclusions of aggregated α-SYN in the brain of PD, MSA, and DLB patients reveal that α-SYN misfolding and aggregation plays a central role in disease pathogenesis. Isolation of LB from the brain of synucleinopathy patients showed that Lewy inclusions are mainly comprised of bundled α-SYN filaments with a typical ultrastructural appearance [48,49]. These large Lewy inclusions have a very densely packed core with radiating aggregated filaments at the outer rim [6]. These filaments are the product of a complex misfolding process that starts from monomeric α-SYN that forms a folding intermediate and which self-associates with other α-SYN molecules into smaller oligomeric variants that structurally convert into larger fibrils.

FOLDING INTERMEDIATES

It is clear that α-SYN embodies a wide conformational repertoire of native folding assemblies that shuttle between unfolded monomeric, partially folded and folded structural states. α-SYN has to partition between these different conformational states without misfolding so that it preserves its functional integrity. The folding of α-SYN thus has to be very closely monitored by protein homeostasis mechanisms to avoid exposure of its hydrophobic sequences, such as its NAC core, since this might lead to misfolding and aggregation. However, since membrane-bound and unfolded α-SYN cannot form aggregates directly, an intermediate configuration has to be adopted in order to transition from an intrinsically disordered or α-helical state to a highly structured oligomeric or fibrillar state. Therefore, α-SYN has to fold into a configuration that is different from both its native and amyloid state and thus a new question arises as to how α-SYN is able to transition between these distinct conformations during the initial steps of protein misfolding?

Like all universal systems, α-SYN will try to reach a folding state that is associated with the lowest amount of free energy. In general terms, this is the resultant of the molecule and the environment with which it interacts, and it is the sum of enthalpic and entropic contributions. In this way, proteins are funneled through a rugged energy landscape with various conformational states governed by a multitude of energetic factors that will lead to the formation of α-SYN intermediates on pathway to its native assembly [50]. The unique combination of hydrophobicity and uncompensated negative charged groups in the C-terminal region that limits the conformational space of α-SYN, allows it to populate an unstructured conformation in physiological conditions[16]. In contrast, variations in solution conditions can lead to partial unfolding of α-SYN or a molten globule-like state. Molten globules are used to describe a state of secondary structure without significant tertiary features as well as compact states with an intermediate secondary structure [51]. Conditions that can lead to such molten globules include changes in salt concentrations, pH, temperature, and differential interactions with various types of surfaces. This could occur in vivo during acidification in organelles resulting in protonation of the C-terminus, via the screening of charged residues by salts or divalent ions that will enhance the solubility of α-SYN or via binding to different types of lipid membranes that will induce partial folding. All these processes have the effect to lower the intramolecular repulsive forces and to decrease the protein-excluded volume, which in turn increases the probability that α-SYN self-associates with other molecules [52,53].

In order to characterize the role of these folding intermediates and determine their importance in disease pathogenesis, there is a need to track individual particles with an intermediate assembly state that sparsely populates the total ensemble of proteins. Only recently, the detection of single α-SYN molecules allowed deciphering the specific folding states of individual α-SYN molecules in solution. By fluorescent labeling of different residues of the α-SYN monomer, it was shown that in conditions with low pH the C-terminal tail undergoes collapse with minor effects on the NAC- and the N-terminal region [54,55]. The use of techniques that rely on bulk detection can then further provide averaged information on the α-SYN folding ensemble. This was, for example, done via NMR techniques where instead of individual α-SYN particles a more homogenous ensemble showed strong contacts between the C-terminus and NAC region with weak and transient contacts between the N- and the C-terminus [56,57]. Compaction of the C-terminal region results in a more rigid monomeric state with less conformational freedom that allows intermolecular associations between different hydrophobic NAC regions resulting into hydrophobic clusters. A collapse of the C-terminus will thus lead to partially folded intermediates or molten globule-like states with a higher propensity to self-associate via the fibrillogenic NAC core of α-SYN [12]. Alternatively, via increasing the concentration of salt, the negative charged residues in the C-terminal tail become screened, which does not necessarily results into compaction, but weakens the electrostatic repulsion between multiple α-SYN molecules. This is also seen after removing or truncation of the C-terminal tail, which leads to an enhancement of the aggregation rates of α-SYN [53,54,58]. Again, this results in a similar effect as observed for low pH conditions, allowing juxtapositioning of α-SYN molecules that facilitate subsequent oligomerization.

Instead of examining conditions with altered levels of pH, salts, or temperature, it is also possible to detect different populations of folding intermediates via single-molecule detection methods under physiological conditions [59]. Here, distinct monomeric assembly states can be subdivided in different classes and include a partially folded intermediate with a certain degree of β-content that is also observed in α-SYN amyloid [60,61]. Although these β-sheet assemblies comprised the smallest fraction of the different monomeric classes observed in these studies, they seem to be important in the aggregation process since shielding the negative charge of α-SYN induces a shift toward these β-curved monomers and point mutations analogous to those observed in clinical variants further shift the folding equilibrium toward this peculiar conformer [59,62,63]. This β-like subpopulation of α-SYN folding intermediates formed under native conditions might thus be aggregation competent and, in contrast to natively unfolded monomers that have a random coil or show weak interactions, the β-curved conformers have the ability to assemble into high-molecular weight assemblies via self-association through their hydrophobic NAC core as observed in conditions with low pH [64].

It thus becomes clear that α-SYN folding is subjected to a stochastic process where a variety of physical forces might favor the formation of partially folded or molten globule-like states. Kinetic barriers that confine α-SYN within its sharply delineated and rugged energy landscape will determine how α-SYN transitions between its different folding states. Pathological conditions lower the kinetic barriers and allow α-SYN to assemble more readily into folding intermediates that rely on transient noncovalent interactions [65]. Alterations within its physicochemical or environmental properties can therefore tip the discrete balance that exists between different monomeric α-SYN conformers. Partially folded or compacted states of α-SYN are therefore problematic since they tend to aggregate more easily [66]. Aberrant behavior or elevated levels of α-SYN and impaired proteostasis, which exist for example during disease conditions or occur simply during aging, will thus lead to a higher tendency of α-SYN to misfold, escape the natural clearance machinery, and form oligomers.

OLIGOMERIC VARIANTS

Early studies on α-SYN aggregation have shown that α-SYN assembles through a process that can be described by a general two-step nucleation model [67–70]. During the initial steps of amyloid formation, α-SYN folding intermediates will self-associate after hydrophobic collapse with the formation of hydrogen bonds accompanied by the appearance of β-sheets [51]. A stochastic process that is concentration dependent governs the association of α-SYN molecules where β-sheets form between adjacent polypeptide backbones [67]. The incorporation of individual α-SYN monomers into larger aggregates hereby occurs via a generic mechanism that mainly involves the main chain atoms and which can even be independent of the composition of the specific amino acid sequence [71,72].

During this process of amyloid formation, small oligomeric assemblies that become gradually enriched in β-sheet content have to structurally reorganize before assembling into an aggregation-competent fibril [70,73,74]. In terms of assembly kinetics, this will translate into a long lag phase during the initial aggregation steps and because of these reactions occurring randomly different discrete oligomeric species will coexist (Fig. 2.2). Several studies have reported pleomorphic oligomeric assemblies during the initial lag phase of α-SYN assembly [75–77]. Although the exact nature of oligomeric species and their role in the aggregation process is hard to determine due to their relative instability and

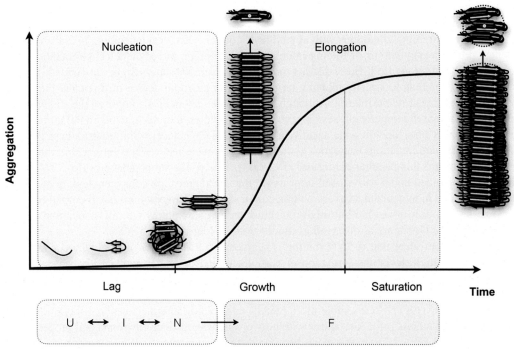

FIGURE 2.2 α-Synuclein Aggregation Kinetics is Characterized by a Nucleation-Dependent Polymerization Reaction

The mechanisms of α-SYN amyloid formation are proposed to occur via a two-step nucleation model that involves stochastic self-assembly of unfolded monomers *(U)* and subsequent structural conversions of transient intermediates *(I)* into a thermodynamically stable aggregation-prone nucleus *(N)*. These structural conversions impose high kinetic barriers and are apparent by a long initial lag phase. Primary nucleation or the formation of a stable nucleus is the rate-limiting step during α-SYN aggregation and is followed by rapid elongation via addition of soluble α-SYN to the freely available ends of the fibril *(F)*. During elongation, nucleation continues via primary but mostly secondary processes where fibril fragmentation will generate new available fibril ends that will favor the incorporation of α-SYN monomers. The self-assembly into fibrils will then continue rapidly until the total pool of free monomers becomes depleted, which translates in a plateau phase or saturation. The sum of these microscopic processes result in a typical sigmoidal curve that represent the kinetics of α-SYN aggregation in vitro.

heterogeneity, recent reports have been able to isolate kinetically trapped oligomers [78,79]. These stable oligomers have a variable degree of hydrophobicity, β-sheet content and stacking with a low tendency to elongate [78,79]. Fibrils of the same size but with a typical β-sheet conformation showed faster elongation rates, which indicates that the structure of these oligomers determines their kinetic stability and that distinct types of oligomers with a certain degree of β-sheet arrangement might convert more easily into stable fibrillar nuclei, which as a result are more transient and less detectable.

Other attempts to classify oligomeric intermediates have shown similar observations [76,79,80]. Early and transient oligomers that form during the lag phase appear to participate in fibril formation [76]. In contrast, oligomers that are generated during later steps persist longer and are even detectable at the end of the reaction [80]. Different oligomeric intermediates show differential β-sheet structures with early oligomeric intermediates showing remarkable similarities in their β-content compared to mature fibrils [76]. In order to form a stable aggregation-prone nucleus, transient oligomeric species require further compaction via reorganization of their three-dimensional structure [81]. This includes the disruption of β-sheets, the repositioning of amino acid side chains, and exposure of hydrophobic variants that were initially buried in the oligomer nucleus. Exposed hydrophobic residues are also observed in amyloid fibrils and the conversion of oligomers into aggregation-potent nuclei, which can arise after spontaneous association or via chemical stabilization by for instance dopamine, can then finally promote fibrillization by seeding soluble α-SYN [79,82]. α-SYN oligomers represent thus a continuum of coexisting heterogeneous and aggregated assemblies, which are on- or off-pathway toward fibril formation (Fig. 2.2).

FIBRILLAR VARIANTS

It is thus clear that a key step in fibril formation involves the structural reorganization and the formation of a stable seeding nucleus through a remarkably slow and large energy landscape that forms the basis of a core filament with one central axis. This protofilament grows bidirectional by incorporating monomeric α-SYN in uninterrupted stretches of β-sheets along the fibril axis [72] resulting into typical unbranched filamentous structures with a diameter that is only a few nanometers, but which extends in length over several microns [83,84]. Consecutive polypeptide sequences in the fibril are now aligned at a characteristic distance of 4.7 Å, which is shown by their β X-ray diffraction pattern [84].

Fibrils have the capacity to recruit soluble monomers, but not oligomeric α-SYN, at their growth-competent ends via an induced-fit mechanism [85,86]. This allows fibrils to elongate, which during the process can break and accelerate aggregation kinetics because they will expose more growth-competent ends that can be used for seeding [66,67]. This process of fragmentation, in some studies referred to as secondary nucleation, will add a last layer of complexity to the aggregation reaction and further explains why aggregation kinetics is often described by a sigmoid curve (Fig. 2.2) [87]. At the end of the reaction, the aggregation curve flattens to reach a plateau phase and reflects the situation where mature fibrils have consumed all available precursors.

Even though aggregation can be described by a common kinetic model, fibrillar polymorphisms can drastically influence aggregation kinetics [23,52,73,78]. At the atomic level, polymorphisms are imprinted in the fibril core through an excessive amount of molecular contacts. Single amino acids can engage in a multifold of interactions with other residues, for example, through hydrophobic contacts so that two opposing β-planes, which include the main chain atoms, intercalate into steric zippers and become stabilized [88,89]. Steric zippers mostly involve smaller hydrophobic amino acid side chains and are located inside a core that is void of water [90]. Larger hydrophobic side chains can also stabilize the nucleus in a similar manner but via the formation of hydrophobic pockets. In addition, electrostatic interactions maintain the fibril its conformation via salt bridges or by salt ladders that run perpendicular with the fibril axis along a single β-sheet plane [91]. Minor changes in the protein's physicochemical properties, such as amino acid sequence, charge, and hydrophobicity but also truncations and post-translational modifications (PTMs) can introduce variations in the structural ensemble and which is irrespective of the aggregation conditions [92].

To this end, different comparative studies assessed the self-assembly between α-SYN WT and its clinical variants A30P, E46K, and A53T, and showed that these variants assemble into fibrils with distinct secondary structures and morphologies [93–95]. Under identical aggregation conditions, the E46K and A53T variants showed accelerated aggregation kinetics as a result of more compact and stable interactions in the fibril core [94,95]. Another study reported increased thermodynamic stability of α-SYN fibrils for the recently identified H50Q clinical variant [96]. Additional differences in aggregation kinetics were furthermore prevalent as a result of decreased solubility of the H50Q variant that translated into increased aggregation rates of one magnitude [96,97].

Collectively, it is this clear that variations within the primary sequence or its interacting environment will dramatically influence the formation of protofibrils [98,99]. Even during subsequent assembly steps that include the complementation of fully assembled protofibrils additional polymorphisms can arise as a result of the differential association of individual protofibrils before entwining into a mature fibril [100]. Therefore, α-SYN is able to aggregate into different polymorphs, which suggests that not only differences during the aggregation process but also the end products per se might be important in terms of their pathogenic potential.

MISFOLDING IN COMPLEX ENVIRONMENTS

α-SYN aggregation is mostly studied in vitro under strictly controlled conditions. However, it is important to consider that aggregation is not just a two-dimensional in vitro process but that aggregation takes place in complex living system, which includes many variables that have different roles during α-SYN aggregation. In a crowded physiological environment, α-SYN interacts with various cellular components, such as molecular chaperones and lipid membranes [43,101–103]. All of these physiological interactors can differentially affect α-SYN its folding state, which can prevent but also promote α-SYN aggregation in vivo, for instance at the membrane.

α-SYN is highly enriched at the membrane and during association it adopts different conformations when it switches from a cytosolic state to a membrane-bound state. It is estimated that about a quarter of total α-SYN is bound to membranes [104] and although the cytosolic form is most abundant, the membrane-bound state has drawn considerable attention. An increasing number of studies are showing that conformational changes of α-SYN during lipid association can significantly influence its aggregation behavior. The different membrane-specific and cytosolic conformations that α-SYN adopts are not mutually exclusive and interchange rapidly during synaptic activity [46,47]. By binding with lipids, α-SYN accommodates a new secondary structure that can include unfolded, partially folded, and multimeric helical structures. It has a binding preference for small vesicles and lipid rafts, and changing the membrane composition not only alters its binding affinity but also increases its local concentration [34,105].

Interestingly, Lewy pathology is a common feature of different lipidoses, such as Gaucher's disease, GM2 gangliosidosis, or Niemann–Pick disease [106–108]. Mutations in the *GBA* gene that encodes for glucocerebrosidase, an enzyme that is active in lysosome and which is involved in lipid metabolism, is found in approximately 25% of the patients that have young onset PD [109]. Whereas homozygous and compound heterozygous mutations in the *GBA* gene give rise to Gaucher's disease, heterozygous mutations are now known as the most important risk factor to develop juvenile PD [110]. Additionally, mutations in the *PLA2G6* gene that encodes for a phospholipase responsible for removing polyunsaturated fatty acids from the lipid membrane, also results in neurodegeneration with extensive Lewy pathology [111,112]. Similarly, the fatty acid composition in the brain of idiopathic PD patients can be severely altered [112] and suggest a relationship between membrane interaction and α-SYN misfolding.

One hypothesis to explain membrane-induced aggregation is via the actions of the membrane on α-SYN folding. Negatively charged lipid membranes have the tendency to attract α-SYN via long-range electrostatic interactions and concentrate α-SYN locally before it even binds with its outer leaflets [22]. After association, this results in a local buildup of α-SYN molecules that are wedged together so closely that they appear as a carpet of interacting molecules, especially at sites where α-SYN is required to induce membrane curvature and drive tubulation [37,113–116]. These membrane-bound forms have a distinctive helical conformation and a concentration that exceeds three magnitudes of that in bulk solution [23].

Environmental perturbations, such as increased levels of α-SYN or changes in lipid composition, further alter α-SYN its membrane structure by promoting the formation of aggregation-competent folding intermediates [23,117]. These assemblies are considered to be isoforms of the typical membrane-bound helical state that in general inhibits aggregation [21,118,119]. The aggregation-prone conformer retains its helicity within the N-terminal region, which keeps α-SYN anchored to the membrane, but is void of secondary structure in the central domain, which only loosely associates with the membrane and remains within the cytosol [119]. These partially folded molecules can subsequently self-associate via their unstructured NAC region that gradually increases in β-content while the N-terminal region remains helical and only reorganizes during later fibrillation steps [119,120]. Similar to the initial association events of aggregation-prone oligomers and amyloid formation in solution, the kinetics of membrane-dependent aggregation can therefore be characterized by a two-step nucleation model [73,78] and requires a folding intermediate to gradually convert into a prenucleus complex that needs to rearrange into a growth-competent nucleus from which sustained aggregation can proceed [23].

Although some discrepancies existed between initial studies concerning the role of membranes in either stimulating or inhibiting α-SYN aggregation, it is now becoming clear that this is the result of different protein-to-lipid ratios and synthetic lipid compositions that can affect the aggregation rate of α-SYN [21–23]. In order to allow α-SYN to aggregate, it is crucial that the lipid–protein ratio is sufficiently low so that a remainder of soluble and unfolded fraction persists in order for aggregation to continue. If the lipid–protein ratio increases, then virtually all α-SYN will bind the membrane and resist aggregation because there is no soluble α-SYN available for seeding [23]. However, these conditions are not physiologically relevant since several thousands of α-SYN molecules exist at high concentrations for on average 300 synaptic vesicles [121]. The amount of α-SYN particles can even increase in disease conditions as a result of increased α-SYN expression or impaired homeostasis mechanisms. Hence, in normal or pathological conditions there will always be an excess of α-SYN protein and therefore only a smaller fraction of α-SYN will be associated with the membrane. Within this fraction, α-SYN can form an aggregation-prone nucleus that induces aggregation via primary nucleation [23]. The remainder unbound fraction will then aggregate on the surface of the lipid vesicle and elongate within the solution on the periphery of the vesicle [23,104,122]. The aggregated filament can subsequently break, dissociate from the membrane, and grow bi- instead of unidirectionally via two aggregation-competent ends and an excess of soluble precursor that is available in the cytosol.

Membrane-dependent aggregation is thus clearly influenced by several physiological processes that can include protein-lipid ratios, the composition of the lipid membrane, and again its physicochemical properties. Variations in membrane interactions were first noted for the A30P clinical variant that showed weakened binding with synthetic and biological membranes upon association [123]. In contrast, the A53T variant shows no clear changes in membrane interaction while the E46K strengthens it [124–126]. Even though these variations have a differential impact on the membrane-binding properties, there is a general consensus in that they all change the membrane-cytosolic equilibrium in which they exist.

Given all these variables that are at play during different steps of the aggregation process it becomes evident that there is no such thing as a prototype oligomer or fibril. The assembly of monomeric α-SYN into high-molecular weight species is shaped by an almost indeterminable amount of variables, each of which influences the aggregate its morphometry.

SYNUCLEIN STRAINS

It is thus clear that conformational conversions and structural variations during the aggregation process give rise to different aggregated polymorphs. In vitro, this can easily lead to batch variability and the detection of multiple secondary structures within a single preparation, which has significantly hampered the characterization of α-SYN aggregates [82,127]. In vivo, this situation is even more complex since various processes will influence the aggregation behavior of α-SYN. For instance, lipid coaggregation can result in tangle-like structures, which are undetectable in parallel conditions where lipids are omitted [117].

Under stringent aggregation and purification procedures it is shown that α-SYN polymerization can lead to the generation of different populations of α-SYN fibrils or strains [85], the importance of which was only very recently acknowledged [128]. Exposure of strains to cell culture or animal models leads to a strain-encoded manner of inheritanceand are paralleled with strain-specific phenotypes in these models [129–133]. It is based on a mechanism of protein-based inheritance and reminds us of prions that have the capacity to self-replicate by imprinting their structural information to their endogenous counterpart [134]. Prions are also found in fungi, yeast, and bacterial cells, and their infectious behavior can be traced back to the structural hierarchy of the aggregated assembly at the atomic level [135].

The existence of strains might finally provide an explanation for the observed heterogeneity that exists between different synucleinopathies, such as PD, DLB, and MSA, which all share the deposition of α-SYN as common histopathological hallmark but differ extensively in their clinical phenotype. Initial characterization of the fibrillar component of neuronal LB and glial cytoplasmic inclusions from PD and MSA patients revealed that these aggregated polymorphs have straight and twisted morphologies, respectively [1,136]. This provided one of the first indications that different structural polymorphs in the brain of synucleinopathy patients might indeed exist [49,136].

More recently, isolation of insoluble α-SYN fractions from the brain of PD and MSA patients showed that α-SYN-enriched fractions of MSA patients were able to induce neurotoxic and pathogenic effects in cellular assays and animal models [132,133,137]. This was in large contrast to fractions that were isolated from the brain of patients with PD, which did not show any effects in the same assays. Another study that assessed the properties of aggregates α-SYN of brain samples obtained from parkinsonian patients found similar observations. Here, brain samples derived from patients with idiopathic PD and patients with LRRK2 mutations were shown to significantly differ in their solubility and biochemical profile [138]. These experiments therefore suggest that different infectious or seeding-potent components exist in the brain of patients with different synucleinopathies and this has important implications in terms of developing biomarkers and strain- or conformation-specific aggregation inhibitors.

GLYCATION OF α-SYN—AN AGE-ASSOCIATED POSTTRANSLATIONAL MODIFICATION

α-SYN is known to undergo several PTMs, some of which are associated with pathology, like phosphorylation at serine 129. Glycation is an age-associated PTM that has been mostly studied in the context of diabetes mellitus, and consists in the nonenzymatic reaction of reducing sugars with phospholipids, nucleotides, and proteins, a process known as Maillard reaction [139,140]. In this reaction, reducing sugars covalently bind to amino-containing compounds forming early glycation products that, eventually, form final irreversible modifications named advanced glycation end products (AGEs) [141,142]. Several carbonyls that are by-products of glucose metabolism display very high reactivity and are the major glycation agents in cells. Among these, methylglyoxal (MGO) is the most reactive, being 20,000-fold more reactive than glucose [143]. MGO reacts with arginine and lysine residues, forming AGEs that cannot be normally processed in the cell.

The first reports of glycation in the context of PD identified AGEs at the periphery of LB in PD patients [144]. AGEs were also observed in newly formed LB in cases of incidental LB disease, suggesting that AGEs may trigger LB formation [145]. Glycation of α-SYN in lysine residues, for example, is thought to block normal ubiquitination and lead to the accumulation of the toxic forms of the protein, causing impairment of protein clearance pathways, such as the ubiquitin proteasome system, or autophagy.

Thus, strategies aimed at interfering with α-SYN glycation hold great potential as putative therapeutics, and need to be further investigated.

MISFOLDING OF α-SYNUCLEIN AS A THERAPEUTIC TARGET
PRESERVING THE NATIVE STATE

Understanding how α-SYN progresses through its complex folding landscape is crucial in order to develop new strategies that might prevent misfolding events. (outlined in Fig. 2.3). A successful strategy could herein lie within the stabilization of its native conformation.

Within its native helical multimeric form, α-SYN is shown to be protected against aggregation [17]. Interestingly, purified α-SYN does not spontaneously form helical multimers in vitro, which means that in order to fold correctly cofactors are required. Chaperones or membrane surfaces are known to be involved in the folding of α-SYN and can aid or stabilize α-SYN multimers. As such, the identification of these components might yield new therapeutic targets.

An example exists for transthyretin tetramers that are known to aggregate and subsequently cause systemic amyloidosis after destabilization of its native state. Transthyretin multimer dissociation, which is the rate-limiting step in transthyretin aggregation, can lead to misfolding and the formation of insoluble toxic aggregates [146]. Kinetic stabilization of native transthyretin tetramers increases the activation energy required for tetramer dissociation and slows down amyloid formation and disease progression [147]. Even though transthyretin aggregation is subjected to downhill polymerization aggregation kinetics, which is clearly different than that of α-SYN, preserving the native state of α-SYN might also prevent its dissociation into monomers and prevent the subsequent formation of misfolded and potentially toxic α-SYN assemblies.

PREVENTING THE FORMATION OF FOLDING INTERMEDIATES

Before or during aggregation, α-SYN needs to adapt a structured, more compact conformation in order to associate with other molecules to form oligomeric or fibrillar complexes. The observation that α-SYN is able to fold from an intrinsically disordered conformation into more a compact state that contains β-sheet structures raised the idea that pharmacological interventions aiming to stabilize

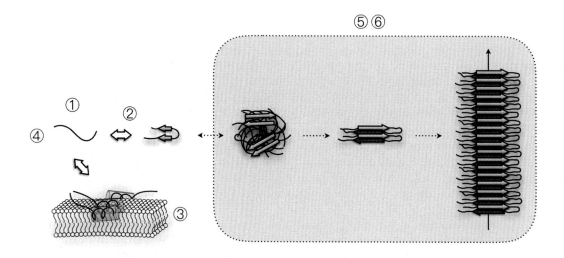

① Preserving the native state

② Preventing the formation of folding intermediates

③ Stabilization of membrane interactions

④ Reducing the expression levels of α-SYN

⑤ Direct inhibition of α-SYN aggregation

⑥ Disaggregation

FIGURE 2.3 Potential Entry Points for Therapeutic Interventions That Prevent, Inhibit or Reduce α-Synuclein Pathogenicity

α-SYN exists natively as an unfolded or a multimeric state that can be cytosolic or membrane bound. These different native states are in equilibrium and destabilization of multimeric assemblies will lead to an increase of unfolded α-SYN that can assemble into folding intermediates. The formation of folding intermediates occurs fast under physiological conditions allowing self-association of α-SYN molecules, which form heterogeneous oligomeric assemblies. In rare conditions, an aggregation-prone nucleus persists and further assembles into toxic fibrils. Different therapeutic strategies that directly target different steps in the misfolding pathway of α-SYN might ameliorate the toxic effects associated with different α-SYN misfolded assemblies. This can occur through stabilization of the native state to prevent dissociation into aggregation-competent unstructured monomers (1) or by preventing the subsequent formation of β-sheet intermediates on pathway to oligomeric self-assembly (2). Alternatively, alterations in membrane composition as observed in different lipidoses, such as GM2 gangliosidosis and Gaucher's disease influence the aggregation behavior of α-SYN and modifying or stabilizing these interactions could further lower the propensity of α-SYN to aggregate (3). A more direct strategy to reduce α-SYN neurotoxicity is by lowering the expression levels of α-SYN via targeting its mRNA transcripts or alternatively by boosting protein clearance via the proteosomal system (4). Finally, directly targeting α-SYN aggregates by preventing growth through β-sheet blockers, stimulating disassembly by disaggregase complexes, promoting clearance by the lysosomal system, or immunization might offer some of the most promising therapeutic strategies that will further lower the toxic burden of α-SYN aggregates (5,6).

the unstructured state or modulate the equilibrium that controls different conformations might lower the probability of encountering a misfolding event that leads to oligomeric association. A comparison between monomeric forms of α-SYN and their propensity to form secondary structures showed that introducing an A53T mutation significantly increased monomeric β-sheet content [63], which correlated with their propensity to aggregate. Introducing N-terminal acetylation, a PTMs of α-SYN that occurs in vivo, has shown to increase helical content in the N-terminal fragment, which in turn altered its membrane-binding and aggregation properties [148–150].

Even though the intrinsically disordered state of α-SYN lacks secondary structure, recent studies have identified molecules that bind unfolded regions of Aβ and other pathogenic proteins [151,152]. Pharmacological chaperoning via binding to a specific conformation of the intrinsically disorder ensemble of α-SYN showed that α-SYN pathogenicity could be rescued in various cellular assays [153]. Limiting the conformational flexibility of α-SYN monomers via the use of β-sheet blockers furthermore showed inhibitory effects on aggregation [154].

However, since diagnosis is usually made in later stages of the disease and the aggregation process, prevention of rapid α-SYN interconversion will no longer suffice. Therefore, it remains to be shown whether this method might also block the templated assembly of α-SYN monomers into fibrillar seeds. A preventive strategy could thus be used in combination with other therapeutic interventions that directly target α-SYN aggregates.

TARGETING AGGREGATION

Lowering α-synuclein expression levels

One of the most promising strategies in tackling synucleinopathies remains the inhibition of α-SYN aggregation either via direct targeting α-SYN aggregate formation or via decreasing α-SYN expression levels. α-SYN fibrils can generate toxicity if they can seed and amplify via conversion of its soluble precursor. Removing or lowering the expression levels of its fundamental building block would therefore provide a good strategy to prevent α-SYN from further aggregating. The idea to lower the expression level of α-SYN stems from observations that α-SYN knock out animals do not show any aberrant pathology, possible via functional redundancy that exist between α-, β-, and γ-SYN [102,155–157]. Moreover, in vivo propagation and aggregation of α-SYN via inoculation of pathological seeds can be completely abolished or ameliorated when these animals lack expression of α-SYN [133] or when α-SYN expression levels are lowered [129].

A popular approach to lower the expression of α-SYN consists out of targeted suppression of the *SNCA* transcript. Initial studies in rodent models that performed acute silencing of α-SYN in dopaminergic neurons of the substantia nigra were effective in lowering α-SYN expression but strikingly, this was accompanied with neurodegeneration and inflammation [158,159]. This is in contrast with more recent studies performed in rats and primates where similar strategies did not induce detrimental effects but which were still effective in lowering the expression levels of α-SYN [160,161]. These apparent conflicting results could be explained by the fact that in the initial studies acute knock down of α-SYN resulted in a drastic reduction of α-SYN expression levels accompanied by viral vector toxicity while in the more recent studies, lower suppression up to only 35–50% was achieved and which still sufficed to obtain neuroprotection. Acute knock down of α-SYN could thus be detrimental to dopaminergic neurons and mild suppression of *SNCA* transcripts might be a more recommendable procedure.

Another strategy to lower α-SYN expression levels is by clearance of aggregated or soluble α-SYN via the natural clearance machinery. Degradation of soluble monomeric α-SYN mainly involves Hsc70 chaperone-mediated delivery of α-SYN to the lysosomes via binding to LAMP2 while degradation of aggregated α-SYN involves autophagophore formation with subsequent lysosomal fusion [162]. Enhancing chaperone-mediated autophagy by viral vector-mediated overexpression of beclin-1, LAMP2, and TFEB, which are all directly involved in the regulation of autophagy, resulted in lowered α-SYN levels with beneficial effects on α-SYN neurotoxicity and dopaminergic survival [163–165]. Similar effects were also observed via pharmacological modulation. Administration of rapamycin, an immunosuppressant that activates autophagy has been shown to effectively counteract α-SYN accumulation and aggregation, and lowered its associated neurotoxic effects in various cellular and animal models [166,167].

Next to autophagy, the proteosomal system might also provide opportunities to lower the levels of α-SYN. The identification of Nedd4 as a ubiquitin ligase that targets α-SYN to the endosomal–lysosomal pathway provided a new entry point to control the expression levels of α-SYN. Enhancing the degradation of α-SYN via Nedd4 overexpression or pharmacological stimulation lowered the aggregation of α-SYN and protected against neurotoxicity and cell loss in various animal models [168–170]. Administration of the small molecule NAB2, that activates Nedd4, reversed the pathological phenotype observed in iPS-derived cells from PD patients [153]. Collectively, these observations indicate that stimulation of degradation pathways and clearance of α-SYN is a promising strategy to stop α-SYN aggregation, which could be achieved via autophagy-activating small molecules, or viral vector-based approaches.

Direct inhibition of α-synuclein aggregation

The conversion of soluble α-SYN and its subsequent aggregation is considered a key event in PD. Considerable efforts have been devoted to identify compounds that inhibit the aggregation of α-SYN. Targeting α-SYN aggregation directly in the central but also the peripheral nervous system via small molecules, aggregate-binding chaperones, or antibodies are some of the most attractive therapeutic strategies to directly bind and inhibit aggregates.

Large library compound screens have already identified several interesting small molecules, such as CLR01, epigallocatachin-3-gallate (EGCG), and Anle-138 that can inhibit α-SYN aggregation in vitro [171–173]. CLR01 is a molecular tweezer that inhibits amyloid aggregation via binding to lysine residues and disrupts hydrophobic and electrostatic interactions in supramolecular assemblies [174]. EGCG not only binds to α-SYN fibrils and inhibits its aggregation but also prevents the structural conversion of native protein into amyloid [172]. This led to the formation of distinct aggregates that were amorphous and nontoxic to cells, probably via binding the C-terminal region, which abolished the interaction of the aggregate with the plasma membrane [175,176]. Anle-138 binds aggregate-specific sequences and not monomers, which leaves another advantage over other strategies since it will not interfere with the protein its physiological function. Testing of Anle-138 and CLR01 in α-SYN transgenic animal models showed a reduction in oligomer formation and α-SYN aggregation [171,177]. Furthermore, Anle-138 was shown to bind distinct prion strains, which from a therapeutic point of view might be important since α-SYN strains may underlie distinct synucleinopathies as well.

Passive or active immunization therapies by antibodies directed against α-SYN or designed to recognize α-SYN via conformational-specific epitopes have also shown encouraging results. Immunization with monoclonal antibodies against α-SYN showed clearance and reduced aggregation that lowered behavioral deficits and α-SYN-induced pathogenicity in various rodent models [178–180].

The successful outcome of this immunization strategy may lie in several reasons. First, α-SYN is an intracellular protein but many studies have shown that misfolded α-SYN becomes secreted [181]. These extracellular assemblies can then propagate between cells and targeting these species might ameliorate the viscous cascade of α-SYN propagation, internalization and finally, aggregation [182]. In addition, and quite unexpectedly, α-SYN-recognizing antibodies also become internalized and colocalize intracellularly with α-SYN in lysosomes [179]. Both passive and active immunization therapies result in activation of the autophagy pathway that increases the clearance of α-SYN. Lastly, immunization not only reduces the levels of α-SYN but it also actively targets aggregated α-SYN. This was shown by experiments where α-SYN pathology caused by intracerebral injections of preformed fibrillar seeds could be rescued by intraperitoneal injections of monoclonal antibodies that recognize misfolded α-SYN [180].

All these promising observation have now led to two commercially funded research initiatives that are currently assessing the safety and efficacy of α-SYN antibodies in clinical trials. AFFiRis is testing a peptide-carrier conjugate vaccine AFFITOPE PD01 whereas Roche (PRX002) is testing the safety of passive immunization against α-SYN.

A third promising method of directly targeting α-SYN aggregation is via the direct disassembly of toxic α-SYN aggregates. Disaggregation of amyloid protein occurs via the normal homeostasis system and is regulated by the concerted action of multiple complexed chaperone proteins of the cellular stress response. Hsp70, Hsp110, and complexed J-protein cochaperones orchestrate together the disassembly of small and large aggregates and form together the metazoan disaggregase complex [183]. Molecular cochaperones can bind misfolded or aggregated proteins and target them for refolding or degradation.

Several studies have attempted to block α-SYN aggregation by modulating the cellular stress response via overexpression of heat shock proteins, such as Hsp70 that are known to directly bind α-SYN fibrils with high affinity. These cochaperones act as antiaggregation barriers during stress situations and although initial studies showed promising results in their capacity to bind and block α-SYN aggregation in vitro, in vivo studies failed to reproduce these results [184,185]. This shows that although modulating the stress response might be an attractive therapeutic target, it remains a complicated strategy probably due to the many different cellular functions with which stress response is involved.

Indeed, these contrasting results could be explained by the regulation of Hsp70 in a more relevant cellular environment where first, the stress-induced Hsp70 is in competition for available binding spots on the fibril with its constitutive counterpart Hsc70 and second, upon binding in their ADP-bound form, ATP will replace ADP and cause rapid dissociation. In vivo, this results in a dynamic release and binding cycle that will counteract Hsp70-dependent aggregate disassembly since both Hsc70 and ATP are abundantly present [186]. Therefore, these cochaperones are likely to be more involved in the initial binding of aggregated assemblies, which results in the subsequent recruitment of the disaggregase complex.

Instead of simply focusing on the single actions of these chaperones, interesting alternatives might thus exist in synthetic peptides that mimic these fibril-binding proteins but which do not dissociate and therefore block subsequent elongation and aggregation or by stimulating the collective actions of the disaggregase complex. Stimulating their disaggregase function could lead to the neutralization of aggregated α-SYN and has indeed been shown to be successful [187]. Yeast cells also have the ability to fragment fibrils via the protein-remodeling Hsp104 disaggregase [188]. Mammalian cells, however, lack Hsp104 but overexpression does ameliorate dopaminergic neurodegeneration in an animal model based on overexpression of α-SYN [189]. Slight modifications in the primary sequence of the Hsp104 disaggregase have yielded large improvements in its ability to degrade α-SYN inclusions [190], and strategies that are aimed in improving the disassembly kinetics or the availability of this complex might be an interesting and new opportunity to antagonize misfolding events and degrade toxic assemblies.

STABILIZING MEMBRANE INTERACTIONS

Alteration in membrane composition results in drastic changes in α-SYN folding and aggregation kinetics, and can give rise to Lewy inclusions. Mutations in the *GBA* gene that leads to improper folding and reduced activity of the GCase enzyme, has an impact on lysosomal activity and the interaction of α-SYN with the membrane and leads to lowered clearance and subsequent accumulation of α-SYN [191,192]. GCase is a multimeric protein and variations in its primary sequence lead to destabilization of its biological and functional state. Therefore, stabilizing its functional state has been proposed as a strategy to reduce its pathogenicity [193]. Two synthetic chaperones, ambroxol and AT2101, have been tested in vitro and enhanced GCase activity via improved folding, transportation through the ER, and integration in the lysosome [194,195]. Overexpression of GCase in a transgenic mouse model of PD indeed reduced the levels α-SYN and improved cognitive defects [191].

Furthermore, biochemical studies have provided evidence that not only changes in the lipid head group but also changes in the acyl chain might alter α-SYN interaction with the membrane altering its propensity to aggregate [105]. It is therefore not surprising that mutations in *PLA2G6* or phospholipase A2, involved in metabolism of polyunsaturated fatty acids that preserves the integrity of the cell membrane also lead to Lewy pathology [111,112]. Several membrane components, such as phospholipids, sphingolipids, and their acyl chains can influence the behavior of α-SYN with the membrane [105]. Interestingly, a double-blinded study that assessed the effects of 2-weekly subcutaneous injections of GM1 glycolipid ganglioside resulted in improved symptomatic effects in PD patients although it is not known whether these effects could also act disease modifying [196].

CONCLUSIONS

α-SYN is a highly dynamic protein that is able to adopt a wide array of distinct structural assemblies. This conformational plasticity is crucial for α-SYN to efficiently fulfill its biological role at the synapse. However, the folding of α-SYN and the environment with which it interacts has to be meticulously regulated to prevent misfolding and subsequent aggregation. Many homeostasis mechanisms maintain α-SYN in its native conformation state, refold α-SYN after aggregate disassembly, or clear large aggregates via lysosomal degradation. Alterations in any of these mechanisms can lead to accumulation and formation of toxic assemblies. Understanding how α-SYN folds and partitions within this large conformational landscape is crucial to develop therapeutic strategies that might prevent, inhibit, or remove toxic aggregates. High-throughput screenings of small molecules that directly inhibit different steps in the aggregation pathway of α-SYN disserve great attention and have led to early successes that are now entering clinical trials. These continuous advances that we make raise optimism that one day we might be able stop the progressive prion-like cascade of α-SYN aggregation and allow us to move from a symptomatic to a curable condition.

REFERENCES

[1] M. Goedert, M.G. Spillantini, K. Del Tredici, H. Braak, 100 years of Lewy pathology, Nat. Rev. Neurol. 9 (2013) 13–24.

[2] F.H. Lewy, Die Entstehung der Einschlußkörper und ihre Bedeutung für die systematische Einordnung der sogenannten Viruskrankheiten, Deutsche Zeitschrift für Nervenheilkunde 124 (1932) 93–110.

[3] K. Ueda, et al. Molecular cloning of cDNA encoding an unrecognized component of amyloid in Alzheimer disease, Proc. Natl. Acad. Sci. USA 90 (1993) 11282–11286.

[4] L. Maroteaux, J.T. Campanelli, R.H. Scheller, Synuclein: a neuron-specific protein localized to the nucleus and presynaptic nerve terminal, J. Neurosci. 8 (1988) 2804–2815.

[5] M.H. Polymeropoulos, et al. Mutation in the alpha-synuclein gene identified in families with Parkinson's disease, Science 276 (1997) 2045–2047.

[6] M.G. Spillantini, et al. Alpha-synuclein in Lewy bodies, Nature 388 (1997) 839–840.

[7] T. Shibayama-Imazu, et al. Cell and tissue distribution and developmental change of neuron specific 14 kDa protein (phosphoneuroprotein 14), Brain Res. 622 (1993) 17–25.

[8] D.F. Clayton, J.M. George, Synucleins in synaptic plasticity and neurodegenerative disorders, J. Neurosci. Res. 58 (1999) 120–129.

[9] J.M. George, H. Jin, W.S. Woods, D.F. Clayton, Characterization of a novel protein regulated during the critical period for song learning in the zebra finch, Neuron 15 (1995) 361–372.

[10] G. Fusco, et al. Direct observation of the three regions in alpha-synuclein that determine its membrane-bound behaviour, Nat. Commun. 5 (2014) 3827.

[11] V.N. Uversky, A.L. Fink, Amino acid determinants of alpha-synuclein aggregation: putting together pieces of the puzzle, FEBS Lett. 522 (2002) 9–13.

[12] H. Han, P.H. Weinreb, P.T. Lansbury Jr., The core Alzheimer's peptide NAC forms amyloid fibrils which seed and are seeded by beta-amyloid: is NAC a common trigger or target in neurodegenerative disease?, Chem. Biol. 2 (1995) 163–169.

[13] W.S. Davidson, A. Jonas, D.F. Clayton, J.M. George, Stabilization of alpha-synuclein secondary structure upon binding to synthetic membranes, J. Biol. Chem. 273 (1998) 9443–9449.

[14] C.C. Jao, A. Der-Sarkissian, J. Chen, R. Langen, Structure of membrane-bound alpha-synuclein studied by site-directed spin labeling, Proc. Natl. Acad. Sci. USA 101 (2004) 8331–8336.

[15] W. Hoyer, D. Cherny, V. Subramaniam, T.M. Jovin, Impact of the acidic C-terminal region comprising amino acids 109-140 on alpha-synuclein aggregation in vitro, Biochemistry 43 (2004) 16233–16242.

[16] V.N. Uversky, J.R. Gillespie, A.L. Fink, Why are "natively unfolded" proteins unstructured under physiologic conditions?, Proteins 41 (2000) 415–427.

[17] T. Bartels, J.G. Choi, D.J. Selkoe, Alpha-synuclein occurs physiologically as a helically folded tetramer that resists aggregation, Nature 477 (2011) 107–110.

[18] E.S. Luth, T. Bartels, U. Dettmer, N.C. Kim, D.J. Selkoe, Purification of alpha-synuclein from human brain reveals an instability of endogenous multimers as the protein approaches purity, Biochemistry 54 (2015) 279–292.

[19] U. Dettmer, A.J. Newman, E.S. Luth, T. Bartels, D. Selkoe, In vivo cross-linking reveals principally oligomeric forms of alpha-synuclein and beta-synuclein in neurons and non-neural cells, J. Biol. Chem. 288 (2013) 6371–6385.

[20] U. Dettmer, A.J. Newman, V.E. von Saucken, T. Bartels, D. Selkoe, KTKEGV repeat motifs are key mediators of normal alpha-synuclein tetramerization: their mutation causes excess monomers and neurotoxicity, Proc. Natl. Acad. Sci. USA 112 (2015) 9596–9601.

[21] M. Zhu, A.L. Fink, Lipid binding inhibits alpha-synuclein fibril formation, J. Biol. Chem. 278 (2003) 16873–16877.

[22] M. Necula, C.N. Chirita, J. Kuret, Rapid anionic micelle-mediated alpha-synuclein fibrillization in vitro, J. Biol. Chem. 278 (2003) 46674–46680.

[23] C. Galvagnion, et al. Lipid vesicles trigger alpha-synuclein aggregation by stimulating primary nucleation, Nat. Chem. Biol. 11 (2015) 229–234.

[24] F.X. Theillet, et al. Structural disorder of monomeric alpha-synuclein persists in mammalian cells, Nature 530 (2016) 45–50.

[25] T.S. Ulmer, A. Bax, N.B. Cole, R.L. Nussbaum, Structure and dynamics of micelle-bound human alpha-synuclein, J. Biol. Chem. 280 (2005) 9595–9603.

[26] R. Bussell Jr., D. Eliezer, A structural and functional role for 11-mer repeats in alpha-synuclein and other exchangeable lipid binding proteins, J. Mol. Biol. 329 (2003) 763–778.

[27] R. Bussell Jr., T.F. Ramlall, D. Eliezer, Helix periodicity, topology, and dynamics of membrane-associated alpha-synuclein, Protein Sci. 14 (2005) 862–872.

[28] P. Borbat, T.F. Ramlall, J.H. Freed, D. Eliezer, Inter-helix distances in lysophospholipid micelle-bound alpha-synuclein from pulsed ESR measurements, J. Am. Chem. Soc. 128 (2006) 10004–10005.

[29] A.C. Ferreon, Y. Gambin, E.A. Lemke, A.A. Deniz, Interplay of alpha-synuclein binding and conformational switching probed by single-molecule fluorescence, Proc. Natl. Acad. Sci. USA 106 (2009) 5645–5650.

[30] E.R. Georgieva, T.F. Ramlall, P.P. Borbat, J.H. Freed, D. Eliezer, The lipid-binding domain of wild type and mutant alpha-synuclein: compactness and interconversion between the broken and extended helix forms, J. Biol. Chem. 285 (2010) 28261–28274.

[31] T. Bartels, et al. The N-terminus of the intrinsically disordered protein alpha-synuclein triggers membrane binding and helix folding, Biophys. J. 99 (2010) 2116–2124.

[32] C.R. Bodner, C.M. Dobson, A. Bax, Multiple tight phospholipid-binding modes of alpha-synuclein revealed by solution NMR spectroscopy, J. Mol. Biol. 390 (2009) 775–790.

[33] D.L. Fortin, et al. Lipid rafts mediate the synaptic localization of alpha-synuclein, J. Neurosci. 24 (2004) 6715–6723.

[34] Z. Martinez, M. Zhu, S. Han, A.L. Fink, GM1 specifically interacts with alpha-synuclein and inhibits fibrillation, Biochemistry 46 (2007) 1868–1877.

[35] S. Ghio, F. Kamp, R. Cauchi, A. Giese, N. Vassallo, Interaction of alpha-synuclein with biomembranes in Parkinson's disease—role of cardiolipin, Prog. Lipid Res. 61 (2016) 73–82.

[36] K. Nakamura, et al. Optical reporters for the conformation of alpha-synuclein reveal a specific interaction with mitochondria, J. Neurosci. 28 (2008) 12305–12317.

[37] C.M. Pfefferkorn, Z. Jiang, J.C. Lee, Biophysics of alpha-synuclein membrane interactions, Biochim. Biophys. Acta. 1818 (2012) 162–171.

[38] J. Burre, M. Sharma, T.C. Sudhof, Alpha-synuclein assembles into higher-order multimers upon membrane binding to promote SNARE complex formation, Proc. Natl. Acad. Sci. USA 111 (2014) E4274–4283.

[39] J. Burre, M. Sharma, T.C. Sudhof, Systematic mutagenesis of alpha-synuclein reveals distinct sequence requirements for physiological and pathological activities, J. Neurosci. 32 (2012) 15227–15242.

[40] C. Betzer, et al. Identification of synaptosomal proteins binding to monomeric and oligomeric alpha-synuclein, PLoS One 10 (2015) e0116473.

[41] A.J. Trexler, E. Rhoades, Alpha-synuclein binds large unilamellar vesicles as an extended helix, Biochemistry 48 (2009) 2304–2306.

[42] J. Diao, et al. Native alpha-synuclein induces clustering of synaptic-vesicle mimics via binding to phospholipids and synaptobrevin-2/VAMP2, Elife. 2 (2013) e00592.

[43] J. Burre, et al. Alpha-synuclein promotes SNARE-complex assembly in vivo and in vitro, Science 329 (2010) 1663–1667.

[44] K.J. Vargas, et al. Synucleins regulate the kinetics of synaptic vesicle endocytosis, J. Neurosci. 34 (2014) 9364–9376.

[45] Y. Lai, et al. Nonaggregated alpha-synuclein influences SNARE-dependent vesicle docking via membrane binding, Biochemistry 53 (2014) 3889–3896.

[46] L. Wang, et al. Alpha-synuclein multimers cluster synaptic vesicles and attenuate recycling, Curr. Biol. 24 (2014) 2319–2326.

[47] D.L. Fortin, et al. Neural activity controls the synaptic accumulation of alpha-synuclein, J. Neurosci. 25 (2005) 10913–10921.

[48] M.G. Spillantini, et al. Filamentous alpha-synuclein inclusions link multiple system atrophy with Parkinson's disease and dementia with Lewy bodies, Neurosci. Lett. 251 (1998) 205–208.

[49] M.G. Spillantini, R.A. Crowther, R. Jakes, M. Hasegawa, M. Goedert, Alpha-synuclein in filamentous inclusions of Lewy bodies from Parkinson's disease and dementia with Lewy bodies, Proc. Natl. Acad. Sci. USA 95 (1998) 6469–6473.

[50] F.U. Hartl, A. Bracher, M. Hayer-Hartl, Molecular chaperones in protein folding and proteostasis, Nature 475 (2011) 324–332.

[51] V.N. Uversky, J. Li, A.L. Fink, Evidence for a partially folded intermediate in alpha-synuclein fibril formation, J. Biol. Chem. 276 (2001) 10737–10744.

[52] A.K. Buell, et al. Solution conditions determine the relative importance of nucleation and growth processes in alpha-synuclein aggregation, Proc. Natl. Acad. Sci. USA 111 (2014) 7671–7676.

[53] Y. Izawa, et al. Role of C-terminal negative charges and tyrosine residues in fibril formation of alpha-synuclein, Brain Behav. 2 (2012) 595–605.

[54] A.J. Trexler, E. Rhoades, Single molecule characterization of alpha-synuclein in aggregation-prone states, Biophys. J. 99 (2010) 3048–3055.

[55] T.L. Yap, C.M. Pfefferkorn, J.C. Lee, Residue-specific fluorescent probes of alpha-synuclein: detection of early events at the N-and C-termini during fibril assembly, Biochemistry 50 (2011) 1963–1965.

[56] K.P. Wu, D.S. Weinstock, C. Narayanan, R.M. Levy, J. Baum, Structural reorganization of alpha-synuclein at low pH observed by NMR and REMD simulations, J. Mol. Biol. 391 (2009) 784–796.

[57] S. McClendon, C.C. Rospigliosi, D. Eliezer, Charge neutralization and collapse of the C-terminal tail of alpha-synuclein at low pH, Protein Sci. 18 (2009) 1531–1540.

[58] W. Li, et al. Aggregation promoting C-terminal truncation of alpha-synuclein is a normal cellular process and is enhanced by the familial Parkinson's disease-linked mutations, Proc. Natl. Acad. Sci. USA 102 (2005) 2162–2167.

[59] M. Sandal, et al. Conformational equilibria in monomeric alpha-synuclein at the single-molecule level, PLoS Biol. 6 (2008) e6.

[60] H.Y. Kim, H. Heise, C.O. Fernandez, M. Baldus, M. Zweckstetter, Correlation of amyloid fibril beta-structure with the unfolded state of alpha-synuclein, Chembiochem. 8 (2007) 1671–1674.

[61] N.C. Maiti, M.M. Apetri, M.G. Zagorski, P.R. Carey, V.E. Anderson, Raman spectroscopic characterization of secondary structure in natively unfolded proteins: alpha-synuclein, J. Am. Chem. Soc. 126 (2004) 2399–2408.

[62] M. Brucale, et al. Pathogenic mutations shift the equilibria of alpha-synuclein single molecules towards structured conformers, Chembiochem. 10 (2009) 176–183.

[63] C. Camilloni, M. Vendruscolo, A relationship between the aggregation rates of alpha-synuclein variants and the beta-sheet populations in their monomeric forms, J. Phys. Chem. B 117 (2013) 10737–10741.

[64] B.I. Giasson, I.V. Murray, J.Q. Trojanowski, V.M. Lee, A hydrophobic stretch of 12 amino acid residues in the middle of alpha-synuclein is essential for filament assembly, J. Biol. Chem. 276 (2001) 2380–2386.

[65] A.I. Bartlett, S.E. Radford, An expanding arsenal of experimental methods yields an explosion of insights into protein folding mechanisms, Nat. Struct. Mol. Biol. 16 (2009) 582–588.

[66] L. Breydo, J.W. Wu, V.N. Uversky, Alpha-synuclein misfolding and Parkinson's disease, Biochim. Biophys. Acta. 1822 (2012) 261–285.

[67] S.J. Wood, et al. alpha-synuclein fibrillogenesis is nucleation-dependent. Implications for the pathogenesis of Parkinson's disease, J. Biol. Chem. 274 (1999) 19509–19512.

[68] B.I. Giasson, K. Uryu, J.Q. Trojanowski, V.M. Lee, Mutant and wild type human alpha-synucleins assemble into elongated filaments with distinct morphologies in vitro, J. Biol. Chem. 274 (1999) 7619–7622.

[69] M. Hashimoto, et al. Human recombinant NACP/alpha-synuclein is aggregated and fibrillated in vitro: relevance for Lewy body disease, Brain Res. 799 (1998) 301–306.

[70] T.R. Serio, et al. Nucleated conformational conversion and the replication of conformational information by a prion determinant, Science 289 (2000) 1317–1321.

[71] C.M. Dobson, The structural basis of protein folding and its links with human disease, Philos. Trans. R. Soc. Lond. B Biol. Sci. 356 (2001) 133–145.

[72] R. Nelson, et al. Structure of the cross-beta spine of amyloid-like fibrils, Nature 435 (2005) 773–778.

[73] N. Cremades, et al. Direct observation of the interconversion of normal and toxic forms of alpha-synuclein, Cell 149 (2012) 1048–1059.

[74] K.C. Luk, E.G. Hyde, J.Q. Trojanowski, V.M. Lee, Sensitive fluorescence polarization technique for rapid screening of alpha-synuclein oligomerization/fibrillization inhibitors, Biochemistry 46 (2007) 12522–12529.

[75] K.A. Conway, et al. Acceleration of oligomerization, not fibrillization, is a shared property of both alpha-synuclein mutations linked to early-onset Parkinson's disease: implications for pathogenesis and therapy, Proc. Natl. Acad. Sci. USA 97 (2000) 571–576.

[76] J. Kaylor, et al. Characterization of oligomeric intermediates in alpha-synuclein fibrillation: FRET studies of Y125W/Y133F/Y136F alpha-synuclein, J. Mol. Biol. 353 (2005) 357–372.

[77] S. Nath, J. Meuvis, J. Hendrix, S.A. Carl, Y. Engelborghs, Early aggregation steps in alpha-synuclein as measured by FCS and FRET: evidence for a contagious conformational change, Biophys. J. 98 (2010) 1302–1311.

[78] S.W. Chen, et al. Structural characterization of toxic oligomers that are kinetically trapped during alpha-synuclein fibril formation, Proc. Natl. Acad. Sci. USA 112 (2015) E1994–E2003.

[79] L. Pieri, K. Madiona, R. Melki, Structural and functional properties of prefibrillar alpha-synuclein oligomers, Sci. Rep. 6 (2016) 24526.

[80] A. Dusa, et al. Characterization of oligomers during alpha-synuclein aggregation using intrinsic tryptophan fluorescence, Biochemistry 45 (2006) 2752–2760.

[81] V.N. Uversky, A protein-chameleon: conformational plasticity of alpha-synuclein, a disordered protein involved in neurodegenerative disorders, J. Biomol. Struct. Dyn. 21 (2003) 211–234.

[82] M. Vilar, et al. The fold of alpha-synuclein fibrils, Proc. Natl. Acad. Sci. USA 105 (2008) 8637–8642.

[83] T.P. Knowles, M. Vendruscolo, C.M. Dobson, The amyloid state and its association with protein misfolding diseases, Nat. Rev. Mol. Cell. Biol. 15 (2014) 384–396.

[84] L.C. Serpell, J. Berriman, R. Jakes, M. Goedert, R.A. Crowther, Fiber diffraction of synthetic alpha-synuclein filaments shows amyloid-like cross-beta conformation, Proc. Natl. Acad. Sci. USA 97 (2000) 4897–4902.

[85] L. Bousset, et al. Structural and functional characterization of two alpha-synuclein strains, Nat. Commun. 4 (2013) 2575.

[86] N. Lorenzen, et al. The role of stable alpha-synuclein oligomers in the molecular events underlying amyloid formation, J. Am. Chem. Soc. 136 (2014) 3859–3868.

[87] T.P. Knowles, et al. An analytical solution to the kinetics of breakable filament assembly, Science 326 (2009) 1533–1537.

[88] Deleted in review.

[89] J.A. Rodriguez, et al. Structure of the toxic core of alpha-synuclein from invisible crystals, Nature 525 (2015) 486–490.

[90] M.R. Sawaya, et al. Atomic structures of amyloid cross-beta spines reveal varied steric zippers, Nature 447 (2007) 453–457.

[91] M.D. Tuttle, G. Comellas, A.J. Nieuwkoop, D.J. Covell, D.A. Berthold, K.D. Kloepper, J.M. Courtney, J.K. Kim, A.M. Barclay, A. Kendall, W. Wan, G. Stubbs, C.D. Schwieters, V.M. Lee, J.M. George, C.M. Rienstra, Soilid-state NMR structure of a pathogenic fibril of full-length human α-synuclein, Nat. Struct. Mol. Biol. 23 (2016) 409–415.

[92] M. Stefani, C.M. Dobson, Protein aggregation and aggregate toxicity: new insights into protein folding, misfolding diseases and biological evolution, J. Mol. Med. 81 (2003) 678–699.

[93] S.B. Nielsen, et al. Wildtype and A30P mutant alpha-synuclein form different fibril structures, PLoS One 8 (2013) e67713.

[94] L.R. Lemkau, et al. Mutant protein A30P alpha-synuclein adopts wild-type fibril structure, despite slower fibrillation kinetics, J. Biol. Chem. 287 (2012) 11526–11532.

[95] L.R. Lemkau, et al. Site-specific perturbations of alpha-synuclein fibril structure by the Parkinson's disease associated mutations A53T and E46K, PLoS One 8 (2013) e49750.

[96] R. Porcari, et al. The H50Q mutation induces a 10-fold decrease in the solubility of alpha-synuclein, J. Biol. Chem. 290 (2015) 2395–2404.

[97] D. Ghosh, et al. The Parkinson's disease-associated H50Q mutation accelerates alpha-synuclein aggregation in vitro, Biochemistry 52 (2013) 6925–6927.

[98] T. Shirahama, A.S. Cohen, High-resolution electron microscopic analysis of the amyloid fibril, J. Cell Biol. 33 (1967) 679–708.

[99] R. Khurana, et al. A general model for amyloid fibril assembly based on morphological studies using atomic force microscopy, Biophys. J. 85 (2003) 1135–1144.

[100] A.D. Dearborn, et al. Alpha-synuclein amyloid fibrils with two entwined, asymmetrically associated, proto-fibrils, J. Biol. Chem. 291 (5) (2015) 2310–2318.

[101] V.M. Nemani, et al. Increased expression of alpha-synuclein reduces neurotransmitter release by inhibiting synaptic vesicle reclustering after endocytosis, Neuron 65 (2010) 66–79.

[102] B. Greten-Harrison, et al. Alpha beta gamma-synuclein triple knockout mice reveal age-dependent neuronal dysfunction, Proc. Natl. Acad. Sci. USA 107 (2010) 19573–19578.

[103] A. Abeliovich, et al. Mice lacking alpha-synuclein display functional deficits in the nigrostriatal dopamine system, Neuron 25 (2000) 239–252.

[104] H.J. Lee, C. Choi, S.J. Lee, Membrane-bound alpha-synuclein has a high aggregation propensity and the ability to seed the aggregation of the cytosolic form, J. Biol. Chem. 277 (2002) 671–678.

[105] S. Kubo, et al. A combinatorial code for the interaction of alpha-synuclein with membranes, J. Biol. Chem. 280 (2005) 31664–31672.

[106] K. Suzuki, et al. Neuronal and glial accumulation of alpha- and beta-synucleins in human lipidoses, Acta. Neuropathol. 114 (2007) 481–489.

[107] Y. Saito, K. Suzuki, C.M. Hulette, S. Murayama, Aberrant phosphorylation of alpha-synuclein in human Niemann-Pick type C1 disease, J. Neuropathol. Exp. Neurol. 63 (2004) 323–328.

[108] K. Wong, et al. Neuropathology provides clues to the pathophysiology of Gaucher disease, Mol. Genet. Metabol. 82 (2004) 192–207.

[109] E. Sidransky, et al. Multicenter analysis of glucocerebrosidase mutations in Parkinson's disease, N. Engl. J. Med. 361 (2009) 1651–1661.

[110] R. Duran, et al. The glucocerebrosidase E326K variant predisposes to Parkinson's disease, but does not cause Gaucher's disease, Mov. Disord. 28 (2013) 232–236.

[111] A. Gregory, et al. Neurodegeneration associated with genetic defects in phospholipase A(2), Neurology 71 (2008) 1402–1409.

[112] C. Paisan-Ruiz, et al. Widespread Lewy body and tau accumulation in childhood and adult onset dystonia-parkinsonism cases with PLA2G6 mutations, Neurobiol. Aging 33 (2012) 814–823.

[113] Z.P. Jiang, M. de Messieres, J.C. Lee, Membrane remodeling by alpha-synuclein and effects on amyloid formation, J. Am. Chem. Soc. 135 (2013) 15970–15973.

[114] R. Sharon, et al. alpha-synuclein occurs in lipid-rich high molecular weight complexes, binds fatty acids, and shows homology to the fatty acid-binding proteins, Proc. Natl. Acad. Sci. USA 98 (2001) 9110–9115.

[115] A.R. Braun, M.M. Lacy, V.C. Ducas, E. Rhoades, J.N. Sachs, Alpha-synuclein-induced membrane remodeling is driven by binding affinity, partition depth, and interleaflet order asymmetry, J. Am. Chem. Soc. 136 (2014) 9962–9972.

[116] A.R. Braun, et al. Alpha-synuclein induces both positive mean curvature and negative Gaussian curvature in membranes, J. Am. Chem. Soc. 134 (2012) 2613–2620.

[117] E. Hellstrand, A. Nowacka, D. Topgaard, S. Linse, E. Sparr, Membrane lipid co-aggregation with alpha-synuclein fibrils, PLoS One 8 (2013) e77235.

[118] L.A. Munishkina, C. Phelan, V.N. Uversky, A.L. Fink, Conformational behavior and aggregation of alpha-synuclein in organic solvents: modeling the effects of membranes, Biochemistry 42 (2003) 2720–2730.

[119] G. Comellas, L.R. Lemkau, D.H. Zhou, J.M. George, C.M. Rienstra, Structural intermediates during alpha-synuclein fibrillogenesis on phospholipid vesicles, J. Am. Chem. Soc. 134 (2012) 5090–5099.

[120] D.P. Smith, et al. Formation of a high affinity lipid-binding intermediate during the early aggregation phase of alpha-synuclein, Biochemistry 47 (2008) 1425–1434.

[121] B.G. Wilhelm, et al. Composition of isolated synaptic boutons reveals the amounts of vesicle trafficking proteins, Science 344 (2014) 1023–1028.

[122] N.P. Reynolds, et al. Mechanism of membrane interaction and disruption by alpha-synuclein, J. Am. Chem. Soc. 133 (2011) 19366–19375.

[123] P.H. Jensen, M.S. Nielsen, R. Jakes, C.G. Dotti, M. Goedert, Binding of alpha-synuclein to brain vesicles is abolished by familial Parkinson's disease mutation, J. Biol. Chem. 273 (1998) 26292–26294.

[124] R. Bussell Jr., D. Eliezer, Effects of Parkinson's disease-linked mutations on the structure of lipid-associated alpha-synuclein, Biochemistry 43 (2004) 4810–4818.

[125] E. Jo, N. Fuller, R.P. Rand, P. St George-Hyslop, P.E. Fraser, Defective membrane interactions of familial Parkinson's disease mutant A30P alpha-synuclein, J. Mol. Biol. 315 (2002) 799–807.

[126] W. Choi, et al. Mutation E46K increases phospholipid binding and assembly into filaments of human alpha-synuclein, FEBS Lett. 576 (2004) 363–368.

[127] H. Heise, et al. Molecular-level secondary structure, polymorphism, and dynamics of full-length alpha-synuclein fibrils studied by solid-state NMR, Proc. Natl. Acad. Sci. USA 102 (2005) 15871–15876.

[128] S.J. Lee, E. Masliah, Neurodegeneration: aggregates feel the strain, Nature 522 (2015) 296–297.

[129] W. Peelaerts, et al. Alpha-synuclein strains cause distinct synucleinopathies after local and systemic administration, Nature 522 (2015) 340–344.

[130] S.B. Prusiner, Biology and genetics of prions causing neurodegeneration, Annu. Rev. Genet. 47 (2013) 601–623.

[131] J.L. Guo, et al. Distinct alpha-synuclein strains differentially promote tau inclusions in neurons, Cell 154 (2013) 103–117.

[132] A.L. Woerman, et al. Propagation of prions causing synucleinopathies in cultured cells, Proc. Natl. Acad. Sci. USA 112 (2015) E4949–4958.

[133] S.B. Prusiner, et al. Evidence for alpha-synuclein prions causing multiple system atrophy in humans with parkinsonism, Proc. Natl. Acad. Sci. USA 112 (2015) E5308–5317.

[134] S.B. Prusiner, Molecular biology of prion diseases, Science 252 (1991) 1515–1522.

[135] F. Chiti, C.M. Dobson, Protein misfolding, functional amyloid, and human disease, Annu. Rev. Biochem. 75 (2006) 333–366.

[136] M.G. Spillantini, M. Goedert, The alpha-synucleinopathies: Parkinson's disease, dementia with Lewy bodies, and multiple system atrophy, Ann. NY Acad. Sci. 920 (2000) 16–27.

[137] J.C. Watts, et al. Transmission of multiple system atrophy prions to transgenic mice, Proc. Natl. Acad. Sci. USA 110 (2013) 19555–19560.

[138] A. Mamais, et al. Divergent alpha-synuclein solubility and aggregation properties in G2019S LRRK2 Parkinson's disease brains with Lewy body pathology compared to idiopathic cases, Neurobiol. Dis. 58 (2013) 183–190.

[139] M. Brownlee, Biochemistry and molecular cell biology of diabetic complications, Nature 414 (2001) 813–820.

[140] H. Vlassara, M.R. Palace, Diabetes and advanced glycation endproducts, J. Intern. Med. 251 (2002) 87–101.

[141] L. M., Action des acides amines sur les sucre: formation des melanoidines par voie methodique, C R Hebd Seances Acad Sci 54 (1912) 66–68.

[142] S.F. Yan, R. Ramasamy, A.M. Schmidt, Mechanisms of disease: advanced glycation end-products and their receptor in inflammation and diabetes complications, Nat. Clin. Pract. Endocrinol. Metab. 4 (2008) 285–293.

[143] P.J. Thornalley, Dicarbonyl intermediates in the maillard reaction, Ann. NY Acad. Sci. 1043 (2005) 111–117.

[144] R. Castellani, M.A. Smith, P.L. Richey, G. Perry, Glycoxidation and oxidative stress in Parkinson disease and diffuse Lewy body disease, Brain Res. 737 (1996) 195–200.

[145] G. Munch, et al. Crosslinking of alpha-synuclein by advanced glycation endproducts—an early pathophysiological step in Lewy body formation?, J. Chem. Neuroanat. 20 (2000) 253–257.

[146] S.M. Johnson, S. Connelly, C. Fearns, E.T. Powers, J.W. Kelly, The transthyretin amyloidoses: from delineating the molecular mechanism of aggregation linked to pathology to a regulatory-agency-approved drug, J. Mol. Biol. 421 (2012) 185–203.

[147] J.D. Gillmore, P.N. Hawkins, Pathophysiology and treatment of systemic amyloidosis, Nat. Rev. Nephrol. 9 (2013) 574–586.

[148] L. Kang, et al. N-terminal acetylation of alpha-synuclein induces increased transient helical propensity and decreased aggregation rates in the intrinsically disordered monomer, Protein Sci. 21 (2012) 911–917.

[149] B. Fauvet, et al. Characterization of semisynthetic and naturally Nalpha-acetylated alpha-synuclein in vitro and in intact cells: implications for aggregation and cellular properties of alpha-synuclein, J. Biol. Chem. 287 (2012) 28243–28262.

[150] A.S. Maltsev, J. Ying, A. Bax, Impact of N-terminal acetylation of alpha-synuclein on its random coil and lipid binding properties, Biochemistry 51 (2012) 5004–5013.

[151] S.J. Metallo, Intrinsically disordered proteins are potential drug targets, Curr. Opin. Chem. Biol. 14 (2010) 481–488.

[152] G. Toth, et al. Targeting the intrinsically disordered structural ensemble of alpha-synuclein by small molecules as a potential therapeutic strategy for Parkinson's disease, PLoS One 9 (2014) e87133.

[153] C.Y. Chung, et al. Identification and rescue of alpha-synuclein toxicity in Parkinson patient-derived neurons, Science 342 (2013) 983–987.

[154] R. Hervas, et al. Common features at the start of the neurodegeneration cascade, Plos Biol. 10 (2012).

[155] S. Chandra, et al. Double-knockout mice for alpha- and beta-synucleins: effect on synaptic functions, Proc. Natl. Acad. Sci. USA 101 (2004) 14966–14971.

[156] D.E. Cabin, et al. Synaptic vesicle depletion correlates with attenuated synaptic responses to prolonged repetitive stimulation in mice lacking alpha-synuclein, J. Neurosci. 22 (2002) 8797–8807.

[157] C.G. Specht, R. Schoepfer, Deletion of the alpha-synuclein locus in a subpopulation of C57BL/6J inbred mice, BMC Neurosci. 2 (2001) 11.

[158] O.S. Gorbatyuk, et al. In vivo RNAi-mediated alpha-synuclein silencing induces nigrostriatal degeneration, Mol. Ther. 18 (2010) 1450–1457.

[159] C.E. Khodr, et al. An alpha-synuclein AAV gene silencing vector ameliorates a behavioral deficit in a rat model of Parkinson's disease, but displays toxicity in dopamine neurons, Brain Res. 1395 (2011) 94–107.

[160] A.L. McCormack, et al. Alpha-synuclein suppression by targeted small interfering RNA in the primate substantia nigra, PLoS One 5 (2010) e12122.

[161] A.D. Zharikov, et al. shRNA targeting alpha-synuclein prevents neurodegeneration in a Parkinson's disease model, J. Clin. Invest. 125 (2015) 2721–2735.

[162] A.M. Cuervo, E.S. Wong, M. Martinez-Vicente, Protein degradation, aggregation, and misfolding, Mov. Disord. 25 (Suppl. 1) (2010) S49–S54.

[163] B. Spencer, et al. Beclin 1 gene transfer activates autophagy and ameliorates the neurodegenerative pathology in alpha-synuclein models of Parkinson's and Lewy body diseases, J. Neurosci. 29 (2009) 13578–13588.

[164] M. Decressac, et al. TFEB-mediated autophagy rescues midbrain dopamine neurons from alpha-synuclein toxicity, Proc. Natl. Acad. Sci. USA 110 (2013) E1817–E1826.

[165] M. Xilouri, et al. Boosting chaperone-mediated autophagy in vivo mitigates alpha-synuclein-induced neuro-degeneration, Brain 136 (2013) 2130–2146.

[166] J.L. Webb, B. Ravikumar, J. Atkins, J.N. Skepper, D.C. Rubinszten, Alpha-synuclein is degraded by both autophagy and the proteasome, J. Biol. Chem. 278 (2003) 25009–25013.

[167] J. Bove, M. Martinez-Vicente, M. Vila, Fighting neurodegeneration with rapamycin: mechanistic insights, Nat. Rev. Neurosci. 12 (2011) 437–452.

[168] S.E. Davies, et al. Enhanced ubiquitin-dependent degradation by Nedd4 protects against alpha-synuclein accumulation and toxicity in animal models of Parkinson's disease, Neurobiol. Dis. 64 (2014) 79–87.

[169] G.K. Tofaris, et al. Ubiquitin ligase Nedd4 promotes alpha-synuclein degradation by the endosomal-lysosomal pathway, Proc. Natl. Acad. Sci. USA 108 (2011) 17004–17009.

[170] D.F. Tardiff, et al. Yeast reveal a "druggable" Rsp5/Nedd4 network that ameliorates alpha-synuclein toxicity in neurons, Science 342 (2013) 979–983.

[171] J. Wagner, et al. Anle138b: a novel oligomer modulator for disease-modifying therapy of neurodegenerative diseases such as prion and Parkinson's disease, Acta. Neuropathol. 125 (2013) 795–813.

[172] D.E. Ehrnhoefer, et al. EGCG redirects amyloidogenic polypeptides into unstructured, off-pathway oligomers, Nat. Struct. Mol. Biol. 15 (2008) 558–566.

[173] S. Prabhudesai, et al. A novel "molecular tweezer" inhibitor of alpha-synuclein neurotoxicity in vitro and in vivo, Neurotherapeutics 9 (2012) 464–476.

[174] S. Sinha, et al. Lysine-specific molecular tweezers are broad-spectrum inhibitors of assembly and toxicity of amyloid proteins, J. Am. Chem. Soc. 133 (2011) 16958–16969.

[175] J. Bieschke, et al. EGCG remodels mature alpha-synuclein and amyloid-beta fibrils and reduces cellular toxicity, Proc. Natl. Acad. Sci. USA 107 (2010) 7710–7715.

[176] N. Lorenzen, et al. How epigallocatechin gallate can inhibit alpha-synuclein oligomer toxicity in vitro, J. Biol. Chem. 289 (2014) 21299–21310.

[177] J. Levin, et al. The oligomer modulator anle138b inhibits disease progression in a Parkinson mouse model even with treatment started after disease onset, Acta. Neuropathol. 127 (2014) 779–780.

[178] E. Masliah, et al. Effects of alpha-synuclein immunization in a mouse model of Parkinson's disease, Neuron 46 (2005) 857–868.

[179] E. Masliah, et al. Passive immunization reduces behavioral and neuropathological deficits in an alpha-synuclein transgenic model of Lewy body disease, PLoS One 6 (2011) e19338.

[180] H.T. Tran, et al. Alpha-synuclein immunotherapy blocks uptake and templated propagation of misfolded alpha-synuclein and neurodegeneration, Cell Rep. 7 (2014) 2054–2065.

[181] S.J. Lee, P. Desplats, C. Sigurdson, I. Tsigelny, E. Masliah, Cell-to-cell transmission of non-prion protein aggregates, Nat. Rev. Neurol. 6 (2010) 702–706.

[182] E.J. Bae, et al. Antibody-aided clearance of extracellular alpha-synuclein prevents cell-to-cell aggregate transmission, J. Neurosci. 32 (2012) 13454–13469.

[183] N.B. Nillegoda, et al. Crucial HSP70 co-chaperone complex unlocks metazoan protein disaggregation, Nature 524 (2015) 247–251.

[184] K.M. Danzer, et al. Heat-shock protein 70 modulates toxic extracellular alpha-synuclein oligomers and rescues trans-synaptic toxicity, FASEB J. 25 (2011) 326–336.

[185] N.R. McFarland, et al. Chronic treatment with novel small molecule Hsp90 inhibitors rescues striatal dopamine levels but not alpha-synuclein-induced neuronal cell loss, PLoS One 9 (2014) e86048.

[186] S. Pemberton, et al. Hsc70 protein interaction with soluble and fibrillar alpha-synuclein, J. Biol. Chem. 286 (2011) 34690–34699.

[187] X. Gao, et al. Human Hsp70 disaggregase reverses Parkinson's-linked alpha-synuclein amyloid fibrils, Mol. Cell. 59 (2015) 781–793.

[188] J.R. Glover, S. Lindquist, Hsp104, Hsp70, and Hsp40: a novel chaperone system that rescues previously aggregated proteins, Cell 94 (1998) 73–82.

[189] C. Lo Bianco, et al. Hsp104 antagonizes alpha-synuclein aggregation and reduces dopaminergic degeneration in a rat model of Parkinson disease, J. Clin. Invest. 118 (2008) 3087–3097.

[190] M.E. Jackrel, J. Shorter, Engineering enhanced protein disaggregases for neurodegenerative disease, Prion 9 (2015) 90–109.

[191] S.P. Sardi, et al. Augmenting CNS glucocerebrosidase activity as a therapeutic strategy for parkinsonism and other Gaucher-related synucleinopathies, Proc. Natl. Acad. Sci. USA 110 (2013) 3537–3542.

[192] K.E. Murphy, et al. Reduced glucocerebrosidase is associated with increased alpha-synuclein in sporadic Parkinson's disease, Brain 137 (2014) 834–848.

[193] S.D. Orwig, et al. Binding of 3,4,5,6-tetrahydroxyazepanes to the acid-beta-glucosidase active site: implications for pharmacological chaperone design for Gaucher disease, Biochemistry 50 (2011) 10647–10657.

[194] A. Zimran, G. Altarescu, D. Elstein, Pilot study using ambroxol as a pharmacological chaperone in type 1 Gaucher disease, Blood Cells Mol. Dis. 50 (2013) 134–137.

[195] I. Bendikov-Bar, G. Maor, M. Filocamo, M. Horowitz, Ambroxol as a pharmacological chaperone for mutant glucocerebrosidase, Blood Cells Mol. Dis. 50 (2013) 141–145.

[196] J.S. Schneider, et al. A randomized, controlled, delayed start trial of GM1 ganglioside in treated Parkinson's disease patients, J. Neurolog. Sci. 324 (2013) 140–148.

NEUROINFLAMMATION AS A THERAPEUTIC TARGET IN NEURODEGENERATIVE DISEASES

3

Richard Gordon, Trent M. Woodruff

The University of Queensland, St. Lucia, QLD, Australia

CHAPTER OUTLINE

Introduction 50
    The Global Burden of Neurodegenerative Diseases and the Challenge
    of Developing Effective Therapeutics 50
Glial Cells in CNS Development, Homeostasis, and Pathology 51
    Microglia 51
    Astrocytes 53
    Oligodendrocytes and Ependymal Cells 54
Chronic Neuroinflammation as a Common Pathophysiological Mediator
in Progressive Neurodegenerative Diseases 54
    Mechanisms of Neuroinflammation-Mediated Neurodegeneration 56
    Alzheimer's Disease 58
    Parkinson's Disease 60
    Amyotrophic Lateral Sclerosis 64
    Huntington's Disease 65
    Frontotemporal Dementia and Lewy Body Dementia 66
Therapeutic Strategies Targeting Neuroinflammation in Progressive
Neurodegenerative Diseases 66
Future Perspectives 69
Conclusions 70
References 70

INTRODUCTION
THE GLOBAL BURDEN OF NEURODEGENERATIVE DISEASES AND THE CHALLENGE OF DEVELOPING EFFECTIVE THERAPEUTICS

Neurodegenerative diseases are characterized by a gradual and progressive loss of neuronal subpopulations in distinct regions of the central nervous system (CNS), which is accompanied by a debilitating spectrum of cognitive, psychiatric, and motor deficits resulting from atrophy of brain structures. Collectively, neurodegenerative diseases exert a major global disease burden, with dementia emerging as the public health challenge of this generation in several developed countries (2013 WHO Global Burden of Disease Survey). Critically, as aging is the strongest risk factor for the most common neurodegenerative conditions [1,2], the global economic and social impact of these diseases on healthcare systems is expected to surge significantly in the coming decades with increasingly aging populations and longer life spans [2,3].

Over 36 million people worldwide are diagnosed with Alzheimer's disease (AD) or Parkinson's disease (PD), the two most common neurodegenerative disorders. The lack of effective disease-modifying treatments and the failure of most clinical trials for new therapies for these diseases, underscores the need to identify new targets for drug discovery to mitigate disease progression. A major challenge in the development of treatment strategies for most progressive neurodegenerative diseases is their complex multifactorial etiology and heterogeneous disease course [4–6]. For the most common neurodegenerative diseases, such as AD, PD, and amyotrophic lateral sclerosis (ALS), the precise causes of disease onset and progression are largely unknown. Additionally, the disease course and severity varies considerably between patients, further complicating the challenge of effective therapeutic intervention. The emerging consensus suggests that progressive neurodegenerative diseases are complex multifactorial disorders with potentially multiple pathogenic mechanisms that trigger neurodegeneration in the aging brain. Several "multiple hit hypotheses" have been proposed to explain the onset of progressive neurodegenerative diseases; however, clear mechanistic evidence remains fragmented [4,6,7]. Although much progress needs to be made in terms of understanding the causes and triggers of neurodegeneration, significant progress has been made over the last two decades in understanding the key pathological mechanisms involved during disease progression. Common pathological mechanisms that have been identified in most progressive neurodegenerative diseases include mitochondrial dysfunction, oxidative stress, proteasomal impairment, and neurotoxic protein misfolding [8,9]. Mounting evidence suggests that there are causal mechanistic links between toxic misfolded protein aggregates and neurodegeneration (see Chapters 1 and 2). In fact, atypical protein assemblies are now considered a hallmark of most neurodegenerative disorders including PD, ALS, and Huntington's disease (HD), although their pathological significance is still debated [10,11]. Further, exciting new evidence suggests that these toxic misfolded protein aggregates can spread spatiotemporally across the CNS by prion-like mechanisms [12–15]. Over the last decade, chronic neuroinflammation driven by CNS glial cells has emerged as a common pathological mechanism that has been shown to drive the progression of neurodegeneration. Persistent neuroinflammation has now been consistently documented in all progressive neurodegenerative diseases, including AD, HD, ALS, and PD [8,16,17]. Mechanistic evidence from clinical, experimental, and epidemiological studies has established that the chronic neuroinflammation, which persists in the CNS for decades, can exacerbate ongoing neurodegeneration and thereby drive disease progression [5,8,18]. Crucially, a wealth of evidence now suggests that chronic neuroinflammation is a common pathological response that occurs in the aging CNS irrespective of the cause of disease onset.

Therefore, it has been proposed that glial-derived neuroinflammation could be an attractive therapeutic target given the multifactorial causes that can trigger neurodegeneration in the aging brain [5,8,19].

GLIAL CELLS IN CNS DEVELOPMENT, HOMEOSTASIS, AND PATHOLOGY

For over a century, glial cells were regarded as passive cells that provided structural support to neurons (the Greek term *glia* meaning "glue"). However, it is now well established that glia are highly dynamic and play a central role in almost every aspect of neuronal function from the regulation of neurotransmission to synapse formation, synaptic scaling, and elimination [20]. There are four major glial cell types in the CNS, namely astrocytes, microglia, oligodendrocytes, and ependymal cells. Together, glial cells comprise approximately half of the total volume of the adult human brain and spinal cord with recent estimates indicating that there are around 86 billion neurons and 85 billion glial cells [21,22]. The glia/neuron ratio, however, differs markedly across the brain being around 3.23 in the cerebral cortex, 0.23 in the cerebellum, and 11.35 in the combined basal ganglia, diencephalons, and brainstem [21]. Nevertheless, a detailed understanding of the pathological functions of glia in general, and astroglia in particular, remains fragmentary, because of a long-lasting prevalence of neurocentric views in neurology and neuropathology. The major glial cell types and their functions are discussed in further detail in the following sections.

MICROGLIA

Microglia are the resident macrophages in the brain and spinal cord and are the primary immunocompetent cells in the CNS. They were first described in 1920 by Rio Hortega, who identified their phagocytic properties and suggested they could function similar to peripheral macrophages. This was confirmed several decades later in the 1980s by Hickey and Kimura when they demonstrated that perivascular microglia are indeed antigen-presenting cells and express high levels of MHC class II. The exact origins of microglia were originally unclear, but are now demonstrated to originate in the yolk sac, derived from primitive macrophages or macrophage progenitors that migrate into the brain early during development [23]. Microglia make up approximately 10%–15% of the total cells in the brain and their numbers and densities vary according to brain region. The hippocampus, basal ganglia, olfactory telencephalon, and substantia nigra have the highest microglial cell density in the CNS [24,25]. In the normal adult brain, microglia are typically present in their so-called "resting state" with a ramified morphology. Classically, these resting microglia were considered to be "inactive" in the healthy physiological state; however, recent in vivo two-photon microscopy has revealed that resting microglia are not dormant cells, but are in fact highly dynamic and efficient in forming processes and continuously scan their neuronal microenvironment [26]. With adjacent microglia constantly surveying overlapping regions of the brain parenchyma, studies estimate that the entire neuronal network of the brain can be scanned within a few hours [26–28]. In response to acute neuronal injury or pathogenic stimuli, microglia undergo a rapid and dynamic transformation to reactive amoeboid forms with concurrent upregulation of various cell surface proteins, including MHC molecules and cytokine and chemokine receptors. Microglial activation is, therefore, the earliest and most evident response to neuronal injury in both acute and chronic neurodegenerative states (Fig. 3.1). In fact, elegant in vivo imaging studies have recently demonstrated that resting microglia can respond to chemoattractants and morphologically realign themselves within

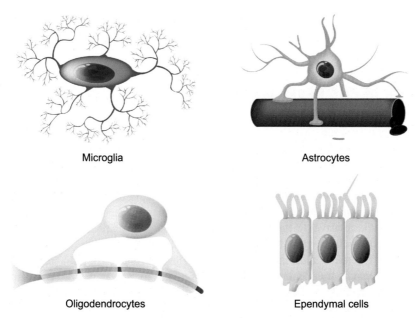

Microglia

Astrocytes

Oligodendrocytes

Ependymal cells

FIGURE 3.1 Major Glial Cell Types Involved in Central Nervous System (CNS) Neuroinflammatory Responses

seconds to minutes. Transcriptional activation of several stimulus-specific sets of response genes also occurs rapidly in activated microglia with functional proteins being produced at around 2 h. As would be necessary, the activation of microglia is both a dynamic and tightly regulated process relying on constant "ON" and "OFF" signals from neurons, astrocytes, and the extracellular milieu, and can be rapidly shut down once homeostasis has been attained [29].

Activated microglia generate a large number of secreted factors including, proteases, proinflammatory mediators, complement factors, cytokines, and chemokines that can alter the local neuronal microenvironment in the brain parenchyma in a stimulus-specific manner. Transient and regulated activation of microglia typically facilitate a microenvironment that promotes neuron survival and homeostasis [30–32]. Activated microglia can respond with the release of antiinflammatory and neurotrophic factors, including BDNF, GDNF, IL10, and transforming growth factor beta 1 (TGFβ1), which can stimulate antiapoptotic survival signaling pathways, including Akt/PI3K in neurons. Microglia have been shown to protect neurons from excitotoxicity by upregulating GLT1, a glutamate uptake protein under conditions where astrocyte glutamate uptake is impaired [33,34]. Microglial activation facilitates phagocytic removal of dead cells at the site of injury and can participate in regeneration of neural connections by synaptic stripping. In the entorhinal cortex, for example, activated microglia facilitate the physical removal of synapses and entire dendritic structures following neuronal injury. The synaptic stripping and phagocytic removal of cellular debris allow for the formation new synaptic connections and restoration of local neuronal homeostasis [35]. Microglia can also directly promote neurogenesis and direct the migration of adult neural stem cells within the brain. They can also negatively regulate neurogenesis by inducing a proinflammatory microenvironment [36]. The basic biology and functional role of microglia in CNS development and homeostasis is still being unraveled with exciting new discoveries, such as the expression of neurotransmitter receptors on microglia and the profound regulation of microglial responses by the gut microbiome [37,38].

ASTROCYTES

Astrocytes are the most abundant and heterogeneous nonneuronal cells in the CNS and have neuroectodermal origins [39]. They are derived from a heterogeneous population of glial progenitor cells in the neuroepithelium of the developing CNS. The radial glia that provide a supportive matrix for neuronal migration and synaptogenesis are thought to differentiate into astrocytes when neuronal development is complete [20,40]. Astrocytes have a characteristic star-shaped morphology based on which they are named. Classically, the primary function of astrocytes was believed to biochemically support the endothelial cells that make up the blood brain barrier (BBB), and, therefore, astrocytes were viewed primarily as static, largely structural entities [39]. Over the last two decades, this static view of astrocytes has been drastically redefined, as their proactive roles in neuronal repair and their dynamic interactions with neuronal and glial cells have been identified. It is now well established that astrocytes can actively regulate the formation, maturation, maintenance, and strength of neuronal synapses [41]. Astrocytes also provide a vital metabolic link between the vasculature and neurons transporting glucose and other metabolites out of the bloodstream [40,42,43]. Astrocytes process glucose through the glycolytic pathway to lactate, which is the primary energy substrate used by neurons [44]. Astrocytes also play a vital role in the propagation of action potentials by taking up neurotransmitters, such as glutamate and γ-aminobutyric acid, and actively regulating extracellular potassium ion concentrations [42,44,45]. Astrocytes can also synthesize and release vital neurotrophic factors, and are involved in the formation of neural and glial scars following neural injury or trauma [46]. This acute astrogliosis response has been found to be essential in terms of both limiting the spread of damage and for the postinjury remodeling of the microenvironment to facilitate recovery of neuronal function by isomorphic astrogliosis [20,43,47].

Over the last decade, the multiple immune functions of astrocytes have been recognized in greater detail with astrocytes being shown to express MHC molecules and pattern recognition receptors (PRRs) for danger-associated molecular patterns (DAMPs), both intracellularly and on their cell surface [39,40]. Compared to microglia, however, astrocytes express fewer PRRs and do so at a lower density. Astrocytes have been shown to contribute to a proinflammatory environment by secreting mediators, including IL6, IL10, IL12, tumor necrosis factor alpha (TNFα), and CXCL10. Although the neuroinflammatory functions of astrocytes have not been as well studied, as microglia and astrocytes have been implicated in most progressive neurodegenerative diseases, including AD, PD, and ALS [5,19]. Chemokines secreted by astrocytes can recruit both peripheral and CNS immune cells to the site of neuronal injury and degeneration [48]. Astrocytes also express cell adhesion molecules (such as ICAM1 and VCAM1) and integrins (LFA1 and VLA4), which can engage lymphocyte cells and facilitate their entry into the CNS. Further, as astrocytes are the gatekeepers of the BBB, they have been implicated in its pathological breakdown, which occurs during progressive neurodegeneration. This crucial step of BBB breakdown has been proposed to facilitate the entry of peripheral immune cells and blood components that can perpetuate CNS neuroinflammation and neurotoxicity [49,50]. Nonetheless, the primary functions of astrocyte activation and inflammatory responses appear to be geared toward limiting CNS damage and tissue remodeling to facilitate neuronal recovery [50]. In this regard, recent studies have highlighted the role of astrocyte degeneration and atrophy in the early stages of neurodegenerative diseases, which could have implications for neuronal loss and disease progression [50].

OLIGODENDROCYTES AND EPENDYMAL CELLS

Oligodendrocytes are specialized glial cells that wrap themselves around neurons present in the CNS. Oligodendrocytes are primarily responsible for maintenance and generation of the myelin sheath that surrounds axons. They also participate in axonal regulation and the sculpting of higher order neuronal circuits [51]. Oligodendrocytes have also been reported to produce a number of neurotrophins, such as nerve growth factor, brain-derived neurotrophic factor (BDNF), and neurotrophin 3, which provide neuronal trophic support [52]. While oligodendrocytes play important roles in the pathology of demyelinating autoimmune diseases, such as multiple sclerosis, their precise roles in progressive neurodegeneration are still being elucidated. Some studies have shown that impaired oligodendrocyte function can result in immune cell infiltration and neuroinflammation. Ependymal cells are specialized glial cells that line the ventricles and the central canal of the spinal cord [53]. The functions of ependymal cells remain are still unclear but they are believed to be important for the directional movement of cerebrospinal fluid (CSF). An important role for the bidirectional flow of CSF is to facilitate transport of nutrients into the brain and the removal of metabolic waste. The unique microvilli present on ependymal cells facilitate CSF movement, and its disruption can lead to hydrocephaly. Recently, an NFκB-driven neuroinflammatory mechanism was implicated in the formation of hydrocephaly using experimental rodent models [54]. However, to date, the role of these cells in progressive neurodegenerative diseases has not been examined in great detail.

CHRONIC NEUROINFLAMMATION AS A COMMON PATHOPHYSIOLOGICAL MEDIATOR IN PROGRESSIVE NEURODEGENERATIVE DISEASES

Chronic neuroinflammation has now been consistently documented in all progressive neurodegenerative diseases, including AD, PD, HD, and ALS [8,16,17]. Within the CNS, the hallmark of neuroinflammation is a persistent and localized activation of the primary immunocompetent glial cells, the microglia, and astrocytes, at the sites of neurodegeneration. The progressive loss of neurons is almost always accompanied by the presence of typical hypertrophic microglia (Fig. 3.2) and dystrophic astrocytes. Perhaps intuitively, for many decades the primary functions of this reactive gliosis was believed to be phagocytic clearance and immune surveillance in the case of activated microglia and neurotrophic support for astrocytes. In fact, this is true in most cases of acute brain injury or infection, where neuroinflammatory responses are transient and self-limiting and typically subside once the infection is cleared or homeostasis is restored at the site of injury. As discussed earlier, the transient and regulated activation of microglia during cases of infection or acute neurotrauma normally facilitates a microenvironment that promotes neuron survival, thereby, enabling recovery of CNS homeostasis [30–32]. Acute and regulated microglial activation can drive antiinflammatory and neurotrophic responses, as well as protect neurons from excitotoxic injury [33,34]. The prohomeostatic role of microglia is further substantiated by in vivo microglial depletion and replenishment studies [55,56]. Therefore, it is generally assumed that acute neuroinflammatory responses are usually beneficial in the CNS, as they tend to limit further injury and contribute to neuronal survival [57].

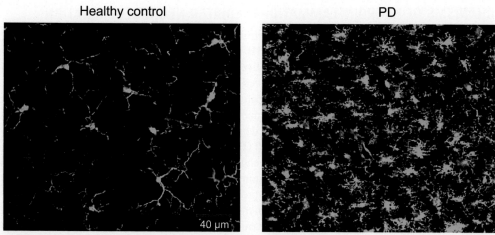

FIGURE 3.2 Reactive Microgliosis During Progressive Neurodegeneration

Microglia labeled with Iba1 in postmortem substantia nigra tissue sections of a Parkinson's disease (PD) patient and an age-matched healthy control. Prominent microglial hypertrophy and increased microglial cell density is evident in the PD brain.

In contrast to the transient and highly regulated activation profile seen during acute neuronal injury, the chronic microglial activation observed in progressive neurodegenerative disorders is morphologically and functionally distinct. Sustained microglial activation at sites of neuronal loss in the CNS is now considered a hallmark of most chronic neurodegenerative disorders. Prominent reactive microgliosis has been documented in postmortem brains of patients in all progressive neurodegenerative disorders [8,19]. These hypertrophic microglia typically display their fully activated amoeboid state and increased microglial cell density (Fig. 3.2) at the sites of neurodegeneration. Reactive microglia rapidly upregulate several cell surface receptors and activation markers, including CD1, CD11b/Mac1 (CR3), lymphocyte function–associated antigen 1 (LFA1), intercellular adhesion molecule 1 (ICAM1 or CD54), and vascular cell adhesion molecule (VCAM1 or CD106). Activated microglia also generate a variety of inflammatory mediators, including cytokines (TNF and interleukins IL1β and IL6) and chemokines [macrophage inflammatory protein (MIP1α), monocyte chemoattractant protein (MCP1), and interferon (IFN) inducible protein 10 (IP10)] that promote inflammation [58]. The microglial reaction observed in chronic neurodegenerative diseases was initially regarded as a compensatory neuroprotective response. However, several lines of evidence from clinical and epidemiological studies, preclinical animal models, and in vitro mechanistic work, now support the view that chronic neuroinflammation and reactive microgliosis can be self-propagating and eventually become neurotoxicleading to progressive neuron loss [5,8]. Reactive microglia generate a large number of secreted factors, including proteases, proinflammatory mediators, cytokines, and chemokines that can alter the local neuronal microenvironment in the brain parenchyma. In a persistently activated state, microglia can augment oxidative stress levels and secrete proinflammatory mediators, which can become neurotoxic over time [19,59].

MECHANISMS OF NEUROINFLAMMATION-MEDIATED NEURODEGENERATION

Although for decades the brain was viewed as an immune-privileged site where inflammation only occurred in the context of direct infection or BBB breakdown, it has now been conclusively established that all endogenous CNS cell types constitutively express specialized PRRs capable of activating an innate immune response to specific host-derived ligands, termed DAMPs [5,60,61]. The highest density of PRRs are expressed on immunocompetent microglial cells, which are capable of selectively recognizing and responding to DAMPs, such as misfolded aggregated proteins, mislocalized nucleic acids, and secreted danger signals from degenerating neurons, such as high-mobility group box 1 protein (HMGB1) or heat shock protein 70 (HSP70) [62]. The emerging paradigm in the case of progressive neurodegenerative diseases is that endogenous DAMPs can directly trigger and sustain neuroinflammation in the CNS and thereby shift immunocompetent cells from their beneficial roles to a chronically reactive state that contributes to the progression of neurodegeneration [5]. This view has fundamentally changed our understanding and therapeutic approach to neurodegenerative disorders from a previously neuron-centric perspective of merely preventing neuronal death or providing trophic support for degenerating neuronal populations. Neuropathological and positon emission tomography (PET)–based neuroimaging studies in patients with neurodegenerative diseases indicate that neuroinflammatory responses may indeed begin prior to significant loss of neuronal populations. The major neuroinflammatory mechanisms and pathways activated in progressive neurodegenerative diseases are discussed in more detail here.

As indicated in Table 3.1, a strikingly common feature of most progressive neurodegenerative diseases is the presence of persistent insoluble protein aggregates and/or inclusion bodies containing misfolded proteins [i.e., amyloid-β, α-synuclein, transactive response DNA-binding protein 43 (TDP43), and mutant huntingtin (mHTT)]. The exact physiological functions of these aggregated misfolded proteins remain unclear and it is still debated if indeed they are protective or neurotoxic and if they actively contribute to disease progression. However, as stated earlier, these insoluble protein aggregates can function as DAMPs to activate PRRs [toll-like receptors (TLRs); NLRP3, nucleotide-binding domain (NOD) leucine-rich repeat (LRR), and pyrin domain–containing 3 (NLRP3); receptor for advanced glycation end products (RAGE), etc.] on microglia and, thereby, perpetuate chronic reactive gliosis and neuroinflammation. Elegant experimental studies using transgenic and knockout mice in preclinical models of neurodegenerative diseases have demonstrated that innate immune neuroinflammatory responses can exacerbate the ongoing neuronal degeneration that has been initiated by apparently diverse mechanisms. This paradigm shift has, over the last decade, led to some exciting advances in understanding the precise nature of pathology in neurodegenerative diseases, and has shifted the focus from neuronal death more broadly to neuron–glial interactions and the CNS microenvironment. Some of the established and emerging mechanisms and pathways by which neuroinflammation has been shown to contribute to neurodegeneration are discussed further.

Oxidative stress, mitochondrial dysfunction, and proteasomal impairment are well-established pathological mechanisms that are known to initiate and drive neuronal degeneration in the aging brain [63]. The human brain microenvironment is believed to be uniquely susceptible to oxidative damage because of its high-energy demands and oxygen consumption, as well as the oxidative metabolic pathways used to generate and catabolize neurotransmitters, such as dopamine [63]. In neuronal cells, the mitochondrial electron transport chain is the primary source of intracellular reactive oxygen species (ROS). In chronically activated microglia on the other hand, excessive activation of the phagocytic NADPH oxidase (NOX2) pathway has emerged as a major proinflammatory

Table 3.1 Danger Associated Molecular Patterns (DAMPs) and Pattern Recognition Receptors (PRRs) in Neurodegenerative Diseases

Disease	Pathological Protein Aggregates	DAMPs	Associated PRRs
AD	Primarily extracellular	Amyloid-β Chromogranin A S100A9	NLRP3
PD	Intracellular and extracellular	α-Synuclein Hsp-70 HMGB1 MMP3	TLR2 and TLR4 Unknown RAGE and TLR2/4 PARs
LBD	Intracellular and extracellular	α-Synuclein oligomers	TLR2/4
ALS	Intracellular and extracellular	Mutant SOD1 FUS/TLS TDP43 C9orf72 dipeptides	NLRP3 TLRs (Unknown)
FTD	Intracellular and extracellular	TDP43	TLR2/4 (NLRP3)
HD	Primarily intracellular	mHTT	TLRs

AD, Alzheimer's disease; ALS, amyotrophic lateral sclerosis; FTD, frontotemporal dementia; FUS/TLS, fused in sarcoma/translocated in liposarcoma; HD, Hunginton's dieae; HMGB1, high-mobility group box 1 protein; LBD Lewy body dementia; mHTT, mutant huntingtin; NLRP3, NOD-, LRR-, and pyrin domain–containing 3; PAR, protease-activated receptor; RAGE, receptor for advanced glycation end products; SOD1, superoxide dismutase 1; TDP43, transactive response DNA-binding protein 43 kDa; TLR, toll-like receptor.

oxidative enzyme in the most prevalent progressive neurodegenerative diseases [8,64,65]. NADPH oxidase catalyzes the production of superoxide from oxygen. Although this pathway is quiescent in resting phagocytes, NADPH oxidase is rapidly upregulated and activated by multiple stimuli, including inflammogens, such as LPS; environmental neurotoxins; and importantly by misfolded protein aggregates functioning as DAMPs [64]. Emerging studies suggest that NADPH oxidase might be a common downstream pathway activated by many microglial PRRs, including NLRP3, TLRs, and RAGE. It has been proposed that as there could be significant similarities and overlap between PAMPs and CNS-derived DAMPs, chronically activated microglia may eventually not be able to differentiate between these signals at the molecular level and, therefore, overreact by activating the NADPH pathways that are primarily utilized to combat infections by innate immune cells [5]. Therefore, targeting NOX2 activation in the CNS could be a potentially useful disease-modifying therapeutic approach for multiple neurodegenerative diseases [65].

Another common feature of neurodegenerative diseases is the excessive production and release of proinflammatory cytokines and chemokines within the CNS [66]. Both the interleukin and TNF family of proinflammatory cytokines are elevated in postmortem brains from patients with neurodegenerative diseases, as well as preclinical animal models [8,67,68]. Mechanistic studies have demonstrated that sustained exposure to TNF can activate the extrinsic apoptotic pathway in vulnerable neuronal populations to drive neurodegeneration [69,70]. Similarly, the overexpression of IL1β drives neurodegeneration in the CNS, while blockade of inflammasome-mediated IL1β production can protect against amyloid-β pathology in mouse models of AD [71]. Several cytokines and chemokines belonging to the TNF and interleukin superfamily are also elevated in the CSF and serum of patients with neurodegenerative diseases, raising the possibility of utilizing them as potential markers of disease onset, progression, or treatment efficacy. However, to date no selective biomarkers have been conclusively identified for progressive neurodegenerative diseases. The complement system is another important arm of the innate immune system that has been shown to play a pathogenic role in progressive neurodegenerative diseases, including AD, PD, HD, and ALS [72–78]. Selected complement factors and receptors are also upregulated in the serum and on circulating immune cells in the periphery of neurodegenerative disease patients, suggesting that complement activation and signaling could be an important link between the CNS and peripheral arms of innate immune system in neurodegenerative diseases [79]. The terminal complement component C5a has been implicated in BBB disruption, which is proposed to be an important step in perpetuation of neuroinflammation by allowing infiltration of peripheral macrophages and potentially other systemic inflammatory mediators into the CNS [74].

Recently, inflammasomes expressed on glial cells have been implicated as key intracellular receptors for DAMPs. The NLR family is a group of intracellular receptors that can form inflammasomes. Assembly of an inflammasome activates caspase-1 and, subsequently, the proteolysis and release of the cytokines IL1β and IL18, as well as pyroptotic cell death [62]. Microglial NLRP3 inflammasome activation occurs in response to amyloid-β aggregates that contribute to neuroinflammation and neuropathology in AD [68,80]; however, it is still unclear whether there is a similar role for inflammasome activation in vivo in synucleopathies, such as PD or Lewy body dementia (LBD). Recently, NLRP3 was shown to be expressed by glial cells in ALS patients and mouse models [81]. However, their functional roles have not yet been elucidated clearly. In addition to the well-characterized NLRP3 inflammasome pathway, studies using in vivo models of stroke and neurotrauma have identified that other inflammasomes, including absent in melanoma 2 (AIM2), NLRP1, and NLRP2 are functional in the CNS. Their specific roles in progressive neurodegenerative diseases, however, remain to be completely elucidated [62,82].

ALZHEIMER'S DISEASE

AD is characterized by progressive memory decline and cognitive impairment and is the most common cause of dementia worldwide affecting over 150 million people [83]. The pathological hallmark of AD is the progressive accumulation of amyloid-β in the brain parenchyma accompanied by neurofibrillary tangles made up of the microtubule-associated protein tau [7,84]. Amyloid-β is generated from its amyloid precursor protein (APP) by proteolytic cleavage in a sequential process mediated by two aspartyl proteases, γ-secretase and β-secretase 1 [85]. Current research studies suggest that reducing the concentration of amyloid-β is vital to prevent its aggregation and pathology in the CNS. Although there are rare genetic mutations that contribute to the production of APP and amyloid-β, the vast majority of cases are sporadic in nature. Emerging evidence suggests that impaired clearance mechanisms in the aging brain could be an important contributing factor to amyloid-β pathology in sporadic AD [86] (see also Chapter 5).

Over the past two decades, substantial epidemiological evidence has emerged supporting a role for neuroinflammation as a pathological contributor to disease progression and indeed a risk factor for the development of AD [84]. Further, the canonical antiinflammatory class, nonsteroidal antiinflammatory drugs (NSAIDs) have been shown to reduce the risk of developing AD in many (but not all) cohort studies. On the other hand factors, such as a history of systemic infections, obesity in midlife, or reduced physical activity, which can contribute to higher levels of systemic inflammatory factors are all associated with an increased risk of developing AD. Recent clinical studies suggest that patients who have experienced severe infections show accelerated cognitive decline and this is positively correlated with peripheral levels of TNFα. Similarly, a US Health and Retirement Study found that severe septicemia can lead to greater cognitive impairment in subsequent years [5,87]. Additionally, survivors of septic shock episodes appear to have cognitive changes that persist over time and changes in electroencephalography measures [5,87]. Other reports have suggested that individuals affected by chronic periodontitis and poor oral health have a slightly higher risk of developing AD later in life, although the mechanisms are poorly understood. More recently, rare variants of the triggering receptor expressed on myeloid cells 2 (TREM2) have been associated with an increased AD risk [88,89]. TREM2 is a cell surface receptor that initiates immune signaling and activation of macrophages by coupling to TYRO protein tyrosine kinase–binding protein (TYROBP). In the brain, TYROBP is expressed mainly by microglia cells, and increased expression of TYROBP has been documented in AD brains and in APP transgenic mice [90]. TREM2 may also be involved in promoting the phagocytic clearance of cellular debris and in the downregulation of inflammatory signaling in response to TLR ligation. Further, in some genetic studies analyzing gene expression networks in late-onset AD, the expression of TYROBP showed the strongest association with disease onset [87].

Elevated levels of proinflammatory mediators, such as complement components, inflammatory cytokines, and chemokines, as well as several eicosanoids, have all been found in the serum or CSF of AD patients. Complement activation is evident in AD patient brains, and blocking proinflammatory complement components, such as C5a, has been shown to reduce pathological outcomes in AD models [91]. More recently, the use of specific PET ligands, such as $[^{11}C](R)$-PK11195 has allowed for the visualization of neuroinflammation in living patients [92]. Large-scale genetic studies of gene networks that are involved in late-onset AD have identified novel genes and regulatory pathways that might be associated with innate immune mechanisms and microglial activation. In particular, several pathways that are involved in phagocytosis and amyloid-β clearance have been implicated in large-scale network analysis studies [93]. These findings are further supported by recent genome-wide association study (GWAS) analyses of sporadic AD, which revealed a set of genes that point to a role for neuroinflammation in AD pathology [94]. For example, variants that are associated with an increased risk of developing AD have been found in genes that encode complement receptor 1 (CR1), myeloid cell–expressed membrane-spanning 4-domains subfamily A member 6A (MS4A6A), putative membrane-spanning 4-domains subfamily A member 4E (MS4A4E), and CD33 [5].

The major reactive oxygen and nitrogen intermediates, such as nitric oxide, peroxynitrite, and intracellular ROS, are well-established downstream mediators of inflammatory pathology in the several neurodegenerative diseases. In AD brains and preclinical mouse models, iNOS is upregulated by microglia and the nitric oxide generated has been shown to cause axonal and synaptic degeneration and, thereby drive, neuronal loss [95]. Critically, excessive nitric oxide production and its reaction product, peroxynitrite, have been shown to drive the nitration of amyloid-β peptides at tyrosine residues, which increase its tendency to form neurotoxic aggregates [96]. Nitrated amyloid-β peptides have now been demonstrated in the brains of AD patients and preclinical mouse models with high concentrations of nitrated peptides localized to the core of senile plaques. Further, knockout mice for the iNOS pathway are protected against some aspects of cognitive decline in AD models, suggesting that this pathway could be a potential therapeutic target [95].

It has recently been proposed that aggregated amyloid-β could mimic a conserved molecular pattern against which the innate immune system has evolved one or more PRRs [68,80]. This hypothesis is supported by the fact that microbial pathogens can express amyloid-like fibrillar proteins (also referred to as curli fibers) [97]. Although there is no direct evidence for this hypothesis, recent reports have shown that the detection of amyloid-β aggregates by PRRs activates the intracellular NLRP3 inflammasome complex, which drives the production of IL1β and IL18. Further, activation of NLRP3-driven caspase-1 is significantly higher in the brains of PD patients, providing direct evidence of inflammasome activation in AD. Remarkably in the APP/PS1 transgenic mouse model of AD, mice that are deficient in NLRP3 are strongly protected against most aspects of amyloid pathology, displaying almost normal cognitive function [68,80]. Besides microglia and astrocytes, other CNS-resident cells, including endothelial cells and oligodendrocytes, have been implicated in the innate immune neuroinflammatory response. For example, oligodendrocytes have been shown to contribute excessive amounts of proinflammatory complement components, which can drive inflammation locally [98,99]. Additionally, a crucial role for infiltrating neutrophils has also been recently uncovered in AD pathology, again demonstrating that peripheral immune responses and systemic inflammation could play a greater role than previously appreciated [100,101]. Finally, an interesting emerging paradigm is the loss of protective antiinflammatory cell surface proteins on neurons during the course of the disease [102]. Fractalkine and the complement regulating proteins CD59 and CD200 have been shown to be reduced in expression during AD pathology [103,104]. The loss of these regulatory protective proteins on neurons has been proposed to contribute to inflammation, thereby driving disease progression. Further studies will be required to gain more insight into this complex regulatory process.

PARKINSON'S DISEASE

Idiopathic PD is a chronic, progressive neurodegenerative disorder characterized by extrapyramidal motor deficits, including akinesia, bradykinesia, rigidity, and postural instability. First described by James Parkinson in 1817 as the shaking palsy, it is now recognized as the second most common neurodegenerative disorder affecting about 1 million people in the United States and over 4 million worldwide [105]. Idiopathic PD has a median onset of around 60 years and its occurrence increases with age. Typically, when motor symptoms manifest at the time of diagnosis, around a 60% loss of dopaminergic (DA) cell bodies in the substantia nigra is evident with about an 80% depletion in striatal dopamine levels [106]. The conspicuous presence of Lewy bodies containing aggregated α-synuclein and other cellular degradation products is a characteristic feature of the disease in humans. Reactive microgliosis is another prominent pathological feature of PD that is evident in clinical subjects and experimental models of the disease. Until recently, the glial reaction accompanying nigral DA degeneration was presumed to be an innocuous consequence of extensive DA degeneration. Over the last two decades, however, a wealth of evidence has accumulated that strongly supports an active role for reactive microgliosis in the pathogenesis of PD, particularly in the progressive phase of the disease.

In postmortem studies of PD patients, extensive microglial activation is well documented in the nigrostriatal system, particularly around the most vulnerable melanin-containing DA neurons of the substantia nigra. Elevated levels of inflammatory mediators are also present in the CSF of PD patients, indicative of an ongoing inflammatory reaction that persists throughout the patient's life [107–110].

Proinflammatory mediators, including TNFα, IL1β, and IFNγ, are markedly elevated in PD brains and their role in the degenerative process has been characterized in experimental animal models [107,111–113]. The activation of caspase-3 and caspase-8 has been reported in PD patients, with caspase activity being required for microglial activation [114–116]. Microglial activation has also been observed in the striatum and the remainder of the basal ganglia by real-time PET imaging in living patients with idiopathic PD [117]. Significantly, the microglial activation is evident in PD patients irrespective of the number of years with the disease, when compared to healthy age-matched controls. Prominent nuclear translocation of NFκB, a crucial regulator of oxidative stress and proinflammatory responses, is evident in brains of PD patients [118]. In support of the function of NFκB in PD, the use of highly specific NFκB inhibitors was shown to be efficacious in protecting against nigrostriatal DA degeneration in mouse models by mitigating the microglial neuroinflammatory responses and oxidative stress. Emerging evidence also suggests that the adaptive arm of the immune response might influence the degenerative process in PD. In postmortem patients, CD8+ and CD4+ T cells, but not B cells, were found in the substantia nigra in the absence of BBB damage [119]. Further, animal studies demonstrate that the T cells enter the brain following neuronal damage during the course of degeneration. Recent reports also suggest that the type of T cell response can both influence microglial activation and contribute to either DA neuron loss or neuroprotection [120–123]. Postmortem analysis of humans who developed rapid-onset Parkinsonism after being exposed to the Parkinsonian toxin 1-methyl-4-phenyl-1,2,3,6-tetrahydropyridine (MPTP) show abundant reactive microgliosis in the substantia nigra, indicative of a sustained neuroinflammatory response that is perpetuated for the life of the disease [124].

Genetic linkage studies in cohorts of PD patients have also uncovered specific single nucleotide polymorphisms in promoter regions of proinflammatory genes that are linked to the development of PD or early-onset PD possibly through transcriptional mechanisms [125,126]. The G−174C SNP in the IL6 promoter is linked to the development of early-onset PD in a study with 258 PD patients and 308 controls. Similar polymorphisms in position −511 of the IL1β gene promoter and position −1031 in the TNF gene promoter have also been reported [127,128]. The experimental evidence and confirmation by metaanalyses for these promoter polymorphism studies, however, is lacking except in the case of TNFα, which has been shown by independent research groups to have a neurotoxic role in PD pathogenesis in different animal models. Epidemiological studies provide further evidence implicating inflammatory mechanisms in PD [129]. Continual use of NSAIDs has been shown to lower the risk of PD by about 46% in human populations and this has been confirmed in animal models of PD [130,131]. More recent epidemiological studies, however, have shown that only certain classes of NSAIDs, such as ibuprofen, lower the risk for PD over a long period of time [132]. Strong epidemiological evidence also supports a role for caffeine consumption in decreasing the risk of developing PD [133]. Some studies have suggested that caffeine may modulate the microglial neuroinflammatory response as a potential mechanism, supporting the epidemiological data [134]. Most of the findings from postmortem patients supporting a role for reactive microgliosis and neuroinflammation in PD pathology have also been verified using animal models of the disease. Reactive microgliosis and localized neuroinflammation in the nigrostriatal system are evident in all neurotoxin models of PD both in rodents and nonhuman primates [135,136]. Further, the magnitude of the microglial neuroinflammatory response is greater in the substantia nigra than in the striatum in animal models, similar to what is seen in PD patients and postmortem human studies.

Abundant evidence over the last decade, ranging from clinical studies, animal models, and mechanistic in vitro experiments, strongly supports a deleterious role for sustained microglial activation

Table 3.2 Microglial Activation Mechanisms Relevant to Dopaminergic (DA) Degeneration

Activation Stimulus	Activation Mechanism	Functional Response in Microglia	References
α-Synuclein	Phagocytosis from DA neurons	ROS generation Cytokine production	[151,152]
Neuromelanin	Phagocytosis from DA neurons	ROS generation NFκB activation Cytokine production	[153,154]
Active MMP3	Secreted from DA neurons Possibly binds to protease receptors	ROS generation Cytokine production	[155,156]
Hsp70	Released from apoptotic DA neurons Activates TLR4	iNOS activation Nitric oxide production Microglial neurotoxicity	[157,158]
RGS10	Modulates Gαi/o signaling in microglia	RGS-deficient microglia show increased activation and DA neurotoxicity	[150,159]
Fractalkine	Negatively regulates microglial activation via fractalkine receptor signaling	Knockout mice show increased microglial activation and DA neurotoxicity	[160,161]
Gangliosides	Component of DA neuron membranes Activate microglia by surface galectins	ROS generation iNOS activation Cytokine production	[162,163]
Paraquat and rotenone	NKκB activation	ROS generation Potentiates microglial neurotoxicity	[164,165]
TNFα	Signals through TNFR1 and TNFR2	ROS generation iNOS activation Nitric oxide production	[166,167]
Diesel exhaust nanoparticles	Phagocytosis by microglia	ROS generation	[168]

Hsp70, Heat shock protein; MMP3, matrix metalloproteinase 3; RGS10, regulator of G-protein signaling 10; ROS, reactive oxygen species; TNFα, tumor necrosis factor α.

during progressive DA neuron loss [19,110,137] (Table 3.2). Although it is now more widely accepted that persistent microglial activation can be neurotoxic, a mechanistic basis for microglial neurotoxicity on DA degeneration remains to be established. The substantia nigra is known to have a relatively higher population of resident microglia than most other regions in the brain [138], and this has been linked to increased DA neuron vulnerability to microglial neurotoxicity. For example, when LPS is injected into the substantial nigra, it induces progressive DA degeneration accompanied by robust reactive microgliosis [139,140]. However, when LPS is injected at equivalent doses in other brain regions, such as the hippocampus and cortex, no neurotoxicity is evident, despite apparently robust microglial activation

[139]. DA neurons in particular are considered to exist in an intrinsically prooxidant microenvironment due to their high-dopamine and -iron contents and their apparent glutathione deficiencies [141–145], making them uniquely vulnerable to elevated levels of oxidative stress resulting from local microglial activation. As would be expected, the microglial activation response in the healthy brain is tightly controlled with inherent protective mechanisms to prevent excessive microglial activation that would be detrimental. Cannabinoids and glucocorticoids have been identified as potent endogenous signaling mechanisms that can negatively regulate the microglial inflammatory response and, thus, protect neurons from deleterious microglial activation [32,140,146–149]. Similarly, the regulator of G-protein signaling 10 (RGS10) and signaling via the fractalkine receptor in microglia have both been identified as endogenous regulatory mechanisms that regulate microglial activation during DA degeneration. Both RGS10 knockout mice and fractalkine receptor knockouts show increased DA degeneration resulting from dysregulated activation of microglia [150].

The most compelling mechanistic evidence for microglial neurotoxicity in the MPTP model comes from multiple studies and independent research groups, which have consistently demonstrated that microglial activation amplifies local ROS production in the substantia nigra to exacerbate DA degeneration. Components of the NADPH oxidase system are elevated following MPTP treatment in the mouse substantia nigra and also in human PD patients [64,169–171]. The membrane-bound component of the NADPH oxidase is highly induced following MPTP treatment and localizes to activated microglia, which were shown to be the primary producers of ROS in vivo using dihydroedthidine histochemistry. Most importantly, NADPH-deficient mice are protected against DA degeneration and microglial oxidative stress in vivo [64,172]. Targeting the NADPH oxidase system using small-molecule inhibitor drugs or with antioxidants that scavenge the ROS species have been effective in protecting the nigrostriatal system from MPTP-induced damage, with some being tested in clinical trials. Microglial iNOS is elevated following MPTP treatment and contributes to local ROS levels in the substantia nigra with knockout mice, demonstrating robust neuroprotection from MPTP toxicity [135,173]. Targeted inhibition of iNOS has also shown efficacy in the MPTP model of PD and in vitro. Signaling through the NFκB pathway regulates the production of several proinflammatory cytokines and chemokines, as well as iNOS and components of the NADPH oxidase system in microglia and other innate immune cells. NFκB activation could, therefore be a common downstream target of multiple activation pathways that result in downstream production of ROS species and proinflammatory mediators. Indeed, targeting NFκB activation using specific NBD peptide inhibitors effectively protects against DA degeneration in the MPTP model, with attenuation of microglial neurotoxicity being a major protective mechanism [174,175]. In some cases almost a complete preservation of the DA neurons in the substantia nigra has been demonstrated, indicating potential for targeting neuroinflammation therapeutically in the progression of PD [172,176–180].

The chronically activated, neurotoxic microglial response that accompanies DA degeneration implies that the inherently protective and neurotrophic process of microglial activation gradually breaks down and is transformed into one, that is, deregulated and progressively neurotoxic in the diseased state. Based on current evidence supporting a pathogenic role for sustained neuroinflammation and microglial neurotoxicity, it has been suggested that in PD, the degenerating DA neurons eventually lose their capacity to produce the signaling mediators that keep microglial activation in check, setting the stage for a vicious cycle of self-propelling neurotoxicity where uncontrolled reactive microgliosis further potentiates DA cell death. This line of thought fits well with the complex, multifactorial etiology that is now thought to underlie the pathogenesis of sporadic PD. It also supports the multiple hit hypothesis that has been proposed recently, wherein transient inflammatory and neurotoxic insults over the course of several decades can initiate a breakdown of regulatory mechanisms with time, resulting in a self-sustained process of chronic microglial activation and progressive DA neurodegeneration. Sustained microglial activation through ROS generation and proinflammatory cytokine responses can augment oxidative stress and apoptosis, which may further exacerbate the vulnerability of DA neurons. Nonetheless, the process by which transient microglial activation becomes dysregulated during DA degeneration and the factors responsible for perpetuating a chronically activated state are yet to be defined and remain an exciting area of current research in PD.

AMYOTROPHIC LATERAL SCLEROSIS

ALS is an incurable progressive neurodegenerative disease that primarily affects the upper and lower motor neurons in the brain stem, spinal cord, and motor cortex. Degeneration of the affected areas typically results in corticospinal tract signs, and atrophy of relevant muscles ultimately leading to almost complete paralysis [181]. The clinical features of ALS include fasciculation, muscle wasting and weakness, increased spasticity, and hyperreflexia. Respiratory complications normally develop in patients with advanced disease, and the cause of death is generally paralysis of the respiratory muscles and diaphragm. ALS is normally fatal, with a median age of onset of 55 years and a survival of 2–5 years after the onset of symptoms [182]. While most cases are known to be of sporadic origin with unknown causative factors, about 5%–10% of ALS cases are caused by autosomal genetic mutations. While multiple genetic loci have been identified in ALS, 20% of familial cases involve a gain-of-function mutation in Cu/Zn superoxide dismutase I (SOD1). Interestingly, in addition to the well-defined role of SOD1 as a critical antioxidant enzyme, recent evidence suggests that its normal function is to protect against protein aggregation, which is a characteristic feature of neurodegenerative diseases [182,183] (Chapter 1). The major pathological mechanisms that have been implicated in ALS include excitotoxic neuronal death, proapoptotic signaling, and metabolic dysfunction. Substantial evidence now suggests that the innate immune system contributes to ALS disease pathology [182]. In line with this, immune cell infiltration and activation is evident at the sites of neurodegeneration in experimental models of the disease [184,185]. Prominent neuroinflammation characterized by reactive gliosis and the accumulation of large numbers of activated microglia and astrocytes is apparent in the affected areas of the CNS and in spinal cords from both human ALS patients and mouse models of the disease [19]. These findings are further substantiated by PET imaging of living patients with ALS using the ligand [^{11}C](R)-PK11195, demonstrating increased microglial activation in affected brain areas. Further, the degree of microglial activation has been correlated with the severity of disease symptoms. Similar to other neurodegenerative diseases, inflammatory mediators, including MCP1, a potent chemotactic stimulus for microglia, have been shown to be elevated in the CSF of ALS patients [79,184]. The proposed mechanisms by which innate immune neuroinflammation may affect the function and survival of motor neurons in ALS include the persistent activation of microglia and astrocytes by mutant SOD1 via PRRs TLR2 and TLR4 and the NLRP3 inflammasome [81]. Selective ablation of SOD1 in microglia has been shown to increase survival in transgenic mice, indicating that mutant SOD contributes to neuronal cell death by affecting microglial responses independently of any direct effects on neurons. The release of proinflammatory cytokines, ROS, and reactive nitrogen species derived from persistently activated microglia and astrocytes have also been proposed to drive motor neuron death [184]. In preclinical mouse models of ALS, antiinflammatory treatments aimed at reducing microglia activation have been shown to increase the life expectancy of mice expressing human mutant SOD1. Major histocompatibility complex molecules and complement receptors are highly expressed by reactive microglia in the primary motor cortex and in the anterior horn of the spinal cords of ALS patients [186,187]. In summary, chronic neuroinflammatory processes in ALS are emerging as important contributors to disease pathology and could be useful disease-modifying therapeutic targets.

HUNTINGTON'S DISEASE

HD is an autosomal dominant progressive neurodegenerative disease caused by the expansion of the trinucleotide CAG in the gene that encodes HTT resulting in mHTT being produced. The mHTT protein possesses a polyglutamine repeat expansion at the amino terminus of the protein, which has been shown to affect the stability, structure, and function of the protein. The primary neuropathology of HD is a progressive loss of medium-spiny neurons within the striatum. Recently, inflammatory responses have been suggested to have pathological roles in HD. Postmortem studies of the brains of patients who had HD have shown increased microgliosis and astrogliosis [18,188,189]. Increased microglial production of complement proteins (C3 and C9), as well as elevated IL1β have been reported in HD patients. Elevated mRNA expression of inflammatory factors, such as CCL2, IL10, IL6, IL8, and TNFα have been reported in the brains of HD patients [18]. Also, the plasma levels of TNFα were found to be elevated in the early stages of HD, with increased levels of TNFα correlating with clinical progression. Similarly, elevated IL6 levels have been detected in patient plasma years before disease onset in mHTT carriers, while CCL4, CCL11, and CCL26 were elevated early during the disease course [190]. PET studies have detected microglial cell activation in the early stages of living patients with HD, and suggest that increased microglial cell activation inversely correlates with reduced neuronal activity [191,192]. Similarly, ex vivo brain slice culture studies have shown that microglia are activated and proliferate in the vicinity of degenerating neurites of mHTT-expressing neurons. In the preclinical R6/2 mouse model of HD, microglia are activated at the early stages of disease and have been shown to accumulate ferritin [193,194]. R6/2 mice are transgenic for exon 1 of the human HTT protein carrying about 150 glutamine repeats and the upstream promoter regions. Patients with HD similarly show an accumulation of ferritin in microglial cells in the early stages of disease. The expression of proinflammatory genes (TNFα, IFNγ, and TGFβ) are also upregulated in the preclinical R6/2 model [188,193]. Inflammasome-driven caspase-1 activation has been documented in HD models, and is proposed to contribute to neuropathology through the generation of cleaved IL1β and pyroptotic cell death mechanisms [5]. As mHTT is also expressed by astrocytes, it has been postulated that these cells might play a role in neuronal dysfunction. This has been supported by knockin studies for mHTT under the control of the human glial fibrillary astrocytic protein (*GFAP*) promoter, which showed neurological deficits, shorter life spans and dysregulated glutamate uptake, which contributes to neuronal excitotoxicity [195–197]. In summary, mHTT can serve as a DAMP in the CNS microenvironment to induce pathological innate immune activation both in the CNS and also in the periphery [188,198]. As persistent neuroinflammation has been linked to disease progression and severity in HD, targeting inflammatory pathways could be a potential therapeutic strategy to slow disease progression.

FRONTOTEMPORAL DEMENTIA AND LEWY BODY DEMENTIA

Frontotemporal dementia (FTD) is a form of progressive neuronal atrophy that involves the progressive loss of cells from both the frontal and temporal cortex. Dementia is often accompanied by changes in personality, behavior, and language skills. FTD is the most prevalent type of dementia after AD. Most FTD patients have intraneuronal inclusions primarily composed of the cytosolic phosphorylated TDP43. Neuroinflammation has been well documented clinically in the pathology of FTD with elevated levels of TNF and TGFβ present in the CSF [199]. Studies using the microglial PET marker [^{11}C] (R)-PK11195 indicate that there is an increased activation of microglial cells in the frontotemporal lobe of patients with FTD early during the disease course [200]. Additionally, *Trem2* variants, which are linked to a greater risk of AD, have now also been associated with an increased risk of FTD [201,202]. Recently, mutations in the progranulin gene have been associated with the development of some forms of FTD [203]. Mechanistic studies have demonstrated that microglial activation is dysregulated in progranulin-deficient mice leading to a neurotoxic proinflammatory phenotype with detrimental effects on neuronal survival [204]. Neuroinflammation has also been reported in the brains of patients suffering from LBD [205]. Increased microglial activation has been reported in LBD patients compared to healthy controls. Pathologically, LBD shares features of both AD and PD, although the extent of reactive gliosis observed in LBD appears to be less compared to AD. Both IL1β and TNFα have been reported in activated microglia in LBD brains, although their roles in this disease remain unclear [206].

THERAPEUTIC STRATEGIES TARGETING NEUROINFLAMMATION IN PROGRESSIVE NEURODEGENERATIVE DISEASES

As discussed earlier, an extensive and growing body of scientific evidence now supports a role for neuroinflammatory mediators in the pathology and progression of neurodegenerative diseases. The sustained neuroinflammatory response in the CNS provides several attractive therapeutic opportunities and drug targets for neuroprotection. Disappointingly, however, several antiinflammatory therapeutics that have shown great promise in preclinical models, have had little to no therapeutic benefit in the clinic. Thus, for targeted therapeutics to be efficacious in the clinic, much progress remains to be made in terms of understanding the precise functions of reactive gliosis. Additionally, the time course and mechanisms by which microglia and astrocytes transform from their seemingly protective neurotrophic roles to a chronically reactive state, that is, detrimental to neuron survival need to be established with better certainty [5,8,19]. More research into the basic biology of neuron–glial interactions would be necessary in this regard. It also remains to be clearly established if neuroinflammation can be the primary causative factor that initiates progressive neurodegeneration as suggested by some studies. There is some evidence from LPS-based models of neuroinflammation, in which neurodegeneration is driven primarily by strong systemic inflammation induced by repeated LPS stimulation [207,208]. However, the relevance of these models to sporadic neurodegeneration of idiopathic etiology remains to be established conclusively. It might be the case that repeated bouts of systemic infection or inflammation over a lifetime may indeed be an initiating factor in some, but not all, cases of neurodegeneration. This is consistent with the proposed concepts of multifactorial causative factors and the multiple hit hypothesis [6,7]. Further, the emerging paradigm of misfolded proteins and protein aggregates serving as persistent DAMPs in the CNS to chronically activate microglia and astrocytes suggests that reactive gliosis could have a greater pathological role at the early stages of neurodegeneration than previously appreciated. There is considerable evidence from clinical studies, preclinical animal models, and in vitro mechanistic studies that neuroinflammatory responses are self-perpetuating once they are initiated [8]. However, the precise mechanisms by which this occurs, including the interactions between microglia, astrocytes, and degenerating neurons, as well as the signaling pathways involved, remain to be elucidated in sufficient detail. Identifying these mechanisms would allow for therapeutic targeting of the positive feedback loops between reactive microglia that are implicated in perpetuating neuroinflammatory responses, which exacerbate ongoing neurodegeneration.

It has been suggested that the major cause for failure of the vast majority of therapeutic strategies targeting neurodegenerative diseases has been the fact that therapeutic interventions in the clinic occur too late in the disease course. At mid to late disease stages postdiagnosis, the damage to the CNS and the associated neuronal loss might be too great to achieve any significant improvement in symptoms. Further, the neuroinflammatory responses might be irreversible due to a neurotoxic CNS microenvironment that is overwhelmed by the presence of misfolded protein aggregates, as is the case with mid to late stages of AD. Therefore, it is critical that therapeutic interventions targeting neuroinflammation are tested as early as possible in the course of the disease. This is often hampered by the lack of definitive diagnostic biomarkers of disease onset or progression. In this regard, PET-based imaging of microglia in the brains of living patients shows much promise to be able to detect neuroinflammation at early stages even before the onset of neurodegeneration. However, more detailed studies would be required to establish definitive correlations between disease stages and PET-based microglial morphometries and phenotypes. Although there has been little success with the use of neuroinflammatory mediators as diagnostic or prognostic markers of neurodegeneration, the use of large-scale proteomic and genomic approaches to identify inflammatory signatures holds promise in this area, and may allow for "presymptomatic" treatment paradigms to be established.

Based on our current knowledge, the primary targets for therapeutic modulation of neuroinflammation are the resident innate immune microglia and astrocytes, as well as infiltrating peripheral immune cells, such as lymphocytes and macrophages. Multiple therapeutic approaches targeting neuroinflammatory responses have been utilized in experimental models of neurodegenerative diseases with varying degrees of success. Despite the apparent failure of many therapeutics in clinical settings, valuable lessons can be learnt from these trials, which can be applied to refine and validate more effective therapeutic approaches or drugs. For example, the lack of efficacy with free radical–scavenging antioxidants in most clinical trials for AD and PD suggests that new therapeutic approaches that block the sources of oxidative stress in neurodegeneration are necessary. To this end, targeting the upregulation of NOX2 (gp91–Phox) and iNOS in chronically activated microglia with CNS-permeable small molecules might be a feasible therapeutic approach [65]. Oxidative stress and mitochondrial dysfunction have been shown to be major pathological mechanisms driving neuronal dysfunction. Both microglial iNOS and NOX2 pathways have been shown to be the major contributors of ROS in the degenerating CNS being upregulated in patients and experimental models. Further as iNOS is associated with a proinflammatory and neurotoxic M1 microglial phenotype, therapeutic approaches that switch microglia to an antiinflammatory M2 phenotype could be neuroprotective. This was elegantly demonstrated recently using NLRP3 knockout mice in transgenic AD models and was suggested as a major mechanism of neuroprotection in these NLRP3-deficient mice [68,80]. It must be mentioned, however, that research over the last decade in macrophage biology has questioned the entire concept of M1 versus M2 phenotypes with several experts no longer utilizing this framework. Nonetheless, despite the concept of microglial phenotypes, the notion of halting or reversing the chronically activated microglial response in the CNS is a promising therapeutic approach [8,58]. Inhibition of upstream signaling pathways regulating microglial activation has been shown to be effective in preclinical models of neurodegeneration. Blockade of NFκB using CNS-permeable inhibitors can protect against DA degeneration in PD models [174]. NFκB is downstream of many PRRs and is the main transcriptional pathway required for upregulation of NOX2 and iNOS, as well as priming the NLRP3 inflammasome in microglia [62,209–211]. Therefore, it represents an attractive therapeutic target as a master regulator of multiple proinflammatory pathways. Similarly, kinases that transduce microglial activation signals downstream of PRRs, such as Fyn kinase, Syk kinase, and PKCδ, could be useful therapeutic targets if CNS-permeable small-molecule inhibitors can be developed [212–214].

Targeting the production of secreted and cell surface inflammatory mediators that drive neuronal dysfunction and death represents another major therapeutic strategy. Dominant negative inhibitors of soluble TNF signaling in the form of small-molecule inhibitors or by viral overexpression have shown efficacy in multiple experimental models of neurodegenerative disorders, including AD and PD [178,215,216]. However, results in clinical trials showed little or no real benefit for patients with TNF inhibitors. Given that there are relatively safe and effective TNF inhibitors, such as Etanercept, that are in use for systemic inflammatory diseases, it could be worthwhile if inhibitors are tested at earlier disease stages or in combination with other antiinflammatory therapeutics [216]. Chronic IL1β production in the CNS has been shown to be neurotoxic to some populations of neurons. Viral overexpression of IL1β at low levels drives DA degeneration in the substantia nigra; whereas, inhibition of the NLRP3-driven IL1β is protective in transgenic AD models [68]. Inhibition of the NLRP3 inflammasome could be an effective means of blocking downstream IL1β production, which has been proposed to be a master regulator of proinflammatory responses in the brain [217]. In 2015, the first small-molecule inhibitor of the NLRP3 inflammasome (MCC950) was published and showed therapeutic efficacy in preclinical mouse model of multiple sclerosis [218]. Recently, it was also demonstrated that endogenous ketone bodies are potent inhibitors of the NLRP3 inflammasome, suggesting that CNS-permeable compounds, such as β-hydroxybutyrate, could have therapeutic value for neuroinflammatory diseases [219]. In experimental models of AD, both IL12 and IL23 have recently been shown to be involved in driving neuropathology with neutralizing antibodies capable of some potential therapeutic benefit when given systemically [220]. IL12 and IL23 were found to be elevated in the CSF and brain tissue of AD patients. However, it remains to be established if these mechanisms are also relevant in other neurodegenerative diseases. Given that there are already FDA-approved IL12p40-neutralizing drugs in use for psoriasis, these drugs may be worth evaluating for AD and potentially other neurodegenerative diseases [221]. Complement activation and deposition around plaques and degenerating neurons have also been demonstrated in multiple neurodegenerative disorders, including AD, PD, HD, and ALS. In models of AD, ALS, and HD, pharmacological inhibition of the receptor for the terminal complement component, C5a has been shown to have positive effects on disease pathology and progression [75,222]. During postnatal development, microglia have a well-defined role in complement-mediated synaptic pruning; therefore, it has also been suggested that a reactivation of this mechanism can occur in neurodegenerative disease to propagate neuronal synaptic removal [223].

PRRs, such as RAGE, Mac1, and TLRs, represent exciting targets in theory based on the emerging paradigm of neuronal DAMPs and misfolded proteins driving neuroinflammation via multiple PRRs. Further, stimulation of PRRs by neuronal DAMPs has been shown to directly link to the

Table 3.3 Selected Clinical Trials for Therapeutics Targeting Neuroinflammation

Target/Mechanism	Therapeutic Agent	Disease	Study Type	Outcome
Amyloid β clearance	Tarenflurbil	AD	Phase III	No benefit
RAGE inhibitor	PF-04494700	AD	Phase II	No benefit
PPARγ agonist	Rosiglitazone	AD and PD	Phase II–III	No benefit or inconclusive
Src/Fyn kinase inhibitor	Saracatinib (AZD0530)	AD	Phase II	Ongoing
Cox2	Celecoxib and Nimesulide	AD and ALS	Phase II/III	No benefit
Calcium channel blocker	Nilvadipine	AD	Phase II/III	Ongoing
ROS	CoenzymeQ and celastrol	AD and ALS	Phase I–III	No benefit
Microglial activation	Minocycline	PD and ALS	Phase II–III	Inconclusive
Astrocytes	ONO-2506	ALS	Phase III	No benefit

activation of downstream proinflammatory pathways, including NOX2, iNOS, and TNFα. Therefore, effectively targeting one or more relevant PRRs in each neurodegenerative disease could be a potential mechanism to shut down reactive gliosis and mitigate the cycle of chronic self-perpetuating neuroinflammation and neurotoxicity. To this end, PF-04494700, an inhibitor of the RAGE showed promise in preclinical models, but results were inconclusive in clinical trials [224]. Statins have been tested in clinical trials for their efficacy in terms of augmenting microglial clearance of amyloid-β, but no therapeutic benefit was found. Recently discovered NLRP3 inhibitors, such as MCC950, also hold promise in this area based on recent findings in transgenic AD models, demonstrating that NLRP3 is a major intracellular PRR that recognizes aggregated misfolded proteins [68]. Again, a better understanding of neuron–glial and microglia–astrocyte interactions in the CNS during neurodegeneration would allow for the development of more specific therapeutic strategies targeting PRRs to achieve neuroprotection. Selected clinical trials with therapeutic agents targeting neuroinflammation are summarized in Table 3.3.

FUTURE PERSPECTIVES

In the last few years, exciting new paradigms have emerged in this field that hold promise for future therapeutic targets and strategies. A direct role for the gut microbiome in modulation of microglial activation and homeostasis has recently been reported suggesting that modulation of the gut microbiota could be of potential therapeutic benefit [38]. Recent reports have in fact confirmed that PD patients have an altered gut microbiome compared to age-matched healthy individuals [225]. Altered sleep patterns have been reported in most progressive neurodegenerative diseases and recent reports have demonstrated that these dysregulated sleep patterns can alter neuroinflammatory responses [226–228]. The precise mechanisms by which sleep and the loss of sleep quality can modulate pro- or antiinflammatory responses in the CNS are yet to be determined. Finally, the role of systemic immune responses in shaping acute and chronic neuroinflammation is being uncovered in greater mechanistic detail. A major paradigm shift that would be relevant in this regard is the recent discovery of functional lymphatic vessels in the brain that can transport fluid and immune cells from the CSF [229]. This system of CNS lymphatic networks that are connected to deep cervical lymph nodes can have enormous implications in terms of the scope for regulation of neuroinflammatory responses and the role of peripheral immune mediators. Already, a previously unknown and unexpected role for neutrophils in the pathology of AD has been uncovered and has a role for peripheral blood monocytes in ALS pathology [101]. It is therefore very likely that exciting new advances would be made in these areas in the coming years that will lead to the development of novel therapeutic strategies.

CONCLUSIONS

Remarkable progress has been made in the last two decades in terms of our fundamental understanding of innate immune responses in the CNS during healthy aging and in neurodegenerative diseases. However, there are still significant knowledge gaps in terms of understanding the entire spectrum of beneficial and pathogenic neuroinflammatory responses that occur during healthy aging and during the onset and progression of neurodegeneration. As greater insight into neuron–glial and glial–glial interactions in the CNS microenvironment emerges, targeted and potentially more effective therapeutic strategies can be pursued. The consensus from several unsuccessful trials targeting neuroinflammation is that the timing of intervention during the disease course appears to be crucial for therapeutic benefit. It is also obvious that at this point, there is still a limited understanding of the complexities of innate immune responses in the CNS and its interplay with systemic immunity. The innate immune system, therefore, provides exciting opportunities for disease-modifying treatments in the CNS that are both innovative and feasible. It is foreseeable, that as our knowledge of the precise underling immune mechanisms advances, more effective disease-modifying therapeutics can be developed to treat these currently intractable neurodegenerative diseases.

REFERENCES

[1] R.J. Ward, et al. The role of iron in brain ageing and neurodegenerative disorders, Lancet Neurol. 13 (10) (2014) 1045–1060.

[2] The Lancet. Joining forces to fight neurodegenerative diseases, Lancet Neurol. 12 (2) (2013) 119.

[3] M. Prince, et al. The global prevalence of dementia: a systematic review and metaanalysis, Alzheimers Dement. 9 (1) (2013) 63–75.

[4] J.A. Obeso, et al. Missing pieces in the Parkinson's disease puzzle, Nat. Med. 16 (6) (2010) 653–661.

[5] M.T. Heneka, M.P. Kummer, E. Latz, Innate immune activation in neurodegenerative disease, Nat. Rev. Immunol. 14 (7) (2014) 463–477.

[6] D. Sulzer, Multiple hit hypotheses for dopamine neuron loss in Parkinson's disease, Trends Neurosci. 30 (5) (2007) 244–250.

[7] X. Zhu, et al. Alzheimer's disease: the two-hit hypothesis, Lancet Neurol. 3 (4) (2004) 219–226.

[8] M.L. Block, L. Zecca, J.S. Hong, Microglia-mediated neurotoxicity: uncovering the molecular mechanisms, Nat. Rev. Neurosci. 8 (1) (2007) 57–69.

[9] M.T. Lin, M.F. Beal, Mitochondrial dysfunction and oxidative stress in neurodegenerative diseases, Nature 443 (7113) (2006) 787–795.

[10] C. Soto, Unfolding the role of protein misfolding in neurodegenerative diseases, Nat. Rev. Neurosci. 4 (1) (2003) 49–60.

[11] Y.S. Eisele, et al. Targeting protein aggregation for the treatment of degenerative diseases, Nat. Rev. Drug Discov. 14 (11) (2015) 759–780.

[12] W. Peelaerts, et al. Alpha-synuclein strains cause distinct synucleinopathies after local and systemic administration, Nature 522 (7556) (2015) 340–344.

[13] K.C. Luk, et al. Intracerebral inoculation of pathological alpha-synuclein initiates a rapidly progressive neurodegenerative alpha-synucleinopathy in mice, J. Exp. Med. 209 (5) (2012) 975–986.

[14] K.C. Luk, et al. Pathological alpha-synuclein transmission initiates Parkinson-like neurodegeneration in nontransgenic mice, Science 338 (6109) (2012) 949–953.

[15] M. Polymenidou, D.W. Cleveland, Prion-like spread of protein aggregates in neurodegeneration, J. Exp. Med. 209 (5) (2012) 889–893.

[16] L. Minghetti, Role of inflammation in neurodegenerative diseases, Curr. Opin. Neurol. 18 (3) (2005) 315–321.

[17] B. Liu, H.M. Gao, J.S. Hong, Parkinson's disease and exposure to infectious agents and pesticides and the occurrence of brain injuries: role of neuroinflammation, Environ. Health Perspect. 111 (8) (2003) 1065–1073.

[18] A. Crotti, C.K. Glass, The choreography of neuroinflammation in Huntington's disease, Trends Immunol. 36 (6) (2015) 364–373.

[19] C.K. Glass, et al. Mechanisms underlying inflammation in neurodegeneration, Cell 140 (6) (2010) 918–934.

[20] B.A. Barres, The mystery and magic of glia: a perspective on their roles in health and disease, Neuron 60 (3) (2008) 430–440.

[21] F.A. Azevedo, et al. Equal numbers of neuronal and nonneuronal cells make the human brain an isometrically scaled-up primate brain, J. Comp. Neurol. 513 (5) (2009) 532–541.

[22] S. Herculano-Houzel, The remarkable, yet not extraordinary, human brain as a scaled-up primate brain and its associated cost, Proc. Natl. Acad. Sci. USA 109 (Suppl. 1) (2012) 10661–10668.

[23] K. Saijo, C.K. Glass, Microglial cell origin and phenotypes in health and disease, Nat. Rev. Immunol. 11 (11) (2011) 775–787.

[24] D. Nayak, T.L. Roth, D.B. McGavern, Microglia development and function, Annu. Rev. Immunol. 32 (2014) 367–402.

[25] D.P. Schafer, B. Stevens, Microglia function in central nervous system development and plasticity, Cold Spring Harb. Perspect. Biol. 7 (10) (2015) pa020545.

[26] A. Nimmerjahn, F. Kirchhoff, F. Helmchen, Resting microglial cells are highly dynamic surveillants of brain parenchyma in vivo, Science 308 (5726) (2005) 1314–1318.

[27] U.K. Hanisch, H. Kettenmann, Microglia: active sensor and versatile effector cells in the normal and pathologic brain, Nat. Neurosci. 10 (11) (2007) 1387–1394.

[28] D. Davalos, et al. ATP mediates rapid microglial response to local brain injury in vivo, Nat. Neurosci. 8 (6) (2005) 752–758.

[29] G.W. Kreutzberg, Microglia: a sensor for pathological events in the CNS, Trends Neurosci. 19 (8) (1996) 312–318.

[30] I. Napoli, H. Neumann, Protective effects of microglia in multiple sclerosis, Exp. Neurol. 225 (1) (2010) 24–28.

[31] A.R. Simard, S. Rivest, Neuroprotective effects of resident microglia following acute brain injury, J. Comp. Neurol. 504 (6) (2007) 716–729.

[32] I. Glezer, A.R. Simard, S. Rivest, Neuroprotective role of the innate immune system by microglia, Neuroscience 147 (4) (2007) 867–883.

[33] M. Persson, et al. Lipopolysaccharide increases microglial GLT-1 expression and glutamate uptake capacity in vitro by a mechanism dependent on TNF-alpha, Glia 51 (2) (2005) 111–120.

[34] I. Shaked, et al. Protective autoimmunity: interferon-gamma enables microglia to remove glutamate without evoking inflammatory mediators, J. Neurochem. 92 (5) (2005) 997–1009.

[35] B.D. Trapp, et al. Evidence for synaptic stripping by cortical microglia, Glia 55 (4) (2007) 360–368.

[36] O. Butovsky, et al. Microglia activated by IL-4 or IFN-gamma differentially induce neurogenesis and oligodendrogenesis from adult stem/progenitor cells, Mol. Cell. Neurosci. 31 (1) (2006) 149–160.

[37] M. Domercq, N. Vazquez-Villoldo, C. Matute, Neurotransmitter signaling in the pathophysiology of microglia, Front. Cell. Neurosci. 7 (2013) 49.

[38] D. Erny, et al. Host microbiota constantly control maturation and function of microglia in the CNS, Nat. Neurosci. 18 (7) (2015) 965–977.

[39] H.K. Kimelberg, M. Nedergaard, Functions of astrocytes and their potential as therapeutic targets, Neurotherapeutics 7 (4) (2010) 338–353.

[40] M.V. Sofroniew, H.V. Vinters, Astrocytes: biology and pathology, Acta Neuropathol. 119 (1) (2010) 7–35.

[41] L.E. Clarke, B.A. Barres, Emerging roles of astrocytes in neural circuit development, Nat. Rev. Neurosci. 14 (5) (2013) 311–321.

[42] F. Tang, et al. Lactate-mediated glia-neuronal signalling in the mammalian brain, Nat. Commun. 5 (2014) 3284.

[43] M.V. Sofroniew, Astrogliosis, Cold Spring Harb. Perspect. Biol. 7 (2) (2015) pa020420.

[44] V. Parpura, P.G. Haydon, Physiological astrocytic calcium levels stimulate glutamate release to modulate adjacent neurons, Proc. Natl. Acad. Sci. USA 97 (15) (2000) 8629–8634.

[45] A.V. Singh, et al. Astrocytes increase ATP exocytosis mediated calcium signaling in response to microgroove structures, Sci. Rep. 5 (2015) 7847.

[46] D. Sun, T.C. Jakobs, Structural remodeling of astrocytes in the injured CNS, Neuroscientist 18 (6) (2012) 567–588.

[47] M.V. Sofroniew, Molecular dissection of reactive astrogliosis and glial scar formation, Trends Neurosci. 32 (12) (2009) 638–647.

[48] E. Hennessy, E.W. Griffin, C. Cunningham, Astrocytes are primed by chronic neurodegeneration to produce exaggerated chemokine and cell infiltration responses to acute stimulation with the cytokines IL-1beta and TNF-alpha, J. Neurosci. 35 (22) (2015) 8411–8422.

[49] B.V. Zlokovic, The blood-brain barrier in health and chronic neurodegenerative disorders, Neuron 57 (2) (2008) 178–201.

[50] N.J. Maragakis, J.D. Rothstein, Mechanisms of disease: astrocytes in neurodegenerative disease, Nat. Clin. Pract. Neurol. 2 (12) (2006) 679–689.

[51] N.J. Abbott, L. Ronnback, E. Hansson, Astrocyte-endothelial interactions at the blood-brain barrier, Nat. Rev. Neurosci. 7 (1) (2006) 41–53.

[52] C.M. Kassmann, et al. Axonal loss and neuroinflammation caused by peroxisome-deficient oligodendrocytes, Nat. Genet. 39 (8) (2007) 969–976.

[53] H.B. Sarnat, Ependymal reactions to injury. A review, J. Neuropathol. Exp. Neurol. 54 (1) (1995) 1–15.

[54] M. Lattke, et al. Nuclear factor kappaB activation impairs ependymal ciliogenesis and links neuroinflammation to hydrocephalus formation, J. Neurosci. 32 (34) (2012) 11511–11523.

[55] F.L. Heppner, et al. Experimental autoimmune encephalomyelitis repressed by microglial paralysis, Nat. Med. 11 (2) (2005) 146–152.

[56] G. Gowing, L. Vallieres, J.P. Julien, Mouse model for ablation of proliferating microglia in acute CNS injuries, Glia 53 (3) (2006) 331–337.

[57] K. Biber, T. Owens, E. Boddeke, What is microglia neurotoxicity (Not)?, Glia 62 (6) (2014) 841–854.

[58] T.C. Frank-Cannon, et al. Does neuroinflammation fan the flame in neurodegenerative diseases?, Mol. Neurodegener. 4 (2009) 47.

[59] M. Aschner, et al. Glial cells in neurotoxicity development, Annu. Rev. Pharmacol. Toxicol. 39 (1999) 151–173.

[60] J. Minkiewicz, J.P. de Rivero Vaccari, R.W. Keane, Human astrocytes express a novel NLRP2 inflammasome, Glia 61 (7) (2013) 1113–1121.

[61] T. Maslanik, et al. The inflammasome and danger associated molecular patterns (DAMPs) are implicated in cytokine and chemokine responses following stressor exposure, Brain Behav. Immun. 28 (2013) 54–62.

[62] J.G. Walsh, D.A. Muruve, C. Power, Inflammasomes in the CNS, Nat. Rev. Neurosci. 15 (2) (2014) 84–97.

[63] S. Saxena, P. Caroni, Selective neuronal vulnerability in neurodegenerative diseases: from stressor thresholds to degeneration, Neuron 71 (1) (2011) 35–48.

[64] D.C. Wu, et al. NADPH oxidase mediates oxidative stress in the 1-methyl-4-phenyl-1,2,3,6-tetrahydropyridine model of Parkinson's disease, Proc. Natl. Acad. Sci. USA 100 (10) (2003) 6145–6150.

[65] H.M. Gao, H. Zhou, J.S. Hong, NADPH oxidases: novel therapeutic targets for neurodegenerative diseases, Trends Pharmacol. Sci. 33 (6) (2012) 295–303.

[66] R.E. Mrak, W.S. Griffin, Glia and their cytokines in progression of neurodegeneration, Neurobiol. Aging 26 (3) (2005) 349–354.

[67] M. Reale, et al. Peripheral cytokines profile in Parkinson's disease, Brain Behav. Immun. 23 (1) (2009) 55–63.

[68] M.T. Heneka, et al. NLRP3 is activated in Alzheimer's disease and contributes to pathology in APP/PS1 mice, Nature 493 (7434) (2013) 674–678.

[69] R. Gordon, et al. Proteolytic activation of proapoptotic kinase protein kinase C delta by tumor necrosis factor alpha death receptor signaling in dopaminergic neurons during neuroinflammation, J. Neuroinflamm. 9 (2012) 82.

[70] M.K. McCoy, M.G. Tansey, TNF signaling inhibition in the CNS: implications for normal brain function and neurodegenerative disease, J. Neuroinflamm. 5 (2008) 45.

[71] M.C. Pott Godoy, et al. Central and systemic IL-1 exacerbates neurodegeneration and motor symptoms in a model of Parkinson's disease, Brain 131 (Pt. 7) (2008) 1880–1894.

[72] T.M. Woodruff, et al. The role of the complement system and the activation fragment C5a in the central nervous system, Neuromol. Med. 12 (2) (2010) 179–192.

[73] I. Farkas, et al. Complement C5a receptor-mediated signaling may be involved in neurodegeneration in Alzheimer's disease, J. Immunol. 170 (11) (2003) 5764–5771.

[74] A. Jacob, J.J. Alexander, Complement and blood-brain barrier integrity, Mol. Immunol. 61 (2) (2014) 149–152.

[75] L.G. Bodea, et al. Neurodegeneration by activation of the microglial complement-phagosome pathway, J. Neurosci. 34 (25) (2014) 8546–8556.

[76] D.T. Briggs, et al. Astrocyte-specific expression of a soluble form of the murine complement control protein Crry confers demyelination protection in the cuprizone model, Glia 55 (14) (2007) 1405–1415.

[77] T. Wyss-Coray, et al. Prominent neurodegeneration and increased plaque formation in complement-inhibited Alzheimer's mice, Proc. Natl. Acad. Sci. USA 99 (16) (2002) 10837–10842.

[78] P. Mukherjee, G.M. Pasinetti, The role of complement anaphylatoxin C5a in neurodegeneration: implications in Alzheimer's disease, J. Neuroimmunol. 105 (2) (2000) 124–130.

[79] S. Mantovani, et al. Elevation of the terminal complement activation products C5a and C5b-9 in ALS patient blood, J Neuroimmunol 276 (1–2) (2014) 213–218.

[80] A. Halle, et al. The NALP3 inflammasome is involved in the innate immune response to amyloid-beta, Nat. Immunol. 9 (8) (2008) 857–865.

[81] S. Johann, et al. NLRP3 inflammasome is expressed by astrocytes in the SOD1 mouse model of ALS and in human sporadic ALS patients, Glia 63 (12) (2015) 2260–2273.

[82] J.G. Walsh, et al. Rapid inflammasome activation in microglia contributes to brain disease in HIV/AIDS, Retrovirology 11 (2014) 35.

[83] M.T. Heneka, D.T. Golenbock, E. Latz, Innate immunity in Alzheimer's disease, Nat. Immunol. 16 (3) (2015) 229–236.

[84] M.T. Heneka, et al. Neuroinflammation in Alzheimer's disease, Lancet Neurol. 14 (4) (2015) 388–405.

[85] L. Crews, E. Masliah, Molecular mechanisms of neurodegeneration in Alzheimer's disease, Hum. Mol. Genet. 19 (R1) (2010) R12–R20.

[86] J.M. Tarasoff-Conway, et al. Clearance systems in the brain-implications for Alzheimer disease, Nat. Rev. Neurol. 11 (8) (2015) 457–470.

[87] F.L. Heppner, R.M. Ransohoff, B. Becher, Immune attack: the role of inflammation in Alzheimer disease, Nat. Rev. Neurosci. 16 (6) (2015) 358–372.

[88] S.C. Jin, et al. Coding variants in TREM2 increase risk for Alzheimer's disease, Hum. Mol. Genet. 23 (21) (2014) 5838–5846.

[89] R. Guerreiro, et al. TREM2 variants in Alzheimer's disease, N. Engl. J. Med. 368 (2) (2013) 117–127.

[90] S.E. Hickman, J. El Khoury, TREM2 and the neuroimmunology of Alzheimer's disease, Biochem. Pharmacol. 88 (4) (2014) 495–498.

[91] M.I. Fonseca, et al. Treatment with a C5aR antagonist decreases pathology and enhances behavioral performance in murine models of Alzheimer's disease, J. Immunol. 183 (2) (2009) 1375–1383.

[92] A. Schuitemaker, et al. Microglial activation in Alzheimer's disease: an (R)-[(1)(1)C]PK11195 positron emission tomography study, Neurobiol. Aging 34 (1) (2013) 128–136.

[93] C. Villegas-Llerena, et al. Microglial genes regulating neuroinflammation in the progression of Alzheimer's disease, Curr. Opin. Neurobiol. 36 (2015) 74–81.

[94] J.C. Lambert, et al. Meta-analysis of 74,046 individuals identifies 11 new susceptibility loci for Alzheimer's disease, Nat. Genet. 45 (12) (2013) 1452–1458.

[95] C. Nathan, et al. Protection from Alzheimer's-like disease in the mouse by genetic ablation of inducible nitric oxide synthase, J. Exp. Med. 202 (9) (2005) 1163–1169.

[96] M.P. Kummer, et al. Nitration of tyrosine 10 critically enhances amyloid beta aggregation and plaque formation, Neuron 71 (5) (2011) 833–844.

[97] J. Miklossy, et al. Alzheimer disease: curly fibers and tangles in organs other than brain, J. Neuropathol. Exp. Neurol. 58 (8) (1999) 803–814.

[98] A.D. Roth, et al. Oligodendrocytes damage in Alzheimer's disease: beta amyloid toxicity and inflammation, Biol. Res. 38 (4) (2005) 381–387.

[99] K. Kobayashi, et al. Apoptosis of astrocytes with enhanced lysosomal activity and oligodendrocytes in white matter lesions in Alzheimer's disease, Neuropathol. Appl. Neurobiol. 28 (3) (2002) 238–251.

[100] S.H. Baik, et al. Migration of neutrophils targeting amyloid plaques in Alzheimer's disease mouse model, Neurobiol. Aging 35 (6) (2014) 1286–1292.

[101] E. Zenaro, et al. Neutrophils promote Alzheimer's disease-like pathology and cognitive decline via LFA-1 integrin, Nat. Med. 21 (8) (2015) 880–886.

[102] C. Lauro, et al. Adenosine A1 receptors and microglial cells mediate CX3CL1-induced protection of hippocampal neurons against Glu-induced death, Neuropsychopharmacology 35 (7) (2010) 1550–1559.

[103] D.G. Walker, et al. Decreased expression of CD200 and CD200 receptor in Alzheimer's disease: a potential mechanism leading to chronic inflammation, Exp. Neurol. 215 (1) (2009) 5–19.

[104] L.B. Yang, et al. Deficiency of complement defense protein CD59 may contribute to neurodegeneration in Alzheimer's disease, J. Neurosci. 20 (20) (2000) 7505–7509.

[105] O. von Bohlen und Halbach, A. Schober, K. Krieglstein, Genes, proteins, and neurotoxins involved in Parkinson's disease, Prog. Neurobiol. 73 (3) (2004) 151–177.

[106] W. Dauer, S. Przedborski, Parkinson's disease: mechanisms and models, Neuron 39 (6) (2003) 889–909.

[107] P.L. McGeer, et al. Reactive microglia are positive for HLA-DR in the substantia nigra of Parkinson's and Alzheimer's disease brains, Neurology 38 (8) (1988) 1285–1291.

[108] M.P. Vawter, et al. TGFbeta1 and TGFbeta2 concentrations are elevated in Parkinson's disease in ventricular cerebrospinal fluid, Exp. Neurol. 142 (2) (1996) 313–322.

[109] R.B. Banati, S.E. Daniel, S.B. Blunt, Glial pathology but absence of apoptotic nigral neurons in long-standing Parkinson's disease, Mov. Disord. 13 (2) (1998) 221–227.

[110] P.S. Whitton, Inflammation as a causative factor in the aetiology of Parkinson's disease, Br. J. Pharmacol. 150 (8) (2007) 963–976.

[111] S. Hunot, et al. Nuclear translocation of NF-kappaB is increased in dopaminergic neurons of patients with parkinson disease, Proc. Natl. Acad. Sci. USA 94 (14) (1997) 7531–7536.

[112] M. Mogi, et al. Caspase activities and tumor necrosis factor receptor R1 (p55) level are elevated in the substantia nigra from Parkinsonian brain, J. Neural Transm. 107 (3) (2000) 335–341.

[113] E.C. Hirsch, S. Hunot, A. Hartmann, Neuroinflammatory processes in Parkinson's disease, Parkinsonism Relat. Disord. 11 (Suppl. 1) (2005) S9–S15.

[114] M.A. Burguillos, et al. Caspase signalling controls microglia activation and neurotoxicity, Nature 472 (7343) (2011) 319–324.

[115] A. Hartmann, et al. Caspase-3: a vulnerability factor and final effector in apoptotic death of dopaminergic neurons in Parkinson's disease, Proc. Natl. Acad. Sci. USA 97 (6) (2000) 2875–2880.

[116] A. Hartmann, et al. Caspase-8 is an effector in apoptotic death of dopaminergic neurons in Parkinson's disease, but pathway inhibition results in neuronal necrosis, J. Neurosci. 21 (7) (2001) 2247–2255.

[117] A. Gerhard, et al. In vivo imaging of microglial activation with [^{11}C](R)-PK11195 PET in idiopathic Parkinson's disease, Neurobiol. Dis. 21 (2) (2006) 404–412.

[118] S. Hunot, et al. Nuclear translocation of NF-kappaB is increased in dopaminergic neurons of patients with Parkinson disease, Proc. Natl. Acad. Sci. USA 94 (14) (1997) 7531–7536.

[119] V. Brochard, et al. Infiltration of CD4+ lymphocytes into the brain contributes to neurodegeneration in a mouse model of Parkinson disease, J. Clin. Invest. 119 (1) (2009) 182–192.

[120] A.D. Reynolds, et al. Regulatory T cells attenuate Th17 cell-mediated nigrostriatal dopaminergic neurodegeneration in a model of Parkinson's disease, J. Immunol. 184 (5) (2010) 2261–2271.

[121] L.M. Kosloski, et al. Adaptive immune regulation of glial homeostasis as an immunization strategy for neurodegenerative diseases, J. Neurochem. 114 (5) (2010) 1261–1276.

[122] A.D. Reynolds, et al. Neuroprotective activities of CD4+CD25+ regulatory T cells in an animal model of Parkinson's disease, J. Leukoc. Biol. 82 (5) (2007) 1083–1094.

[123] E.J. Benner, et al. Therapeutic immunization protects dopaminergic neurons in a mouse model of Parkinson's disease, Proc. Natl. Acad. Sci. USA 101 (25) (2004) 9435–9440.

[124] J.W. Langston, et al. Evidence of active nerve cell degeneration in the substantia nigra of humans years after 1-methyl-4-phenyl-1,2,3,6-tetrahydropyridine exposure, Ann. Neurol. 46 (4) (1999) 598–605.

[125] A. Hakansson, et al. Interaction of polymorphisms in the genes encoding interleukin-6 and estrogen receptor beta on the susceptibility to Parkinson's disease, Am. J. Med. Genet. 133B (1) (2005) 88–92.

[126] R. Kruger, et al. Genetic analysis of immunomodulating factors in sporadic Parkinson's disease, J. Neural Transm. 107 (5) (2000) 553–562.

[127] M. Nishimura, et al. Tumor necrosis factor gene polymorphisms in patients with sporadic Parkinson's disease, Neurosci. Lett. 311 (1) (2001) 1–4.

[128] M. Nishimura, et al. Influence of interleukin-1beta gene polymorphisms on age-at-onset of sporadic Parkinson's disease, Neurosci. Lett. 284 (1–2) (2000) 73–76.

[129] M.A. Schwarzschild, et al. Neuroprotection by caffeine and more specific A2A receptor antagonists in animal models of Parkinson's disease, Neurology 61 (11 Suppl. 6) (2003) S55–S61.

[130] H. Chen, et al. Nonsteroidal anti-inflammatory drugs and the risk of Parkinson disease, Arch. Neurol. 60 (8) (2003) 1059–1064.

[131] H. Chen, et al. Nonsteroidal antiinflammatory drug use and the risk for Parkinson's disease, Ann. Neurol. 58 (6) (2005) 963–967.

[132] A.D. Wahner, et al. Nonsteroidal anti-inflammatory drugs may protect against Parkinson disease, Neurology 69 (19) (2007) 1836–1842.

[133] H. Checkoway, et al. Parkinson's disease risks associated with cigarette smoking, alcohol consumption, and caffeine intake, Am. J. Epidemiol. 155 (8) (2002) 732–738.

[134] H.M. Brothers, Y. Marchalant, G.L. Wenk, Caffeine attenuates lipopolysaccharide-induced neuroinflammation, Neurosci. Lett. 480 (2) (2010) 97–100.

[135] I. Kurkowska-Jastrzebska, et al. The inflammatory reaction following 1-methyl-4-phenyl-1,2,3,6-tetrahydropyridine intoxication in mouse, Exp. Neurol. 156 (1) (1999) 50–61.

[136] A. Czlonkowska, et al. Microglial reaction in MPTP (1-methyl-4-phenyl-1,2,3,6-tetrahydropyridine) induced Parkinson's disease mice model, Neurodegeneration 5 (2) (1996) 137–143.

[137] R. Gordon, et al. Protein kinase C delta upregulation in microglia drives neuroinflammatory responses and dopaminergic neurodegeneration in experimental models of Parkinson's disease, Neurobiol. Dis. 93 (2016) 96–114.

[138] L.J. Lawson, et al. Heterogeneity in the distribution and morphology of microglia in the normal adult mouse brain, Neuroscience 39 (1) (1990) 151–170.

[139] W.G. Kim, et al. Regional difference in susceptibility to lipopolysaccharide-induced neurotoxicity in the rat brain: role of microglia, J. Neurosci. 20 (16) (2000) 6309–6316.

[140] S. Nadeau, S. Rivest, Glucocorticoids play a fundamental role in protecting the brain during innate immune response, J. Neurosci. 23 (13) (2003) 5536–5544.

[141] M.J. Zigmond, T.G. Hastings, R.G. Perez, Increased dopamine turnover after partial loss of dopaminergic neurons: compensation or toxicity?, Parkinsonism Relat. Disord. 8 (6) (2002) 389–393.

[142] D.A. Loeffler, et al. Effects of enhanced striatal dopamine turnover in vivo on glutathione oxidation, Clin. Neuropharmacol. 17 (4) (1994) 370–379.

[143] G.D. Zeevalk, L.P. Bernard, W.J. Nicklas, Role of oxidative stress and the glutathione system in loss of dopamine neurons due to impairment of energy metabolism, J. Neurochem. 70 (4) (1998) 1421–1430.

[144] K. Nakamura, W. Wang, U.J. Kang, The role of glutathione in dopaminergic neuronal survival, J. Neurochem. 69 (5) (1997) 1850–1858.

[145] P. Jenner, C.W. Olanow, Oxidative stress and the pathogenesis of Parkinson's disease, Neurology 47 (6 Suppl. 3) (1996) S161–S170.

[146] F. Ros-Bernal, et al. Microglial glucocorticoid receptors play a pivotal role in regulating dopaminergic neurodegeneration in parkinsonism, Proc. Natl. Acad. Sci. USA 108 (16) (2011) 6632–6637.

[147] G.A. Cabral, L. Griffin-Thomas, Cannabinoids as therapeutic agents for ablating neuroinflammatory disease, Endocr. Metab. Immune Disord. Drug Targets 8 (3) (2008) 159–172.

[148] S. Rivest, Cannabinoids in microglia: a new trick for immune surveillance and neuroprotection, Neuron 49 (1) (2006) 4–8.

[149] M.F. McCarty, Down-regulation of microglial activation may represent a practical strategy for combating neurodegenerative disorders, Med. Hypotheses 67 (2) (2006) 251–269.

[150] J.K. Lee, et al. Regulator of G-protein signaling 10 promotes dopaminergic neuron survival via regulation of the microglial inflammatory response, J. Neurosci. 28 (34) (2008) 8517–8528.

[151] E.J. Lee, et al. Alpha-synuclein activates microglia by inducing the expressions of matrix metalloproteinases and the subsequent activation of protease-activated receptor-1, J. Immunol. 185 (1) (2010) 615–623.

[152] W. Zhang, et al. Aggregated alpha-synuclein activates microglia: a process leading to disease progression in Parkinson's disease, FASEB J. 19 (6) (2005) 533–542.

[153] H. Wilms, et al. Activation of microglia by human neuromelanin is NF-kappaB dependent and involves p38 mitogen-activated protein kinase: implications for Parkinson's disease, FASEB J. 17 (3) (2003) 500–502.

[154] W. Zhang, et al. Neuromelanin activates microglia and induces degeneration of dopaminergic neurons: implications for progression of Parkinson's disease, Neurotox. Res. 19 (1) (2011) 63–72.

[155] Y.S. Kim, et al. A pivotal role of matrix metalloproteinase-3 activity in dopaminergic neuronal degeneration via microglial activation, FASEB J. 21 (1) (2007) 179–187.

[156] Y.S. Kim, et al. Matrix metalloproteinase-3: a novel signaling proteinase from apoptotic neuronal cells that activates microglia, J. Neurosci. 25 (14) (2005) 3701–3711.

[157] L. Stefano, et al. The surface-exposed chaperone, Hsp60, is an agonist of the microglial TREM2 receptor, J. Neurochem. 110 (1) (2009) 284–294.

[158] S. Lehnardt, et al. A vicious cycle involving release of heat shock protein 60 from injured cells and activation of toll-like receptor 4 mediates neurodegeneration in the CNS, J. Neurosci. 28 (10) (2008) 2320–2331.

[159] J.L. Waugh, et al. Regional, cellular, and subcellular localization of RGS10 in rodent brain, J. Comp. Neurol. 481 (3) (2005) 299–313.

[160] S. Shan, et al. NEW evidences for fractalkine/CX3CL1 involved in substantia nigral microglial activation and behavioral changes in a rat model of Parkinson's disease, Neurobiol. Aging 32 (3) (2011) 443–458.

[161] A.E. Cardona, et al. Control of microglial neurotoxicity by the fractalkine receptor, Nat. Neurosci. 9 (7) (2006) 917–924.

[162] K.J. Min, et al. Gangliosides activate microglia via protein kinase C and NADPH oxidase, Glia 48 (3) (2004) 197–206.

[163] H. Pyo, et al. Gangliosides activate cultured rat brain microglia, J. Biol. Chem. 274 (49) (1999) 34584–34589.

[164] X.F. Wu, et al. The role of microglia in paraquat-induced dopaminergic neurotoxicity, Antioxid. Redox Signal. 7 (5–6) (2005) 654–661.

[165] H.M. Gao, et al. Synergistic dopaminergic neurotoxicity of the pesticide rotenone and inflammogen lipopolysaccharide: relevance to the etiology of Parkinson's disease, J. Neurosci. 23 (4) (2003) 1228–1236.

[166] M.K. McCoy, et al. TNF: a key neuroinflammatory mediator of neurotoxicity and neurodegeneration in models of Parkinson's disease, Adv. Exp. Med. Biol. 691 (2011) 539–540.

[167] C. Barcia, et al. IFN-gamma signaling, with the synergistic contribution of TNF-alpha, mediates cell specific microglial and astroglial activation in experimental models of Parkinson's disease, Cell Death Dis. 2 (2011) e142.

[168] M.L. Block, et al. Nanometer size diesel exhaust particles are selectively toxic to dopaminergic neurons: the role of microglia, phagocytosis, and NADPH oxidase, FASEB J. 18 (13) (2004) 1618–1620.

[169] W. Zhang, et al. Neuroprotective effect of dextromethorphan in the MPTP Parkinson's disease model: role of NADPH oxidase, FASEB J. 18 (3) (2004) 589–591.

[170] H.M. Gao, et al. Critical role of microglial NADPH oxidase-derived free radicals in the in vitro MPTP model of Parkinson's disease, FASEB J. 17 (13) (2003) 1954–1956.

[171] W. Zhang, et al. Microglial PHOX and Mac-1 are essential to the enhanced dopaminergic neurodegeneration elicited by A30P and A53T mutant alpha-synuclein, Glia 55 (11) (2007) 1178–1188.

[172] L. Qian, et al. NADPH oxidase inhibitor DPI is neuroprotective at femtomolar concentrations through inhibition of microglia over-activation, Parkinsonism Relat. Disord. 13 (Suppl. 3) (2007) S316–S320.

[173] G.T. Liberatore, et al. Inducible nitric oxide synthase stimulates dopaminergic neurodegeneration in the MPTP model of Parkinson disease, Nat. Med. 5 (12) (1999) 1403–1409.

[174] A. Ghosh, et al. Selective inhibition of NF-kappaB activation prevents dopaminergic neuronal loss in a mouse model of Parkinson's disease, Proc. Natl. Acad. Sci. USA 104 (47) (2007) 18754–18759.

[175] A. Ghosh, et al. Simvastatin inhibits the activation of p21ras and prevents the loss of dopaminergic neurons in a mouse model of Parkinson's disease, J. Neurosci. 29 (43) (2009) 13543–13556.

[176] F. Zhang, et al. Resveratrol protects dopamine neurons against lipopolysaccharide-induced neurotoxicity through its anti-inflammatory actions, Mol. Pharmacol. 78 (3) (2010) 466–477.

[177] M.C. Hernandez-Romero, et al. Simvastatin prevents the inflammatory process and the dopaminergic degeneration induced by the intranigral injection of lipopolysaccharide, J. Neurochem. 105 (2) (2008) 445–459.

[178] M.K. McCoy, et al. Blocking soluble tumor necrosis factor signaling with dominant-negative tumor necrosis factor inhibitor attenuates loss of dopaminergic neurons in models of Parkinson's disease, J. Neurosci. 26 (37) (2006) 9365–9375.

[179] W. Zhang, et al. 3-hydroxymorphinan is neurotrophic to dopaminergic neurons and is also neuroprotective against LPS-induced neurotoxicity, FASEB J. 19 (3) (2005) 395–397.

[180] T. Wang, et al. Protective effect of the SOD/catalase mimetic MnTMPyP on inflammation-mediated dopaminergic neurodegeneration in mesencephalic neuronal-glial cultures, J. Neuroimmunol. 147 (1–2) (2004) 68–72.

[181] W. Robberecht, T. Philips, The changing scene of amyotrophic lateral sclerosis, Nat. Rev. Neurosci. 14 (4) (2013) 248–264.

[182] A.E. Renton, A. Chio, B.J. Traynor, State of play in amyotrophic lateral sclerosis genetics, Nat. Neurosci. 17 (1) (2014) 17–23.

[183] A.J. Pratt, et al. Aggregation propensities of superoxide dismutase G93 hotspot mutants mirror ALS clinical phenotypes, Proc. Natl. Acad. Sci. USA 111 (43) (2014) E4568–E4576.

[184] T. Philips, W. Robberecht, Neuroinflammation in amyotrophic lateral sclerosis: role of glial activation in motor neuron disease, Lancet Neurol. 10 (3) (2011) 253–263.

[185] O. Butovsky, et al. Modulating inflammatory monocytes with a unique microRNA gene signature ameliorates murine ALS, J. Clin. Invest. 122 (9) (2012) 3063–3087.

[186] L.A. Lampson, P.D. Kushner, R.A. Sobel, Major histocompatibility complex antigen expression in the affected tissues in amyotrophic lateral sclerosis, Ann. Neurol. 28 (3) (1990) 365–372.

[187] J.D. Lee, et al. Dysregulation of the complement cascade in the hSOD1G93A transgenic mouse model of amyotrophic lateral sclerosis, J. Neuroinflamm. 10 (2013) 119.

[188] A. Crotti, et al. Mutant huntingtin promotes autonomous microglia activation via myeloid lineage-determining factors, Nat. Neurosci. 17 (4) (2014) 513–521.

[189] A. Silvestroni, et al. Distinct neuroinflammatory profile in post-mortem human Huntington's disease, Neuroreport 20 (12) (2009) 1098–1103.

[190] Y.F. Tai, et al. Microglial activation in presymptomatic Huntington's disease gene carriers, Brain 130 (Pt. 7) (2007) 1759–1766.

[191] T.C. Andrews, et al. Huntington's disease progression. PET and clinical observations, Brain 122 (Pt. 12) (1999) 2353–2363.

[192] N. Pavese, et al. Microglial activation correlates with severity in Huntington disease: a clinical and PET study, Neurology 66 (11) (2006) 1638–1643.

[193] M. Bjorkqvist, et al. The R6/2 transgenic mouse model of Huntington's disease develops diabetes due to deficient beta-cell mass and exocytosis, Hum. Mol. Genet. 14 (5) (2005) 565–574.

[194] D.A. Simmons, et al. Ferritin accumulation in dystrophic microglia is an early event in the development of Huntington's disease, Glia 55 (10) (2007) 1074–1084.

[195] J.Y. Shin, et al. Expression of mutant huntingtin in glial cells contributes to neuronal excitotoxicity, J. Cell. Biol. 171 (6) (2005) 1001–1012.

[196] J. Bradford, et al. Mutant huntingtin in glial cells exacerbates neurological symptoms of Huntington disease mice, J. Biol. Chem. 285 (14) (2010) 10653–10661.

[197] J. Bradford, et al. Expression of mutant huntingtin in mouse brain astrocytes causes age-dependent neurological symptoms, Proc. Natl. Acad. Sci. USA 106 (52) (2009) 22480–22485.

[198] S. Mantovani, et al. Motor deficits associated with Huntington's disease occur in the absence of striatal degeneration in BACHD transgenic mice, Hum. Mol. Genet. 25 (9) (2016) 1780–1791.

[199] G. Pasqualetti, D.J. Brooks, P. Edison, The role of neuroinflammation in dementias, Curr. Neurol. Neurosci. Rep. 15 (4) (2015) 17.

[200] J. Zhang, Mapping neuroinflammation in frontotemporal dementia with molecular PET imaging, J. Neuroinflamm. 12 (2015) 108.

[201] R.J. Guerreiro, et al. Using exome sequencing to reveal mutations in TREM2 presenting as a frontotemporal dementia-like syndrome without bone involvement, JAMA Neurol. 70 (1) (2013) 78–84.

[202] B. Borroni, et al. Heterozygous TREM2 mutations in frontotemporal dementia, Neurobiol. Aging 35 (4) (2014) 934e7–934e10.

[203] M. Baker, et al. Mutations in progranulin cause tau-negative frontotemporal dementia linked to chromosome 17, Nature 442 (7105) (2006) 916–919.

[204] L.H. Martens, et al. Progranulin deficiency promotes neuroinflammation and neuron loss following toxin-induced injury, J. Clin. Invest. 122 (11) (2012) 3955–3959.

[205] A. Surendranathan, J.B. Rowe, J.T. O'Brien, Neuroinflammation in Lewy body dementia, Parkinsonism Relat. Disord. 21 (12) (2015) 1398–1406.

[206] W.S. Griffin, et al. Interleukin-1 mediates Alzheimer and Lewy body pathologies, J. Neuroinflamm. 3 (2006) 5.

[207] H.M. Gao, et al. Neuroinflammation and alpha-synuclein dysfunction potentiate each other, driving chronic progression of neurodegeneration in a mouse model of Parkinson's disease, Environ. Health Perspect. 119 (6) (2011) 807–814.

[208] L. Qin, et al. Systemic LPS causes chronic neuroinflammation and progressive neurodegeneration, Glia 55 (5) (2007) 453–462.

[209] K. Schroder, R. Zhou, J. Tschopp, The NLRP3 inflammasome: a sensor for metabolic danger?, Science 327 (5963) (2010) 296–300.

[210] J. Tschopp, K. Schroder, NLRP3 inflammasome activation: the convergence of multiple signalling pathways on ROS production?, Nat. Rev. Immunol. 10 (3) (2010) 210–215.

[211] Y.H. Youm, et al. Canonical Nlrp3 inflammasome links systemic low-grade inflammation to functional decline in aging, Cell Metab. 18 (4) (2013) 519–532.

[212] D. Zhang, et al. Neuroprotective effect of protein kinase C delta inhibitor rottlerin in cell culture and animal models of Parkinson's disease, J. Pharmacol. Exp. Ther. 322 (3) (2007) 913–922.

[213] D. Paris, et al. The spleen tyrosine kinase (Syk) regulates Alzheimer amyloid-beta production and Tau hyperphosphorylation, J. Biol. Chem. 289 (49) (2014) 33927–33944.

[214] N. Panicker, et al. Fyn kinase regulates microglial neuroinflammatory responses in cell culture and animal models of Parkinson's disease, J. Neurosci. 35 (27) (2015) 10058–10077.

[215] A.S. Harms, et al. Delayed dominant-negative TNF gene therapy halts progressive loss of nigral dopaminergic neurons in a rat model of Parkinson's disease, Mol. Ther. 19 (1) (2011) 46–52.

[216] C.J. Barnum, et al. Peripheral administration of the selective inhibitor of soluble tumor necrosis factor (TNF) XPro(R)1595 attenuates nigral cell loss and glial activation in 6-OHDA hemiparkinsonian rats, J. Parkinsons Dis. 4 (3) (2014) 349–360.

[217] J.B. Koprich, et al. Neuroinflammation mediated by IL-1beta increases susceptibility of dopamine neurons to degeneration in an animal model of Parkinson's disease, J. Neuroinflamm. 5 (2008) 8.

[218] R.C. Coll, et al. A small-molecule inhibitor of the NLRP3 inflammasome for the treatment of inflammatory diseases, Nat. Med. 21 (3) (2015) 248–255.

[219] Y.H. Youm, et al. The ketone metabolite beta-hydroxybutyrate blocks NLRP3 inflammasome-mediated inflammatory disease, Nat. Med. 21 (3) (2015) 263–269.

[220] J. Vom Berg, et al. Inhibition of IL-12/IL-23 signaling reduces Alzheimer's disease-like pathology and cognitive decline, Nat. Med. 18 (12) (2012) 1812–1819.

[221] M. Papatriantafyllou, Immunotherapy: immunological bullets against Alzheimer's disease, Nat. Rev. Drug Discov. 12 (1) (2013) 24.

[222] T.M. Woodruff, K.S. Nandakumar, F. Tedesco, Inhibiting the C5-C5a receptor axis, Mol. Immunol. 48 (14) (2011) 1631–1642.

[223] A.H. Stephan, B.A. Barres, B. Stevens, The complement system: an unexpected role in synaptic pruning during development and disease, Annu. Rev. Neurosci. 35 (2012) 369–389.

[224] M.N. Sabbagh, et al. PF-04494700, an oral inhibitor of receptor for advanced glycation end products (RAGE), in Alzheimer disease, Alzheimer Dis. Assoc. Disord. 25 (3) (2011) 206–212.

[225] F. Scheperjans, et al. Gut microbiota are related to Parkinson's disease and clinical phenotype, Mov. Disord. 30 (3) (2015) 350–358.

[226] A.C. Keene, W.J. Joiner, Neurodegeneration: paying it off with sleep, Curr. Biol. 25 (6) (2015) R234–R236.

[227] B. Zhu, et al. Sleep disturbance induces neuroinflammation and impairment of learning and memory, Neurobiol. Dis. 48 (3) (2012) 348–355.

[228] J.P. Wisor, M.A. Schmidt, W.C. Clegern, Evidence for neuroinflammatory and microglial changes in the cerebral response to sleep loss, Sleep 34 (3) (2011) 261–272.

[229] A. Louveau, et al. Structural and functional features of central nervous system lymphatic vessels, Nature 523 (7560) (2015) 337–341.

STEM CELLS IN NEURODEGENERATION: MIND THE GAP

4A

Christel Claes*,**, Joke Terryn*,†,‡, Catherine M. Verfaillie*

*KU Leuven Stem Cell Institute, Leuven, Belgium
**VIB Center for the Biology of Disease, Leuven, Belgium
†VIB Vesalius Research Center, Leuven, Belgium
‡University Hospitals Leuven, Leuven, Belgium

CHAPTER OUTLINE

Introduction 81
Part I: Stem Cells 82
   Genome Editing of Stem Cells 83
   The Generation of Specific Cell Types 84
Part II: Stem Cells as Regenerative Therapy 85
   Clinical Translation and Immunology 88
   Keeping Track of Cell Transplants 90
Part III: Stem Cells to Model Neurodegenerative Diseases 91
   iPSC Characteristics: The Role of Development in Neurodegenerative Diseases 91
   iPSC Technology: Disease Modeling and Drug Screening 92
   Future Prospects 96
References 96

INTRODUCTION

Pluripotency and self-renewal, two essential features of stem cells, might make your head spin thinking about the possibilities they create. Could they be an unlimited source of "spare body parts?" Stem cell therapy has already proven what it is worth for hematological diseases, but would that also be possible for other organs? And what about our most precious organ? Are they capable of rescuing a brain in decay? Unlike hematological stem cell transplants where the host tissue is destroyed and replaced by donor tissue, mending a brain creates additional challenges.

In this chapter, we'll discuss the challenges and strategies for the use of pluripotent stem cells in neurodegeneration. The cell therapy trials that have been conducted in Parkinson's disease (PD) and Alzheimer's disease (AD) will serve as an example to address the major questions concerning cell therapy. Additionally, we will discuss how disease modeling can aid in the development of therapies for neurodegeneration. Induced pluripotent stem cell (iPSC) technology has opened new opportunities to crack neurodegeneration, as patient-derived iPSCs are used for disease modeling. Stem cell models shed light on the mechanisms underlying neurodegeneration and provide a platform for drug screening. Stem cell therapy for neurodegenerative diseases (NDDs) is still in its infancy, but a bright future lies ahead. The challenge will be to allow the technology to mature and not rush into clinical trials.

PART I: STEM CELLS

Stem cell research continues to evolve. Where initially the focus lied on deriving and studying the properties of stem cells, current research is focused on applying stem cell technology to model and treat diseases. One major breakthrough underlying this shift was the discovery by the Yamanaka group in 2006 that somatic cells could be reprogrammed to become pluripotent cells. The discovery opened up new possibilities for disease modeling and regenerative medicine, now that patient-derived stem cell lines could be generated. The Nobel Prize committee was quick to recognize the significance of this research and awarded the Nobel Prize for Physiology or Medicine in 2012 jointly to Sir John Gurdon and Dr. Shinya Yamanaka. Forty years prior, Sir Gurdon had taken the first steps toward reprogramming. In 1962, he succeeded in cloning a frog by transplanting the nucleus of a somatic intestinal cell from a tadpole to an enucleated frog egg [1]. This experiment proved that an adult cell nucleus still contains all the information needed to generate all the cell types and that certain factors in the egg cell are capable of reprogramming a differentiated cell nucleus. It was only in 1996, when Dolly, the first cloned sheep [2], was born that cloning by somatic cell nuclear transfer became widely recognized. Apart from animal cloning, two additional breakthroughs in science made way for Yamanaka's work [3,4]: the discovery that transcription factors can function as master regulators controlling cell fate [5], and the development of cultured embryonic stem cell (ESC) lines in 1981 [6]. Provided with the tools and culture conditions to influence cell fate, the Yamanaka group succeeded to reprogram mouse somatic cells by the overexpression of four selected transcription factors, *Oct4*, *Sox2*, *Klf4*, and *c-Myc* (OSKM) using retroviral vectors [7] (Fig. 4A.1). The resulting pluripotent cells were named iPSCs. As even short-term overexpression can activate endogenous pluripotency genes and thus reprogram cells to the pluripotent stage, nonintegrating methods have been developed to reprogram cells. Reprogramming can be achieved among others by transfection of synthetic mRNA, by protein transfection, or by using nonintegrating viral vectors, such as Sendai viral vectors [8]. Furthermore, addition of small molecules that modify the epigenome or activate specific signaling pathways can enhance reprogramming [8].

How do these reprogrammed cells compare to ESCs? ESC lines are derived from the inner cell mass of an embryo and are considered to be a more natural cell type, representing in mouse, an early preimplantation developmental stage, and in humans an early postimplantation stage. However, harvesting these cells does destroy the embryo, even if it has now been shown that ESCs can also be obtained by culturing a single morula–derived cell, which then theoretically would allow the embryo to develop further [9]. Nevertheless, as ESCs have an unlimited capacity to expand in culture, only a limited amount of ESC lines are currently in use for research and therapeutic purposes. Comparing iPSC and ESC lines showed that they have slightly different methylation and gene expression patterns; however, whether these differences are of importance remains unclear [3,10]. Patient-derived stem cells do have the benefit that they can be generated from the patient her/himself, therefore, being autologous when considered for cell therapy, and posing fewer ethical concerns. However, to date, the

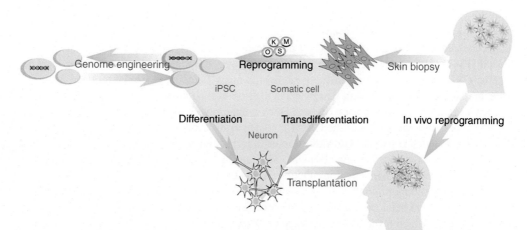

FIGURE 4A.1 Schematic Overview of the Possible Ways to Manipulate Cells In Vitro and In Vivo

Induced pluripotent stem cells (iPSC) are typically derived from fibroblasts (obtained by skin biopsy) by introducing the four classical Yamanaka factors: *Oct4*, *Sox2*, *Klf4*, and *c-Myc* (OSKM), preferably by nonintegrating methods. The derived iPSCs can then be further genetically engineered if needed. Alternatively allogeneic embryonic stem cells (ESCs) or iPSCs can be used for differentiation toward the desired cell type for transplantation. Transdifferentiation methods omit the pluripotent stage, which is a faster, yet less versatile, approach. In vivo reprogramming methods focus on the in situ conversion of somatic cells toward the preferred cell type. Adult stem cells (ASCs) have been excluded from this schematic overview, as they are restricted in differentiation capacity.

process of generating iPSCs and fully characterizing the generated cells is still labor intensive, costly, and time consuming, making the goal of personalized medicine harder to obtain.

Another source of stem cells are so-called adult stem cells (ASCs). ASCs are obtained from an organism after the embryo stage, and can be found in many tissues of the body. They are multipotent, meaning they can only differentiate to a limited amount of cell types, usually limited to the cell types of the tissue of origin. The best-studied ASC is the hematopoietic stem cell, which is being used clinically. Stem cells can also be found in other tissues with a fast turnover, including the skin and the epithelium of the gastrointestinal tract. Another kind of ASCs is the mesenchymal stem cell (MSC) that can be harvested from bone marrow, adipose tissue, dental pulp, umbilical cord blood, and other tissues. In addition, stem cells have been identified in tissues with a slow turnover, such as the adult central nervous system and skeletal muscle. Although they have a more limited range of use, as they are multipotent and not pluripotent, ASCs are attractive candidates for cell therapy. For example, MSC possess the unique property to modulate the immune system and promote vascularization and neurogenesis through secreted factors [11,12]. They can be easily harvested from adult tissue without ethical concerns, but compared to ESCs and iPSCs with an unlimited expansion capacity; ASCs can only be expanded for 5–10 passages in vitro.

GENOME EDITING OF STEM CELLS

Genome editing to introduce or correct a mutation in hiPSC or hESC is a valuable method to demonstrate genotype–phenotype causal relationships in human genetic disorders. In addition, this would allow correcting the genetic defect in autologous iPSCs from patients with a genetic disorder, before grafting the cells in vivo, for regenerative therapy. To evaluate gene function throughout the years, either zinc finger nuclease (ZFN), transcription activator–like effector nucleases (TALEN), or more recently clustered regularly interspaced short palindromic repeats (CRISPR)–Cas9 technologies have been used [13–15]. In particular, CRISPR–Cas9 technology offers a user-friendly approach to target stem cells. CRISPR loci are found in the genomes of many bacteria and most archaea, where they underlie an adaptive immune system that protects the host cell against invasive nucleic acids, such as viral genomes. Upon invasion of bacteria by a DNA-containing element, for example, a virus, a short part of its DNA is incorporated into the CRISPR locus. Transcription of this locus produces CRISPR RNAs (crRNAs), which associate with Cas proteins to target and cleave the invading DNA. Modifying this natural system via expression of a 20-nucleotide long single-guide RNA (sgRNA), Cas9 nucleases can be directed to specific genomic regions. Double-strand DNA breaks, produced by these nucleases, can be either repaired by nonhomologous end joining, which causes a so-called indel or a mismatch repair, leading to mutation and hence a knockout of the gene function, or the site can be repaired by homologous recombination, which involves the exchange of DNA sequences provided by a similar template and can be used to modify or add sequence. In addition, CRISPR technologies keep expanding, including Cas9 nickase or Cas9–FokI nucleases. The latter Cas9 nucleases are mutated to result in only one active nuclease domain. By a combination of two sgRNA and two mutated Cas9 enzymes, a double-strand break will only occur when both enzymes are cut in proximity of one another, lowering the chance of off-target effects [16,17]. As an example of genome editing, various PD-associated abnormalities, including α-synuclein deposition, have been observed in LRRK2 iPSC–derived neurons, as well as in isogenic control iPSC in which this mutation was introduced. In addition this phenotype can be rescued by genetic correction of this mutation [18].

THE GENERATION OF SPECIFIC CELL TYPES

Once the desired stem cell line has been generated, the cells need to be differentiated toward the relevant cell type for disease modeling or transplantation. As stem cells differentiate, they pass through distinctive developmental stages and become more specialized along the way. The differentiation process mimics embryonic development and is controlled through the sequential addition of specific growth factors and morphogens, mimicking signals the cells would see throughout development. For example, neural differentiation protocols typically commence with a neural induction phase, generating neural stem cells (NSCs) comparable to those of the neural plate. Without further patterning, these neural cells would mature to their default anterior fate and become telencephalic neurons [19]. By the addition of morphogens early in the differentiation process, it is, however, possible to navigate through the neural tube and alter cell identity. Increasing concentration of Wnts and retinoic acid will lead to the generation of more caudal neural precursors [20,21]. Combined activation of the Sonic hedgehog pathway in these neural precursors will induct a ventral fate, priming them for the generation of spinal cord motor neurons [21,22]. Apart from regional patterning, neural precursors are also subjected to temporal patterning in vitro and in vivo [20]. The human cerebral cortex develops in an inside–out fashion, where early-born neurons form the deeper layers of the cortex and late-born neurons shape the upper layers. After neurogenesis the neural precursor cells switch to gliogenesis. As stem cell differentiation protocols mimic embryonic development, generation of the desired cell type generally takes weeks to even months.

Transdifferentiating somatic cells to the desired cell type has been proven to be a faster approach, as there is a direct conversion without going through the developmental stages and without the prior need to reset the somatic cell to the pluripotent stage. Several attempts have shown that it is possible to generate induced neuronal cells by overexpressing a selected set of transcription factors; however, the yield and purity of the generated neurons is relatively low. For a concise review of in vitro transdifferentiation techniques on mouse and human fibroblasts, refer to the article by Yang et al. [23]. Similar strategies have also led to the development of in vivo reprogramming techniques, focusing on the conversion of resident glial cells to the desired cell type. Niu et al. successfully transdifferentiated adult glial cells to neurons by lentiviral overexpression of *SOX2* under the *GFAP* promotor [24]. This technique, however, generated a heterogenous pool of induced neurons. Targeting specific subtypes of glial cells might allow a better control over this process. Overexpression of a single transcription factor, *NeuroD1*, led to the conversion of astrocytes into glutamatergic neurons and to the generation of glutamatergic and GABAergic neurons when oligodendrocyte progenitor cells were targeted [25]. As the field of transdifferentiation evolved, some cross-pollination occurred, introducing transdifferentiation techniques to improve differentiation protocols for ESCs. Zhang et al. were the first to describe a highly efficient protocol to convert pluripotent stem cells to a homogenous population of excitatory cortical neurons by overexpression of Neurogenin-2 by a lentiviral vector [26].

Current available differentiation and transdifferentiation protocols allow for the generation of a multitude of cell types that are well suited for disease modeling. However, few protocols succeed in generating a pure cell population, which is a requirement for cell therapy. Persistence of stem cells in the transplanted population holds the risk of teratoma formation, and contamination of cultures with undesired cell types might lead to unwanted side effects. For example, the presence of serotonergic neurons in transplanted embryonic grafts in PD patients has been associated with the occurrence of graft-induced dyskinesia [27]. Further purification of in vitro–generated cell types can be achieved by the use of fluorescence-activated cell sorting (FACS) or magnetic-activated cell sorting (MACS), but this does negatively affect cell survival and further complicates the production process. In vivo reprogramming techniques will require alternative strategies to control cell faith, for example, more fine-tuned transdifferentiation protocols.

Another obstacle in the development of efficient therapeutic paradigms resides in the diffuse and progressive nature of NDDs and the influence of this hostile environment on transplanted cells. Studies with human neural precursor cells in AD showed that exposing them to high doses of secreted amyloid precursor protein (APP) in vitro or by grafting them in the brain of APP mice in vivo, promoted glial differentiation [28]. Direct implantation of primary rat NSCs into injured or intact spinal cord also lead chiefly to the generation of glial cells, with only a fraction of them differentiating into neurons [29]. To obtain successful transplantations, we first need to increase our understanding of the specification and differentiation mechanisms and the influence of the host environment in controlling the fate of transplanted cells.

PART II: STEM CELLS AS REGENERATIVE THERAPY

The first clinical transplantation trials in neurodegeneration focused on PD. A disease clinically marked by motor symptoms: tremor, rigidity, slowing of movement (bradykinesia), and loss of postural reflexes leads to a stooped posture (see also Chapter 6). Nonmotor symptoms, such as hypersalivation, orthostatic hypotension, and, in a later stage of the disease, cognitive decline contribute to the disability. The disease starts to manifest when patients have an estimated loss of about 30% of the dopaminergic

(DA) neurons in the substantia nigra of the mesencephalon, which causes reduced innervation of the striatum and, therefore, loss of control over motor function [30]. Oral PD medication aims at restoring the dopamine levels by intake of a dopamine precursor or dopamine agonist. Medication improves rigidity and slowness of movement but fails to treat the postural imbalance and nonmotor symptoms. As the disease progresses, patients become more sensitive to the fluctuations in dopamine levels and start to experience sudden wearing off or excessive movements (medication-induced dyskinesia) [27]. This can be avoided by continuous duodenal administration of L-DOPA or by opting for deep brain stimulation, a technique modulating pathological network activity in the basal ganglia. Unfortunately symptoms that are resistant to DA medication are also resistant to deep brain stimulation [31]. Ideally, stem cell therapy would provide additional clinical benefits over the current treatment options.

Between 1990 and 2003, several trials tested if transplanting fetal mesencephalic tissue in the basal ganglia of PD patients would provide benefit. Although placebo-controlled trials failed to prove benefit, subgroups of patients, mainly the ones with milder disease at the time of grafting, did improve significantly [27]. Graft function was evident from PET studies measuring ^{18}F-DOPA uptake, although striatal ^{18}F-DOPA levels never reached normal levels [27,32]. Additional evidence for graft function came from autopsy studies on a few of patients that had died from unrelated causes. These studies showed that the grafted cells had survived and had formed synapses within the basal ganglia, demonstrated by staining for the dopamine transporter [27,33,34]. Unexpectedly, however, α-synuclein accumulation and aggregation was also detected in some patients. Some of the aggregates were heavily ubiquitinated, resembling Lewy bodies, the anatomopathological hallmark of PD [33]. This raised questions on how sustainable the treatment really would be, as the disease had seemed to spread from the surrounding area into the graft. In addition, one patient who had been clinically diagnosed with PD, revealed at the time of autopsy to be affected by multisystem atrophy [34], raising questions on how patient selection had occurred before initiation of the trial. The most serious side effect of the cell therapy was the development of off-medication dyskinesia in about half of the patients after an average latency of 5 months. Possible explanations are aberrant synaptic connectivity, dysfunction of the implanted neurons, or contamination of the graft with serotonergic neurons [27,35,36]. Apart from the medical problems, ethical problems hindered the development of the therapy, as for each graft multiple aborted human embryos were required. Other cell therapy trials in PD have been conducted, grafting neuroendocrine cells (adrenal or carotid body), retinal pigment epithelium cells (RPECs) capable of secreting dopamine, or DA neurons from xenografts [34,37]. Although many of these trials were negative, they provided information on patient selection and revealed obstacles and questions to address when considering cell replacement therapy for neurodegeneration in general.

Instead of paralyzing current research, the complexity of the matter led to the generation of an international working group that designed and guides the ongoing TransEuro trial for PD. The first patients included in this EU-funded trial received their fetal mesencephalic grafts in 2015 [38]. Although future trials will most likely avoid the use of fetal tissue, this trial will serve as a proof of principle and an important stepping-stone, allowing the field to evolve. Furthermore, these collaborative efforts have also contributed to the generation of a global consortium, G-force PD [38]. As multiple questions remain to be answered, joining forces is the only way to efficiently take these therapies forward.

Achieving the goal of replacing lost neurons and restoring network activity has proven to be more challenging to achieve in AD. Unlike PD that mainly affects the basal ganglia, the network dysfunction in AD is much more widespread and multiple neuronal systems are affected. The complex pathogenesis has also led to a variety of experimental cell therapy approaches for tackling this neurodegenerative disease. AD, accounting for two-thirds of all dementia cases, clinically translates to progressive memory loss in combination with deterioration of other cognitive functions (e.g., attention, language, or spatial orientation). In 1906, Alois Alzheimer described the neuropathology of the disease that was to bear his name [39]. Ever since, our understanding of AD has grown starting with a link to genetics. Autosomal dominant mutations of APP, PS1, and PS2 genes that cause early-onset AD have been very informative, leading to the "amyloid cascade hypothesis" [40–42]. This hypothesis has been the basis for several disease-modifying therapeutic approaches for AD, focusing on amyloid peptides as the main target (see also Chapter 5). In the past 5 years, there have been six amyloidocentric programs that completed phase III clinical testing. None met their primary outcome measures, although one, solanezumab, showed encouraging results in a prespecified secondary outcome measure [43]. Thus, the AD field has struggled to move drugable targets from preclinical research into effective therapies. Standard pharmacological treatment is currently still restricted to cholinesterase inhibitors. Considering this cholinergic hypothesis in AD, a first reflex could be to try to restore the amount of cholinergic neurons in the brain. However, this hypothesis has lost some of its significance, partially because of the lack of major clinical effects when using cholinergic drugs, and because restoring the widespread cholinergic network might be less feasible. A second cell type that has been implicated in the pathogenesis of AD in mice models and humans are the inhibitory interneurons. These neurons inherently possess a great capacity to migrate. During embryonic development these cortical GABAergic neurons are generated in the medial ganglionic eminence and migrate tangentially throughout the brain. Transplantation experiments using embryonic medial ganglionic eminence–derived interneurons, and targeting the hippocampus of aged apolipoprotein E4 (apoE4) mice, indicated that they were able to rescue learning and memory [44]. As other cell types do not inherently pose the capacity to migrate extensively and because the degenerating brain, compared to embryonic development, might lack the necessary environmental cues to lead the cells, additional strategies to promote migration are being evaluated. Ladewig et al. demonstrated that chemotactic interactions between transplanted neural precursor cells and the derived neurons limited their migration capacity. The group showed that by administering antibodies interfering with the FGF2 and VEGF signaling pathways migration could be enhanced [45]. However, we should keep in mind that forced stimulation of migration could lead to aberrant regeneration of the entire network and clinical symptoms, such as abnormal sensations. Altered striatal network activity has been suggested as one of the possible causes underlying graft-induced dyskinesia in PD [27].

Aside from the early-onset familial AD (FAD), the cause of late-onset sporadic AD, which affects 90% of all AD cases, remains unresolved. So far, rare risk factors have been uncovered with ApoE4 carriers and mutations in triggering receptor expressed on myeloid cells 2 (TREM2) conferring to the highest risk (4- to 10-fold) in developing the disease at an older age. These among other risk factors point in the direction of neuroinflammation [46–48] (see also Chapter 3). To target neuroinflammation in the pathogenesis of AD, treatment with human MSCs is being evaluated (https://clinicaltrials.gov/ct2/show/NCT01297218), where MSCs are injected in the ventricles of the brain. Studies in mouse models showed beneficial effects, and currently phase II trials are ongoing to show efficacy in human patients [49].

Another therapeutic strategy relies on the fact that stem cells can be easily genetically modified to carry new or corrected genes and are able to deliver compounds to prevent degeneration [50]. For example, earlier studies in AD showed that nerve growth factor (NGF) is able to prevent neuronal death and improve memory in animal models of aging, excitotoxicity, and amyloid toxicity, and could be useful for treating neuronal degeneration and cell death in the AD brain [51,52]. However, NGF is not able to cross the blood–brain barrier due to its size and polarity, and thus cannot be administered peripherally. To overcome this difficulty, human genomic engineered NSCs, instead of human fibroblasts, have been successfully used to deliver NGF in learning-deficit AD model rats. Transplanted NGF human NSCs were found throughout the brain, differentiated into neurons and astrocytes, and enhanced cognitive function [53].

An alternative to introducing cells by transplantation is reprogramming cells in vivo. Cell loss is usually emphasized in neurodegeneration, as the net result is atrophy but the parallel process of reactive gliosis shouldn't be ignored. Reversing this secondary gliosis back to neural tissue by in vivo reprogramming has been successfully attempted in AD models [25]. It is highly experimental and difficult to control, but still a method worth pursuing due to its general applicability in neurodegeneration.

Though versatile in approach, the therapeutic strategy of cell therapy in AD is mainly bidirectional. One goal is to modulate the environment, by downregulating inflammation, altering disease pathogenesis directly, or by inducing the activation of endogenous stem cells. A second approach is through cell autonomous effects to replenish the lost neurons or glial cells.

CLINICAL TRANSLATION AND IMMUNOLOGY

The use of hPSC-derived cells in the clinic remains limited so far. To ensure the success of clinical trials one must respect some essential restriction factors, such as safety regulations and understanding the regenerative capacity of cell transplants. First of all, the safety issues of hPSC-derived cells need to be fully addressed and it must be demonstrated by preclinical data that they were produced under good manufacturing practice conditions. In vitro culture conditions and passage number are known to contribute to an increased incidence of mutations in hPSC lines. Therefore, safety regulations include assays for chromosomal stability and genome-wide mutations. The importance of this is shown by an ongoing trial in Japan for macular degeneration that was recently arrested because mutations were discovered in iPSC lines [54]. In addition, understanding the regenerative capacity, meaning mechanism of action, of any cell therapy is crucial to the success of clinical trials. Transplanted hPSC–derived cells have to integrate, mature, and function following transplantation and correctly blend in with the help of endogenous cells. Sufficient preclinical data is needed to prevent premature translation to the clinic. The International Society for Stem Cell Research (ISSCR) recently updated its guidelines for stem cell research and clinical translation, addressing all challenges, from manufacturing to ethical or regulatory issues. They emphasize oversight and transparency to drive progress [55].

Once the desired cells have been generated and are ready to be transplanted, the next goal is to ensure their survival. Allogeneic transplantation inevitably entails a risk of rejection, provoked by mismatching of HLA antigens. The HLA system comprises of a set of genes encoding for cell surface proteins that directly interact with the immune system. HLA mismatching will alert the immune system of the "non-self" nature of the transplanted cells and provoke an immune reaction, either an antibody response or cytotoxic response depending on the HLA type [56–58]. The opposite reaction, graft versus host disease, can occur as well, when the transplant contains T cells capable of attacking the host. This can be a problem in hematological but not in stem cell–derived cell transplants, as they do not contain T cells [58]. With the help of immunosuppressive drugs, partial HLA mismatches can be tolerated. However, these drugs are intrinsically toxic, increase the risk for opportunistic infections, and lower immunological surveillance for tumor cells.

Even though there have been a number of transplantation trials for NDDs, there is no clear consensus yet on whether the use of immunosuppressive agents is warranted and for how long they should be administered.

The lack of clear guidelines is a consequence of the belief that the brain is an immunologically privileged organ and that immunosuppressive therapy or donor matching might not be necessary. The presence of the blood–brain barrier shields the brain from circulating antibodies, immune mediators, and immune cells [59]. The brain parenchyma itself is marked by low-HLA expression, few antigen-presenting cells, and the absence of a classical lymphatic system. Recently, however, these dogmas have been subjected for review. Louveau et al. demonstrated the existence of a meningeal perisinu-soidal lymphatic system, providing a route for antigens and immune cells from the cerebrospinal fluid to the periphery [60]. Vascular insults like a stroke, a common comorbidity in the aging population, can also breach the blood–brain barrier, leading to an influx of immune components. Furthermore the brain itself is not devoid of immune reactivity. Even risk factors for NDDs point in the direction of the immune system (see also Chapter 3). For example, recently discovered variants in TREM2, a receptor expressed on myeloid cells, are associated with a 4- to 10-fold increased risk for developing AD at an older age [46–48]. Therefore, one can no longer disregard the role of the immune system in the pathogenesis of neurodegeneration, and, therefore likely also, in transplantation studies.

When transplanting stem cell–derived cells, the classical immunological questions need to be addressed. However, the use of stem cells on its own provides new opportunities to tackle potential problems.

First of all, the origin of the transplanted cells needs to be considered. Cell therapy trials using allogeneic cells, fetal tissue, or hESC-based trials have faced the problem of graft failure and immune rejection. Using patient-derived cells—iPSCs or ASCs–avoids the possibility of rejection. Unexpectedly, there have been reports mentioning an immune reaction to iPSC-derived teratomas when transplanted into syngeneic mice, but those reactions could be explained by the lack of central immune tolerance toward pluripotency markers [61]. When cells are differentiated to cell types natural to the host environment, such immune reactions did not occur [62].

Abandonment of the use of allogeneic cells and embracing personalized (iPSC-based) cell therapy is what would be expected. However, the opposite trend is manifesting itself. Two major iPSC-based trials in Japan, one treating macular degeneration and the other PD, are switching from autologous patient–derived iPSCs to allogeneic iPSCs. Reasons for halting one trial was the discovery of point mutations in a patient-derived iPSC line rendering the line useless. Concurrently, there was a change in Japanese legislation only allowing the use of allogeneic, HLA-matched, cryobanked iPSC cells (www.ipscell.com/2015/11/parkinsons-ips-cell-trial-in-japan-switching-to-allogeneic/). This decision has most likely been influenced by the joint efforts of the Yamanaka and Okano group in creating an HLA-matched stem cell and neural precursor bank [4]. The generation of these cell banks was pushed forward by the need for a rapidly available source of NSCs for their use in spinal cord injury (SCI). Stem cell trials for SCI demonstrated that rapid transplantation after SCI was a necessity to ensure a beneficial functional outcome [4]. As an HLA-matched stem cell bank could aid in the development of cell therapies for many different diseases, the efforts needed to generate the bank largely outweigh the time and cost to manufacture a clinical grade iPSC line for every individual patient requiring cell therapy.

A second approach to tackle the immune response is the combined grafting of cells with immune-regulating properties and neuronal cells to minimize the need for immunosuppressive drugs. In the field of hematopoietic stem cell transplantation, MSCs are widely investigated as a cotransplant to improve grafting. MSCs are ideally suited as they express low levels of MHC class I molecules and are known to suppress T-cell activation and proliferation [11]. In the field of neurodegeneration and neuroinflammation,

MSCs are currently being explored for a number of diseases including AD, Huntington's disease, and multiple sclerosis [12,37]. Another possibility would be to cotransplant regulatory T cells (Tregs), cells known to tamper the immune response [63]. Piquet et al. suggested the combined grafting of RPECs, a cell type secreting dopamine and known to have immunosuppressive properties. Grafting of RPECs alone did not prove to be beneficial in a double-blind trial (the STEPS trial), but combined grafting could still be advantageous [34].

As stem cells can easily be genomically engineered, the deletion of MHC antigens could also prevent or minimize the risk of immune rejection [56], with as a drawback, however, the increased chance of escaping immune surveillance in the setting of viral infections or tumor formation.

KEEPING TRACK OF CELL TRANSPLANTS

Apart from preventing an immune response, it is also necessary to monitor the survival of the transplanted cells. One of the PD patients that had received a fetal mesencephalic graft improved significantly after transplantation, but then deteriorated again after 10 years. Three years after deterioration, she died due to cardiac arrest. Anatomopathological evaluation of the graft region showed extensive microglial invasion. Possibly this immune response was the reason for graft failure 3 years earlier, but it could also be due to an immune response in relation to systemic illness in the course of her life [33]. A method for cell tracking could have provided interesting information on graft survival and correlation with clinical function.

For PD, graft function can be indirectly assessed with nuclear imaging of the DA system, but a better and more generally applicable system would be to label all the transplanted cells, regardless of cell type.

Direct labeling of cells with MRI or radionuclide probes before transplantation allows short-term cell tracking, but provides insufficient data on cell survival. Cell division dilutes all probes, and radionuclide probes decay and MRI probes can be taken up by macrophages after cell death leading to false localization [64].

Gathering long-term survival information requires genetically engineered stem cell lines stably expressing a reporter gene. The first study testing reporter gene–based imaging of therapeutic cells was conducted by Yaghoubi et al. [65]. In 2006 a patient diagnosed with glioblastoma multiforme was infused with autologous genetically modified T cells. The cells expressed a reporter, thymidine kinase, from the herpes simplex virus 1 capable of phosphorylating a radiolabeled probe and consequently trapping it inside the cell, allowing repeated imaging [65]. Aside from functioning as a PET imaging reporter gene, the gene also functioned as a suicide gene. In this trial the gene was randomly integrated in the host genome. However, to minimize the risks involved with genome engineering, site-specific incorporation of the transgene in a safe harbor locus is recommended.

At least in clinical trials, diagnostic cell tracking would provide the necessary information to evaluate and guide the need for immunotherapy, to monitor cell migration, etc.

To conclude, stem cell therapy is still in its infancy. As research and clinical trials continue, hopefully many remaining questions will be answered. The goal of curing neurodegeneration and completely mending the decaying brain might be hard to reach. Alternative or complementary strategies should continue to be developed, which might include neuroprotective therapies, neurosurgical approaches, and cognitive and physical rehabilitation [66]. In this respect, disease modeling with iPSCs might turn out to be the breakthrough the neurodegenerative field has been waiting for.

PART III: STEM CELLS TO MODEL NEURODEGENERATIVE DISEASES

Translational research for NDDs chiefly depends on animal models. Nonetheless, therapies developed using mouse models mostly fail in clinical trials. In this respect, transcriptome studies demonstrate that mouse models rarely mimic the transcriptome of human NDDs. For instance in murine studies consistent trends toward preserved mitochondrial function, protein catabolism, DNA repair responses, and chromatin maintenance is seen, highlighting the inability of mouse models to recapitulate the multifactorial pathophysiology of human NDDs and need for human models to take center stage [67]. The advent of hiPSC [68] creates the possibility to generate neural cells from fibroblasts or blood cells from patients with these NDDs, recreating cells in vitro that contain the exact same genetic background as these patients. This way, human iPSC technology allows us to study known and unknown disease modifiers, which contribute to the manifestation of disease.

IPSC CHARACTERISTICS: THE ROLE OF DEVELOPMENT IN NEURODEGENERATIVE DISEASES

As NDDs commonly appear in or beyond the fifth to sixth decade of life, it was initially believed that iPSC technology would not be suitable to model these diseases. However, for a number of familial forms of these diseases, there is mounting evidence that the underlying cellular or molecular pathological process may be present throughout life. Clinicians are now starting to recognize the link between neurodevelopment and neurodegeneration. For example, there is an increased frequency of learning disability, especially dyslexia, in patients with frontotemporal dementia (FTD) and their first-degree relatives [69]. FTD is a dementia marked by behavior and language deficits with a much earlier onset compared to AD dementia (see also Chapter 8). Mutations in MAPT (tau), C9Oorf72, and progranulin (PGRN) explain half of the cases and have linked FTD and amyotrophic lateral sclerosis (ALS) [70,71]. Fifteen percent of FTD patients develop features of ALS, and up to 50% of ALS patients exhibit abnormal neuropsychological testing, indicative of frontal lobe dysfunction [72]. A search for modifier genes is in progress to explain why some patients predominantly develop motor neuron degeneration, while others manifest cortical degeneration. Recently, dyslexia susceptibility genes were also identified as risk factors for the development of language deficits in FTD [73]. To further investigate this neurodevelopmental link, iPSC models are ideally suited, as they allow sequential monitoring of the developmental and degenerative phase. For example, iPSC-derived neurons from patients with FTD carrying the $GRN^{IVS1+5G>C}$ mutation showed a developmental phenotype in vitro, as the production of cortical neurons was reduced (Fig. 4A.2) [74]. The mutation, however, did not affect differentiation toward motor neurons. Interestingly, transcriptome analysis revealed deregulation of the Wnt signaling pathway, and treatment with a Wnt inhibitor improved cortical differentiation [74].

A potential problem with current iPSC technologies in modeling late-onset NDDs is the difficulty in obtaining age-related phenotypes in a relatively short timeframe. iPSC technologies face the challenge that the somatic cells generated from undifferentiated iPSC remain immature for long periods. For instance, the group of Livesey demonstrated an 80-day, three-stage process that recapitulates cortical development, in which human iPSCs are first differentiated to cortical stem and progenitor cells that then generate cortical projection neurons in a stereotypical temporal order before maturing functionally [75]. Maturation leads to the active firing of action potentials, synaptogenesis, and the formation of neural circuits in vitro [75].

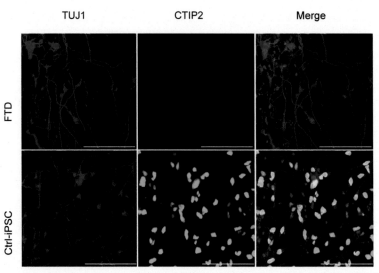

FIGURE 4A.2 Generation of Cortical Neurons From Frontotemporal Dementia (FTD)–iPSC and Control Lines

Immunostainings show double positivity of the derived neurons at D40 for the neural marker, TUJ1, and the cortical marker, CTIP2. FTD lines show less positive cells for this markers compared to control cell lines. Scale bar, 100 μm.

Adapted with permission from S. Raitano, L. Ordovàs, L. De Muynck, W. Guo, I. Espuny-Camacho, M. Geraerts, S. Khurana, K. Vanuytsel, B.I. Tóth, T. Voets, R. Vandenberghe, Restoration of progranulin expression rescues cortical neuron generation in an induced pluripotent stem cell model of frontotemporal dementia, Stem Cell Rep. 4 (1) (2015) 16–24. [74].

Studer's group recently succeeded in inducing rapid cell aging by misexpression of a truncated form of lamin A that is associated with premature aging [76]. This method made it possible to induce aging in PD patient iPSC–derived DA neurons. This resulted in disease-related phenotypes, including severe dendrite degeneration, progressive loss of TH-positive cells, abnormal mitochondria, and Lewy body precursor-like inclusions, which are difficult to identify using conventional neuronal differentiation methods for iPSCs [76]. Notably, they did not observe typical Lewy bodies in DA neurons induced from PD patient–derived iPSCs, even though α-synuclein accumulated in progerin-expressing DA neurons in vitro and Lewy bodies were prominent in the brain autopsy of the patient. Still, progerin-induced aging is a useful method to investigate the features of age-related diseases by iPSC-based disease modeling, while their application to other age-related diseases needs to be verified.

IPSC TECHNOLOGY: DISEASE MODELING AND DRUG SCREENING

Familial forms of NDDs are modeled by either using iPSC from patients with a known mutation, or by introducing a mutation in human immortalized cells and ESC lines. This allows us to compare the phenotypic outcome and conclude whether the mutation alone is sufficient to cause this disease, or if the environment and/or genetic background also contributes due to presence of certain unknown modifiers. For instance, less than 10% of all AD cases are caused by familial mutations. Thus, additional risk factors should be taken into account to fully model sporadic forms. Reprogramming of iPSC-derived from patients carrying low-frequency, high-risk genes and coculture of several cell types present within the brain, will be key to model the sporadic forms.

Once a robust disease phenotype has been established, this can serve as a model to screen and evaluate candidate therapeutic drugs. Candidate drugs have already been tested for their efficacy in iPSC-derived disease-relevant cell types, such as neurons and astrocytes derived from familial and sporadic AD patients [77,78], motor neurons from familial and sporadic ALS [79,80], and DA neurons derived from PD patients with either LRRK2 or PINK1 mutations [81]. These models are suitable for screening, as the in vitro observations resemble the in vivo pathophysiology.

In the following paragraphs we will elaborate on the progress made with iPSC technology for following NDDs: ALS, AD, and PD.

Amyotrophic lateral sclerosis

A recent characterization of iPSC-derived motor neurons from patients with ALS emphasizes the applicability of iPSC to the clinic. ALS is a relentlessly progressive disease marked by the loss of motor neurons in the central and peripheral nervous system, with a median survival of 3 years. Seventy percent of patients present with limb-onset disease, marked by muscle atrophy, fasciculations, spasticity, and consequent loss of fine and gross motor skills. Bulbar-onset disease is characterized by dysarthria, tongue wasting, and dysphagia [82]. As muscle weakness progresses, respiratory function declines, leading to respiratory failure and death. In the treatment of ALS only riluzole has been shown to improve survival by 3–6 months [82,83]. As the prognosis of ALS remains grim, additional disease-modifying drugs are desperately needed (see also Chapter 10).

Among a growing number of genes implicated in familial ALS, mutations in four genes account for the majority of cases. Superoxide dismutase 1 (SOD1) was the first to be discovered, later on missense mutations in genes encoding two RNA/DNA-binding proteins, TAR DNA–binding protein-43 (TDP-43) and fused-in-sarcoma (FUS), and the GGGGCC hexanucleotide expansion in C9orf72 were identified [82,84]. The discovery of TDP-43 as a component in the ubiquitin-positive, tau-negative insoluble protein aggregates in neurons and glia, represents a major breakthrough in FTD/ALS research [85,86].

The characterization of iPSC-derived motor neurons from ALS patients harboring SOD1, C9orf72, and FUS mutations, has brought patients closer to a potential therapy. Through means of multielectrode array and patch clamp recording, iPSC-derived motor neurons show hyperexcitability, which is as well detected in these patients [87]. Moreover, the genetically corrected, but otherwise isogenic, $SOD1^{+/+}$ stem cell line does not display this phenotype. In addition, diseased iPSC-derived motor neurons showed reduced delayed rectifier potassium current amplitudes compared to the control, which was hypothesized to underlie their hyperexcitability. To prove this hypothesis, the Kv7 channel activator, retigabine, was added to the cells. Retigabine was able to block hyperexcitability and improve motor neuron survival by suppressing ALS neuron spontaneous firing, with an EC_{50} of 1.5 µM, a concentration consistent with its pharmacological activity as an antiepileptic agent and similar to its EC_{50} for Kv7 channels. As this drug is already in the market to treat epilepsy, it is now being tested for its suitability in the treatment for ALS.

The discovery of the repeat expansion in C9orf72 as the most prevalent genetic cause of ALS caused a wave of excitement in the field [84]. However, the development of a mouse model that could reproduce the anatomopathological changes, neuronal loss, and behavior deficits proved to be a struggle [88]. ALS patients with this repeat expansion usually carry hundreds to thousands of repeats. Due to the repetitiveness, this genetic deficit is hard to introduce in animal models, limiting the number of introduced repeats to less than hundred. Different disease mechanisms have been suggested. Loss of function, RNA toxicity, and accumulation of toxic proteins due to repeat-associated non-ATG translation (RAN translation) have been implicated in the pathogenesis. Modeling ALS with iPSC-derived neurons from C9orf72 patients bypasses the problem animal models encounter. Two studies already showed the use of antisense oligonucleotides to reverse RNA toxicity. These antisense oligonucleotides target the hexanucleotide repeat expansions, which results in reduced RNA foci and rescue the increased vulnerability of neurons to stress-mediated cytotoxicity. These results show that RNA toxicity, rather than loss of function of C9orf72, is a major contributing factor to the pathogenesis [89]. However, despite the fact that iPSC-derived neurons recapitulate the key neuropathological features of these mutations and allow further characterization of RNA foci, it remains unclear how these phenotypes contribute to human disease [89,90]. This issue can be addressed by performing transcriptome analysis on iPSC-derived neurons. For instance, transcriptome analysis of iPSC-derived motor neurons of a C9orf72 ALS patient revealed deregulation of genes that are involved in cell adhesion, synaptic transmission, and neural differentiation [89,91]. Second, C9orf72 iPSC–derived neurons exhibit elevated sensitivity to stress-mediated cytotoxicity [89]. To conclude, C9orf72 iPSC–derived neurons provide an efficient tool to reveal disease mechanisms and subsequently identify potential therapeutics.

Another problem the ALS field has encountered is finding a model that recapitulates the TDP-43 protein pathology, a common feature in sporadic ALS. The cause–effect relationship between TDP translocation (loss of function) and protein aggregates in the pathogenesis of ALS remains unclear. Animal models either overexpress mutant TDP-43 or eliminate its expression and, therefore, do not represent the human situation where both disease mechanisms are present [92]. Moreover, none of the rodent models that express TDP-43 mutations show definitive TDP-43 proteinopathy that recapitulates human disease. Given these limitations, iPSC-derived motor neurons from these patients pose an ideal and highly relevant model to further study this phenomenon.

Egawa et al. demonstrated that TDP-43M337V iPSC–derived motor neurons, when exposed to arsenite to induce oxidative stress, show increased insoluble TDP-43 in the cytoplasm and reduced cell survival [79]. These iPSC-derived motor neurons also revealed misregulation of genes that are implicated in RNA metabolism and cytoskeleton functions; in particular the expression of neurofilament medium and light chains was significantly reduced [79]. These results are similar as those reported in SOD1D90V iPSC–derived motor neurons in ALS, suggesting that different genetic mutations may share the same target genes leading to an ALS phenotype. This model can be used as a tool to identify candidates to promote survival of neurons [79]. Further studies on iPSC will focus on the identification of several compounds to inhibit the formation of TDP-43 aggregates.

As ALS is marked by loss of motor neurons in the central and peripheral nervous system, the focus of attention has been directed at motor neurons. However, there are reports of astrocyte involvement and neuroinflammation. Given the pluripotent nature of iPSCs, they provide a convenient tool to determine the contribution of glial cells expressing mutant proteins.

Animal studies focusing on SOD1 mutations have implicated astrocytes as key players in motor neuron degeneration. Graft-derived SOD1G93A astrocytes induced host motor neuron ubiquitination and death, forelimb motor and respiratory dysfunction, reactive astrocytosis, and reduced GLT-1 transporter expression in wild type (WT) animals. The SOD1G93A astrocyte–induced motor neuron death seemed in part mediated by host microglial activation. These findings show that mSOD1 astrocytes alone can induce WT motor neuron death and associated pathological changes in vivo.

However, coculture experiments of iPSC-derived astrocytes carrying the M337V TDP-43 mutation, showed that mutant TDP-43 astrocytes do not adversely affect survival of cocultured neurons. Mutant astrocytes did, however, exhibit increased levels of TDP-43, subcellular mislocalization of TDP-43, and decreased cell survival. These observations reveal that TDP-43 proteinopathies may just cause a neuronal cell–autonomous pathogenesis in cell culture [93].

Alzheimer's disease

FAD iPSC–derived neurons with APP/PSEN1 mutations demonstrated a bias toward production of amyloid beta 42 relative to amyloid beta 40, as seen in vivo, as well as higher levels of phosphorylated tau, as observed in mutated hESC lines. Nonetheless, no plaques or tangles were observed using 2D stem cell models [77,78]. Therefore, researchers started to develop 3D culture models to more closely recreate the in vivo environment. By culturing genetically engineered human immortalized NSC (overexpression mutations in APP and PS1) in human matrigel as a 3D support matrix, Tanzi and coworkers were able to demonstrate both key events of AD, production of plaques and tangles, after 6 weeks of culture. Human matrigel was chosen, as it more closely resembles the stiffness and extracellular matrix of the brain [94]. In addition, matrigel limited the diffusion of secreted amyloid, triggering plaque formation. The model was also consistent with the "amyloid hypothesis" stating that the accumulation of amyloid plaques and oligomers results in hyperphosphorylated Tau and tangles, ultimately leading to neurodegeneration. The study could recapitulate this hypothesis by adding a γ-secretase inhibitor, which not only reduced the production of amyloid beta 42, but also caused lower amounts of pTau, suggesting that accumulations of pTau are induced by amyloid beta 42 [94].

Parkinson's disease

An excellent example of a familial model for PD is described by Okano's group, where they reprogrammed fibroblasts obtained from PARK2 PD patients to iPSCs. Subsequently, these iPSCs, as well as control iPSCs and ESCs were differentiated toward DA neurons [95]. The transcriptome, metabolome, proteome, and mitochondrial homeostasis of the normal and diseased iPSC–derived DA neurons were then evaluated. Increased oxidative stress and dysfunctional mitochondria are implicated in the pathogenesis of PD. To test whether this is represented in vitro, oxidative stress was measured in PARK2 iPSC–derived neurons, and shown to be increased in diseased versus control neurons. In addition, metabolomics studies indicated that mitochondria were dysfunctional in PARK2 iPSC–derived neurons, with abnormal mitochondrial morphology and impaired mitochondrial turnover. Examination of iPSC-derived neuronal cells from a PARK2 patient revealed that α-synuclein had accumulated in a similar manner as in the postmortem brain of this patient. This emphasizes the beauty of iPSC-derived neurons, able to represent the patient pathology in vitro [95].

In correlation with Reinhardt et al. [18] another paper shows the generation of two patient-derived iPSC lines: one carrying a mutation in the LRRK2 gene and the second containing a triplication of the SNCA gene. DA neurons derived from these lines showed increased expression of oxidative stress response genes and α-synuclein protein. Ubiquitin-positive puncta were detected in the SNCA DA neurons and in addition compared to WT DA neurons, mutant DA neurons were more sensitive to caspase-3 activation upon addition of hydrogen peroxide, of which the SNCA iPSC line was the most sensitive. These in vitro findings correlate with the data of Reinhardt et al. [18] and shows resemblance with early phenotypes linked to PD [96].

FUTURE PROSPECTS

iPSCs are extensively used to model familial disorders. Nonetheless, more attention should be paid to study sporadic disease causes, which account for 90% of all disease cases (AD and PD). In this respect, more low-frequency high-risk genes discovered through genome-wide association studies should be studied through means of iPSCs. In addition sporadic cases without a known cause or risk factor can only be modeled through iPSCs, whereas animal models are restricted to diseases with known genetic causes. This will be crucial to identify the molecular etiology underlying sporadic diseases. Also, it will be essential to develop large-scale automated systems for the production and differentiation of iPSCs, to include a sufficient number of patients.

Studies characterizing iPSCs derived from somatic cells of patients provide an opportunity to investigate the gradual changes occurring during the disease course, from the asymptomatic phase to the later stages when the pathology has become prominent. Such studies could help to develop an appropriate preemptive neuroprotective treatment, including small molecules, gene therapy, or cell therapy, which could be started early in the asymptomatic phase. For example, AD usually has a long progression of more than 30 years that consists of an asymptomatic phase of ~20 years, a mild cognitive impairment phase of ~10 years, and a dementia phase of unlimited length. Amyloid plaques form and continue to enlarge in the asymptomatic phase, and there is already substantial neuronal loss and brain atrophy in the mild cognitive impairment phase [97]. Thus, if diagnosis were possible in the asymptomatic phase, it would provide a great advantage by enabling the use of treatments to prevent dementia. The development of preemptive treatments for late-onset NDDs would be enormously important in rapidly aging countries, such as Japan. However, to date, iPSC technologies have not yet saved any patient's life. Continuous efforts through the cooperation of basic stem cell biology, system biology, clinical investigation of the diseases, translational research, pharmaceutical science, and regulatory science will be necessary to make it possible for iPSC-based research to really contribute to human health.

REFERENCES

[1] J.B. Gurdon, The developmental capacity of nuclei taken from intestinal epithelium cells of feeding tadpoles, J. Embryol. Exp. Morphol. 10 (1962) 622–640.

[2] I. Wilmut, A.E. Schnieke, J. McWhir, A.J. Kind, K.H. Campbell, Viable offspring derived from fetal and adult mammalian cells, Nature 385 (1997) 810–813.

[3] S. Yamanaka, Induced pluripotent stem cells: past, present, and future, Stem Cell 10 (2012) 678–684.

[4] H. Okano, S. Yamanaka, iPS cell technologies: significance and applications to CNS regeneration and disease, Mol. Brain 7 (2014) 1–12.

[5] R.L. Davis, H. Weintraub, A.B. Lassar, Expression of a single transfected cDNA converts fibroblasts to myoblasts, Cell 51 (1987) 987–1000.

[6] M.J. Evans, M.H. Kaufman, Establishment in culture of pluripotential cells from mouse embryos, Nature 292 (1981) 154–156.

[7] K. Takahashi, et al. Induction of pluripotent stem cells from adult human fibroblasts by defined factors, Cell 131 (2007) 861–872.

[8] F. González, S. Boué, J.C.I. Belmonte, Methods for making induced pluripotent stem cells: reprogramming à la carte, Nat. Rev. Genet. 12 (2011) 231–242.

[9] Y. Chung, et al. Human embryonic stem cell lines generated without embryo destruction, Cell Stem Cell 2 (2008) 113–117.

[10] P. Cahan, G.Q. Daley, Origins and implications of pluripotent stem cell variability and heterogeneity, Nature 14 (2013) 357–368.

[11] S.A. Jacobs, V.D. Roobrouck, C.M. Verfaillie, S.W. Van Gool, Immunological characteristics of human mesenchymal stem cells and multipotent adult progenitor cells, Immunol. Cell Biol. 91 (2013) 32–39.

[12] N. Joyce, et al. Mesenchymal stem cells for the treatment of neurodegenerative disease, Regen. Med. 5 (2010) 933–946.

[13] F.D. Urnov, E.J. Rebar, M.C. Holmes, H.S. Zhang, P.D. Gregory, Genome editing with engineered zinc finger nucleases, Nat. Rev. Genet. 11 (2010) 636–646.

[14] J.C. Miller, et al. A TALE nuclease architecture for efficient genome editing, Nat. Biotech. 29 (2011) 143–148.

[15] S.W. Cho, S. Kim, J.M. Kim, J.-S. Kim, Targeted genome engineering in human cells with the Cas9 RNA-guided endonuclease, Nat. Biotech. 31 (2013) 230–232.

[16] B. Shen, et al. Efficient genome modification by CRISPR-Cas9 nickase with minimal off-target effects, Nat. Methods 11 (2014) 399–402.

[17] S.Q. Tsai, et al. Dimeric CRISPR RNA-guided FokI nucleases for highly specific genome editing, Nat. Biotech. 32 (2014) 569–576.

[18] P. Reinhardt, et al. Genetic correction of a LRRK2 mutation in human iPSCs links Parkinsonian neurodegeneration to ERK-dependent changes in gene expression, Cell Stem Cell 12 (2013) 354–367.

[19] N. Gaspard, et al. An intrinsic mechanism of corticogenesis from embryonic stem cells, Nature 455 (2008) 351–357.

[20] N. Gaspard, P. Vanderhaeghen, Mechanisms of neural specification from embryonic stem cells, Curr. Opin. Neurobiol. 20 (2010) 37–43.

[21] Y. Maury, et al. Combinatorial analysis of developmental cues efficiently converts human pluripotent stem cells into multiple neuronal subtypes, Nat. Biotech. 33 (2015) 89–96.

[22] B.-Y. Hu, S.-C. Zhang, Differentiation of spinal motor neurons from pluripotent human stem cells, Nat. Protoc. 4 (2009) 1295–1304.

[23] N. Yang, Y.H. Ng, Z.P. Pang, T.C. Südhof, M. Wernig, Induced neuronal cells: how to make and define a neuron, Stem Cell 9 (2011) 517–525.

[24] W. Niu, et al. SOX2 reprograms resident astrocytes into neural progenitors in the adult brain, Stem Cell Rep. 4 (2015) 780–794.

[25] Z. Guo, et al. In vivo direct reprogramming of reactive glial cells into functional neurons after brain injury and in an Alzheimer's disease model, Stem Cell 14 (2014) 188–202.

[26] Y. Zhang, et al. Rapid single-step induction of functional neurons from human pluripotent stem cells, Neuron 78 (2013) 785–798.

[27] C.W. Olanow, et al. Clinical pattern and risk factors for dyskinesias following fetal nigral transplantation in Parkinson's disease: a double blind video-based analysis, Mov. Disord. 24 (2008) 336–343.

[28] Y.-D. Kwak, et al. Amyloid precursor protein regulates differentiation of human neural stem cells, Stem Cells Dev. 15 (2006) 381–389.

[29] Q.L. Cao, et al. Pluripotent stem cells engrafted into the normal or lesioned adult rat spinal cord are restricted to a glial lineage, Exp. Neurol. 167 (2001) 48–58.

[30] J.M. Fearnley, A.J. Lees, Ageing and Parkinson's disease: substantia nigra regional selectivity, Brain 114 (Pt. 5) (1991) 2283–2301.

[31] OVID-DS. Deep brain stimulation in Parkinson's disease, (2010) 1–13.

[32] D.J. Brooks, Neuroimaging in Parkinson's disease, Neurotherapeutics 1 (2004) 243–254.

[33] J.H. Kordower, Y. Chu, R.A. Hauser, T.B. Freeman, C.W. Olanow, Lewy body–like pathology in long-term embryonic nigral transplants in Parkinson's disease, Nat. Med. 14 (2008) 504–506.

[34] A.L. Piquet, K. Venkiteswaran, N.I. Marupudi, M. Berk, T. Subramanian, The immunological challenges of cell transplantation for the treatment of Parkinson's disease, Brain Res. Bull. 88 (2012) 320–331.

[35] O. Lindvall, Z. Kokaia, A. Martinez-Serrano, Stem cell therapy for human neurodegenerative disorders—how to make it work, Nat. Med. 10 (2004) S42–S50.

[36] M. Politis, K. Wu, C. Loane, N.P. Quinn, D.J. Brooks, S. Rehncrona, A. Bjorklund, O. Lindvall, P. Piccini, Serotonergic neurons mediate dyskinesia side effects in Parkinson's patients with neural transplants, Sci. Transl. Med. 2 (38) (2010) 38ra46.

[37] S.B. Dunnett, A.E. Rosser, Challenges for taking primary and stem cells into clinical neurotransplantation trials for neurodegenerative disease, Neurobiol. Dis. 61 (2014) 79–89.

[38] R.A. Barker, L. Studer, E. Cattaneo, J. Takahashi, G-Force PD: a global initiative in coordinating stem cell-based dopamine treatments for Parkinson's disease, NPJ Parkinsons Dis. 1 (2015) 15015–15017.

[39] A. Alzheimer, On a peculiar disease of the cerebral cortex, Alzheimer Dis. Assoc. Disord. 1 (1) (1987) 3–8.

[40] J.A. Hardy, G.A. Higgins, Alzheimer's disease: the amyloid cascade hypothesis, Science 256 (1992) 184–185.

[41] D.J. Selkoe, The molecular pathology of Alzheimer's disease, Neuron 6 (1991) 487–498.

[42] G.G. Glenner, C.W. Wong, Alzheimer's disease and Down's syndrome: sharing of a unique cerebrovascular amyloid fibril protein, Biochem. Biophys. Res. Commun. 122 (1984) 1131–1135.

[43] E. Karran, J. Hardy, A critique of the drug discovery and phase 3 clinical programs targeting the amyloid hypothesis for Alzheimer disease, Ann. Neurol. 76 (2014) 185–205.

[44] L.M. Tong, H. Fong, Y. Huang, Stem cell therapy for Alzheimer's disease and related disorders: current status and future perspectives, Exp. Mol. Med. 47 (2015) e151–e1518.

[45] J. Ladewig, P. Koch, O. Brüstle, Auto-attraction of neural precursors and their neuronal progeny impairs neuronal migration, Nature 17 (2013) 24–26.

[46] R. Guerreiro, et al. TREM2 variants in Alzheimer's disease, N. Engl. J. Med. 368 (2013) 117–127.

[47] T. Jonsson, et al. Variant of TREM2 associated with the risk of Alzheimer's disease, N. Engl. J. Med. 368 (2013) 107–116.

[48] E.H. Corder, et al. Gene dose of apolipoprotein E type 4 allele and the risk of Alzheimer's disease in late onset families, Science 261 (1993) 921–923.

[49] J.K. Lee, et al. Intracerebral transplantation of bone marrow-derived mesenchymal stem cells reduces amyloid-beta deposition and rescues memory deficits in Alzheimer's disease mice by modulation of immune responses, Stem Cell 28 (2010) 329–343.

[50] S.U. Kim, J. de Vellis, Stem cell-based cell therapy in neurological diseases: a review, J. Neurosci. Res. 87 (2009) 2183–2200.

[51] D.F. Emerich, et al. Implants of polymer-encapsulated human NGF-secreting cells in the nonhuman primate: rescue and sprouting of degenerating cholinergic basal forebrain neurons, J. Comp. Neurol. 349 (1994) 148–164.

[52] M.L. Hemming, et al. Reducing amyloid plaque burden via ex vivo gene delivery of an Abeta-degrading protease: a novel therapeutic approach to Alzheimer disease, PLoS Med. 4 (2007) e262.

[53] H.J. Lee, et al. Human neural stem cells genetically modified to express human nerve growth factor (NGF) gene restore cognition in the mouse with ibotenic acid-induced cognitive dysfunction, Cell Transplant. 21 (2012) 2487–2496.

[54] A. Trounson, N.D. DeWitt, Pluripotent stem cells progressing to the clinic, Nature 17 (2016) 194–200.

[55] G.Q. Daley, et al. Setting global standards for stem cell research and clinical translation: the 2016 ISSCR guidelines, Stem Cell Rep. 6 (2016) 787–797.

[56] J.A. Bradley, E.M. Bolton, R.A. Pedersen, Stem cell medicine encounters the immune system, Nat. Rev. Immunol. 2 (2002) 859–871.

[57] M.A. Ayala García, B. González Yebra, A.L. López Flores, E. Guaní Guerra, The major histocompatibility complex in transplantation, J. Transplant. 2012 (2012) 1–7.

[58] M. Turner, et al. Toward the development of a global induced pluripotent stem cell library, Cell Stem Cell 13 (2013) 382–384.

[59] L.L. Muldoon, et al. Immunologic privilege in the central nervous system and the blood–brain barrier, J. Cereb. Blood Flow Metab. 33 (2012) 13–21.

[60] A. Louveau, et al. Structural and functional features of central nervous system lymphatic vessels, Nature 523 (2015) 337–341.

[61] T. Zhao, Z.-N. Zhang, Z. Rong, Y. Xu, Immunogenicity of induced pluripotent stem cells, Nature 474 (2011) 212–215.

[62] P. Guha, J.W. Morgan, G. Mostoslavsky, N.P. Rodrigues, A.S. Boyd, Lack of immune response to differentiated cells derived from syngeneic induced pluripotent stem cells, Stem Cell 12 (2013) 407–412.

[63] G.E. Tullis, K. Spears, M.D. Kirk, Immunological barriers to stem cell therapy in the central nervous system, Stem Cells Int. 2014 (2014) 1–12.

[64] M. Rodriguez-Porcel, J.C. Wu, S.S. Gambhir, Molecular imaging of stem cells, StemBook, Harvard Stem Cell Institute, Cambridge, MA, (2008).

[65] S.S. Yaghoubi, et al. Noninvasive detection of therapeutic cytolytic T cells with 18F-FHBG PET in a patient with glioma, Nat. Clin. Pract. Oncol. 6 (2009) 53–58.

[66] E. Cattaneo, L. Bonfanti, Therapeutic potential of neural stem cells: greater in people's perception than in their brains?, Front. Neurosci. 8 (2014).

[67] T.C. Burns, M.D. Li, S. Mehta, A.J. Awad, A.A. Morgan, Mouse models rarely mimic the transcriptome of human neurodegenerative diseases: a systematic bioinformatics-based critique of preclinical models, Eur. J. Pharmacol. 759 (2015) 101–117.

[68] K. Takahashi, S. Yamanaka, Induction of pluripotent stem cells from mouse embryonic and adult fibroblast cultures by defined factors, Cell 126 (2006) 663–676.

[69] E. Rogalski, N. Johnson, S. Weintraub, M. Mesulam, Increased frequency of learning disability in patients with primary progressive aphasia and their first-degree relatives, Arch. Neurol. 65 (2008) 1–9.

[70] C. Lomen-Hoerth, T. Anderson, B. Miller, The overlap of amyotrophic lateral sclerosis and frontotemporal dementia, Neurology 59 (2002) 1077–1079.

[71] A. Sieben, et al. The genetics and neuropathology of frontotemporal lobar degeneration, Acta Neuropathol. 124 (2012) 353–372.

[72] C. Lomen-Hoerth, et al. Are amyotrophic lateral sclerosis patients cognitively normal?, Neurology 60 (2003) 1094–1097.

[73] D. Paternico, et al. Dyslexia susceptibility genes influence brain atrophy in frontotemporal dementia, Neurol. Genet. 1 (2015) e24–e124.

[74] S. Raitano, L. Ordovàs, L. De Muynck, W. Guo, I. Espuny-Camacho, M. Geraerts, S. Khurana, K. Vanuytsel, B.I. Tóth, T. Voets, R. Vandenberghe, T. Cathomen, L. Van Den Bosch, P. Vanderhaeghen, P. Van Damme, C.M. Verfaillie, Restoration of progranulin expression rescues cortical neuron generation in an induced pluripotent stem cell model of frontotemporal dementia, Stem Cell Rep. 4 (1) (2015) 16–24.

[75] Y. Shi, P. Kirwan, F.J. Livesey, Directed differentiation of human pluripotent stem cells to cerebral cortex neurons and neural networks, Nat. Protoc. 7 (2012) 1836–1846.

[76] J.D. Miller, et al. Human iPSC-based modeling of late-onset disease via progerin-induced aging, Cell Stem Cell 13 (2013) 691–705.

[77] T. Kondo, et al. Modeling Alzheimer's disease with iPSCs reveals stress phenotypes associated with intracellular Aβ and differential drug responsiveness, Cell Stem Cell 12 (2013) 487–496.

[78] M.A. Israel, et al. Probing sporadic and familial Alzheimer's disease using induced pluripotent stem cells, Nature 482 (2013) 216–220.

[79] N. Egawa, et al. Drug screening for ALS using patient-specific induced pluripotent stem cells, Sci. Transl. Med. 4 (2012) 145ra104.

[80] M.F. Burkhardt, et al. A cellular model for sporadic ALS using patient-derived induced pluripotent stem cells, Mol. Cell. Neurosci. 56 (2013) 355–364.

[81] O. Cooper, et al. Pharmacological rescue of mitochondrial deficits in iPSC-derived neural cells from patients with familial Parkinson's disease, Sci. Transl. Med. 4 (2012) 141ra90.

[82] M.C. Kiernan, et al. Amyotrophic lateral sclerosis, Lancet 377 (2011) 942–955.

[83] L. Lacomblez, G. Bensimon, P.N. Leigh, P. Guillet, V. Meininger, Dose-ranging study of riluzole in amyotrophic lateral sclerosis. Amyotrophic lateral sclerosis/riluzole study group II, Lancet 347 (1996) 1425–1431.

[84] M. DeJesus-Hernandez, et al. Expanded GGGGCC hexanucleotide repeat in noncoding region of C9ORF72 causes chromosome 9p-linked FTD and ALS, Neuron 72 (2011) 245–256.

[85] T. Arai, et al. TDP-43 is a component of ubiquitin-positive tau-negative inclusions in frontotemporal lobar degeneration and amyotrophic lateral sclerosis, Biochem. Biophys. Res. Commun. 351 (2006) 602–611.

[86] M. Neumann, et al. Ubiquitinated TDP-43 in frontotemporal lobar degeneration and amyotrophic lateral sclerosis, Science 314 (2006) 130–133.

[87] B.J. Wainger, et al. Intrinsic membrane hyperexcitability of amyotrophic lateral sclerosis patient-derived motor neurons, Cell Rep. 7 (2014) 1–11.

[88] J. Chew, et al. C9ORF72 repeat expansions in mice cause TDP-43 pathology, neuronal loss, and behavioral deficits, Science 348 (2015) 1151–1154.

[89] C.J. Donnelly, et al. RNA toxicity from the ALS/FTD C9ORF72 expansion is mitigated by antisense intervention, Neuron 80 (2013) 415–428.

[90] S. Almeida, et al. Modeling key pathological features of frontotemporal dementia with C9ORF72 repeat expansion in iPSC-derived human neurons, Acta Neuropathol. 126 (2013) 385–399.

[91] D. Sareen, et al. Targeting RNA foci in iPSC-derived motor neurons from ALS patients with a C9ORF72 repeat expansion, Sci. Transl. Med. 5 (2013) 208ra149.

[92] E.B. Lee, V.M.Y. Lee, J.Q. Trojanowski, Gains or losses: molecular mechanisms of TDP43-mediatedneurodegeneration, Nat. Rev. Neurosci. 13 (1) (2012) 38–50.

[93] A. Serio, et al. Astrocyte pathology and the absence of non-cell autonomy in an induced pluripotent stem cell model of TDP-43 proteinopathy, Proc. Natl. Acad. Sci. USA 110 (2013) 4697–4702.

[94] S.H. Choi, et al. A three-dimensional human neural cell culture model of Alzheimer's disease, Nature 515 (2014) 274–278.

[95] Y. Imaizumi, et al. Mitochondrial dysfunction associated with increased oxidative stress and α-synuclein accumulation in PARK2 iPSC-derived neurons and postmortem brain tissue, Mol. Brain 5 (2012) 35.

[96] B. Byers, H.-L. Lee, R. Reijo Pera, Modeling Parkinson's disease using induced pluripotent stem cells, Curr. Neurol. Neurosci. Rep. 12 (2012) 237–242.

[97] A. Iwata, T. Iwatsubo, Disease-modifying therapy for Alzheimer's disease: challenges and hopes, Neurol. Clin. Neurosci. 1 (2013) 49–54.

THE POTENTIAL OF STEM CELLS IN TACKLING NEURODEGENERATIVE DISEASES

4B

Sarah Libbrecht
KU Leuven, Leuven, Belgium

CHAPTER OUTLINE

Endogenous Stem Cells as a Therapeutic Target 101
   What are Endogenous Stem Cells? 101
   Adult Neurogenesis in a Neurodegenerative Environment 104
   Adult Brain Regeneration to Tackle Neurodegeneration 108
References 110

ENDOGENOUS STEM CELLS AS A THERAPEUTIC TARGET
WHAT ARE ENDOGENOUS STEM CELLS?
The mechanism of adult neurogenesis

In the 1990s, Reynolds demonstrated that cells could be isolated from the brain with the potency of self-renewal through proliferation and differentiation into a variety of cell types [1]. From then on, it was established that adult neurogenesis (AN) persists at least at two sites in the brain: the subventricular zone (SVZ) of the lateral ventricles and the subgranular zone of the dentate gyrus (DG) in the hippocampus (HC). These two neurogenic regions harbor activated neural stem cells (NSCs) that migrate, differentiate, and finally integrate in the existing network.

The subventricular zone

In the adult murine SVZ, NSCs have been identified as slowly dividing radial glia-like cells (type B cells) separated from the ventricle by a layer of ependymal cells (Fig. 4B.1). They generate rapid amplifying cells (type C cells), which in turn give rise to neuroblasts (type A cells). The majority of these cells migrate through a rostral migratory stream (RMS) to the olfactory bulb (OB) where they radially migrate in the bulbar layers. There they differentiate into two types of interneurons that modulate the activity of mitral cells projecting to the olfactory cortex. The majority of newborn cells differentiate into granule neurons and a minority into periglomerular neurons [2]. It is established that only half of the cells arriving in the OB, survive and integrate into the existing network [3]. The survival of these newborn neurons depends at least partially on sensory and other synaptic inputs during a critical period

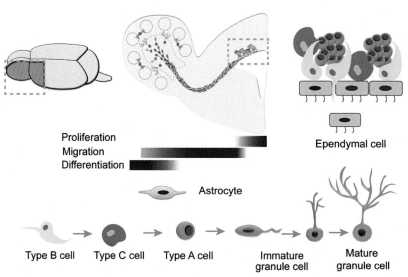

Proliferation
Migration
Differentiation

Ependymal cell

Astrocyte

Type B cell   Type C cell   Type A cell   Immature   Mature
                                          granule cell   granule cell

FIGURE 4B.1 The Rodent Subventricular Zone

In the rodent brain, stem cells reside along the ependymal wall of the lateral ventricles. These type B cells give rise to rapid amplifying type C cells which form neuroblasts. These neuroblasts migrate through the rostral migratory stream (RMS), a hollow tube of astrocytes, to the olfactory bulb (OB) where they migrate into the bulbar layers and differentiate into interneurons.

(between 15 and 45 days) after neuronal birth [3]. Once functionally integrated, the addition of new neurons is thought to play a role in maintaining the existing network, tuning odor information processing, olfactory learning, and memory formation [4].

Additionally, NSCs in the SVZ generate a small population of newborn neurons that migrates toward the corpus callosum, the striatum, and the fornix where they differentiate into oligodendrocytes [5].

The dentate gyrus

In the DG, a population of radial glia-like cells (type B cells) acts as NSCs. They generate actively amplifying nonradial progenitors (type C cells), which in turn give rise to neuroblasts (type A cells) that predominantly differentiate into immature neurons which migrate into the DG granule cell layer (Fig. 4B.2). There, they differentiate into granule cells that extend dendrites into the molecular layer to receive glutamatergic synaptic input from the entorhinal cortex. They contribute to the surrounding network by projection of glutamatergic axons into the hilus of the DG toward CA3 [6]. In contradiction to the SVZ, the major critical period for cell survival in the DG is the transition from amplifying progenitors to neuroblasts. About half of the newborn cells undergo apoptosis during the first 4 days of their life [7]. In line with the SVZ, a second input-dependent critical period of survival takes place at 3 weeks after birth and full integration is achieved after 4 weeks.

Adult HC neurogenesis is believed to be essential for spatial-navigation learning, long-term spatial memory retention, trace and contextual fear conditioning, and presumably in pattern separation, a process by which similar patterns of inputs can be discerned [8].

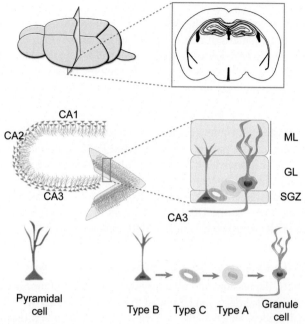

FIGURE 4B.2 The Rodent Dentate Gyrus

In the rodent brain, stem cells reside in the dentate gyrus of the hippocampus, more specific in the subgranular zone *(SGZ)*. Type B cells residing in this zone give rise to type C cells which divide and form neuroblasts. These migrate a short distance into the granular layer *(GL)* where they mature into granule cells. *ML*, molecular layer.

Stem cell niche modulation

Both intrinsic and extrinsic mechanisms regulate different aspects of AN [9]. Extrinsic factors include morphogens, growth factors, neurotrophins, cytokines, hormones, and neurotransmitters. These factors come into contact with the microenvironment through the CSF, the vasculature, and the extracellular matrix. Neurotransmitters provide important niche signals through release at the synapse or diffusion after nonsynaptic secretion. Further, intrinsic factors most likely provide some of the tight control on the neurogenesis process. Cell-intrinsic roles for Notch, Cdk5, neurotrophin receptors, neurotransmitter receptors, and much more have already been proposed. Last but not least, epigenetic mechanisms take part in the regulation of AN by DNA methylation, histone modification, and the action of noncoding RNAs.

Human relevance

The subventricular zone

The group of Alvarez-Buylla identified proliferating stem cells in the human SVZ as GFAP$^+$ astrocytes that parallel the ependymal cell layer [10]. Although the existence of neuronal precursors in the SVZ was confirmed by others [11], the migration of these neuroblasts to the OB was heavily debated as subsequent studies could not confirm the presence of neuroblasts in the RMS nor a robust RMS reaching the OB [12,13]. Using a birth-dating technique based on measuring ^{14}C generated during nuclear bomb testing in the Cold War, the Group of Frisén doubted the existence of OB neurogenesis [14]. Very recently, it was shown that, unexpectedly, the NSCs of the human SVZ give rise to interneurons in the adjacent striatum. This unique pattern of neurogenesis in the adult human brain compared to rodents might result from a different expression of guidance molecules rerouting the neuroblasts to another brain region [15]. Contradicting studies speculate that this ectopic rerouting of SVZ neuroblasts is only a transient response upon brain damage and arises from local astrocytes [16,17].

The dentate gyrus
Direct evidence for the existence of human adult HC neurogenesis came from a pioneering study in 1998 [18]. During that time, cancer patients often received BrdU, a thymidine analog, for diagnostic purposes. Immunohistochemical analysis of the DG of these patients revealed the presence of dividing cells in five individuals up to 72 years of age. Indirect evidence was obtained by the analysis of doublecortin expression in a large cohort of individuals ranging from 1 day to 100 years old revealing doublecortin-positive neuroblasts in the DG up to 85 years of age [19]. Recently, ^{14}C measurements confirmed the existence of HC neurogenesis in the adult human brain [20].

ADULT NEUROGENESIS IN A NEURODEGENERATIVE ENVIRONMENT
Symptoms, such as anxiety, depression, and olfactory dysfunction, that precede the clinical symptoms of several neurodegenerative disorders, can be linked to known functions of AN. Alterations in AN are observed in both animal models and patients suffering from Alzheimer's disease (AD), Parkinson's disease (PD), and Huntington's disease (HD).

Parkinson's disease
PD is the most common movement disorder characterized by a degeneration of dopaminergic neurons in the substantia nigra. One of the pathological hallmarks of the disease is the formation of α-synuclein inclusions or aggregates. The most prominent clinical features of PD are motor symptoms including bradykinesia, tremor, rigidity, and postural instability. Also nonmotor symptoms, such as olfactory deficits, autonomic dysfunction, depression, cognitive deficits, and sleep disorders, are related to PD (see also Chapter 6) and can be linked to impaired neurogenesis [4,21]. PD animal models, in which AN is studied, include acute lesion models and transgenic animal models.

Acute lesion models
Acute lesion models are toxin-induced models that predominantly replicate the dopaminergic neurodegeneration in the substantia nigra and the denervation in the striatum. The two most used models are based on the administration of the neurotoxins 6-hydroxy-dopamine (6-OHDA) and 1-methyl-4-phenyl-1,2,3,6-tetrahydropyridine (MPTP) that selectively and rapidly destroy catecholaminergic neurons which leads to dopaminergic cell death and striatal dopamine depletion [22]. Dopamine can bind several receptors (D1–D5) of which D2-like receptors are expressed on type C cells [23]. In rodents and monkeys, it has been reported that a loss of dopamine has a negative influence on these type C cells leading to a decrease in SVZ proliferation correlating with the level of striatal dopaminergic denervation [24–29]. Administration of levodopa, a dopamine precursor, or pramipexole, a dopamine receptor agonist, rescued the proliferation level in the SVZ demonstrating the role of dopamine [24,26]. Subsequent to decreased SVZ proliferation, newborn neuron arrival in the OB was declined [24] while apoptosis of migrating neuroblasts was induced [30,31]. The study of Winner et al. refined this

observation as a transient decrease of BrdU$^+$ newborn neurons in the granule cell layer while BrdU$^+$ cells increased in the glomerular layer indicating a compensatory mechanism of increased newborn neuron survival [25]. On top of that, they observed a shift in the differentiation to the dopaminergic phenotype suggesting a role for dopamine on proliferation, survival, and differentiation.

Despite the numerous reports of an impaired SVZ neurogenesis, a handful studies contest this. A study of van den Berge et al. observed no change in proliferation after MPTP administration, which might be explained by the insufficient dopaminergic denervation [32]. On the other hand, an increased SVZ proliferation has been reported as well in the MPTP lesion model [33,34].

In contrast to multiple observations that couple dopamine depletion to SVZ neurogenesis deficits, the effect on HC neurogenesis is less defined. Both a decreased [24] and an increased [33] proliferation have been reported.

Transgenic animal models

Although most of the PD cases appear sporadically (about 90%), 5%–10% of the cases can be linked to genetic mutations in several genes of which SNCA and LRRK2 are the most common [35].

SNCA encodes for the protein α-synuclein, which plays a central role in the pathophysiology of PD. Several mutations, duplications, and triplications in the gene have been associated with autosomal dominant forms. Furthermore, a role for α-synuclein in AN has been proposed. Overexpression of human WT α-synuclein decreased newborn neuron arrival in the DG and OB due to an increased apoptosis and decreased survival rate without altering SVZ and DG proliferation [36–38]. The proof of a cell-autonomous influence of α-synuclein overexpression came from studies in which specific α-synuclein overexpression in neuronal progenitors using retroviral vectors delayed migration of the neuroblasts toward the OB and decreased dendritic outgrowth and spine formation in the DG [39,40]. The overexpression of mutant α-synuclein resulted in a decreased proliferation and survival, similar as seen with WT overexpression [37,41]. A53T α-synuclein overexpression also resulted in a decreased SVZ proliferation [37]. In α-synuclein KO mice, neuronal differentiation and newborn neuron arrival was increased in the DG [40].

LRRK2 mutations are the most common cause of familial PD [35]. Overexpression of G2019S LRKK2 resulted in SVZ and HC neurogenesis deficits including decreased proliferation and survival. Further, the newborn neurons were characterized by a reduced dendritic arborization and spine density [42]. On the other hand, LRKK2 KO mice revealed no effect on proliferation and survival but a delayed maturation was noticed by a higher relative percentage of immature to mature neuroblasts. Furthermore, these neuroblasts presented themselves with longer dendrites and a more complex branching pointing toward a role for LRRK2 in neurite outgrowth [43].

In summary, it seems that the influence on AN is highly dependent on the animal model. In lesion models characterized by distinctive striatal neurodegeneration and dopamine depletion, mostly SVZ proliferation is affected while genetic mutations in SNCA or LRRK2 manifest on later stages including predominantly survival deficits for SNCA and morphological alterations in case of LRRK2.

Human patients

Only a handful studies on human PD samples have been conducted and led to inconsistent results. On one hand, studies have reported a decreased cell proliferation in the SVZ [24,27,44] and the DG [24]. On the other hand, van den Berge et al. were not able to find a difference in proliferation of SVZ progenitors in a cohort of 10 PD patients [32].

Alzheimer's disease

AD, which is currently the most common neurodegenerative disorder, is characterized by progressive degeneration of neuronal populations in the neocortex and the limbic system. The cardinal pathological hallmarks are the accumulation of toxic neurofibrillary tangles of hyperphosphorylated tau and the β-amyloid (Aβ) peptide plaques leading to synaptic dysfunction and neuronal death. Patients suffering from AD manifest with olfactory deficits, memory impairment, cognitive, and functional decline (see also Chapter 5). Similar as to PD, some of these can be linked to a compromised neurogenesis, which probably takes place before the clinical onset of the disease [45].

Different experimental animal models have been developed based on the overexpression of genetically altered genes linked to familial forms [45]. However, the emergence of these different animal models in different mouse strains combined with various detection methods have resulted in inconsistent reports of the pathological effect on AN.

Transgenic animal models

The most aggressive familial form of AD is caused by a mutation in presenilin 1 (PSEN1) which encodes a part of the y-secretase complex. Both y- and β-secretases are responsible for the production of the amyloid Aβ peptides by proteolysis of amyloid precursor proteins (APP). PSEN1 transgenic animals replicate the altered Aβ peptide production, neuronal toxicity, and manifest with an altered DG neurogenesis [46–50]. Upon PSEN1 WT overexpression, conflicting studies reported both a decreased and an unimpaired DG cell proliferation [48–50]. In case of mutant overexpression, different outcomes have been reported with different mutant forms ranging from an unaffected to a decreased or increased proliferation [48–51]. A study performed by Chevallier et al. showed an increased proliferation in A246E PSEN1 mice [51] while alterations in the survival rate have been reported in the case of P117L PSEN1 overexpression [48,49]. On the other hand, a PSEN2 universal KO combined with specific KO of PSEN1 in the forebrain resulted in a clear increase in proliferation and an enhanced survival and differentiation in early stages of neurodegeneration [47]. In later stages, the proliferation increase was less pronounced and the survival unaffected.

Aside from PSEN1 transgenic animal models, most of the studies have been performed on transgenic animals expressing the mutant APP. Mostly a decreased neurogenesis in the DG was found [52–56] while in some reports a decreased SVZ neurogenesis [55,57,58] was noted. Furthermore, in a triple transgenic animal harboring the three mutant genes APP, PSEN1, and tau, the decreased proliferation in SVZ and DG correlated to the amount of Aβ plaques and the amount of Aβ+ cells in the HC [59,60]. Contradictory, some groups reported an enhanced neurogenesis in the SVZ and/or the DG [61,62]. When AN is studied in early stages of the disease, before the onset of Aβ plaques, this increased proliferation was confirmed but counteracted by a decreased survival which finally led to a decreased neuronal differentiation [62]. Besides proliferation deficits, APP mutant overexpression has been associated with morphological deficit, such as reduced dendritic length, branching, and spine number [53,62].

In summary, the results are highly variable and unclear which might be due to the genetic backgrounds of the animals and the different mutations used resulting in different phenotypes and rates of disease progression. Further, most of these models replicate the Aβ deposition but not all animal models represent the HC neuronal cell loss [52,56,57] or the cognitive decline [52,57,59,62].

Human patients

A first study by Jin et al. reported an increase in DG neurogenesis in a cohort of 14 severe senile AD patients [63]. Although this might indicate a compensatory mechanism, this report was contradicted by two independent studies. In a group of seven severe AD patients a decrease in DG neurogenesis was observed [64] while in a group of presenile patients, no differences could be found [65]. Furthermore, in a more profound study performed by Li et al., an increased neurogenesis was described but it seemed that the newly formed cells could not mature correctly into functional neurons, which led to a decreased amount of newborn neurons [66]. This might be the explanation that reconciles the seemingly contradictory reports published earlier. Besides the DG, also in the SVZ, a reduction in progenitor cells has been observed [67].

Huntington's disease

HD is an autosomal dominant neurodegenerative disorder caused by a CAG repeat expansion in the huntingtin gene resulting in a progressive cell death of striatal interneurons. HD patients suffer from progressive involuntary choreatic movements, bradykinesia, cognitive decline, and psychiatric syndromes [68] (see also Chapter 9). Both in animal models as in patients, deficits in AN have been detected and could be implicated in the disease progression.

Animal models

The knowledge on the relation between AN and HD in animal models originates mostly from two mouse models, referred to as R6/1 and R6/2. Both models carry the exon 1 of the huntingtin gene with highly expanded CAG repeats (115 or 150). The R6/2 model is the most severe with an early age of onset and the most widespread occurrence of huntingtin inclusions in the brain [69]. In both these animal models, a deficit in DG proliferation has been described by several groups [70–74]. Except for the study performed by Lazic et al. [71], the survival and the differentiation potential was unaffected [72,73]. In a YAC128 mouse model, in which the full huntingtin gene with 128 CAG repeats is expressed, the proliferation deficit was confirmed and an impaired differentiation and morphology of immature neurons was demonstrated [72,75].

Concerning SVZ neurogenesis, there is less of a consensus. However, it seems that striatal neurodegeneration is necessary to induce an increased proliferation [76–78]. Besides the disagreement concerning the SVZ proliferation state, these studies seem to consolidate on the fact that newborn neuroblasts migrate to the diseased striatum. Which state of SVZ progenitor cell is attracted to the diseased striatum, is still under debate [79]. Whether these migrated newborn neurons are able to mature completely and functionally integrate into the circuit is still unclear. The differences observed in this integration potential could be related to the severity of disease progression. If the pathology in the striatum becomes severe, the necessary support for the maturation and integration of newborn neurons might be lacking.

Patients

In contradiction to rodent HD models, in which predominantly a deficit in the DG was demonstrated, mostly SVZ-related impairments were noticed in patient postmortem brain. Curtis et al., reported an increased SVZ proliferation correlating with the severity of the disease [80,81]. Furthermore, a migration of these newborn neurons in the striatum and a differentiation to the neuronal or glial phenotype was observed [15,82]. These studies might indicate that upon striatal neurodegeneration, the SVZ proliferation is increased in an attempt to compensate for the striatal cell loss. Despite the migration of these newborn neurons into the striatum, the affected tissue is not able to sustain the maturation and functional integration in the network to the same degree as healthy tissue.

Summary

A consensus on the state of AN as well as the implications on disease progression is still far away. The discrepancies are at least partially caused by the use of different transgenic mouse models and study protocols. Furthermore, we should keep in mind that transgenic animal models do not represent all the pathological features. Especially for AN, present results indicate that the degree of neuronal cell loss is an important factor known to vary among animal models.

ADULT BRAIN REGENERATION TO TACKLE NEURODEGENERATION

As progressive neuronal loss is the cardinal hallmark of neurodegenerative diseases, trying to ameliorate this disease by neuronal regeneration seems to be a logical step. The age-related aspect of these diseases implicates that these young stem cells would still be unaffected by the pathologic genetic alterations for a significant time span. Indeed, Kohl et al. found that in the R6/2 transgenic HD mouse model, huntingtin aggregates could be observed in mature neurons but not in the B and C cells of the SVZ suggesting a noncell autonomous mechanism of aggregated huntingtin on newly generated neurons [76]. Further, some of the disease symptoms are correlated to AN and thus reversing AN deficits might alleviate these symptoms.

Increasing the endogenous pool of stem cells
Environmental enrichment

It is established that the environment and the existing neuronal network has a major influence on AN. Furthermore, it has been observed that rehabilitation following brain damage is critically aided by physical and intellectual exercise and extensive sensory stimulation. At least for AD and HD, some beneficial effects of environmental enrichment have been reported. In a study of Valero et al., short-term environmental enrichment increased HC neurogenesis in young APP transgenic mice and rescued HC-dependent learning and memory deficits and the dendritic morphology of newborn neurons [53].

Pharmaco- and gene therapy

AN is a process strictly controlled by a chunk of factors, of which some of them are able to cross the blood brain barrier making them attractive for pharmacotherapy. The mechanistic insight behind these factors is still being unraveled, hence, only a handful of studies have attempted pharmacotherapy in animal models with a beneficial outcome.

The fibroblast growth factor 2 (FGF-2) has been shown to stimulate SVZ proliferation and migration to the striatum in both the R6/2 animal model of HD [83] and the 6-OHDA PD model [84]. In the former, newborn neurons were able to differentiate into medium sized spiny neurons. Further, the huntingtin inclusion load was reduced and the motor function and average survival increased. In the 6-OHDA model, both intraventricular injections of FGF-2 and epidermal growth factor 2 [84] and striatal administration of transforming growth factor could induce cell proliferation and migration but not dopaminergic differentiation [85].

The beneficial effect of the brain-derived neurotrophic factor BDNF has been extensively studied by the Goldmand's lab in the R6/2 HD murine model and in squirrel monkeys. They found that viral vector-mediated expression of noggin and BDNF in the ependymal cells of the SVZ resulted in the recruitment of newborn neurons to the striatum. These neurons were able to differentiate into medium spiny neurons and functionally integrate in the striatopallidal pathway resulting in a delayed deterioration of motor function and substantially increased survival [86,87]. Taken into consideration that overexpression of BDNF in the SVZ progenitors itself could not replicate these benefits, underlines the importance of paracrine regulation on SVZ progenitors [88].

Next, overexpression of transcription factors, such as NeuroD1 enhanced the neuronal fate, maturation, differentiation, and synaptic integration of newborn neurons leading to a rescue of dendritic spine density and spatial memory [89]. Further, compounds acting on the Wnt3a pathways have shown to be beneficial as well. Chronic overexpression or activation of Wnt3a was able to enhance DG neurogenesis and rescue behavioral impairments in triple transgenic AD mice [90,91].

Cell transplantation

The use of the endogenous stem cell pool as a source for transplantation overcomes the ethical problems associated with fetal tissue and the detrimental response of the innate immune system. The feasibility of SVZ progenitor cell isolation, subsequent neurosphere culture and differentiation followed by transplantation in the affected region has already been demonstrated [92,93]. Nevertheless, no studies have been performed using this stem cell source in a clinically relevant setting. However, Lévesque et al. were able to isolate dormant stem cells from cortical samples of a PD patient, expand them in vitro and differentiate them to dopaminergic and GABAergic neurons [94]. Three months post retransplantation of this cell suspension into the patient, they could observe a significant increase in dopamine uptake within the transplanted putamen and an improved motor function up to 5 years. The success of this treatment holds promise for the use of the active stem cells residing in the neurogenic niches.

As a model system

Before neuronal progenitors residing in the neurogenic niches can be used as a stem cell source, several remaining questions concerning the AN cascade should be addressed. First, we have to grasp, how to direct the newly generated neurons to the lesioned area, second how to differentiate them into a specific neuronal fate, and third how functional synaptogenesis and long-term survival is induced.

One of the hurdles of cell therapy today is the limited ability of grafted neurons to migrate through the adult brain from the site of injury. Therefore, mechanistic insight in the regulation of neuroblast migration in rodents could be valuable for the design of cell therapy. To ameliorate the differentiation of the migrated newborn neurons to the appropriate cell type, the identification of genetic programs controlling this neuronal differentiation is warranted.

Third, subsequent integration is another bottleneck. In rodents, it has been demonstrated that the environmental cues and the interplay with the existing network play a pivotal role [3]. If you combine this with the observation of the beneficial outcome of brain training on rehabilitation, knowledge of the experience-dependent plasticity of newborn neurons may provide a scientific base for cell transplantation-based rehabilitation therapy. Furthermore, these studies should also be conducted in diseased brain as the interaction between new neurons and the preexisting neuronal circuits might be disturbed.

Summary

There is still a long road ahead before cell replacement therapy for neurodegenerative diseases can become a reality. Besides the limited knowledge to manipulate NSCs, there are some caveats that could still hamper the cell therapy benefits. For example, the diseased environment could have an effect on the engraftment. In case of PD, it has been observed that α-synuclein aggregates spread to the engrafted cells indicating that disease factors in the host can affect the grafted cells [95]. Therefore, a combinational therapy of stem cell replacement and drug therapy seems to hold the most promise.

REFERENCES

[1] B.a. Reynolds, W. Tetzlaff, S. Weiss, A multipotent EGF-responsive striatal embryonic progenitor cell produces neurons and astrocytes, J. Neurosci. 12 (11) (1992) 4565–4574.

[2] A. Alvarez-buylla, J.M. Garcı, Neurogenesis in adult subventricular zone, J. Neurosci. 22 (3) (2002) 629–634.

[3] M. Yamaguchi, K. Mori, Critical period for sensory experience-dependent survival of newly generated granule cells in the adult mouse olfactory bulb, Proc. Natl. Acad. Sci. USA 102 (27) (2005) 9697–9702.

[4] F. Lazarini, P.-M. Lledo, Is adult neurogenesis essential for olfaction?, Trends Neurosci. 34 (1) (2011) 20–30.

[5] B. Menn, J.M. Garcia-Verdugo, C. Yaschine, O. Gonzalez-Perez, D. Rowitch, A. Alvarez-Buylla, Origin of oligodendrocytes in the subventricular zone of the adult brain, J. Neurosci. 26 (30) (2006) 7907–7918.

[6] G. Kempermann, D. Gast, G. Kronenberg, M. Yamaguchi, F.H. Gage, Early determination and long-term persistence of adult-generated new neurons in the hippocampus of mice, Development 130 (2) (2003) 391–399.

[7] A. Sierra, J.M. Encinas, J.J.P. Deudero, et al. Microglia shape adult hippocampal neurogenesis through apoptosis-coupled phagocytosis, Cell Stem Cell. 7 (4) (2010) 483–495.

[8] W. Deng, J.B. Aimone, F.H. Gage, New neurons and new memories: how does adult hippocampal neurogenesis affect learning and memory?, Nat. Rev Neurosci. 11 (5) (2010) 339–350.

[9] G. Ming, H. Song, Adult neurogenesis in the mammalian central nervous system, Annu. Rev. Neurosci. 28 (2005) 223–250.

[10] N. Sanai, A.D. Tramontin, A. Quiñónes-Hinojosa, et al. Unique astrocye ribbon in adult human brain contains neural stem cells but lacks chain migration, Nature 427 (6976) (2004) 740–744.

[11] M.a Curtis, M. Kam, U. Nannmark, et al. Human neuroblasts migrate to the olfactory bulb via a lateral ventricular extension, Science 315 (5816) (2007) 1243–1249.

[12] N. Sanai, T. Nguyen, R.A. Ihrie, et al. Corridors of migrating neurons in the human brain and their decline during infancy, Nature 478 (7369) (2011) 382–386.

[13] C. Wang, F. Liu, Y.-Y. Liu, et al. Identification and characterization of neuroblasts in the subventricular zone and rostral migratory stream of the adult human brain, Cell Res. 21 (11) (2011) 1534–1550.

[14] O. Bergmann, J. Liebl, S. Bernard, et al. The age of olfactory bulb neurons in humans, Neuron 74 (4) (2012) 634–639.

[15] A. Ernst, K. Alkass, S. Bernard, et al. Neurogenesis in the striatum of the adult human brain, Cell 156 (5) (2014) 1072–1083.

[16] J.P. Magnusson, C. Göritz, J. Tatarishvili, et al. A latent neurogenic program in astrocytes regulated by Notch signaling in the mouse, Science 346 (6206) (2014) 237–241.

[17] C. Wang, Y. You, D. Qi, et al. Human and monkey striatal interneurons are derived from the medial ganglionic eminence but not from the adult subventricular zone, J. Neurosci. 34 (33) (2014) 10906–10923.

[18] P.S. Eriksson, E. Perfilieva, T. Björk-Eriksson, et al. Neurogenesis in the adult human hippocampus, Nat. Med. 4 (11) (1998) 1313–1317.

[19] R. Knoth, I. Singec, M. Ditter, et al. Murine features of neurogenesis in the human hippocampus across the lifespan from 0 to 100 years, PLoS One 5 (1) (2010) e8809.

[20] K.L. Spalding, O. Bergmann, K. Alkass, et al. Dynamics of hippocampal neurogenesis in adult humans, Cell 153 (6) (2013) 1219–1227.

[21] J.-M. Revest, D. Dupret, M. Koehl, et al. Adult hippocampal neurogenesis is involved in anxiety-related behaviors, Mol. Psychiatry 14 (10) (2009) 959–967.

[22] A. Schober, Classic toxin-induced animal models of Parkinson's disease: 6-OHDA and MPTP, Cell Tissue Res. 318 (1) (2004) 215–224.

[23] A. Borta, G.U. Höglinger, Dopamine and adult neurogenesis, J. Neurochem. 100 (3) (2007) 587–595.

[24] G.U. Höglinger, P. Rizk, M.P. Muriel, et al. Dopamine depletion impairs precursor cell proliferation in Parkinson disease, Nat. Neurosci. 7 (7) (2004) 726–735.

[25] B. Winner, M. Geyer, S. Couillard-Despres, et al. Striatal deafferentation increases dopaminergic neurogenesis in the adult olfactory bulb, Exp. Neurol. 197 (1) (2006) 113–121.

[26] B. Winner, P. Desplats, C. Hagl, et al. Dopamine receptor activation promotes adult neurogenesis in an acute Parkinson model, Exp. Neurol. 219 (2) (2009) 543–552.

[27] G.C. O'Keeffe, P. Tyers, D. Aarsland, J.W. Dalley, R.A. Barker, M.A. Caldwell, Dopamine-induced proliferation of adult neural precursor cells in the mammalian subventricular zone is mediated through EGF, Proc. Natl. Acad. Sci. 106 (21) (2009) 8754–8759.

[28] S.a Baker, K.A. Baker, T. Hagg, Dopaminergic nigrostriatal projections regulate neural precursor proliferation in the adult mouse subventricular zone, Eur. J. Neurosci. 20 (2) (2004) 575–579.

[29] N. Freundlieb, Dopaminergic substantia nigra neurons project topographically organized to the subventricular zone and stimulate precursor cell proliferation in aged primates, J. Neurosci. 26 (8) (2006) 2321–2325.

[30] X.J. He, H. Nakayama, M. Dong, et al. Evidence of apoptosis in the subventricular zone and rostral migratory stream in the MPTP mouse model of Parkinson disease, J. Neuropathol. Exp. Neurol. 65 (9) (2006) 873–882.

[31] X.J. He, H. Yamauchi, K. Uetsuka, H. Nakayama, Neurotoxicity of MPTP to migrating neuroblasts: studies in acute and subacute mouse models of Parkinson's disease, Neurotoxicology 29 (3) (2008) 413–420.

[32] S.A. van den Berge, M.E. van Strien, J.A. Korecka, et al. The proliferative capacity of the subventricular zone is maintained in the parkinsonian brain, Brain 134 (Pt 11) (2011) 3249–3263.

[33] J. Peng, L. Xie, K. Jin, D.A. Greenberg, J.K. Andersen, Fibroblast growth factor 2 enhances striatal and nigral neurogenesis in the acute 1-methyl-4-phenyl-1,2,3,6-tetrahydropyridine model of Parkinson's disease, Neuroscience 153 (3) (2008) 664–670.

[34] P.M. Aponso, R.L.M. Faull, B. Connor, Increased progenitor cell proliferation and astrogenesis in the partial progressive 6-hydroxydopamine model of Parkinson's disease, Neuroscience 151 (4) (2008) 1142–1153.

[35] S. Lesage, A. Brice, Parkinson's disease: from monogenic forms to genetic susceptibility factors, Hum. Mol. Genet. 18 (R1) (2009) 48–59.

[36] B. Winner, D.C. Lie, E. Rockenstein, et al. Human wild-type alpha-synuclein impairs neurogenesis, J. Neuropathol. Exp. Neurol. 63 (11) (2004) 1155–1166.

[37] L. Crews, H. Mizuno, P. Desplats, et al. Alpha-synucleinalters Notch-1 expression and neurogenesis in mouse embryonic stem cells and in the hippocampus of transgenic mice, J. Neurosci. 28 (16) (2008) 4250–4260.

[38] V.E.L. May, S. Nuber, F. Marxreiter, O. Riess, B. Winner, J. Winkler, Impaired olfactory bulb neurogenesis depends on the presence of human wild-type alpha-synuclein, Neuroscience 222 (2012) 343–355.

[39] M. Tani, H. Hayakawa, T. Yasuda, et al. Ectopic expression of α-synuclein affects the migration of neural stem cells in mouse subventricular zone, J. Neurochem. 115 (4) (2010) 854–863.

[40] B. Winner, M. Regensburger, S. Schreglmann, et al. Role of -synuclein in adult neurogenesis and neuronal maturation in the dentate gyrus, J. Neurosci. 32 (47) (2012) 16906–16916.

[41] F. Marxreiter, S. Nuber, M. Kandasamy, et al. Changes in adult olfactory bulb neurogenesis in mice expressing the A30P mutant form of alpha-synuclein, Eur. J. Neurosci. 29 (5) (2009) 879–890.

[42] B. Winner, H.L. Melrose, C. Zhao, et al. Adult neurogenesis and neurite outgrowth are impaired in LRRK2 G2019S mice, Neurobiol. Dis. 41 (3) (2011) 706–716.

[43] M. Paus, Z. Kohl, N.M.B. Ben Abdallah, D. Galter, F. Gillardon, J. Winkler, Enhanced dendritogenesis and axogenesis in hippocampal neuroblasts of LRRK2 knockout mice, Brain Res. 1497 (2013) 85–100.

[44] S.S. O'Sullivan, M. Johnson, D.R. Williams, et al. The effect of drug treatment on neurogenesis in Parkinson's disease, Mov. Disord. 26 (1) (2011) 45–50.

[45] L. Crews, E. Rockenstein, E. Masliah, APP transgenic modeling of Alzheimer's disease: mechanisms of neurodegeneration and aberrant neurogenesis, Brain Struct. Funct. 214 (2–3) (2010) 111–126.

[46] S.H. Choi, K. Veeraraghavalu, O. Lazarov, et al. Non-cell-autonomous effects of presenilin 1 variants on enrichment-mediated hippocampal progenitor cell proliferation and differentiation, Neuron 59 (4) (2008) 568–580.

[47] Q. Chen, A. Nakajima, S.H. Choi, X. Xiong, S.S. Sisodia, Y.-P. Tang, Adult neurogenesis is functionally associated with AD-like neurodegeneration, Neurobiol. Dis. 29 (2) (2008) 316–326.

[48] P.H. Wen, X. Shao, Z. Shao, et al. Overexpression of wild type but not an FAD mutant presenilin-1 promotes neurogenesis in the hippocampus of adult mice, Neurobiol. Dis. 10 (2002) 8–19.

[49] P.H. Wen, P.R. Hof, X. Chen, et al. The presenilin-1 familial Alzheimer disease mutant P117L impairs neurogenesis in the hippocampus of adult mice, Exp. Neurol. 188 (2) (2004) 224–237.

[50] R. Wang, K.T. Dineley, J.D. Sweatt, H. Zheng, Presenilin 1 familial Alzheimer's disease mutation leads to defective associative learning and impaired adult neurogenesis, Neuroscience 126 (2) (2004) 305–312.

[51] N.L. Chevallier, S. Soriano, D.E. Kang, E. Masliah, G. Hu, E.H. Koo, Perturbed neurogenesis in the adult hippocampus associated with presenilin-1 A246E mutation, Am. J. Pathol. 167 (1) (2005) 151–159.

[52] M.H. Donovan, U. Yazdani, R.D. Norris, D. Games, D.C. German, A.J. Eisch, Decreased adult hippocampal neurogenesis in the PDAPP mouse model of Alzheimer's disease, J. Comp. Neurol. 495 (1) (2006) 70–83.

[53] J. Valero, J. España, A. Parra-Damas, E. Martín, J. Rodríguez-Álvarez, C.A. Saura, Short-term environmental enrichment rescues adult neurogenesis and memory deficits in APPSw,Ind transgenic mice, PLoS One 6 (2) (2011) e16832.

[54] C. Zhang, E. McNeil, L. Dressler, R. Siman, Long-lasting impairment in hippocampal neurogenesis associated with amyloid deposition in a knock-in mouse model of familial Alzheimer's disease, Exp. Neurol. 204 (1) (2007) 77–87.

[55] M. Demars, Y.-S. Hu, A. Gadadhar, O. Lazarov, Impaired neurogenesis is an early event in the etiology of familial Alzheimer's disease in transgenic mice, J. Neurosci. Res. 88 (10) (2010) 2103–2117.

[56] A. Faure, L. Verret, B. Bozon, et al. Impaired neurogenesis, neuronal loss, and brain functional deficits in the APPxPS1-Ki mouse model of Alzheimer's disease, Neurobiol. Aging 32 (3) (2011) 407–418.

[57] N.J. Haughey, A. Nath, S.L. Chan, C. Borchard a, M.S. Rao, M.P. Mattson, Disruption of neurogenesis by amyloid b-peptide, and perturbed neural progenitor cell homeostasis, in models of Alzheimer' s disease, J. Neurochem. 83 (2002) 1509–1524.

[58] N.J. Haughey, D. Liu, Nath a, C. Borchard a, M.P. Mattson, Disruption of neurogenesis in the subventricular zone of adult mice, and in human cortical neuronal precursor cells in culture, by amyloid beta-peptide: implications for the pathogenesis of Alzheimer's disease, Neuromolecular Med. 1 (2) (2002) 125–135.

[59] J.J. Rodríguez, V.C. Jones, M. Tabuchi, et al. Impaired adult neurogenesis in the dentate gyrus of a triple transgenic mouse model of Alzheimer's disease, PLoS One 3 (8) (2008) e2935.

[60] J.J. Rodríguez, V.C. Jones, A. Verkhratsky, Impaired cell proliferation in the subventricular zone in an Alzheimer's disease model, Neuroreport 20 (10) (2009) 907–912.

[61] K. Jin, V. Galvan, L. Xie, et al. Enhanced neurogenesis in Alzheimer's disease transgenic (PDGF-APPSw,Ind) mice, Proc. Natl. Acad. Sci. USA 101 (36) (2004) 13363–13367.

[62] A. Krezymon, K. Richetin, H. Halley, et al. Modifications of hippocampal circuits and early disruption of adult neurogenesis in the Tg2576 mouse model of Alzheimer's disease, PLoS One 8 (9) (2013) 9–11.

[63] K. Jin, A.L. Peel, X.O. Mao, et al. Increased hippocampal neurogenesis in Alzheimer's disease, Proc. Natl. Acad. Sci. USA 101 (1) (2004) 343–347.

[64] L. Crews, A. Adame, C. Patrick, et al. Increased BMP6 levels in the brains of Alzheimer's disease patients and APP transgenic mice are accompanied by impaired neurogenesis, J. Neurosci. 30 (37) (2010) 12252–12262.

[65] K. Boekhoorn, M. Joels, P.J. Lucassen, Increased proliferation reflects glial and vascular-associated changes, but not neurogenesis in the presenile Alzheimer hippocampus, Neurobiol. Dis. 24 (1) (2006) 1–14.

[66] B. Li, H. Yamamori, Y. Tatebayashi, et al. Failure of neuronal maturation in Alzheimer disease dentate gyrus, J. Neuropathol. Exp. Neurol. 67 (1) (2008) 78–84.

[67] I. Ziabreva, E. Perry, R. Perry, et al. Altered neurogenesis in Alzheimer's disease, J. Psychosom. Res. 61 (3) (2006) 311–316.

[68] F.O. Walker, Huntington's disease, Lancet 369 (9557) (2007) 218–228.

[69] L. Mangiarini, K. Sathasivam, M. Seller, et al. Exon I of the HD gene with an expanded CAG repeat is sufficient to cause a progressive neurological phenotype in transgenic mice, Cell 87 (3) (1996) 493–506.

[70] S.E. Lazic, H. Grote, R.J. Armstrong, et al. Decreased hippocampal cell proliferation in R6/1 Huntington's mice, Neuroreport 15 (5) (2004) 811–813.

[71] S.E. Lazic, H.E. Grote, C. Blakemore, et al. Neurogenesis in the R6/1 transgenic mouse model of Huntington's disease: effects of environmental enrichment, Eur. J. Neurosci. 23 (7) (2006) 1829–1838.

[72] J.M.A.C. Gil, P. Mohapel, I.M. Araújo, et al. Reduced hippocampal neurogenesis in R6/2 transgenic Huntington's disease mice, Neurobiol. Dis. 20 (3) (2005) 744–751.

[73] W. Phillips, Abnormalities of neurogenesis in the R6/2 mouse model of Huntington's disease are attributable to the in vivo microenvironment, J. Neurosci. 25 (50) (2005) 11564–11576.

[74] Z. Kohl, M. Kandasamy, B. Winner, et al. Physical activity fails to rescue hippocampal neurogenesis deficits in the R6/2 mouse model of Huntington's disease, Brain Res. 1155 (2007) 24–33.

[75] J.M. Simpson, J. Gil-Mohapel, M.A. Pouladi, et al. Altered adult hippocampal neurogenesis in the YAC128 transgenic mouse model of Huntington disease, Neurobiol. Dis. 41 (2) (2011) 249–260.

[76] Z. Kohl, M. Regensburger, R. Aigner, et al. Impaired adult olfactory bulb neurogenesis in the R6/2 mouse model of Huntington's disease, BMC Neurosci. 11 (2010) 114.

[77] a.S. Tattersfield, R.J. Croon, Y.W. Liu, A.P. Kells, R.L.M. Faull, B. Connor, Neurogenesis in the striatum of the quinolinic acid lesion model of Huntington's disease, Neuroscience 127 (2) (2004) 319–332.

[78] C.M.C. Batista, T.E. Kippin, S. Willaime-morawek, K. Shimabukuro, W. Akamatsu, D. van der Kooy, Neurobiology of disease a progressive and cell non-autonomous increase in striatal neural stem cells in the Huntington's disease R6/2 mouse, Neurobiol. Dis. 26 (41) (2006) 10452–10460.

[79] M. Kandasamy, M. Rosskopf, K. Wagner, et al. Reduction in subventricular zone-derived olfactory bulb neurogenesis in a rat model of Huntington's disease is accompanied by striatal invasion of neuroblasts, PLoS One 10 (2) (2015) e0116069.

[80] M.a Curtis, E.B. Penney, A.G. Pearson, et al. Increased cell proliferation and neurogenesis in the adult human Huntington's disease brain, Proc. Natl. Acad. Sci. USA 100 (15) (2003) 9023–9027.

[81] M.A. Curtis, E.B. Penney, J. Pearson, M. Dragunow, B. Connor, R.L.M. Faull, The distribution of progenitor cells in the subependymal layer of the lateral ventricle in the normal and Huntington's disease human brain, Neuroscience 132 (2005) 777–788.

[82] A. Ernst, J. Frisén, Adult neurogenesis in humans—common and unique traits in mammals, PLOS Biol. 13 (1) (2015) e1002045.

[83] K. Jin, M. LaFevre-Bernt, Y. Sun, et al. FGF-2 promotes neurogenesis and neuroprotection and prolongs survival in a transgenic mouse model of Huntington's disease, Proc. Natl. Acad. Sci. USA 102 (50) (2005) 18189–18194.

[84] B. Winner, E. Rockenstein, D.C. Lie, et al. Mutant alpha-synuclein exacerbates age-related decrease of neurogenesis, Neurobiol. Aging 29 (6) (2008) 913–925.

[85] O. Cooper, Intrastriatal transforming growth factor delivery to a model of Parkinson's disease induces proliferation and migration of endogenous adult neural progenitor cells without differentiation into dopaminergic neurons, J. Neurosci. 24 (41) (2004) 8924–8931.

[86] S.R. Cho, A. Benraiss, E. Chmielnicki, A. Samdani, A. Economides, S.a Goldman, Induction of neostriatal neurogenesis slows disease progression in a transgenic murine model of Huntington disease, J. Clin. Invest. 117 (10) (2007) 2889–2902.

[87] A. Benraiss, M.J. Toner, Q. Xu, et al. Sustained mobilization of endogenous neural progenitors delays disease progression in a transgenic model of huntington's disease, Cell Stem Cell. 12 (6) (2013) 787–799.

[88] V. Reumers, C.M. Deroose, O. Krylyshkina, et al. Noninvasive and quantitative monitoring of adult neuronal stem cell migration in mouse brain using bioluminescence imaging, Stem Cells. 26 (9) (2008) 2382–2390.

[89] K. Richetin, C. Leclerc, N. Toni, et al. Genetic manipulation of adult-born hippocampal neurons rescues memory in a mouse model of Alzheimer's disease, Brain 138 (2) (2015) 440–455.

[90] A. Shruster, D. Offen, Targeting neurogenesis ameliorates danger assessment in a mouse model of Alzheimer's disease, Behav. Brain Res. 261 (2014) 193–201.

[91] S.K. Tiwari, S. Agarwal, B. Seth, et al. Curcumin-loaded nanoparticles potently induce adult neurogenesis and reverse cognitive deficits in Alzheimer's disease model via canonical Wnt/beta-catenin pathway, ACS Nano. 8 (1) (2014) 76–103.

[92] E.M. Vazey, K. Chen, S.M. Hughes, B. Connor, Transplanted adult neural progenitor cells survive, differentiate and reduce motor function impairment in a rodent model of Huntington's disease, Exp. Neurol. 199 (2006) 384–396.

[93] K. Visnyei, K.J. Tatsukawa, R.I. Erickson, et al. Neural progenitor implantation restores metabolic deficits in the brain following striatal quinolinic acid lesion, Exp. Neurol. 197 (2) (2006) 465–474.

[94] M.F. Lévesque, T. Neuman, M. Rezak, Therapeutic microinjection of autologous adult human neural stem cells and differentiated neurons for Parkinson's disease: five-year post-operative outcome, Open Stem Cell J. (2009) 20–29.

[95] J.-Y. Li, E. Englund, J.L. Holton, et al. Lewy bodies in grafted neurons in subjects with Parkinson's disease suggest host-to-graft disease propagation, Nat. Med. 14 (5) (2008) 501–503.

PRECLINICAL MODELS OF ALZHEIMER'S DISEASE FOR IDENTIFICATION AND PRECLINICAL VALIDATION OF THERAPEUTIC TARGETS: FROM FINE-TUNING STRATEGIES FOR VALIDATED TARGETS TO NEW VENUES FOR THERAPY

5

Bruno Vasconcelos*, Matthew Bird*, Ilie-Cosmin Stancu*, Dick Terwel**, Ilse Dewachter[†]

*Institute of Neuroscience, Catholic University of Louvain, Brussels, Belgium
**reMYND NV, Leuven, Belgium
[†]Research Group Physiology, University of Hasselt, Belgium

CHAPTER OUTLINE

Introduction 116
    Alzheimer's Disease: Clinical Features, Pathology, and Genetics Leading
    to the Formulation of the Amyloid Cascade Hypothesis 116
    Transgenic Models Recapitulating Amyloid Pathology as Preclinical
    Models to Identify Targets for Anti-Aβ Therapies...118
    Transgenic Models Recapitulating Tau-Pathology as Preclinical Models
    to Identify Targets for Anti-Tau-Directed Therapies..124
    Transgenic Mouse Models Recapitulating Prion-Like Spreading and
    Propagation of Tau-Pathology: New Venues for Therapeutic Strategies 128
    Transgenic Mouse Models Recapitulating Aβ-to-Tau Axis: New Venues
    to Identify Novel Therapeutic Targets Aiming at the Molecular and
    Cellular Mechanisms Linking Aβ and Tau...132
Conclusions..137
References 138

INTRODUCTION

Alzheimer's disease (AD) is the predominant form of dementia that affects approximately 36 million people worldwide [1,2], representing a major societal and economic problem. Currently, only symptomatic treatment is available, including acetylcholinesterase inhibitors (donepezil, rivastigmine, and galantamine) and an NMDA antagonist (memantine), alleviating the symptoms but unable to halt the disease process. In this work we review the use and continuous optimization of animal models for identification and preclinical validation of therapeutic strategies and their subsequent fine-tuning. First, we will summarize the identification, validation, and optimization of amyloid-β (Aβ)-directed targets in animal models against the general background of AD pathology, clinical features, and genetics. However, results from—failing—clinical trials raised the notion that disease-modifying therapies, once the disease has started, will require multitargeted therapies aiming at, besides Aβ, downstream targets in the amyloid cascade, most notably the protein tau, driving preclinical modeling and target identification aiming at tau. More recently, animal models allowing analysis of mechanisms involved in prion-like spreading and Aβ-induced tau-pathology have gained interest in exploration of novel therapeutic approaches, further discussed in this work.

ALZHEIMER'S DISEASE: CLINICAL FEATURES, PATHOLOGY, AND GENETICS LEADING TO THE FORMULATION OF THE AMYLOID CASCADE HYPOTHESIS

Clinical features

The earliest symptoms of the disease are episodic deficits in short-term memory, while subsequently progressive impairment of declarative and nondeclarative memory and other cognitive symptoms appear. The patients' condition progressively deteriorates over 6–8 years, during which memory loss and cognitive impairment becomes more pronounced. Patients also develop neuropsychiatric symptoms, which can vary from depression, apathy to agitation, aggression, anxiety, disinhibition, etc. [3–5]. In the final stages of the disease, patients are unable to perform activities of daily living.

Neuropathology

Neuropathologically, AD is characterized by the presence of extracellular β-amyloid plaques and neurofibrillary tangles (NFTs), composed of predominantly Aβ peptides and hyperphosphorylated tau, respectively. Aβ is an amyloidogenic peptide, which is derived by endoproteolytic processing of its precursor, the amyloid precursor protein (APP), a type I transmembrane protein. Subsequent cleavage of APP by β- and γ-secretase, results in the generation of Aβ peptides. APP is first cleaved by β-secretase to generate C99, and further cleaved by γ-secretase to generate Aβ [6]. Conversely, cleavage of APP by α-secretase occurs in the middle of the Aβ-sequence, precluding Aβ-peptide formation. NFTs are composed of aggregated hyperphosphorylated tau that accumulates in cell bodies and apical dendrites. Tau also aggregates as neuropil threads in distal dendrites, and hyperphosphorylated tau is also present in dystrophic neurites surrounding the senile plaques. The microtubule-associated protein tau, is highly enriched in neurons, particularly in microtubule dense axons. It is encoded by the microtubule-associated protein tau (MAPT) gene, and as its name suggests, binds to and stabilizes microtubules. NFTs in AD comprise all six tau-isoforms, both with three or four repeat (3R and 4R) domains and zero, one, or two N-terminal inserts [7], whereas in different tauopathies either 3R and/or 4R tau-forms can accumulate. Spatio–temporal progression of the respective Aβ- and tau-pathology has been character-ized in detail and is staged using prototypical Braak and Thal staging [8,9]. In contrast to amyloid pa-thology, tau-pathology is known to progress in a very characteristic pattern, which correlates strongly with the appearance of the symptoms and disease progression, thereby allowing its use as a diagnostic tool. Six Braak stages are defined and range from no NFTs, to NFTs present in entorhinal cortex (EC) and closely related areas (Braak stage I/II), to NFTs abundantly present in hippocampus and amygdala, slightly extending in association cortex (Braak stage III/IV). Finally, in stages V/VI, NFTs are detected throughout the neocortex and ultimately in primary motor and sensory areas [10,11]. Although the ap-pearance of amyloid pathology correlates less with disease symptoms, its progression has been staged as "Thal" stages. Aβ deposits are first exclusively present in the neocortex and in allocortical brain regions (Thal I and II), subsequently in diencephalic nuclei, the striatum, and the cholinergic nuclei of the basal forebrain (Thal III). In phase IV, several brain stem nuclei become affected, and finally (Thal V) cerebellar Aβ deposits appear [9]. Besides the characteristic signature of amyloid plaques and NFTs, AD brains are pathologically characterized by brain atrophy and severe neuronal and synaptic loss, in addition to reactive astro- and microgliosis [12,13]. The most prominently affected brain struc-ture in AD is the hippocampus, followed by degeneration of the association cortices and subcortical structures, including the amygdala and nucleus basalis of Meynert.

Genetics

Genetically, AD patients fall into two groups based on the presence or absence of mutations autosomal dominantly linked to AD, denoted early-onset familial (EOFAD), and the far more common sporadic late-onset (LOAD, ≥99% of cases) AD cases, respectively [14]. Although the pathology of the disease appears largely similar in both groups, the age of onset is not, with disease symptoms usually affect-ing EOFAD patients well before age 65, and much later in the LOAD group, at an average age of 70. Mutations known to cause AD within the EOFAD group are clustered near the (α-, β-, and γ-)secretase cleavage sites of APP (see for an updated list: http://www.molgen.ua.ac.be/ADMutations) [14–17]. In addition, APP duplication is also causally linked to EOFAD. Conversely, a protective A673T mutation in APP, located near the β-secretase cleavage site has been identified [18], further underscoring the role of APP and its processing in the disease process. The majority of mutations linked to EOFAD are actu-ally found in either presenilin (PS) 1 or 2. Their identification as the catalytic subunit of the γ-secretase complex, which responsible for the generation of the C-terminal cleavage of Aβ, further implicated alterations in the biogenesis of APP, which the precursor of Aβ, as the culprit [19–24]. For the LOAD cases, APOE has been identified as the most important risk factor, with ApoE4 conferring increased risk and associated with accumulation of cerebral Aβ [14]. Recently, rare mutations in the prodomain of ADAM10, identified as the major α-secretase in neurons in vivo [25], have been identified to cose-gregate with LOAD and were shown to decrease α-secretase with a concomitant increase in Aβ [26]. The analysis of risk factors for LOAD, has recently been complemented by genome-wide association approaches, which have indicated several additional AD susceptibility loci, pointing to potentially implicated pathways, which remain to be mechanistically analyzed [14,27]. No mutations have been identified in MAPT, the gene encoding the tau-protein, in AD. However, several mutations in MAPT have been autosomal dominantly linked to frontotemporal dementia (FTD) (see also Chapter 8) with Parkinsonism linked to chromosome 17 and related tauopathies [28–31], indicating the causal role of tau-alterations in neurodegenerative processes.

The amyloid cascade hypothesis

The identification of fully penetrant autosomal dominant mutations (see for an updated list: http://www. molgen.ua.ac.be/ADMutations) in APP and in PS1 and 2, in combination with the clinical symptoms and pathological features in EOFAD patients, led to the formulation of the amyloid cascade hypothesis [13,32]. This hypothesis states that accumulation of Aβ peptides causes the initiation of a cascade of events responsible for the development of the clinical symptoms and pathological features of AD. Hence, accumulation of amyloid peptides induces a cascade that leads to tau-hyperphosphorylation, aggregation, and astro- and microgliosis, eventually leading to brain atrophy, neuronal, and synapse loss [32,33]. This process is associated with the appearance of the clinical symptoms in a characteristic way. The amyloid cascade hypothesis has since its formulation been subject to some criticism, but has ever since remained the predominant hypothesis, and major framework for development and identification of disease-modifying therapies. Some adjustments in interpretation have been performed, including increased emphasis on soluble oligomeric Aβ forms [32,33] and the importance of downstream targets, particularly tau [34]. The amyloid cascade hypothesis, originally based on combined genetic, pathological, and clinical data, is further supported by detailed biomarker analysis in patients [35] and data from preclinical in vitro and in vivo models [36] (discussed further). However, and importantly, its final proof, that is, a successful clinical trial decreasing Aβ peptide accumulation, is eagerly awaited. Some cautious optimism has recently emerged with regard to some Aβ-directed therapies with a trend toward positive outcomes, not yet reaching overall significance (discussed further). Generally, however, the notion is held that Aβ-targeting therapies need to start early in the disease process, and that for later stages combined therapies need to be considered [34].

TRANSGENIC MODELS RECAPITULATING AMYLOID PATHOLOGY AS PRECLINICAL MODELS TO IDENTIFY TARGETS FOR ANTI-Aβ THERAPIES

APP and APP/PS1 transgenic models

The central role of Aβ formulated in the amyloid cascade hypothesis for the pathogenesis of AD, instigated the generation and characterization of a plethora of AD mouse models, which are briefly summarized further (in a general, but clearly nonexhaustive way). Transgenic mice overexpressing EOFAD APP mutations, have been reported to recapitulate several features reminiscent of AD, including learning and memory disturbances [37] and neuropsychiatric alterations, such as increased aggression and anxiety [38–52]. Learning and memory deficits correlated with defects in synaptic plasticity, measured by long-term potentiation (LTP) [43,53,54]. Furthermore, seizures and premature death were often demonstrated in several APP-overexpressing models [43,52,55]. Importantly, the brains of EOFAD mutant APP-overexpressing mice, display increased intra-and extracellular Aβ; progressively develop amyloid deposition in the vasculature of the brain and in the parenchyma, associated with astro- and microgliosis, and often with synaptic loss; and display either no or limited degree of neuronal loss. Senile plaques, with a dense core and surrounded by dystrophic neurites containing hyperphosphorylated tau, which is associated with reactive gliosis, are detected in addition to more diffuse amyloid deposits. The age of onset of amyloid plaque development depends on the level of APP overexpression and the mutations used, determining Aβ-peptide concentration and $A\beta_{42}/A\beta_{40}$ ratio. Relative development of angiopathy versus parenchymal plaque deposition was also demonstrated to depend on the relative ratios of more and less amyloidogenic amyloid peptides, with increasing vascular deposition correlating with more soluble Aβ forms [49]. Besides extracellular forms of Aβ, either soluble or aggregated in plaques or vascular deposits, intracellular Aβ forms exist. Within the amyloidogenic processing pathway, Aβ can be generated in intracellular compartments. But Aβ can also accumulate intracellularly in a variety of intracellular organelles, including late endosomes and multivesicular bodies [56]. Aβ peptides of varying length are generated, as well as several posttranslationally modified forms of Aβ have been demonstrated in animal models (ranging from pyro-Glu to phosphorylated or N-terminally truncated) [56–63]. Recently, APP Swedish/Iberian knockin mice have been reported, which display increased concentrations of various Aβ peptides, age-dependent development of Aβ-pathology, neuroinflammation, synaptic loss, and memory impairment, indicating that these features are not dependent on APP overexpression per se, but related to the EOFAD mutation [64]. Particularly, the faithful recapitulation of amyloid plaque pathology provided the basis for the use of these models for target identification for anti-Aβ-directed strategies (Fig. 5.1).

Secretases and their validation as therapeutic targets

The generated models recapitulating amyloid pathology, Aβ-related cognitive deficits, and synaptic dysfunction have provided crucial tools for the in vivo validation of the different secretases in neurons, following their biochemical or in vitro identification, and for their validation and continuous fine-tuning as therapeutic targets in preclinical models. Initially, mutant APP-overexpressing mice were used to analyze the effects of EOFAD PS1 and PS2 mutations in vivo [67,71]. Double-transgenic APP/mutant PS1 (and 2) mice displayed an increased ratio of $A\beta_{42}/A\beta_{40}$, resulting in dramatically aggravated and accelerated amyloid pathology in vivo, in addition to aggravated phenotypic changes [38,39,67,68]. This further instigated the identification of presenilins as essential component of the γ-secretase complex (further comprising aph1, pen2 and nicastrin), by demonstrating inhibited Aβ production and accumulation of C-terminal fragments in PS1 deficient mice, in combination with the demonstration of its catalytic activity [20,21,72]. A panoply of models coexpressing APP and PS1 has been generated to create models with robust and early-amyloid pathology and associated features, including transgenic mice harboring five EOFAD mutations, that is, 5xFAD mice [38,39,65,67,68]. In vitro and in vivo studies led to the formulation of some caveats for using γ-secretase inhibitors, based on the fact that many different substrates of γ-secretase exist (importantly comprising Notch) [73] and the potential detrimental effects of accumulation of C-terminal fragments [53,74]. The latter led to strategies to circumvent this problem by the use of γ-secretase modulators, shifting Aβ-production toward shorter, less-amyloidogenic forms of Aβ. More recently, still more fine-tuned γ-secretase–directed therapies are being developed, which include the targeting of different forms of the γ-secretase complex [75], or the selective targeting of their carboxypeptidase-like and endopeptidase activities, providing attractive strategies [22,24]. In the same vein, following in vitro identification of ADAM10 as a potential α-secretase, ADAM10 was validated and identified in APP transgenic mice as the major α-secretase–exerting enzyme in neurons, thereby presenting as a potential therapeutic target [25]. Indeed, neuronal overexpression of the α-secretase (ADAM10) in APP transgenic mice, increased APPsα production with a concomitant reduction in Aβ concentration, and ameliorated behavioral deficits with a concomitant rescue of LTP impairment [25].

Similarly, following identification of BACE as a β-secretase–cleaving enzyme in vitro [76,77], overexpression of BACE1 in APP transgenic mice has been shown to increase the rate of Aβ plaque deposition in the parenchyma, while conversely, ablation of the β-secretase in APP mutant mice [78,79] resulted in reduced Aβ production and plaque deposition, alleviating memory deficits. These findings positioned BACE as an attractive therapeutic target. However, the identification of a panoply of β-secretase substrates, and identification of its physiological role in BACE1-deficient mice raised

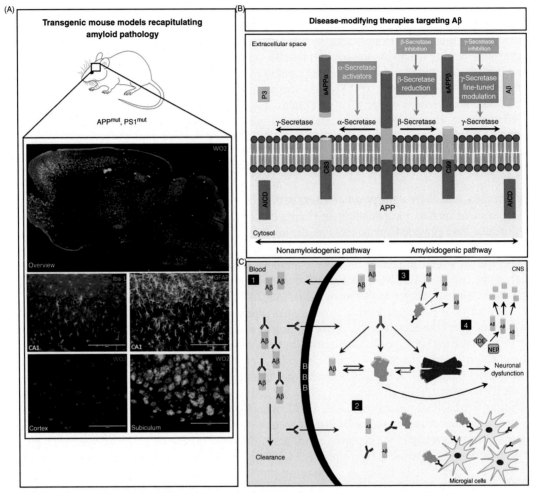

FIGURE 5.1 Transgenic Mouse Models Recapitulating Amyloid Pathology and Their Use to Identify Anti-Aβ-Directed Therapeutic Strategies

(A) Overview of a sagittal section of a 9-month-old transgenic mouse—APP$^{KM670/671NL; I716V; V717I}$, PS1$^{M146L; L286V}$—[65,66] stained with specific anti-Aβ-antibody *(WO2)* is presented in the upper left panel. Middle panels present microglial *(Iba-1)* (left) and astroglial *(GFAP)* (right) stainings, indicating microglial and astroglial inflammation in this model. The lower panels present detailed stainings of Aβ *(WO2)* in the cortex and subiculum area demonstrating high amyloid load in this transgenic mouse model at 9 months of age (scale bars: CA1 subiculum, 200 μm; cortex, 400 μm). A variety of models recapitulating amyloid pathology have been generated [38–40,42–52,55,65,67,68]. (B) Schematic presentation of disease-modifying treatments targeting Aβ processing (based on Ref. [69]). (C) Schematic presentation of disease-modifying treatments aiming at clearance and degradation of different forms of Aβ, ranging from monomers to fibrils (based on Ref. [70]). Immunotherapy strategy can trigger multiple responses leading to the clearance of Aβ. Antibodies can promote the peripheral clearance of circulating Aβ *(1)*, promote microglial cell–mediated clearance *(2)*, or directly block the formation of the toxic Aβ species and desegregate Aβ aggregates *(3)*. On the other hand, Aβ-degrading enzymes, like neprilysin *(NEP)* and insulin-degrading enzyme *(IDE)* can also promote the clearance of Aβ peptides (4). *AICD,* Amyloid precursor protein intracellular domain; *APP,* amyloid precursor protein; *CNS,* central nervous system; *sAPP,* soluble amyloid precursor protein.

similar concerns for tolerance and safety. For BACE1 this includes, neuregulin-dependent myelination, axon guidance, neurogenesis, muscle spindle formation, and neuronal network functions [77,80–82]. Furthermore, schizophrenia endophenotypes have been identified in BACE1-deficient mice [83]. Interestingly, in this respect, a 50% reduction of BACE1, in heterozygous BACE1 knockout mice was demonstrated to be sufficient to decrease amyloid plaque formation and cognitive deficits in APP mice [79]. Interestingly, pharmacological reduction of BACE activity has been shown to be effective in APP mice [84] without major side effects, opening venues for clinical trials. Indeed several clinical trials aiming at BACE inhibition are ongoing, at various stages (Phase I–III), and β-secretase reduction seems to be quite well tolerated. The outcomes of the clinical trials are eagerly awaited, while the time point of starting the treatment will need to be considered, and critical evaluation of target engagement and the clinical trial setup will need to be evaluated in detail. A detailed description of clinical trials targeting the different secretases is discussed in Refs. [85–87] and beyond the scope of this chapter. Preclinical animal models have been crucial for the validation of the identity of secretases in vivo and their validation as therapeutic targets. The aforementioned and related models continue to be crucial to analyze and validate more fine-tuned treatment strategies, and to develop and evaluate compounds with minimal side effects associated with secretase modulation to increase efficiency and limit side effects in clinical trials.

Clearance of Aβ by immunotherapy

Increasing Aβ catabolism or clearance from the brain represents the complementary strategy to inhibit production and lower Aβ concentrations. Aβ peptides can be cleared by enzymatic degradation, by efflux via brain drainage pathways, by active clearance by glial cells, or immunotherapy directed toward the removal of accumulating amyloid peptides [64]. Different enzymes capable of degrading Aβ have been identified, including neprilysin [88], insulin-degrading enzyme (IDE), and angiotensin-converting enzyme. The role of IDE on Aβ degradation was further confirmed in crosses of APP mice with IDE-deficient mice [89] and IDE-overexpressing mice [90], resulting in increased Aβ concentrations and in rescue of amyloid pathology in APP transgenic mice, respectively. Similarly, neprilysin-deficient mice displayed impaired Aβ degradation [88,91], while viral overexpression of neprilysin resulted in increased Aβ degradation, as reflected in an ameliorated amyloid pathology [92–94]. Modulation of the endolysosomal–autophagosomal processes has also been proposed, not only as a disease-contributing mechanism, but also as a potential therapeutic target [56,95–98]. Recently, PS2 and FAD mutant PS1 were shown to selectively cleave late endosomal/lysosomal–localized substrates, thereby generating a prominent pool of intracellular Aβ that contains longer Aβ [99]. Accelerating lysosomal degradation of APP by, for instance, TFEB can reduce Aβ accumulation [100] and may open venues for therapy [100], which is currently a new emerging theme. Notably, a very attractive approach for Aβ removal is provided by Aβ-directed immunotherapy. The proof of concept for Aβ-immunotherapy was provided in a seminal paper of Schenk et al., demonstrating the striking prevention of amyloid burden, neuritic dystrophy, and associated gliosis following immunization in mutant APP transgenic mice, by intraperitoneal immunization with aggregated Aβ [101]. Treatment of older APP mice also markedly reduced the extent and progression of these AD-like neuropathologies. Subsequently, the active immunization approach was complemented with "passive" immunization, by direct administration of anti-Aβ-antibodies, also capable of clearing amyloid pathology [102,103]. A variety of follow-up papers have further corroborated these initial findings, using diverse immunization approaches, demonstrating that immunotherapy efficiently decreased amyloid plaque load and Aβ-associated features in APP and

APP/PS1 mice. It was, furthermore, demonstrated that depending on the peptide or antibody used for immunization, different forms of Aβ could be cleared [47,104,105], ranging from soluble oligomeric forms to plaque-associated Aβ. Aβ-directed therapy decreased Aβ, concomitantly alleviated gliosis, cognitive impairment, and early—but not late—forms of tau-pathology [106] and synaptic deficits [107]. Clearance of soluble Aβ oligomeric forms, rather than amyloid plaques, were thereby associated with improved cognitive performance and synaptic plasticity deficits [47]. While the effects of active and passive Aβ-immunization on Aβ removal is well demonstrated and accepted, the mechanisms involved are not fully resolved and may be a combination of several processes. These include glia-mediated clearance, antibody binding and blocking the formation of the toxic Aβ species, or peripheral Aβ clearance [peripheral sink hypothesis not necessitating crossing of the blood–brain barrier (BBB) by the anti-Aβ-antibodies] (reviewed in Ref. [85]). The success of immunization strategies in preclinical models subsequently led to the first active immunization trial (AN1792), using a preaggregated synthetic Aβ$_{1-42}$ peptide. The trial was, however, prematurely terminated, owing to 6% of the patients developing meningoencephalitis, who received the vaccine [108]. Although unsuccessful in significantly halting the cognitive decline and the disease process, some patients displayed indications of mildly slowed cognitive decline, correlating to increased titers of Aβ-antibodies [85]. Most importantly, this trial has yielded the crucial information that Aβ-directed immunotherapy can effectively, but partially, clear Aβ plaques in patients, which is a remarkable and interesting finding. It must be noted, however, that NFTs were not cleared and unfortunately no significant overall rescue of cognitive decline was demonstrated. Different arguments have been raised to explain these results, including targeting of nontoxic Aβ forms and delayed administration in the disease process [108], when the pathogenetic cascade initiated might be self-propagating. However, this study provided crucial information for further fine-tuning of immunization therapies—including decreasing risk for meningoencephalitis and targeting toxic/causal Aβ-forms—and follow-up in preclinical and clinical trials, reviewed in detail elsewhere [85]. Two recent anti-Aβ-directed phase III clinical trials using passive immunization unfortunately failed to show significant clinical benefit [70,85], although some minor effects in subpopulations were noted. Again late administration needs to be taken into account. Some interesting results from a different immunization trial using aducanumab, a fully human monoclonal antibody against Aβ, have been presented (http://www.alzforum.org/therapeutics/aducanumab). The 3- and 10-mg/kg aducanumab dosing, resulted in decreased plaque load associated with significant decreased cognitive decline, albeit in a small population, requiring further confirmation in a larger cohort. A detailed discussion of clinical trials based on the promising results in preclinical models are beyond the scope of this chapter and an excellent review has been published on this topic [70]. These results raise hope, although they do not provide the eagerly awaited proof of the amyloid cascade hypothesis in clinical trials, yet.

Identification of the exact toxic form(s) of Aβ and aggregation inhibitors

The identification of the exact toxic form(s) of Aβ is central in the identification of therapeutic strategies and has attracted much attention to optimize current strategies. The nature of the Aβ species is far more complex than alluded to earlier, considering Aβ-peptides can vary in length; can exist as intra- and extracellular forms, and in multiple physical interconverting forms ranging from aggregated or soluble; and in monomeric, oligomeric, and fibrillar forms. Several oligomeric forms have been reported to exhibit deregulating effects on synaptic plasticity, including the dodecameric forms and oligomeric forms (particularly trimeric) [12,109]. Furthermore, the length of the Aβ peptide also varies, not only by variations at the C-terminal truncation site, but also by variations in the N-terminal, as truncated forms have been identified [6,59,62]. Furthermore, Aβ-peptides undergo posttranslational modifications, including phosphorylation and N-terminal modifications [58,61,110]. Particularly, the N-terminally modified, pyroglutamic acid Aβ has been characterized in detail, as a particularly toxic and aggregating form in vitro and in vivo [58]. It will be important to identify the exact toxic forms, which need to be eliminated or decreased in clinical trials, and also to understand the conversion between toxic and nontoxic forms of Aβ. This is not only important in the context of immunotherapy, but also in the context of several therapeutic approaches aiming at interfering with Aβ-aggregation. The latter requires a clear-cut identification of the toxic form(s) of Aβ and knowledge and control of their interconversion.

Combination of different anti-Aβ strategies and other approaches

Combination of different anti-Aβ therapies may turn out more efficacious and present less side effects. In this respect, combination of a BACE inhibitor and immunotherapy has been analyzed preclinically in APP$_{V717I}$ transgenic mice [111]. This model is suited for testing efficacy of BACE inhibitors, as the V717I mutation leaves the BACE cleavage site intact. BACE inhibition alone reduced Aβ42 and Aβ40 levels in the brain and slightly reduced plaque burden. Immunotherapy alone reduced plaque burden and brain Aβ42, but not Aβ40, levels. The combination was more effective, especially at the lower dosage of BACE inhibitor used, and may allow for a better therapy or usage of safer dosages. Other approaches aiming at decreasing amyloid load, which have been identified and validated in APP transgenic mice, include GPR3, PKC modulators, Aβ aggregation inhibitors, etc. A more recent, and novel approach, to increase Aβ clearance in AD mouse models used scanning ultrasound (SUS). SUS was shown to be associated with an extensive internalization of Aβ into the lysosomes of activated microglia in mouse brains, with no concomitant increase observed in the number of microglia in APP$_{KM670/671NL}$ mice. SUS was also associated with strong improvements in learning and memory in this model [112]. As an alternative strategy, gene therapy is being explored to modulate Aβ production and clearance. Disease-modifying effects of recombinant adeno-associated viral vector (rAAV) gene delivery in transgenic AD models have been obtained [113]. Furthermore, detailed analysis in preclinical models—recapitulating amyloid pathology—of the pathways that have been indicated by risk factors identified in genome-wide association studies may reveal novel or additional pathways for combinational therapies.

Limitations of APP and APP/PS1 preclinical models

APP and APP/PS1 transgenic models quite faithfully recapitulate amyloid plaque pathology, but do not faithfully develop all aspects of AD. In this respect, the early stages of the disease process should be studied in more detail in AD patients, and faithfully recapitulated in AD models. This may include different early aspects of the disease process, such as early defects in endolysosomal dysfunction [56,95,98]. Most importantly, APP and APP/PS1 models display only limited neurodegeneration and no NFTs were detected in any of these models. Only limited effects on tau-phosphorylation have been reported, most robustly reflected in AT8-positive dystrophic neurites surrounding amyloid plaques (reviewed in Ref. [36]). Despite relatively successful results in preclinical AD models, no successful clinical trials have yet emerged. Successful trials may require models with a more complete recapitulation of the disease, with combined amyloid and tau-pathology. Furthermore, these results raised the notion that a better identification of the toxic form of Aβ is required. In addition to effective therapies at the moment for diagnosis, combinational therapeutic strategies need to be considered, aiming at downstream targets of Aβ, most particularly tau, generally considered as the executor of the disease process, similar as in related tauopathies. Hence, to generate more complete models and to identify targets for combined therapies, APP and APP/PS1 transgenic mice have been complemented with tau transgenic mice to mimic tau-pathology and tau-related aspects.

TRANSGENIC MODELS RECAPITULATING TAU-PATHOLOGY AS PRECLINICAL MODELS TO IDENTIFY TARGETS FOR ANTI-TAU-DIRECTED THERAPIES

Tau-transgenic models recapitulating tau-pathology

Initially transgenic mice overexpressing wild type (WT) tau have been generated for modeling tau-pathology in the context of AD, in view of the lack of MAPT mutations associated with AD. However, initial models overexpressing WT human tau, either 3R or 4R tau, did not develop NFTs [114–117], but instead presented with axonopathy and/or somatodendritic tau-mislocalization and hyperphosphorylation. Limited NFT pathology was finally detected in very old 3R WT tau transgenic mice [118,119]. An alternative approach overexpressing all six isoforms of human WT tau resulted in successful development of NFTs associated with a pronounced neurodegenerative phenotype reflected in brain atrophy [120,121]. To facilitate the development of NFTs, transgenic mice overexpressing clinical mutants of tau, autosomal dominantly linked to tauopathies, have been generated [115,122–133]. This resulted in models that develop clear-cut NFTs, which stain with ThioS, Congo Red, and silver, and can enable identification of filamentous tau by electron microscopy. These models displayed different degrees of hyperphosphorylated and sarkosyl-insoluble tau, different degrees of gliosis, synaptic dysfunction, and synaptic and neuronal loss or brain atrophy. Depending on the expression level and the expression pattern of the transgene, tau transgenic mice presented with behavioral abnormalities ranging from motoric phenotypes (clasping, motoric deficits observed in rotarod, inverted grid hanging, etc.) to cognitive phenotypes and/or a neurodegenerative phenotype, leading to premature death. The behavioral phenotype, furthermore, correlated with the brain regions affected by pathological tau, similarly as observed in different tauopathies. Inducible/repressible tau–expressing mice were used to analyze reversibility of tau-related phenotypes. Mice expressing high levels of repressible human 4R0N tauP301L variant developed progressive age-related NFTs, neuronal loss, brain atrophy, and cognitive decline [129]. Suppression of tau expression rescued memory function and stabilized neuron numbers, while not leading to NFT clearance, indicating dissociation between NFT formation and neuronal loss and cognitive decline in this model and indicating reversibility of certain aspects of tau-dependent toxicity [129]. Furthermore, NFT-containing neurons have been demonstrated, using in vivo two-photon imaging, to be functionally integrated in neuronal networks in vivo [134]. Along the same vein, transgenic mice overexpressing either full-length proaggregant tau or proaggregant tau microtubule–binding domain, further support a role for early aggregating forms of tau, but not necessarily NFTs as causal forms of cognitive dysfunction [135,136]. Hence, although the process of tau-aggregation seems to be associated with toxicity, the exact toxic form(s) of tau and tau* still need to be identified in detail, similar to amyloid peptides.

Tau transgenic mice were subsequently used to generate more complete models of AD with combined amyloid and tau-pathology, and to identify and validate targets for tau-directed therapies [137]. The executive role of tau in neurodegenerative processes and its role as a therapeutic target have become generally accepted. This role is supported by several arguments. First, NFTs are a diagnostic hallmark of AD, and the progression of tau-pathology correlates very strongly with the disease progression, allowing its use as a diagnostic criterion, which is less strong for amyloid pathology [138–140]. In addition, tau-pathology is a characteristic feature of different neurodegenerative disorders, grouped as tauopathies. But most importantly, the identification of mutations in MAPT, causally associated with the development of several tauopathies, demonstrated that tau dysfunction is sufficient to induce a neurodegenerative process in tauopathies. The therapeutic role of tau gained further interest by the demonstration that Aβ-directed immunization trials, resulting in clearance of amyloid pathology but not NFTs, which did not result in significant cognitive benefits. Although several arguments need to be taken into account, this lack of success correlated with the lack of significant clearance of NFTs [141,142] (Fig. 5.2).

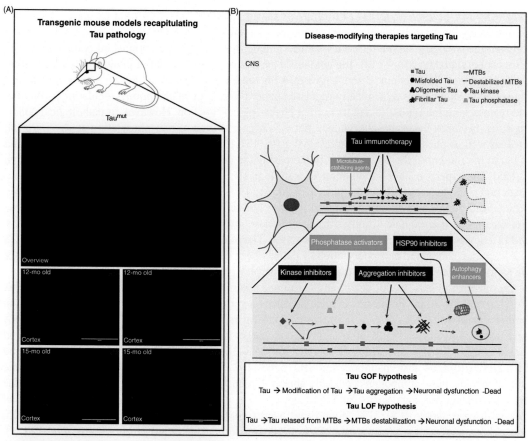

FIGURE 5.2 Transgenic Mouse Models Recapitulating Tau-Pathology and Their Use to Identify Anti-Tau-Directed Therapeutic Strategies

(A) Overview of a sagittal section of a 12-month-old tau transgenic mouse—tau^{P301S}—[66,132] stained with anti-P-tau antibody (P-S202/T205; AT8) is presented (upper panel). Middle panels present stainings of cortex of 12-month-old mice stained with AT8 antibody and lower panels present stainings of mice of 15 months of age (scale bars: left panels, 400 µm; right panels, 200 µm). Tau-pathology and phenotypic changes appear at ±12 months in this model. This is later than initially reported, and could be due to genetic drift, strain background, or housing conditions. Low tau-pathology is detectable in these mice at 12 months of age, and increases over time. A variety of models recapitulating Tau pathology have been generated [115,118–127,129–133,143]. (B) Schematic presentation of disease-modifying therapies targeting tau (based on Ref. [137]). *GOF*, Gain of function; *LOF*, loss of function; *MTBs*, microtubules.

Targeting tau-phosphorylation and other posttranslational modifications of tau

Of the many posttranslational modifications of tau, tau-phosphorylation has—not surprisingly—received the strongest attention, in view of the finding that NFTs in AD brains are composed of hyperphosphorylated tau [144–150]. Tau contains a particularly high content of putative phosphorylation sites (45 serines, 35 threonines, and 4 tyrosines for the longest 441 amino acid human tau-isoform). Phosphorylation of tau is dynamically regulated under physiological conditions, regulating its binding and release from microtubules and its normal physiological function. However, in pathological conditions, tau becomes hyperphosphorylated, that is, a higher degree of phosphorylation at physiological sites, as well as phosphorylation at additional sites [151,152]. Phosphorylation decreases the binding of tau to microtubules, thereby potentially increasing the pool of soluble tau available for aggregation [7,153]. The hyperphosphorylation of tau in pathological conditions pointed toward deregulation of kinases and phosphatases in the pathogenesis of AD [154]. A number of kinases and phosphatases have, hence, been pursued, including glycogen synthase kinase-3 (GSK-3) [155], cyclin-dependent kinase 5 (CDK5), and microtubule affinity–regulating kinase 4 (MARK4) [156–159] as kinases, and PP2A and PP2B, also called calcineurin, as phosphatases. The role of GSK-3β, which colocalizes with NFTs in AD, in the pathogenic phosphorylation of tau has been particularly well studied. Overexpression of GSK-3β in adult WT mice was able to induce hyperphosphorylation and somatodendritic localization of tau, reactive gliosis, neural death, and learning impairments [116,160,161]. Similarly, the overexpression of GSK-3β in tau transgenic mice produced a very robust induction of NFTs in the forebrain, further identifying GSK-3β as a tau protein kinase in neurons in vivo [155]. A number of groups have accordingly pursued GSK-3β as a therapeutic target in the disease. The GSK-3β inhibitors tested included lithium chloride (LiCl, not specific for GSK-3β), SRN-003-556 (also inhibits CDK5 and ERK2), CHIR98104, alsterpaullone, and SB216763 [162–170]. When administered in AD transgenic animal models, these compounds generally appear to reduce levels of insoluble tau, and ameliorate the disease burden to varying degrees, but fail to reduce levels of NFTs. CDK5 also appears to have a central role in disease. Noble et al. [171] found that overexpression of the CDK5 regulatory subunit, p25, was able to increase tau-phosphorylation and induce neurodegeneration in one mouse model [171]. In another model the massive neurodegeneration could be demonstrated as well, but in the absence of increased tau-phosphorylation [172]. Although modulating tau-phosphorylation seems as an attractive therapeutic target, the panoply of physiologically crucial signaling pathways targeted by these different kinases and phosphatases, provides a strong drawback for therapeutic strategies because of side effects. Furthermore, it must be considered that tau-phosphorylation is implicated in the regulation of the physiological functions of tau [7,152]. Hence very fine-tuned and balanced dosing would be required, which might be brain region specific and even patient specific, potentially hampering their use as therapeutic targets. Besides tau-phosphorylation, other posttranslational modifications, including glycosylation, O-GlcNAcylation, nitration, and acetylation, have been reported and affect pathological changes in tau [173–180]. In addition, tau is subject to ubiquitination, truncation, prolyl isomerization, association with heparin sulfate proteoglycans, glycation, and modification by advanced glycation end products [154], presenting potential modulators of tau and tau-pathology.

Microtubule-stabilizing agents

The pathogenetic role of tau may be due to either a toxic gain of function or a loss of function [137]. AD and related tauopathies are characterized by the presence of aggregates of hyperphosphorylated tau. Hyperphosphorylation of tau induces a reduction in the ability of tau to bind microtubules.

Furthermore, aggregation of tau may deplete tau from the microtubules and several MAPT mutations linked to tauopathies cause an imbalance between 3R and 4R tau forms, with different binding capacity to microtubules. The physiological role of tau in microtubule stabilization is regulated by a dynamic balance between phosphorylation and dephosphorylation and the availability of 3R and 4R tau-isoforms. Together this has fueled therapeutic strategies using microtubule-stabilizing drugs to compensate for a loss of function [137]. Paclitaxel, a microtubule-stabilizing drug used in cancer therapy, rescued fast axonal transport deficits, in addition to motor deficits observed in transgenic mice overexpressing WT tau [118,181]. Intriguingly, treatment with a different microtubule-stabilizing agent, epothilone D, in different tau transgenic models (rTg4510 and PS19) overexpressing mutant tau, rescued tau-pathology and cognitive function, although the mechanism remains to be understood in detail [182–184]. Also NAP, a microtubule-stabilizing agent, has been evaluated in preclinical models. NAP decreased Aβ levels and tau-phosphorylation, although pleiotropic actions of NAP are probably involved in these effects [185–187]. These findings have subsequently translated in clinical trials (http://clinicaltrials. gov/ct2/show/NCT01492374 and http://www.clinicaltrials.gov/ct2/show/NCT01056965), for which the results are awaited.

Targeting intracellular tau-degradation

Cells can clear damaged or abnormally folded cytosolic proteins by two means: by the ubiquitin proteasome system (UPS), which targets proteins for degradation by ubiquitination; or by autophagy, linked to lysosomal degradation [188]. Considering the evidence that suggests both of these pathways can degrade hyperphosphorylated tau, the prospect of enhancing these activities in AD to clear pathological forms of tau is an attractive one [97,189,190]. One such approach to promote proteolysis has been not to focus on the proteolytic machinery itself, but rather the role of chaperones in stabilizing and protecting hyperphosphorylated tau from degradation by the proteasome [143,191,192]. This approach has been successfully demonstrated in in vitro and in vivo models with inhibitors of the chaperone Hsp90, which resulted in reduced levels of hyperphosphorylated tau at a number of pathologically relevant sites [191–193]. Noteworthy, it was recently demonstrated that insoluble tau impairs the UPS [194]. Administration of an agent that activates cAMP–protein kinase A (PKA) signaling led to attenuation of proteasome dysfunction, resulting in decreased tau-aggregation and improved cognitive performance in tau transgenic mice [194]. The UPS seems to be a potentially efficient way of dealing with monomeric tau, but is unlikely to clear higher order structures, such as oligomers and fibrils. In this respect, the autophagy system is a much more attractive system with the capacity to remove much larger aggregates of tau, such as those in AD [96,195]. Rapamycin, or its analog, temsirolimus, is one such compound that is known to stimulate macroautophagy, and has been investigated in this context [196]. Herein, temsirolimus was able to reduce levels of hyperphosphorylated tau and to partially rescue spatial learning defects in a tau transgenic model of the disease.

Targeting tau by immunization therapy

Inspired by the strong effects of immunotherapy on Aβ clearance in preclinical models, tau-directed immunization has been addressed in tau transgenic mice [70,197,198]. A seminal paper demonstrated that active immunization, using a phospho-tau peptide as an antigen [containing the paired helical filament (PHF)-1 epitope pSer396/pSer404] [199] in presymptomatic mice, resulted in reduction of tau-pathology and a concomitant improvement in sensorimotor tasks. Active immunization resulting in clearance of tau-pathology was further confirmed in a variety of tau transgenic mice using different immunogens, including PHF and phospho-tau peptides [200–202]. However, active immunization using either full-length recombinant tau in nontransgenic mice, or using different phospho-peptides in transgenic mice and WT mice indicated the possibility of adverse effects of immunization encompassing neuroinflammation and cognitive deficits [203,204]. These findings provide a caveat for active immunization against a physiologically relevant protein. In addition, "prion-like" properties of tau—discussed in detail further—need to be taken into account. Hence, specific targeting of pathological forms is required for active tau-directed immunization strategies lead to promising results in preclinical models (as noted earlier). The potential caveats of active immunization have fueled strategies of passive immunization therapy, as a potentially more safe therapeutic strategy. Administration of different antibodies targeting different pathological tau-epitopes and conformers in different tau transgenic models diminished tau-pathology and associated phenotypic traits (motor deficits, cognition, or premature death) in vivo [205–208]. Mechanistically, the initial demonstration of effective clearance of tau-pathology by (active and passive) immunization was rather surprising in view of the fact that tau is an intraneuronal protein, and hence antibodies are not only required to cross the BBB, but also need to clear an intraneuronal cytosolic protein. Subsequent findings indicated successful BBB crossing of anti-tau-antibodies and antibody translocation by low-affinity Fc receptors followed by binding of pathological tau in the endosomal–lysosomal system [209], providing a potential mechanism of action. Alternatively, antibodies have also been proposed to target propagation of tau-pathology by interfering with a tau-seeding process, encompassing an extracellular form of tau [208], which is discussed further. While these are plausible explanations, the exact mechanism underlying tau-immunization needs to be resolved in detail. Research to develop more efficient and safe active and passive vaccination strategies is ongoing and receives ample attention in preclinical models recapitulating tau-pathology and associated features. Promising active and passive immunotherapy strategies continue to move rapidly from mouse models to the clinic (reviewed in Ref. [198]). These include the AC Immune by AG, KU Leuven, Fred van Leuven, Janssen Pharmaceuticals [210], and AADvac1 by Axon Neuroscience and Michal Novak [211].

TRANSGENIC MOUSE MODELS RECAPITULATING PRION-LIKE SPREADING AND PROPAGATION OF TAU-PATHOLOGY: NEW VENUES FOR THERAPEUTIC STRATEGIES

A variety of neurodegenerative disorders is characterized by the progressive accumulation and aggregation of misfolded proteins (see also Chapter 1). As stated earlier, tau-pathology displays a very characteristic spatio–temporal pattern, strongly correlating with disease progression. The characteristic spreading and propagation of misfolding of proteins is reminiscent for prion disease, and prion-like properties have been demonstrated for different proteins involved in neurodegenerative disorders [212]. The prion paradigm indicated an infectious proteinaceous agent, prion, as the culprit of the pathogenetic neurodegenerative process in prion disease [213]. Propagation of seeded misfolding of the endogenous protein is the cause of disease progression. Interestingly, the latter process has been reiterated in the context of protein misfolding in neurodegenerative disorders. As infectivity is not demonstrated in these cases, the term has been taken to mean proteinaceous nucleating particles or the term prion-like is used to exclude infectivity. Setting aside the semantic debate about the use of the term "prion" or "prion-like," the mechanisms of cell-to-cell transmission of misfolded proteins and propagation of protein misfolding have important therapeutic implications for many neurodegenerative disorders, including for AD and tauopathies, meriting its in-depth analysis.

Demonstration of prion-like spreading of tau-pathology in preclinical tauopathy models in vivo: proof of concept

Over the last few years, accumulating evidence has indicated prion-like propagation of tau-pathology. Noteworthy, prion-like propagation of tau-aggregation, which occurs intracellularly, requires a cell-to-cell transmission of tau-misfolding. In vitro studies delivered the proof of concept that tau-aggregates can be taken up by cells, induce aggregation of natively unfolded intracellular tau, and demonstrated transfer of tau-aggregation to cocultured cells [182,184,214,215]. Propagation of tau-pathology in vivo was first demonstrated in a milestone study by the injection of brain lysates of mice bearing NFTs into human WT–expressing mice that do not develop NFTs during life [216]. These injected mice developed tau-pathology at the site of injection and in regions remote from, but functionally connected to, the injection site [216,217]. This has been further demonstrated by injection of brain extracts from patients with different tauopathies into tau transgenic mice. In this case, tau-pathology developed in the characteristic cell types of the respective tauopathy and tau-aggregates were reminiscent of the respective tauopathy from which the brain lysates were derived [218,219]. Prion-like spreading of tau-pathology was evident in two other studies using mice overexpressing human tauP301L restricted to the EC [133,220]. In these studies, tau-pathology was demonstrated to develop in the EC and, several months later, in the hippocampus. Both studies excluded transgene expression leakage as an explanation for the induction of tau-pathology in the hippocampus, favoring the mechanism of cell-to-cell propagation. Importantly, injection of preaggregated synthetic tau and tau-fragments in tau transgenic mice resulted in the development and spreading of tau-pathology, demonstrating unequivocally that misfolded tau per se, and not some other unknown factor present in brain lysates, was responsible for the development and spreading of tau-pathology [221]. Induction of tau-aggregation by misfolded tau was also achieved in nontransgenic mice, although to a limited extent and later than in transgenic mice [222]. Recently, we demonstrated the propagation of tau-pathology out of EC following injection of synthetic tau-seeds, along functional connections, reminiscent of propagation of tau-pathology observed in AD [223]. Focal injection of minor amounts of misfolded tau, without changing the expression of tau, was sufficient to induce tau-aggregation and propagation through neuronal networks. Interestingly, we demonstrated that propagation of tau-pathology in functionally different circuitries correlated with different behavioral deficits (motor or cognitive deficits), depending on the site of injection. Our findings not only demonstrated that prion-like propagation of tau-pathology is associated with neuronal dysfunction, but are also reminiscent for the heterogeneity of the clinical presentation of tauopathies associated with the same mutation. More recently, identification and maintenance of different tau-strains in vitro and in vivo have been demonstrated, further highlighting the prion-like character of propagation of tau-pathology [219,224].

Targeting extracellular release and uptake of misfolded tau, acting as tau-seeds for prion-like propagation of tau-pathology

The demonstration of prion-like propagation in in vitro and in vivo models provides a compelling mechanism for the progression of tau-pathology in AD. Mechanistic insight in prion-like spreading of tau-aggregation may yield novel therapeutic targets. Prion-like propagation suggests the possible existence of extracellular forms of tau-seeds. The clearance of extracellular aggregates and the inhibition of their release and their uptake into neighboring cells could, hence, prove effective treatments for AD (Fig. 5.3). Although tau is classically considered as an intracellular protein, tau is found in CSF of AD patients, serving as biomarker [35]. Although this was originally considered to be released from dying

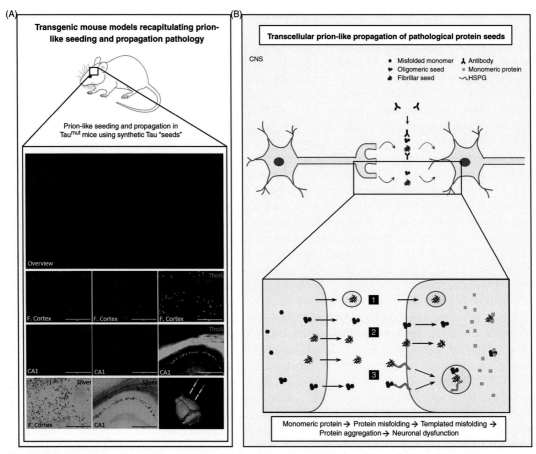

FIGURE 5.3 Transgenic Mouse Models Recapitulating Prion-Like Seeding and Propagation of Tau-Pathology and Their Use to Identify Novel Venues for AD Therapy

(A) Overview of a sagittal section of a tau transgenic mouse—tau^{P301S}—stereotactically injected with tau-seeds in the frontal cortex and hippocampus [223] at the age of 3 months and stained with AT8 antibody, 6 months postinjection (upper panel). Abundant tau-pathology develops at the site of injection and in remote, but functionally connected, brain regions (contralateral side, data not shown). Detailed AT8, ThioS, and Gallyas silver stainings are presented in the middle and lower panels, demonstrating the presence of mature NFTs (scale bars: F.cortex CA1, 400 µm; middle panel AT8 stainings, 200 µm). The model is described in detail in Ref. [223]. A variety of models recapitulating prion-like seeding and propogation of pathology have been generated [133,216–224]. (B) Schematic presentation of the mechanisms involved in prion-like seeding and propagation of pathology and the venues for new therapeutic strategies (based on Ref. [233]). Seeds accumulating in one neuron can be transmitted from neuron to neuron by release associated with vesicles *(1)* or naked release *(2)* with subsequent internalization in receptor neurons. They can also interact with heparan sulfate proteoglycans (HSPGs) *(3)* present on the cell membrane of receptor neurons, triggering their internalization by macropinocytosis. *CA1*, cornus ammonis region I of hippocampus proper; *F. cortex*, frontal cortex.

tangle-bearing neurons, tau is also detected in healthy individuals, and was found in the interstitial fluid of WT mice [225], suggesting that tau occurs extracellularly in physiological conditions. This was further corroborated by in vitro studies demonstrating that neuronal activity and AMPA receptor stimulation stimulates tau-release from primary cortical neurons [226]. Tau-release or secretion has subsequently been demonstrated in a variety of neuronal and nonneuronal cells, for WT, and also over-expressed mutant tau [183,227–232]. Several mechanisms have been proposed, including tau-release by direct secretion [228,230], associated to vesicles [232], and associated to exosomes [229,231]. It is not known if aggregated tau is released from cells and neurons by the same mechanism as monomeric tau. Further detailed research is ongoing to identify mechanisms of tau-"secretion" and the exact forms of tau released by these mechanisms, as interfering with these processes presents targets for therapy, which need further validation in preclinical models of prion-like propagation of tau-pathology.

Another necessary step, and maybe more suitable for drug design to inhibit propagation of tau-pathology, is the internalization of tau-seeds by acceptor cells. In this respect, the mechanism of uptake, as well as the physical form of tau [monomeric oligomeric, fibrillary, and posttranslationally modified (truncated, phosphorylated, etc.)]—for uptake and subsequent propagation—must be considered. Several in vitro and in vivo studies demonstrated uptake of oligomeric forms, as well as fibrillary forms of tau, with less efficient uptake of monomeric tau [182,184,214,234,235]. An in vivo study demonstrated uptake of both oligomeric and PHF-tau, but oligomeric tau had a higher propagation potential [236]. Different studies have started to address the mechanism of tau-seed uptake in cells, implicating bulk endocytosis [182,235,237] and more specifically macropinocytosis, a subtype of fluid-phase bulk endocytosis in the uptake of aggregated tau in neurons [237]. This process was found to be initiated by the binding of aggregated tau to heparan sulfate proteoglycans (HSPGs) on the cell membrane leading to actin restructuring and formation of large intracellular vesicles (macropinosome). Tau-aggregates were subsequently found to be released from the vesicles, leading to seeded aggregation of monomeric intracellular tau by a templated misfolded aggregation mechanism. Binding to HSPGs was prevented in this study by heparin, chlorate, heparinase, and genetic knockdown of a key HSPG synthetic enzyme, EXT1, in vitro and using a heparin mimetic, F6, which lead to blocking of neuronal uptake of stereotactically injected tau-fibrils in vivo [237]. Another study, also demonstrated that exosomes could bind HSPGs, triggering their internalization by macropinocytosis [238]. This could facilitate the internalization of seeds associated to exosomes or lacking the binding motifs to HSPGs. The macropinocytosis mechanism of tau-aggregate uptake represents a new therapeutic opportunity, as the blockage of aggregate binding to these surface proteins or interference in their proper maturation could prevent pathological seed uptake in healthy neurons [239].

Immunotherapy targeting extracellular misfolded forms of tau to inhibit prion-like propagation of tau-pathology

Immunotherapy approaches targeting Aβ [101,102,107,240,241] and tau [201,242–244] rescued behavioral deficits and reduced neuropathology in preclinical models. These findings resulted in a panoply of follow-up preclinical trials for optimization and clinical trials, as discussed earlier. The concept of prion-like propagation of tau-pathology and the concept of extracellular forms of tau are interesting in the context of tau-directed immunotherapy. Extracellular forms of tau provide an easier target for antibodies, and antibodies can be directed toward forms of tau involved in propagation of tau-pathology. In this respect, Yanamandra et al., performed a screening for anti-tau monoclonal antibodies inhibiting tau-seeding activity of tauP301S brain lysates using an in vitro assay [208,214]. The identified anti-tau monoclonal antibodies were subsequently evaluated in vivo, and were found to reduce tau-pathology, tau-seeding activity, and improve cognitive deficits [208]. Continued efforts may target immunotherapy to forms of tau with high-seeding and high-propagation potential. In this respect, the existence of different tau-strains with different propagation potentials, needs to be taken into consideration and can be used for therapeutic design [224].

Targeting tau-misfolding and aggregation to inhibit prion-like propagation of tau-pathology

As highlighted in the different tau-directed therapeutic strategies, the identification of the toxic forms of tau and the most potent propagating forms of tau are important research areas. Another effective treatment for diseases with underlying prion-like mechanism could be the prevention of protein aggregation. Most therapies developed until now target the protein aggregates by disrupting them without changing the overall native protein stability. This strategy presents limitations, as the disruption of large protein aggregates could lead to the formation of multiple additional nucleating centers of smaller sizes, but still capable of seeding and propagating protein aggregation, further aggravating pathology. One possible way to circumvent this limitation is the design of molecules that would increase the stability of the native protein or an inert intermediate form of aggregation. This strategy was successful applied for transthyretin, stabilizing a tetramer form of this protein and consequently inhibiting fibril formation [245].

Over the last few years an increasing interest in prion-like mechanisms as a potential contributing mechanism in neurodegenerative diseases fueled the development of animal models that robustly recapitulate this process. Although prion-like propagation of protein misfolding is a compelling mechanism for the characteristic spreading of disease-associated protein aggregation (including tau-pathology), its actual contribution in the pathogenesis of AD and related disorders remains to be proven. Identification of strategies interfering with this process in preclinical models is the first step to analyze its relevance in the disease process and may provide venues for AD and different neurodegenerative diseases characterized by pathological protein aggregation.

TRANSGENIC MOUSE MODELS RECAPITULATING Aβ-TO-TAU AXIS: NEW VENUES TO IDENTIFY NOVEL THERAPEUTIC TARGETS AIMING AT THE MOLECULAR AND CELLULAR MECHANISMS LINKING Aβ AND TAU

Animal models of AD robustly recapitulate Aβ-induced tau-pathology

Models recapitulating amyloid and tau-pathology separately have been crossed to generate more complete models of AD. Importantly these models have reproducibly demonstrated Aβ-induced tau-pathology, thereby not only lending support to the amyloid cascade hypothesis, but also providing models to study the mechanisms of Aβ-induced tau-pathology, which is generally considered as a key pathogenetic event in AD.

As described earlier, mouse models overexpressing APP or APP/PS1, on an endogenous murine tau background, did not develop mature NFTs. Only subtle changes in tau-phosphorylation and dystrophic neurites surrounding amyloid plaques containing hyperphosphorylated tau were detected [40,43,51,67,246]. In these models the lack of NFTs is probably due to the relative resistance of murine tau to aggregation within the lifetime of the mice or the different properties of murine tau compared to human tau, particularly the expression of different tau-isoforms, as well as the total concentration of tau. However, three milestone papers demonstrated the development of NFTs induced by Aβ in transgenic mice overexpressing human tau [106,247,248], thereby lending support to the amyloid cascade hypothesis. In the first study, injection of preaggregated synthetic Aβ in the brain of tauP301L mice resulted in the development of NFTs at the injection site and in remote areas, but functionally connected to the site of injection [247]. Although NFTs were detected in rather limited numbers, this study points toward a role of extracellularly added Aβ in the induction of intracellular tau-aggregation. In the second study, investigators crossed mutant APP transgenic mice with tauP301L mice and verified that the female double-transgenic mice showed increased NFT burden compared to the parental tau-strain [248]. In the third study, authors elegantly demonstrated, in a double-transgenic mouse model with combined amyloid and tau-pathology, reduced early pathological changes on tau by anti-Aβ immunization [106]. These findings, that is, Aβ-induced tau-pathology, were further confirmed and elaborated in a variety of different models by different groups using different strategies [41,51,66,106,155,185,247–256]. Different mouse models with a more robust Aβ-induced tau-pathology were generated by crossing tau transgenic mice with mutant APP/PS1 transgenic mice or by injection of tau transgenic mice with Aβ-enriched brain lysates of mice or patients [66,155,251,252,254,255]. The accumulation of phosphorylated and pathological tau by APP/amyloid peptides was demonstrated more recently in brains of mice overexpressing WT human tau [253,256]. Importantly, in these models amyloid pathology was not affected or at least not aggravated [255] by tau-pathology, while Aβ induced tau-pathology in all models [66,155,252,254,255] (Fig. 5.4).

Molecular and cellular mechanisms of Aβ-induced tau-pathology

Particularly the models with a very robust induction of tau-pathology, provide an attractive tool for mechanistic analysis of Aβ-induced tau-pathology. In this chapter we focus on three potential mechanisms, acting together or independently, that (nonexclusively) could explain Aβ-induced tau-pathology and that are supported by the majority of the published data. The first mechanism is the direct interaction of Aβ with neurons, leading to altered signaling pathways inducing tau-alterations. Aβ has been reported to bind to receptors on the cell membrane of neurons [257–270] or, due to its sticky nature, directly interact with the cell membrane and surface proteins [257]. The second mechanism that could explain Aβ-induced tau-pathology would be the activation of glia cells by Aβ, leading to inflammation and ultimately tau-alterations [271–273]. Finally, the third mechanism could be the direct cross-seeding of tau-aggregation by Aβ seeds [274]. This later mechanism gained increasing interest over the years, as cross-seeding was already demonstrated between different proteins [275–278]. Pinpointing the molecular and cellular mechanisms of Aβ-induced tau-pathology may open new venues for therapy.

Neuronal signaling pathways downstream of Aβ as a potential mechanism of Aβ-induced tau-pathology

Initially it was considered that in AD Aβ peptides accumulate extracellularly in senile plaques and conversely that NFTs accumulate intracellularly in neurons. So the interaction between the both would require neuronal signaling effects of Aβ leading to tau-phosphorylation. This notion was weakened by the identification of intracellular Aβ species and extracellular tau-species. Nevertheless, many studies demonstrated the binding of Aβ to different classes of receptors, such as alpha7 nicotinic acetylcholine receptors (α7 nAChR), NMDA, and AMPA receptors (directly or indirectly), the ephrin-type B2 receptor (EphB2), insulin receptors, the receptor for advanced glycation end products (RAGE), the prion protein receptor (PrP receptor), the mouse paired immunoglobulin-like receptor (PirB), and its

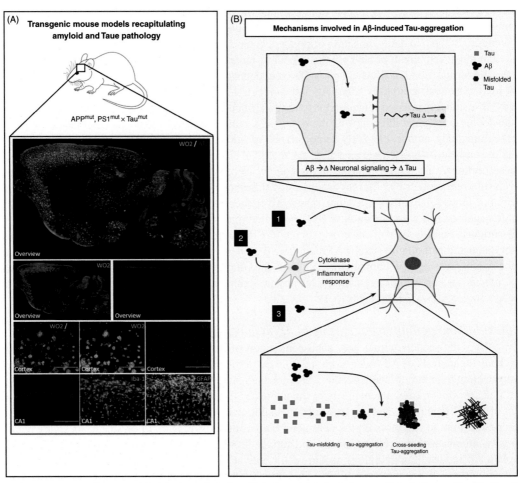

FIGURE 5.4 Transgenic Mouse Models Recapitulating Aβ-Induced Tau-Pathology and Their Use to Identify Novel Venues for AD Therapy

(A) A double-transgenic mouse model that develops combined amyloid and tau-pathology was obtained by crossing the APP$^{KM670/671NL; I716V; V717I}$, PS1$^{M146L; L286V}$ mice with tau^{P301S} mice [66]. Overview of sagittal sections from a 9-month-old double-transgenic mouse stained with WO2 and AT8 antibody (upper panel). Lower and middle panels present detailed stainings from cortex and CA1 region of the hippocampus stained with WO2 and AT8. Stainings with Iba-1 and GFAP are also presented showing reactive microglia and astrocytes, respectively (scale bars: 200 μm). The model is described in detail in Ref. [66]. A variety of models recapitulating amyloid and Tau-pathology have been generated [51,66,106,155,185,247–256]. (B) Schematic presentation of potential mechanisms involved in Aβ-induced tau-aggregation, providing new venues for identifying targets for AD therapy (based on Ref. [36]). Multiple mechanisms were proposed to explain Aβ-induced tau-pathology, including the interaction of Aβ with cellular receptors inducing signaling pathways that lead to tau-alterations *(1)*, Aβ-induced inflammation with downstream effects on tau *(2)*, and the cross-seeding mechanism of Aβ and tau, accelerating preexistent symptomatically silent tau-pathology with possible change of tau-strain *(3)*. These mechanisms are not necessarily mutually exclusive.

human counterpart, leukocyte immunoglobulin-like receptor (LilrB2) [257–270]. In addition, nonspecific binding of amyloid peptides to membranes or membrane proteins has to be considered besides their specific binding to receptors [257]. Monomeric and oligomeric forms of Aβ were demonstrated to induce synaptic and cognitive defects by binding α7 nAChR with high affinity [258,279]. Binding of Aβ peptide to α7 nAChR increased tau-phosphorylation (S202, T181, and T231) in vitro using a neuroblastoma cell line and ex vivo using synaptosomes, representing a receptor potentially involved in Aβ-induced tau-pathology. Aβ was, furthermore, found to impair synaptic plasticity in different APP transgenic models and in different experiments using application of extracellular Aβ to hippocampal slices, promoting a shift from synaptic potentiation (LTP) to synaptic depression (LTD) [12,43,66,280–283]. These findings were further elaborated by highlighting downstream effects of Aβ on NMDA receptor function and NMDA signaling [282–284]. Aβ-derived diffusible ligands were, thereby, found to bind synaptic sites in primary neuronal cultures, partially colocalizing with NMDA receptors [282,284–286]. Several studies tried to explain the impaired NMDA function and the LTP to LTD shift following exposure of neurons to Aβ. Alterations in synaptic localization of NR2B receptors [284,285], conformational changes of the NMDA receptors, a switch in NMDAR composition from GluN2B to GluN2A [286], or changes in downstream signaling cascades or indirectly via mGluR5 [287] have been proposed to be involved. Several other studies also linked Aβ-induced excitotoxicity [288] to NMDA receptors and tau. Importantly, many phenotypic features of APP transgenic mice, such as excitotoxicity, seizures, and premature death were reported to be rescued by tau-deficiency [55,289,290]. Latter studies demonstrated that tau-dependent targeting of fyn kinase couples the NMDA receptor to excitotoxicity and that mistargeting of fyn kinase decreases Aβ toxicity [281,288,291–294], highlighting a role for the Aβ–fyn–tau triad in synaptic and cognitive dysfunction [290]. Alternatively, a role for GSK-3β as a regulatory switch between LTD and LTP [295], and as a potential link between Aβ and tau has been proposed. Aβ-induced reduction of NMDA-dependent LTP was linked to increased tau-phosphorylation and GSK3 activation in hippocampal slices [290], and Aβ-induced deficits in LTP were rescued by GSK-3β inhibition [296]. A potential role of GSK-3β in Aβ-induced tau-pathology was further suggested by increased GSK-3β activation in two different models of Aβ-induced tau-pathology [66,155]. Together these findings have yielded important insights, yet the unequivocal identification of molecular mechanisms of the binding of Aβ to receptors or in a sticky way, its effect on downstream signaling pathways, and the exact contribution of these pathways to disease needs further research. As stated earlier, although amyloid plaques and soluble forms of Aβ are extracellular, Aβ is generated in intracellular compartments within the amyloidogenic pathway, and has been shown to accumulate and aggregate also intracellularly. Aβ can thereby disturb the normal function of the endolysosomal–autophagic pathway, resulting in proteopathic stress and altered signaling pathways that can give rise to tau-alterations (this is reviewed in detail in Refs. [56,98,194]), providing a potential alternative pathway of Aβ-induced tau-pathology. The exact molecular mechanism(s) need to be identified and validated in different models, which may give rise to new therapeutic applications.

Aβ-induced inflammation as a potential mechanism of Aβ-induced tau-pathology

In addition to the main pathological hallmarks of AD, that is, amyloid plaques and NFTs, the brains of patients also present astro- and microgliosis [13,297,298] despite not being directly used as diagnostic criteria (see also Chapter 3). Importantly, amyloid plaque deposition closely correlates with astro- and microgliosis [13,297,298] and in vitro and in vivo models also recapitulate very robust astro- and microgliosis induced by Aβ [51,271–273]. Different studies demonstrated that tau-pathology is dramatically increased by acute and chronic inflammatory insults that induce astro- and microgliosis [299–301]. So, it is reasonable to hypothesize that inflammation may contribute to Aβ-induced tau-pathology. In favor of this hypothesis, some studies demonstrated that blockage of IL-1 signaling, using an IL-1R antibody attenuated tau-pathology in triple-transgenic mice [300], while increasing IL-1β-exacerbated tau-pathology [302]. Further studies need to be performed for the validation of this hypothesis using different animal models. Nevertheless this represents an important area of research and opportunity for therapeutic strategies to interfere with Aβ-induced tau-pathology.

Heterotypic seeding of tau-aggregation by preaggregated Aβ as a potential mechanism of Aβ-induced tau-pathology

As described earlier, prion-like propagation of pathogenic aggregation of tau has recently been demonstrated [216,218,221,303] and is considered as an attractive mechanism for the characteristic spreading of tau-pathology described by Braak and Braak [8]. Prion-like mechanisms have been demonstrated for aggregating proteins in different neurodegenerative disorder as a potentially contributing mechanism including in AD, Parkinson's disease, Huntington's disease, and tauopathies [212,303,304] (see also Chapter 1). The existence of mixed pathologies and the existence of a continuum of intermediate diseases raises the interesting possibility of cross-seeding between different prion-like proteins. Intriguingly in this respect, α-synuclein and tau both possess prion-like features [305–308] and cross-seeding between both has been demonstrated in vitro and in vivo [275,276]. Several studies indicated that Aβ peptides also display prion-like seeding properties [212,304,309], enabling initiation and propagation of amyloid plaque formation to remote brain regions. This raises the interesting question whether Aβ could cross-seed tau-aggregation. As discussed earlier, Aβ peptides induce tau-pathology in a variety of models, which occurs also in remote but functionally connected brain regions [66,106,247,248,251]. Aβ peptides were also reported to directly bind tau [310] and Aβ-induced tau-aggregation was modeled in silico [311,312], setting the stage for a possible cross-seeding mechanism. We recently demonstrated that preaggregated Aβ seeds were capable of inducing tau-aggregation by a cross-seeding mechanism, using a well-characterized cellular tau-aggregation model and a cell-free tau-fibrillization assay [274]. Stereotactic injections of the Aβ-induced tau-aggregates (heterotypic seeds) in tau transgenic mice also resulted in the development of robust tau-aggregation at the site of injection and in the spreading of tau-pathology to areas remote from the injection site (contralateral side), further supporting these newly identified "heterotypic seeds" as strongly propagating. Aβ-induced heterotypic cross-seeding of tau-fibrillization provides a compelling mechanism for the Aβ-induced propagation of tau-pathology beyond EC as observed in AD patients at the onset of the disease process. Aβ-aggregates that strongly accumulate in isocortical regions of early phase (pre-) AD patients' brains could, thereby, be the driving force for spreading of tau-pathology beyond EC. Aβ-seeds could cross-seed tau in the EC via functional connections, thereby changing the existing tau-strain to a more aggressive and propagating strain, that would spread beyond this region and follow the characteristic pattern observed in AD for tau-pathology development [8]. Preferential homotypic seeding of amyloid pathology in isocortical regions, explains the initial spatial paradox between amyloid and tau-pathology in AD. However, increasing amyloid pathology would increase the probability of cross-seeding of incipient tau-pathology in the EC; thereby, changing a silent tau-pathology to an aggressively spreading one, as observed in AD, correlating with the presentation of AD symptoms. In view of these results, a cross-seeding mechanism by preaggregated Aβ to seed tau-aggregation should be considered as an attractive new mechanism involved in AD.

CONCLUSIONS

In this chapter, without being exhaustive, we describe the generation and continuous optimization of animal models for AD and their use for preclinical therapeutic target evaluation. These models have been instrumental in the identification of therapeutic targets that are currently being tested in patients, and are still used for further refinement of efficacy of these strategies and minimizing adverse side effects. These animal models were also used to study the actions of downstream effectors of Aβ, particularly on tau, recapitulating robust Aβ-induced tau-pathology. Recent findings, furthermore, indicate prion-like spreading of tau-pathology as a compelling mechanism for the propagation of tau-pathology in AD, presenting novel venues for therapy.

Future prospects in animal modeling should emphasize the use of models with combined amyloid and tau-pathology, for evaluation of therapeutic strategies for AD, as the disease is defined by the cooccurrence of plaques and tangles as an obligatory signature. Furthermore, future prospects should include—besides the existing models, which are all very valuable—models with combined amyloid and tau-pathology on WT tau background also. Future models should also include, besides models with moderate overexpression, knockin models. In this respect, it must be noted, however, that duplication of APP causes EOFAD, and overexpression of APP is considered to be involved in trisomy 21. Hence, overexpression of APP (to a reasonable extent) could be considered similar to an additional EOFAD mutation. Future prospects should also focus on early stages of the disease process, which should be documented in detail in patients, and animal models recapitulating these early amyloid phases should be generated. In this respect also, animal models with selective formation of different forms of Aβ and tau, will be useful to analyze their respective roles in the disease process and to evaluate their translatability to patient trials. Most importantly, and as a general rule, target identification and validation should be performed, by comparative analysis in different models. In fact the existence of many different models—with and without combined pathologies, with and without PS1 mutations, with selective expression or formation of Aβ* and tau* forms (their respective toxic forms), overexpression or not, etc.—is an extremely strong asset, which should be exploited more. Mechanistic insights and therapeutic strategies should be evaluated and compared in different models, and the results interpreted in the context of the in-depth knowledge of these models.

It must be noted, however, that no successful clinical trials have resulted yet from these analyses. Many factors need to be taken into account for these failures. Successful preclinical trials in animal models require trial optimization, in terms of optimization of the most efficient doses, treatment timing (preventive, curative, dosing per day, etc.), and treatment paradigms (i.c.v., i.c., and i.p.), which require optimization and reiteration and is much more difficult in patients. Patient stratification is another complicating factor, which is omitted in animal models, with an identical genetic background and well-characterized phenotype. Most importantly, the setup of the clinical trial and the reason for failure (side effects, lack of target engagement, time of starting treatment, patient stratification, etc.) needs to be analyzed in detail [313,314]. Importantly, future studies should focus more on detailed analysis of the translational value of findings in different animal models based on the outcomes of these clinical trials. The results obtained in animal models should not be bluntly extrapolated to patients, but interpreted with caution. However, and as a positive note to end, results obtained in immunization studies, showing clearance of amyloid pathology but no clearance of NFTs in mice, were translated in patient studies, showing clearance of amyloid pathology [108], but not of NFTs. At least in terms of pathology, this example shows a certain degree of translational value from mice to men, despite the fact that the models are artificial, and require the use of different mutations, with or without an overexpression, to mimic the respective pathologies. In conclusion, future studies should use comparative analysis in different models, preferably with combined amyloid and tau-pathology, and should be cautiously interpreted to obtain the results.

To conclude, animal models have contributed significantly to the identification of therapeutic targets in AD, and will continue to be useful for optimization, fine-tuning, and novel target identification. In the future it will be important to further optimize the animal models, and particularly to study in detail and optimize translation of results between animal models and patient studies. Availability and accessibility of results in clinical trials will be crucial for this analysis.

REFERENCES

[1] A. Wimo, B. Winblad, L. Jonsson, An estimate of the total worldwide societal costs of dementia in 2005, Alzheimers Dement. 3 (2007) 81–91.

[2] Alzheimer's Association2013 Alzheimer's disease facts and figures, Alzheimers Dement. 9 (2013) 208–245.

[3] A.D. Hutchinson, J.L. Mathias, Neuropsychological deficits in frontotemporal dementia and Alzheimer's disease: a meta-analytic review, J. Neurol. Neurosurg. Psychiatry 78 (2007) 917–928.

[4] B. Dubois, H.H. Feldman, C. Jacova, J.L. Cummings, S.T. Dekosky, P. Barberger-Gateau, A. Delacourte, G. Frisoni, N.C. Fox, D. Galasko, S. Gauthier, H. Hampel, G.A. Jicha, K. Meguro, J. O'Brien, F. Pasquier, P. Robert, M. Rossor, S. Salloway, M. Sarazin, L.C. De Souza, Y. Stern, P.J. Visser, P. Scheltens, Revising the definition of Alzheimer's disease: a new lexicon, Lancet Neurol. 9 (2010) 1118–1127.

[5] G.M. McKhann, D.S. Knopman, H. Chertkow, B.T. Hyman, C.R. Jack Jr., C.H. Kawas, W.E. Klunk, W.J. Koroshetz, J.J. Manly, R. Mayeux, R.C. Mohs, J.C. Morris, M.N. Rossor, P. Scheltens, M.C. Carrillo, B. Thies, S. Weintraub, C.H. Phelps, The diagnosis of dementia due to Alzheimer's disease: recommendations from the National Institute on Aging-Alzheimer's Association workgroups on diagnostic guidelines for Alzheimer's disease, Alzheimers Dement. 7 (2011) 263–269.

[6] C. Haass, C. Kaether, G. Thinakaran, S. Sisodia, Trafficking and proteolytic processing of APP, Cold Spring Harb. Perspect. Med. 2 (2012) a006270.

[7] E.M. Mandelkow, E. Mandelkow, Biochemistry and cell biology of tau protein in neurofibrillary degeneration, Cold Spring Harb. Perspect. Med. 2 (2012) a006247.

[8] H. Braak, E. Braak, Neuropathological stageing of Alzheimer-related changes, Acta Neuropathol. 82 (1991) 239–259.

[9] D.R. Thal, U. Rub, M. Orantes, H. Braak, Phases of A beta-deposition in the human brain and its relevance for the development of AD, Neurology 58 (2002) 1791–1800.

[10] B.T. Hyman, Amyloid-dependent and amyloid-independent stages of Alzheimer disease, Arch. Neurol. 68 (2011) 1062–1064.

[11] B.T. Hyman, C.H. Phelps, T.G. Beach, E.H. Bigio, N.J. Cairns, M.C. Carrillo, D.W. Dickson, C. Duyckaerts, M.P. Frosch, E. Masliah, S.S. Mirra, P.T. Nelson, J.A. Schneider, D.R. Thal, B. Thies, J.Q. Trojanowski, H.V. Vinters, T.J. Montine, National Institute on Aging-Alzheimer's Association guidelines for the neuropathologic assessment of Alzheimer's disease, Alzheimers Dement. 8 (2012) 1–13.

[12] D.J. Selkoe, Alzheimer's disease is a synaptic failure, Science 298 (2002) 789–791.

[13] D.J. Selkoe, Alzheimer's disease, Cold Spring Harb. Perspect. Biol. 3 (2011) a004457.

[14] L. Bertram, C.M. Lill, R.E. Tanzi, The genetics of Alzheimer disease: back to the future, Neuron 68 (2010) 270–281.

[15] A. Goate, M.C. Chartier-Harlin, M. Mullan, J. Brown, F. Crawford, L. Fidani, L. Giuffra, A. Haynes, N. Irving, L. James, et al. Segregation of a missense mutation in the amyloid precursor protein gene with familial Alzheimer's disease, Nature 349 (1991) 704–706.

[16] M. Cruts, C. Van Broeckhoven, Molecular genetics of Alzheimer's disease, Ann. Med. 30 (1998) 560–565.

[17] R.E. Tanzi, L. Bertram, Twenty years of the Alzheimer's disease amyloid hypothesis: a genetic perspective, Cell 120 (2005) 545–555.

[18] T. Jonsson, J.K. Atwal, S. Steinberg, J. Snaedal, P.V. Jonsson, S. Bjornsson, H. Stefansson, P. Sulem, D. Gudbjartsson, J. Maloney, K. Hoyte, A. Gustafson, Y. Liu, Y. Lu, T. Bhangale, R.R. Graham, J. Huttenlocher, G. Bjornsdottir, O.A. Andreassen, E.G. Jonsson, A. Palotie, T.W. Behrens, O.T. Magnusson, A. Kong, U. Thorsteinsdottir, R.J. Watts, K. Stefansson, A mutation in APP protects against Alzheimer's disease and age-related cognitive decline, Nature 488 (2012) 96–99.

[19] M. Citron, D. Westaway, W. Xia, G. Carlson, T. Diehl, G. Levesque, K. Johnson-Wood, M. Lee, P. Seubert, A. Davis, D. Kholodenko, R. Motter, R. Sherrington, B. Perry, H. Yao, R. Strome, I. Lieburburg, J. Rommens, S. Kim, D. Schenk, P. Fraser, P. St George Hyslop, D.J. Selkoe, Mutant presenilins of Alzheimer's disease increase production of 42-residue amyloid beta-protein in both transfected cells and transgenic mice, Nat. Med. 3 (1997) 67–72.

[20] B. De Strooper, P. Saftig, K. Craessaerts, H. Vanderstichele, G. Guhde, W. Annaert, K. Von Figura, F. Van Leuven, Deficiency of presenilin-1 inhibits the normal cleavage of amyloid precursor protein, Nature 391 (1998) 387–390.

[21] M.S. Wolfe, W. Xia, B.L. Ostaszewski, T.S. Diehl, W.T. Kimberly, D.J. Selkoe, Two transmembrane aspartates in presenilin-1 required for presenilin endoproteolysis and gamma-secretase activity, Nature 398 (1999) 513–517.

[22] M. Takami, Y. Nagashima, Y. Sano, S. Ishihara, M. Morishima-Kawashima, S. Funamoto, Y. Ihara, Gamma-secretase: successive tripeptide and tetrapeptide release from the transmembrane domain of beta-carboxyl terminal fragment, J. Neurosci. 29 (2009) 13042–13052.

[23] O. Holmes, S. Paturi, W. Ye, M.S. Wolfe, D.J. Selkoe, Effects of membrane lipids on the activity and processivity of purified gamma-secretase, Biochemistry 51 (2012) 3565–3575.

[24] M. Szaruga, S. Veugelen, M. Benurwar, S. Lismont, D. Sepulveda-Falla, A. Lleo, N.S. Ryan, T. Lashley, N.C. Fox, S. Murayama, H. Gijsen, B. De Strooper, L. Chavez-Gutierrez, Qualitative changes in human gamma-secretase underlie familial Alzheimer's disease, J. Exp. Med. 212 (2015) 2003–2013.

[25] R. Postina, A. Schroeder, I. Dewachter, J. Bohl, U. Schmitt, E. Kojro, C. Prinzen, K. Endres, C. Hiemke, M. Blessing, P. Flamez, A. Dequenne, E. Godaux, F. Van Leuven, F. Fahrenholz, A disintegrin-metalloproteinase prevents amyloid plaque formation and hippocampal defects in an Alzheimer disease mouse model, J. Clin. Invest. 113 (2004) 1456–1464.

[26] J. Suh, S.H. Choi, D.M. Romano, M.A. Gannon, A.N. Lesinski, D.Y. Kim, R.E. Tanzi, ADAM10 missense mutations potentiate beta-amyloid accumulation by impairing prodomain chaperone function, Neuron 80 (2013) 385–401.

[27] International Genomics of Alzheimer's Disease ConsortiumConvergent genetic and expression data implicate immunity in Alzheimer's disease, Alzheimers Dement. 11 (2015) 658–671.

[28] M. Hutton, C.L. Lendon, P. Rizzu, M. Baker, S. Froelich, H. Houlden, S. Pickering-Brown, S. Chakraverty, A. Isaacs, A. Grover, J. Hackett, J. Adamson, S. Lincoln, D. Dickson, P. Davies, R.C. Petersen, M. Stevens, E. de Graaff, E. Wauters, J. van Baren, M. Hillebrand, M. Joosse, J.M. Kwon, P. Nowotny, L.K. Che, J. Norton, J.C. Morris, L.A. Reed, J. Trojanowski, H. Basun, L. Lannfelt, M. Neystat, S. Fahn, F. Dark, T. Tannenberg, P.R. Dodd, N. Hayward, J.B. Kwok, P.R. Schofield, A. Andreadis, J. Snowden, D. Craufurd, D. Neary, F. Owen, B.A. Oostra, J. Hardy, A. Goate, J. van Swieten, D. Mann, T. Lynch, P. Heutink, Association of missense and 5′-splice-site mutations in tau with the inherited dementia FTDP-17, Nature 393 (1998) 702–705.

[29] P. Poorkaj, T.D. Bird, E. Wijsman, E. Nemens, R.M. Garruto, L. Anderson, A. Andreadis, W.C. Wiederholt, M. Raskind, G.D. Schellenberg, Tau is a candidate gene for chromosome 17 frontotemporal dementia, Ann. Neurol. 43 (1998) 815–825.

[30] M.G. Spillantini, J.R. Murrell, M. Goedert, M.R. Farlow, A. Klug, B. Ghetti, Mutation in the tau gene in familial multiple system tauopathy with presenile dementia, Proc. Natl. Acad. Sci. USA 95 (1998) 7737–7741.

[31] M.S. Wolfe, Tau mutations in neurodegenerative diseases, J. Biol. Chem. 284 (2009) 6021–6025.

[32] J. Hardy, D.J. Selkoe, The amyloid hypothesis of Alzheimer's disease: progress and problems on the road to therapeutics, Science 297 (2002) 353–356.

[33] J.A. Hardy, G.A. Higgins, Alzheimer's disease: the amyloid cascade hypothesis, Science 256 (1992) 184–185.

[34] E.S. Musiek, D.M. Holtzman, Three dimensions of the amyloid hypothesis: time, space and 'wingmen', Nat. Neurosci. 18 (2015) 800–806.

[35] C.R. Jack Jr., D.S. Knopman, W.J. Jagust, R.C. Petersen, M.W. Weiner, P.S. Aisen, L.M. Shaw, P. Vemuri, H.J. Wiste, S.D. Weigand, T.G. Lesnick, V.S. Pankratz, M.C. Donohue, J.Q. Trojanowski, Tracking pathophysiological processes in Alzheimer's disease: an updated hypothetical model of dynamic biomarkers, Lancet Neurol. 12 (2013) 207–216.

[36] I.C. Stancu, B. Vasconcelos, D. Terwel, I. Dewachter, Models of beta-amyloid induced Tau-pathology: the long and "folded" road to understand the mechanism, Mol. Neurodegener. 9 (2014) 51.

[37] K.H. Ashe, Learning and memory in transgenic mice modeling Alzheimer's disease, Learn Mem. 8 (2001) 301–308.

[38] D.R. Borchelt, G. Thinakaran, C.B. Eckman, M.K. Lee, F. Davenport, T. Ratovitsky, C.M. Prada, G. Kim, S. Seekins, D. Yager, H.H. Slunt, R. Wang, M. Seeger, A.I. Levey, S.E. Gandy, N.G. Copeland, N.A. Jenkins, D.L. Price, S.G. Younkin, S.S. Sisodia, Familial Alzheimer's disease-linked presenilin 1 variants elevate Abeta1-42/1-40 ratio in vitro and in vivo, Neuron 17 (1996) 1005–1013.

[39] D.R. Borchelt, T. Ratovitski, J. van Lare, M.K. Lee, V. Gonzales, N.A. Jenkins, N.G. Copeland, D.L. Price, S.S. Sisodia, Accelerated amyloid deposition in the brains of transgenic mice coexpressing mutant presenilin 1 and amyloid precursor proteins, Neuron 19 (1997) 939–945.

[40] C. Sturchler-Pierrat, D. Abramowski, M. Duke, K.H. Wiederhold, C. Mistl, S. Rothacher, B. Ledermann, K. Burki, P. Frey, P.A. Paganetti, C. Waridel, M.E. Calhoun, M. Jucker, A. Probst, M. Staufenbiel, B. Sommer, Two amyloid precursor protein transgenic mouse models with Alzheimer disease-like pathology, Proc. Natl. Acad. Sci. USA 94 (1997) 13287–13292.

[41] K. Hsiao, Transgenic mice expressing Alzheimer amyloid precursor proteins, Exp. Gerontol. 33 (1998) 883–889.

[42] P.F. Chapman, G.L. White, M.W. Jones, D. Cooper-Blacketer, V.J. Marshall, M. Irizarry, L. Younkin, M.A. Good, T.V. Bliss, B.T. Hyman, S.G. Younkin, K.K. Hsiao, Impaired synaptic plasticity and learning in aged amyloid precursor protein transgenic mice, Nat. Neurosci. 2 (1999) 271–276.

[43] D. Moechars, I. Dewachter, K. Lorent, D. Reverse, V. Baekelandt, A. Naidu, I. Tesseur, K. Spittaels, C.V. Haute, F. Checler, E. Godaux, B. Cordell, F. Van Leuven, Early phenotypic changes in transgenic mice that overexpress different mutants of amyloid precursor protein in brain, J. Biol. Chem. 274 (1999) 6483–6492.

[44] G. Chen, K.S. Chen, J. Knox, J. Inglis, A. Bernard, S.J. Martin, A. Justice, L. McConlogue, D. Games, S.B. Freedman, R.G. Morris, A learning deficit related to age and beta-amyloid plaques in a mouse model of Alzheimer's disease, Nature 408 (2000) 975–979.

[45] L. Mucke, E. Masliah, G.Q. Yu, M. Mallory, E.M. Rockenstein, G. Tatsuno, K. Hu, D. Kholodenko, K. Johnson-Wood, L. McConlogue, High-level neuronal expression of abeta 1-42 in wild-type human amyloid protein precursor transgenic mice: synaptotoxicity without plaque formation, J. Neurosci. 20 (2000) 4050–4058.

[46] M.A. Chishti, D.S. Yang, C. Janus, A.L. Phinney, P. Horne, J. Pearson, R. Strome, N. Zuker, J. Loukides, J. French, S. Turner, G. Lozza, M. Grilli, S. Kunicki, C. Morissette, J. Paquette, F. Gervais, C. Bergeron, P.E. Fraser, G.A. Carlson, P.S. George-Hyslop, D. Westaway, Early-onset amyloid deposition and cognitive deficits in transgenic mice expressing a double mutant form of amyloid precursor protein 695, J. Biol. Chem. 276 (2001) 21562–21570.

[47] J.C. Dodart, K.R. Bales, K.S. Gannon, S.J. Greene, R.B. Demattos, C. Mathis, C.A. Delong, S. Wu, X. Wu, D.M. Holtzman, S.M. Paul, Immunization reverses memory deficits without reducing brain Abeta burden in Alzheimer's disease model, Nat. Neurosci. 5 (2002) 452–457.

[48] I.H. Cheng, J.J. Palop, L.A. Esposito, N. Bien-Ly, F. Yan, L. Mucke, Aggressive amyloidosis in mice expressing human amyloid peptides with the Arctic mutation, Nat. Med. 10 (2004) 1190–1192.

[49] M.C. Herzig, D.T. Winkler, P. Burgermeister, M. Pfeifer, E. Kohler, S.D. Schmidt, S. Danner, D. Abramowski, C. Sturchler-Pierrat, K. Burki, S.G. van Duinen, M.L. Maat-Schieman, M. Staufenbiel, P.M. Mathews, M. Jucker, Abeta is targeted to the vasculature in a mouse model of hereditary cerebral hemorrhage with amyloidosis, Nat. Neurosci. 7 (2004) 954–960.

[50] J.P. Cleary, D.M. Walsh, J.J. Hofmeister, G.M. Shankar, M.A. Kuskowski, D.J. Selkoe, K.H. Ashe, Natural oligomers of the amyloid-beta protein specifically disrupt cognitive function, Nat. Neurosci. 8 (2005) 79–84.

[51] C. Duyckaerts, M.C. Potier, B. Delatour, Alzheimer disease models and human neuropathology: similarities and differences, Acta Neuropathol. 115 (2008) 5–38.

[52] J. Götz, L.M. Ittner, Animal models of Alzheimer's disease and frontotemporal dementia, Nat. Rev. Neurosci. 9 (2008) 532–544.

[53] I. Dewachter, D. Reverse, N. Caluwaerts, L. Ris, C. Kuiperi, C. Van den Haute, K. Spittaels, L. Umans, L. Serneels, E. Thiry, D. Moechars, M. Mercken, E. Godaux, F. Van Leuven, Neuronal deficiency of presenilin 1 inhibits amyloid plaque formation and corrects hippocampal long-term potentiation but not a cognitive defect of amyloid precursor protein [V717I] transgenic mice, J. Neurosci. 22 (2002) 3445–3453.

[54] D.M. Walsh, D.J. Selkoe, Deciphering the molecular basis of memory failure in Alzheimer's disease, Neuron 44 (2004) 181–193.

[55] E.D. Roberson, B. Halabisky, J.W. Yoo, J. Yao, J. Chin, F. Yan, T. Wu, P. Hamto, N. Devidze, G.Q. Yu, J.J. Palop, J.L. Noebels, L. Mucke, Amyloid-beta/Fyn-induced synaptic, network, and cognitive impairments depend on tau levels in multiple mouse models of Alzheimer's disease, J. Neurosci. 31 (2011) 700–711.

[56] A. Peric, W. Annaert, Early etiology of Alzheimer's disease: tipping the balance toward autophagy or endosomal dysfunction?, Acta Neuropathol. 129 (2015) 363–381.

[57] C.G. Almeida, R.H. Takahashi, G.K. Gouras, Beta-amyloid accumulation impairs multivesicular body sorting by inhibiting the ubiquitin-proteasome system, J. Neurosci. 26 (2006) 4277–4288.

[58] S. Schilling, U. Zeitschel, T. Hoffmann, U. Heiser, M. Francke, A. Kehlen, M. Holzer, B. Hutter-Paier, M. Prokesch, M. Windisch, W. Jagla, D. Schlenzig, C. Lindner, T. Rudolph, G. Reuter, H. Cynis, D. Montag, H.U. Demuth, S. Rossner, Glutaminyl cyclase inhibition attenuates pyroglutamate Abeta and Alzheimer's disease-like pathology, Nat. Med. 14 (2008) 1106–1111.

[59] O. Wirths, H. Breyhan, H. Cynis, S. Schilling, H.U. Demuth, T.A. Bayer, Intraneuronal pyroglutamate-Abeta 3–42 triggers neurodegeneration and lethal neurological deficits in a transgenic mouse model, Acta Neuropathol. 118 (2009) 487–496.

[60] G.K. Gouras, D. Tampellini, R.H. Takahashi, E. Capetillo-Zarate, Intraneuronal beta-amyloid accumulation and synapse pathology in Alzheimer's disease, Acta Neuropathol. 119 (2010) 523–541.

[61] S. Kumar, J. Walter, Phosphorylation of amyloid beta (Abeta) peptides—a trigger for formation of toxic aggregates in Alzheimer's disease, Aging 3 (2011) 803–812.

[62] O. Wirths, T.A. Bayer, Intraneuronal Abeta accumulation and neurodegeneration: lessons from transgenic models, Life Sci. 91 (2012) 1148–1152.

[63] T.A. Bayer, O. Wirths, Focusing the amyloid cascade hypothesis on N-truncated Abeta peptides as drug targets against Alzheimer's disease, Acta Neuropathol. 127 (2014) 787–801.

[64] S. Saito, M. Ihara, New therapeutic approaches for Alzheimer's disease and cerebral amyloid angiopathy, Front. Aging Neurosci. 6 (2014) 290.

[65] H. Oakley, S.L. Cole, S. Logan, E. Maus, P. Shao, J. Craft, A. Guillozet-Bongaarts, M. Ohno, J. Disterhoft, L. Van Eldik, R. Berry, R. Vassar, Intraneuronal beta-amyloid aggregates, neurodegeneration, and neuron loss in transgenic mice with five familial Alzheimer's disease mutations: potential factors in amyloid plaque formation, J. Neurosci. 26 (2006) 10129–10140.

[66] I.C. Stancu, L. Ris, B. Vasconcelos, C. Marinangeli, L. Goeminne, V. Laporte, L.E. Haylani, J. Couturier, O. Schakman, P. Gailly, N. Pierrot, P. Kienlen-Campard, J.N. Octave, I. Dewachter, Tauopathy contributes to synaptic and cognitive deficits in a murine model for Alzheimer's disease, FASEB J. 28 (2014) 2620–2631.

[67] L. Holcomb, M.N. Gordon, E. Mcgowan, X. Yu, S. Benkovic, P. Jantzen, K. Wright, I. Saad, R. Mueller, D. Morgan, S. Sanders, C. Zehr, K. O'Campo, J. Hardy, C.M. Prada, C. Eckman, S. Younkin, K. Hsiao, K. Duff, Accelerated Alzheimer-type phenotype in transgenic mice carrying both mutant amyloid precursor protein and presenilin 1 transgenes, Nat. Med. 4 (1998) 97–100.

[68] I. Dewachter, J. Van Dorpe, L. Smeijers, M. Gilis, C. Kuiperi, I. Laenen, N. Caluwaerts, D. Moechars, F. Checler, H. Vanderstichele, F. Van Leuven, Aging increased amyloid peptide and caused amyloid plaques in brain of old APP/V717I transgenic mice by a different mechanism than mutant presenilin1, J. Neurosci. 20 (2000) 6452–6458.

[69] F.M. LaFerla, K.N. Green, S. Oddo, Intracellular amyloid-beta in Alzheimer's disease, Nat. Rev. Neurosci. 8 (2007) 499–509.

[70] T. Wisniewski, F. Goni, Immunotherapeutic approaches for Alzheimer's disease, Neuron 85 (2015) 1162–1176.

[71] C. Schmitz, B.P. Rutten, A. Pielen, S. Schafer, O. Wirths, G. Tremp, C. Czech, V. Blanchard, G. Multhaup, P. Rezaie, H. Korr, H.W. Steinbusch, L. Pradier, T.A. Bayer, Hippocampal neuron loss exceeds amyloid plaque load in a transgenic mouse model of Alzheimer's disease, Am. J. Pathol. 164 (2004) 1495–1502.

[72] B. De Strooper, Aph-1, Pen-2, and Nicastrin with Presenilin generate an active gamma-secretase complex, Neuron 38 (2003) 9–12.

[73] B. De Strooper, W. Annaert, Novel research horizons for presenilins and gamma-secretases in cell biology and disease, Annu. Rev. Cell. Dev. Biol. 26 (2010) 235–260.

[74] B. De Strooper, T. Iwatsubo, M.S. Wolfe, Presenilins and gamma-secretase: structure, function, and role in Alzheimer Disease, Cold Spring Harb. Perspect. Med. 2 (2012) a006304.

[75] L. Serneels, J. Van Biervliet, K. Craessaerts, T. Dejaegere, K. Horre, T. Van Houtvin, H. Esselmann, S. Paul, M.K. Schafer, O. Berezovska, B.T. Hyman, B. Sprangers, R. Sciot, L. Moons, M. Jucker, Z. Yang, P.C. May, E. Karran, J. Wiltfang, R. D'Hooge, B. De Strooper, Gamma-secretase heterogeneity in the Aph1 subunit: relevance for Alzheimer's disease, Science 324 (2009) 639–642.

[76] R. Vassar, B.D. Bennett, S. Babu-Khan, S. Kahn, E.A. Mendiaz, P. Denis, D.B. Teplow, S. Ross, P. Amarante, R. Loeloff, Y. Luo, S. Fisher, J. Fuller, S. Edenson, J. Lile, M.A. Jarosinski, A.L. Biere, E. Curran, T. Burgess, J.C. Louis, F. Collins, J. Treanor, G. Rogers, M. Citron, Beta-secretase cleavage of Alzheimer's amyloid precursor protein by the transmembrane aspartic protease BACE, Science 286 (1999) 735–741.

[77] R. Vassar, P.H. Kuhn, C. Haass, M.E. Kennedy, L. Rajendran, P.C. Wong, S.F. Lichtenthaler, Function, therapeutic potential and cell biology of BACE proteases: current status and future prospects, J. Neurochem. 130 (2014) 4–28.

[78] M. Ohno, E.A. Sametsky, L.H. Younkin, H. Oakley, S.G. Younkin, M. Citron, R. Vassar, J.F. Disterhoft, BACE1 deficiency rescues memory deficits and cholinergic dysfunction in a mouse model of Alzheimer's disease, Neuron 41 (2004) 27–33.

[79] L. McConlogue, M. Buttini, J.P. Anderson, E.F. Brigham, K.S. Chen, S.B. Freedman, D. Games, K. Johnson-Wood, M. Lee, M. Zeller, W. Liu, R. Motter, S. Sinha, Partial reduction of BACE1 has dramatic effects on Alzheimer plaque and synaptic pathology in APP transgenic mice, J. Biol. Chem. 282 (2007) 26326–26334.

[80] M. Willem, A.N. Garratt, B. Novak, M. Citron, S. Kaufmann, A. Rittger, B. De Strooper, P. Saftig, C. Birchmeier, C. Haass, Control of peripheral nerve myelination by the beta-secretase BACE1, Science 314 (2006) 664–666.

[81] L. Zhou, S. Barao, M. Laga, K. Bockstael, M. Borgers, H. Gijsen, W. Annaert, D. Moechars, M. Mercken, K. Gevaert, B. De Strooper, The neural cell adhesion molecules L1 and CHL1 are cleaved by BACE1 protease in vivo, J. Biol. Chem. 287 (2012) 25927–25940.

[82] S. Barao, D. Moechars, S.F. Lichtenthaler, B. De Strooper, BACE1 physiological functions may limit its use as therapeutic target for Alzheimer's disease, Trends Neurosci. 39 (2016) 158–169.

[83] A.V. Savonenko, T. Melnikova, F.M. Laird, K.A. Stewart, D.L. Price, P.C. Wong, Alteration of BACE1-dependent NRG1/ErbB4 signaling and schizophrenia-like phenotypes in BACE1-null mice, Proc. Natl. Acad. Sci. USA 105 (2008) 5585–5590.

[84] P.C. May, R.A. Dean, S.L. Lowe, F. Martenyi, S.M. Sheehan, L.N. Boggs, S.A. Monk, B.M. Mathes, D.J. Mergott, B.M. Watson, S.L. Stout, D.E. Timm, E. Smith Labell, C.R. Gonzales, M. Nakano, S.S. Jhee, M. Yen, L. Ereshefsky, T.D. Lindstrom, D.O. Calligaro, P.J. Cocke, D. Greg Hall, S. Friedrich, M. Citron, J.E. Audia, Robust central reduction of amyloid-beta in humans with an orally available, non-peptidic beta-secretase inhibitor, J. Neurosci. 31 (2011) 16507–16516.

[85] D. Schenk, G.S. Basi, M.N. Pangalos, Treatment strategies targeting amyloid beta-protein, Cold Spring Harb. Perspect. Med. 2 (2012) a006387.

[86] R. Yan, R. Vassar, Targeting the beta secretase BACE1 for Alzheimer's disease therapy, Lancet Neurol. 13 (2014) 319–329.

[87] B. De Strooper, L. Chavez Gutierrez, Learning by failing: ideas and concepts to tackle gamma-secretases in Alzheimer's disease and beyond, Annu. Rev. Pharmacol. Toxicol. 55 (2015) 419–437.

[88] N. Iwata, S. Tsubuki, Y. Takaki, K. Watanabe, M. Sekiguchi, E. Hosoki, M. Kawashima-Morishima, H.J. Lee, E. Hama, Y. Sekine-Aizawa, T.C. Saido, Identification of the major Abeta1-42-degrading catabolic pathway in brain parenchyma: suppression leads to biochemical and pathological deposition, Nat. Med. 6 (2000) 143–150.

[89] W. Farris, S. Mansourian, Y. Chang, L. Lindsley, E.A. Eckman, M.P. Frosch, C.B. Eckman, R.E. Tanzi, D.J. Selkoe, S. Guenette, Insulin-degrading enzyme regulates the levels of insulin, amyloid beta-protein, and the beta-amyloid precursor protein intracellular domain in vivo, Proc. Natl. Acad. Sci. USA 100 (2003) 4162–4167.

[90] M.A. Leissring, W. Farris, A.Y. Chang, D.M. Walsh, X. Wu, X. Sun, M.P. Frosch, D.J. Selkoe, Enhanced proteolysis of beta-amyloid in APP transgenic mice prevents plaque formation, secondary pathology, and premature death, Neuron 40 (2003) 1087–1093.

[91] N. Iwata, S. Tsubuki, Y. Takaki, K. Shirotani, B. Lu, N.P. Gerard, C. Gerard, E. Hama, H.J. Lee, T.C. Saido, Metabolic regulation of brain Abeta by neprilysin, Science 292 (2001) 1550–1552.

[92] R.A. Marr, E. Rockenstein, A. Mukherjee, M.S. Kindy, L.B. Hersh, F.H. Gage, I.M. Verma, E. Masliah, Neprilysin gene transfer reduces human amyloid pathology in transgenic mice, J. Neurosci. 23 (2003) 1992–1996.

[93] N. Iwata, H. Mizukami, K. Shirotani, Y. Takaki, S. Muramatsu, B. Lu, N.P. Gerard, C. Gerard, K. Ozawa, T.C. Saido, Presynaptic localization of neprilysin contributes to efficient clearance of amyloid-beta peptide in mouse brain, J. Neurosci. 24 (2004) 991–998.

[94] N. Iwata, M. Sekiguchi, Y. Hattori, A. Takahashi, M. Asai, B. Ji, M. Higuchi, M. Staufenbiel, S. Muramatsu, T.C. Saido, Global brain delivery of neprilysin gene by intravascular administration of AAV vector in mice, Sci. Rep. 3 (2013) 1472.

[95] R.A. Nixon, D.S. Yang, Autophagy failure in Alzheimer's disease—locating the primary defect, Neurobiol. Dis. 43 (2011) 38–45.

[96] R.A. Nixon, Autophagy in neurodegenerative disease: friend, foe or turncoat?, Trends Neurosci. 29 (2006) 528–535.

[97] R.A. Nixon, J. Wegiel, A. Kumar, W.H. Yu, C. Peterhoff, A. Cataldo, A.M. Cuervo, Extensive involvement of autophagy in Alzheimer disease: an immuno-electron microscopy study, J. Neuropathol. Exp. Neurol. 64 (2005) 113–122.

[98] R.A. Nixon, The role of autophagy in neurodegenerative disease, Nat. Med. 19 (2013) 983–997.

[99] R. Sannerud, C. Esselens, P. Ejsmont, R. Mattera, L. Rochin, A.K. Tharkeshwar, G. De Baets, V. De Wever, R. Habets, V. Baert, W. Vermeire, C. Michiels, A.J. Groot, R. Wouters, K. Dillen, K. Vints, P. Baatsen, S. Munck, R. Derua, E. Waelkens, G.S. Basi, M. Mercken, M. Vooijs, M. Bollen, J. Schymkowitz, F. Rousseau, J.S. Bonifacino, G. Van Niel, B. De Strooper, W. Annaert, Restricted location of PSEN2/

gamma-secretase determines substrate specificity and generates an intracellular Abeta pool, Cell 166 (2016) 193–208.

[100] Q. Xiao, P. Yan, X. Ma, H. Liu, R. Perez, A. Zhu, E. Gonzales, D.L. Tripoli, L. Czerniewski, A. Ballabio, J.R. Cirrito, A. Diwan, J.M. Lee, Neuronal-targeted TFEB accelerates lysosomal degradation of APP, reducing Abeta generation and amyloid plaque pathogenesis, J. Neurosci. 35 (2015) 12137–12151.

[101] D. Schenk, R. Barbour, W. Dunn, G. Gordon, H. Grajeda, T. Guido, K. Hu, J. Huang, K. Johnson-Wood, K. Khan, D. Kholodenko, M. Lee, Z. Liao, I. Lieberburg, R. Motter, L. Mutter, F. Soriano, G. Shopp, N. Vasquez, C. Vandevert, S. Walker, M. Wogulis, T. Yednock, D. Games, P. Seubert, Immunization with amyloid-beta attenuates Alzheimer-disease-like pathology in the PDAPP mouse, Nature 400 (1999) 173–177.

[102] F. Bard, C. Cannon, R. Barbour, R.L. Burke, D. Games, H. Grajeda, T. Guido, K. Hu, J. Huang, K. Johnson-Wood, K. Khan, D. Kholodenko, M. Lee, I. Lieberburg, R. Motter, M. Nguyen, F. Soriano, N. Vasquez, K. Weiss, B. Welch, P. Seubert, D. Schenk, T. Yednock, Peripherally administered antibodies against amyloid beta-peptide enter the central nervous system and reduce pathology in a mouse model of Alzheimer disease, Nat. Med. 6 (2000) 916–919.

[103] D. Schenk, Amyloid-beta immunotherapy for Alzheimer's disease: the end of the beginning, Nat. Rev. Neurosci. 3 (2002) 824–828.

[104] L.A. Kotilinek, B. Bacskai, M. Westerman, T. Kawarabayashi, L. Younkin, B.T. Hyman, S. Younkin, K.H. Ashe, Reversible memory loss in a mouse transgenic model of Alzheimer's disease, J. Neurosci. 22 (2002) 6331–6335.

[105] I. Klyubin, D.M. Walsh, C.A. Lemere, W.K. Cullen, G.M. Shankar, V. Betts, E.T. Spooner, L. Jiang, R. Anwyl, D.J. Selkoe, M.J. Rowan, Amyloid beta protein immunotherapy neutralizes Abeta oligomers that disrupt synaptic plasticity in vivo, Nat. Med. 11 (2005) 556–561.

[106] S. Oddo, L. Billings, J.P. Kesslak, D.H. Cribbs, F.M. Laferla, Abeta immunotherapy leads to clearance of early, but not late, hyperphosphorylated tau aggregates via the proteasome, Neuron 43 (2004) 321–332.

[107] C. Janus, J. Pearson, J. Mclaurin, P.M. Mathews, Y. Jiang, S.D. Schmidt, M.A. Chishti, P. Horne, D. Heslin, J. French, H.T. Mount, R.A. Nixon, M. Mercken, C. Bergeron, P.E. Fraser, P. St George-Hyslop, D. Westaway, A beta peptide immunization reduces behavioural impairment and plaques in a model of Alzheimer's disease, Nature 408 (2000) 979–982.

[108] P.H. St George-Hyslop, J.C. Morris, Will anti-amyloid therapies work for Alzheimer's disease?, Lancet 372 (2008) 180–182.

[109] S. Lesne, M.T. Koh, L. Kotilinek, R. Kayed, C.G. Glabe, A. Yang, M. Gallagher, K.H. Ashe, A specific amyloid-beta protein assembly in the brain impairs memory, Nature 440 (2006) 352–357.

[110] D.R. Thal, J. Walter, T.C. Saido, M. Fandrich, Neuropathology and biochemistry of Abeta and its aggregates in Alzheimer's disease, Acta Neuropathol. 129 (2015) 167–182.

[111] H. Jacobsen, L. Ozmen, A. Caruso, R. Narquizian, H. Hilpert, B. Jacobsen, D. Terwel, A. Tanghe, B. Bohrmann, Combined treatment with a BACE inhibitor and anti-Abeta antibody gantenerumab enhances amyloid reduction in APPLondon mice, J. Neurosci. 34 (2014) 11621–11630.

[112] G. Leinenga, J. Götz, Scanning ultrasound removes amyloid-beta and restores memory in an Alzheimer's disease mouse model, Sci. Transl. Med. 7 (2015) 278ra33.

[113] P. Nilsson, N. Iwata, S. Muramatsu, L.O. Tjernberg, B. Winblad, T.C. Saido, Gene therapy in Alzheimer's disease—potential for disease modification, J. Cell. Mol. Med. 14 (2010) 741–757.

[114] J. Gotz, A. Probst, M.G. Spillantini, T. Schafer, R. Jakes, K. Burki, M. Goedert, Somatodendritic localization and hyperphosphorylation of tau protein in transgenic mice expressing the longest human brain tau isoform, EMBO J. 14 (1995) 1304–1313.

[115] J.P. Brion, G. Tremp, J.N. Octave, Transgenic expression of the shortest human tau affects its compartmentalization and its phosphorylation as in the pretangle stage of Alzheimer's disease, Am. J. Pathol. 154 (1999) 255–270.

[116] K. Spittaels, C. Van den Haute, J. Van Dorpe, K. Bruynseels, K. Vandezande, I. Laenen, H. Geerts, M. Mercken, R. Sciot, A. Van Lommel, R. Loos, F. Van Leuven, Prominent axonopathy in the brain and spinal cord of transgenic mice overexpressing four-repeat human tau protein, Am. J. Pathol. 155 (1999) 2153–2165.

[117] A. Probst, J. Gotz, K.H. Wiederhold, M. Tolnay, C. Mistl, A.L. Jaton, M. Hong, T. Ishihara, V.M. Lee, J.Q. Trojanowski, R. Jakes, R.A. Crowther, M.G. Spillantini, K. Burki, M. Goedert, Axonopathy and amyotrophy in mice transgenic for human four-repeat tau protein, Acta Neuropathol. 99 (2000) 469–481.

[118] T. Ishihara, M. Hong, B. Zhang, Y. Nakagawa, M.K. Lee, J.Q. Trojanowski, V.M. Lee, Age-dependent emergence and progression of a tauopathy in transgenic mice overexpressing the shortest human tau isoform, Neuron 24 (1999) 751–762.

[119] T. Ishihara, B. Zhang, M. Higuchi, Y. Yoshiyama, J.Q. Trojanowski, V.M. Lee, Age-dependent induction of congophilic neurofibrillary tau inclusions in tau transgenic mice, Am. J. Pathol. 158 (2001) 555–562.

[120] C. Andorfer, Y. Kress, M. Espinoza, R. de Silva, K.L. Tucker, Y.A. Barde, K. Duff, P. Davies, Hyperphosphorylation and aggregation of tau in mice expressing normal human tau isoforms, J. Neurochem. 86 (2003) 582–590.

[121] C. Andorfer, C.M. Acker, Y. Kress, P.R. Hof, K. Duff, P. Davies, Cell-cycle reentry and cell death in transgenic mice expressing nonmutant human tau isoforms, J. Neurosci. 25 (2005) 5446–5454.

[122] J. Lewis, E. Mcgowan, J. Rockwood, H. Melrose, P. Nacharaju, M. Van Slegtenhorst, K. Gwinn-Hardy, M. Paul Murphy, M. Baker, X. Yu, K. Duff, J. Hardy, A. Corral, W.L. Lin, S.H. Yen, D.W. Dickson, P. Davies, M. Hutton, Neurofibrillary tangles, amyotrophy and progressive motor disturbance in mice expressing mutant (P301L) tau protein, Nat. Genet. 25 (2000) 402–405.

[123] J. Gotz, F. Chen, R. Barmettler, R.M. Nitsch, Tau filament formation in transgenic mice expressing P301L tau, J. Biol. Chem. 276 (2001) 529–534.

[124] F. Lim, F. Hernandez, J.J. Lucas, P. Gomez-Ramos, M.A. Moran, J. Avila, FTDP-17 mutations in tau transgenic mice provoke lysosomal abnormalities and Tau filaments in forebrain, Mol. Cell. Neurosci. 18 (2001) 702–714.

[125] B. Allen, E. Ingram, M. Takao, M.J. Smith, R. Jakes, K. Virdee, H. Yoshida, M. Holzer, M. Craxton, P.C. Emson, C. Atzori, A. Migheli, R.A. Crowther, B. Ghetti, M.G. Spillantini, M. Goedert, Abundant tau filaments and nonapoptotic neurodegeneration in transgenic mice expressing human P301S tau protein, J. Neurosci. 22 (2002) 9340–9351.

[126] K. Tanemura, M. Murayama, T. Akagi, T. Hashikawa, T. Tominaga, M. Ichikawa, H. Yamaguchi, A. Takashima, Neurodegeneration with tau accumulation in a transgenic mouse expressing V337M human tau, J. Neurosci. 22 (2002) 133–141.

[127] Y. Tatebayashi, T. Miyasaka, D.H. Chui, T. Akagi, K. Mishima, K. Iwasaki, M. Fujiwara, K. Tanemura, M. Murayama, K. Ishiguro, E. Planel, S. Sato, T. Hashikawa, A. Takashima, Tau filament formation and associative memory deficit in aged mice expressing mutant (R406W) human tau, Proc. Natl. Acad. Sci. USA 99 (2002) 13896–13901.

[128] B. Zhang, M. Higuchi, Y. Yoshiyama, T. Ishihara, M.S. Forman, D. Martinez, S. Joyce, J.Q. Trojanowski, V.M. Lee, Retarded axonal transport of R406W mutant tau in transgenic mice with a neurodegenerative tauopathy, J. Neurosci. 24 (2004) 4657–4667.

[129] K. Santacruz, J. Lewis, T. Spires, J. Paulson, L. Kotilinek, M. Ingelsson, A. Guimaraes, M. Deture, M. Ramsden, E. Mcgowan, C. Forster, M. Yue, J. Orne, C. Janus, A. Mariash, M. Kuskowski, B. Hyman, M. Hutton, K.H. Ashe, Tau suppression in a neurodegenerative mouse model improves memory function, Science 309 (2005) 476–481.

[130] D. Terwel, R. Lasrado, J. Snauwaert, E. Vandeweert, C. Van Haesendonck, P. Borghgraef, F. Van Leuven, Changed conformation of mutant Tau-P301L underlies the moribund tauopathy, absent in progressive, nonlethal axonopathy of Tau-4R/2N transgenic mice, J. Biol. Chem. 280 (2005) 3963–3973.

[131] K. Schindowski, A. Bretteville, K. Leroy, S. Begard, J.P. Brion, M. Hamdane, L. Buee, Alzheimer's disease-like tau neuropathology leads to memory deficits and loss of functional synapses in a novel mutated tau transgenic mouse without any motor deficits, Am. J. Pathol. 169 (2006) 599–616.

[132] Y. Yoshiyama, M. Higuchi, B. Zhang, S.M. Huang, N. Iwata, T.C. Saido, J. Maeda, T. Suhara, J.Q. Trojanowski, V.M. Lee, Synapse loss and microglial activation precede tangles in a P301S tauopathy mouse model, Neuron 53 (2007) 337–351.

[133] A. de Calignon, M. Polydoro, M. Suarez-Calvet, C. William, D.H. Adamowicz, K.J. Kopeikina, R. Pitstick, N. Sahara, K.H. Ashe, G.A. Carlson, T.L. Spires-Jones, B.T. Hyman, Propagation of tau pathology in a model of early Alzheimer's disease, Neuron 73 (2012) 685–697.

[134] K.V. Kuchibhotla, S. Wegmann, K.J. Kopeikina, J. Hawkes, N. Rudinskiy, M.L. Andermann, T.L. Spires-Jones, B.J. Bacskai, B.T. Hyman, Neurofibrillary tangle-bearing neurons are functionally integrated in cortical circuits in vivo, Proc. Natl. Acad. Sci. USA 111 (2014) 510–514.

[135] M.M. Mocanu, A. Nissen, K. Eckermann, I. Khlistunova, J. Biernat, D. Drexler, O. Petrova, K. Schonig, H. Bujard, E. Mandelkow, L. Zhou, G. Rune, E.M. Mandelkow, The potential for beta-structure in the repeat domain of tau protein determines aggregation, synaptic decay, neuronal loss, and coassembly with endogenous Tau in inducible mouse models of tauopathy, J. Neurosci. 28 (2008) 737–748.

[136] K. Hochgrafe, A. Sydow, E.M. Mandelkow, Regulatable transgenic mouse models of Alzheimer disease: onset, reversibility and spreading of Tau pathology, FEBS J. 280 (2013) 4371–4381.

[137] K.R. Brunden, J.Q. Trojanowski, V.M. Lee, Advances in tau-focused drug discovery for Alzheimer's disease and related tauopathies, Nat. Rev. Drug Discov. 8 (2009) 783–793.

[138] P.V. Arriagada, J.H. Growdon, E.T. Hedley-Whyte, B.T. Hyman, Neurofibrillary tangles but not senile plaques parallel duration and severity of Alzheimer's disease, Neurology 42 (1992) 631–639.

[139] C. Bancher, H. Braak, P. Fischer, K.A. Jellinger, Neuropathological staging of Alzheimer lesions and intellectual status in Alzheimer's and Parkinson's disease patients, Neurosci. Lett. 162 (1993) 179–182.

[140] R.D. Terry, The pathogenesis of Alzheimer disease: an alternative to the amyloid hypothesis, J. Neuropathol. Exp. Neurol. 55 (1996) 1023–1025.

[141] E. Masliah, L. Hansen, A. Adame, L. Crews, F. Bard, C. Lee, P. Seubert, D. Games, L. Kirby, D. Schenk, Abeta vaccination effects on plaque pathology in the absence of encephalitis in Alzheimer disease, Neurology 64 (2005) 129–131.

[142] J.A. Nicoll, E. Barton, D. Boche, J.W. Neal, I. Ferrer, P. Thompson, C. Vlachouli, D. Wilkinson, A. Bayer, D. Games, P. Seubert, D. Schenk, C. Holmes, Abeta species removal after abeta42 immunization, J. Neuropathol. Exp. Neurol. 65 (2006) 1040–1048.

[143] H. Zhang, F. Burrows, Targeting multiple signal transduction pathways through inhibition of Hsp90, J. Mol. Med. 82 (2004) 488–499.

[144] J.P. Brion, A.M. Couck, E. Passareiro, J. Flament-Durand, Neurofibrillary tangles of Alzheimer's disease: an immunohistochemical study, J. Submicrosc. Cytol. 17 (1985) 89–96.

[145] A. Delacourte, A. Defossez, Alzheimer's disease: Tau proteins, the promoting factors of microtubule assembly, are major components of paired helical filaments, J. Neurol. Sci. 76 (1986) 173–186.

[146] I. Grundke-Iqbal, K. Iqbal, M. Quinlan, Y.C. Tung, M.S. Zaidi, H.M. Wisniewski, Microtubule-associated protein tau. A component of Alzheimer paired helical filaments, J. Biol. Chem. 261 (1986) 6084–6089.

[147] Y. Ihara, N. Nukina, R. Miura, M. Ogawara, Phosphorylated tau protein is integrated into paired helical filaments in Alzheimer's disease, J. Biochem. 99 (1986) 1807–1810.

[148] K.S. Kosik, C.L. Joachim, D.J. Selkoe, Microtubule-associated protein tau (tau) is a major antigenic component of paired helical filaments in Alzheimer disease, Proc. Natl. Acad. Sci. USA 83 (1986) 4044–4048.

[149] M. Goedert, C.M. Wischik, R.A. Crowther, J.E. Walker, A. Klug, Cloning and sequencing of the cDNA encoding a core protein of the paired helical filament of Alzheimer disease: identification as the microtubule-associated protein tau, Proc. Natl. Acad. Sci. USA 85 (1988) 4051–4055.

[150] C.M. Wischik, M. Novak, P.C. Edwards, A. Klug, W. Tichelaar, R.A. Crowther, Structural characterization of the core of the paired helical filament of Alzheimer disease, Proc. Natl. Acad. Sci. USA 85 (1988) 4884–4888.

[151] E. Kopke, Y.C. Tung, S. Shaikh, A.C. Alonso, K. Iqbal, I. Grundke-Iqbal, Microtubule-associated protein tau. Abnormal phosphorylation of a non-paired helical filament pool in Alzheimer disease, J. Biol. Chem. 268 (1993) 24374–24384.

[152] D.P. Hanger, B.H. Anderton, W. Noble, Tau phosphorylation: the therapeutic challenge for neurodegenerative disease, Trends Mol. Med. 15 (2009) 112–119.

[153] D. Terwel, I. Dewachter, F. Van Leuven, Axonal transport, tau protein, and neurodegeneration in Alzheimer's disease, Neuromol. Med. 2 (2002) 151–165.

[154] F. Chen, D. David, A. Ferrari, J. Gotz, Posttranslational modifications of tau—role in human tauopathies and modeling in transgenic animals, Curr. Drug Targets 5 (2004) 503–515.

[155] D. Terwel, D. Muyllaert, I. Dewachter, P. Borghgraef, S. Croes, H. Devijver, F. Van Leuven, Amyloid activates GSK-3beta to aggravate neuronal tauopathy in bigenic mice, Am. J. Pathol. 172 (2008) 786–798.

[156] J.C. Augustinack, A. Schneider, E.M. Mandelkow, B.T. Hyman, Specific tau phosphorylation sites correlate with severity of neuronal cytopathology in Alzheimer's disease, Acta Neuropathol. 103 (2002) 26–35.

[157] D. Matenia, E.M. Mandelkow, The tau of MARK: a polarized view of the cytoskeleton, Trends Biochem. Sci. 34 (2009) 332–342.

[158] W. Yu, J. Polepalli, D. Wagh, J. Rajadas, R. Malenka, B. Lu, A critical role for the PAR-1/MARK-tau axis in mediating the toxic effects of Abeta on synapses and dendritic spines, Hum. Mol. Genet. 21 (2012) 1384–1390.

[159] G.J. Gu, H. Lund, D. Wu, A. Blokzijl, C. Classon, G. von Euler, U. Landegren, D. Sunnemark, M. Kamali-Moghaddam, Role of individual MARK isoforms in phosphorylation of tau at Ser(2)(6)(2) in Alzheimer's disease, Neuromol. Med. 15 (2013) 458–469.

[160] J.J. Lucas, F. Hernandez, P. Gomez-Ramos, M.A. Moran, R. Hen, J. Avila, Decreased nuclear beta-catenin, tau hyperphosphorylation and neurodegeneration in GSK-3beta conditional transgenic mice, EMBO J. 20 (2001) 27–39.

[161] F. Hernández, J. Borrell, C. Guaza, J. Avila, J.J. Lucas, Spatial learning deficit in transgenic mice that conditionally over-express GSK-3beta in the brain but do not form tau filaments, J. Neurochem. 83 (2002) 1529–1533.

[162] A.E. Aplin, J.S. Jacobsen, B.H. Anderton, J.M. Gallo, Effect of increased glycogen synthase kinase-3 activity upon the maturation of the amyloid precursor protein in transfected cells, Neuroreport 8 (1997) 639–643.

[163] M. Pérez, F. Hernandez, F. Lim, J. Diaz-Nido, J. Avila, Chronic lithium treatment decreases mutant tau protein aggregation in a transgenic mouse model, J. Alzheimers Dis. 5 (2003) 301–308.

[164] C.J. Phiel, C.A. Wilson, V.M. Lee, P.S. Klein, GSK-3alpha regulates production of Alzheimer's disease amyloid-beta peptides, Nature 423 (2003) 435–439.

[165] W. Noble, E. Planel, C. Zehr, V. Olm, J. Meyerson, F. Suleman, K. Gaynor, L. Wang, J. Lafrancois, B. Feinstein, M. Burns, P. Krishnamurthy, Y. Wen, R. Bhat, J. Lewis, D. Dickson, K. Duff, Inhibition of glycogen synthase kinase-3 by lithium correlates with reduced tauopathy and degeneration in vivo, Proc. Natl. Acad. Sci. USA 102 (2005) 6990–6995.

[166] Y.H. Chong, Y.J. Shin, E.O. Lee, R. Kayed, C.G. Glabe, A.J. Tenner, ERK1/2 activation mediates Abeta oligomer-induced neurotoxicity via caspase-3 activation and tau cleavage in rat organotypic hippocampal slice cultures, J. Biol. Chem. 281 (2006) 20315–20325.

[167] T. Engel, P. Goni-Oliver, J.J. Lucas, J. Avila, F. Hernandez, Chronic lithium administration to FTDP-17 tau and GSK-3beta overexpressing mice prevents tau hyperphosphorylation and neurofibrillary tangle formation, but pre-formed neurofibrillary tangles do not revert, J. Neurochem. 99 (2006) 1445–1455.

[168] A. Caccamo, S. Oddo, L.X. Tran, F.M. Laferla, Lithium reduces tau phosphorylation but not A beta or working memory deficits in a transgenic model with both plaques and tangles, Am. J. Pathol. 170 (2007) 1669–1675.

[169] M.P. Mazanetz, P.M. Fischer, Untangling tau hyperphosphorylation in drug design for neurodegenerative diseases, Nat. Rev. Drug Discov. 6 (2007) 464–479.

[170] E. Rockenstein, M. Torrance, A. Adame, M. Mante, P. Bar-On, J.B. Rose, L. Crews, E. Masliah, Neuroprotective effects of regulators of the glycogen synthase kinase-3beta signaling pathway in a transgenic model of Alzheimer's disease are associated with reduced amyloid precursor protein phosphorylation, J. Neurosci. 27 (2007) 1981–1991.

[171] W. Noble, V. Olm, K. Takata, E. Casey, O. Mary, J. Meyerson, K. Gaynor, J. Lafrancois, L. Wang, T. Kondo, P. Davies, M. Burns, Veeranna, R. Nixon, D. Dickson, Y. Matsuoka, M. Ahlijanian, L.F. Lau, K. Duff, Cdk5 is a key factor in tau aggregation and tangle formation in vivo, Neuron 38 (2003) 555–565.

[172] D. Muyllaert, D. Terwel, A. Kremer, K. Sennvik, P. Borghgraef, H. Devijver, I. Dewachter, F. Van Leuven, Neurodegeneration and neuroinflammation in cdk5/p25-inducible mice: a model for hippocampal sclerosis and neocortical degeneration, Am. J. Pathol. 172 (2008) 470–485.

[173] S.D. Yan, S.F. Yan, X. Chen, J. Fu, M. Chen, P. Kuppusamy, M.A. Smith, G. Perry, G.C. Godman, P. Nawroth, et al. Non-enzymatically glycated tau in Alzheimer's disease induces neuronal oxidant stress resulting in cytokine gene expression and release of amyloid beta-peptide, Nat. Med. 1 (1995) 693–699.

[174] F. Liu, T. Zaidi, K. Iqbal, I. Grundke-Iqbal, R.K. Merkle, C.X. Gong, Role of glycosylation in hyperphosphorylation of tau in Alzheimer's disease, FEBS Lett. 512 (2002) 101–106.

[175] F. Liu, K. Iqbal, I. Grundke-Iqbal, G.W. Hart, C.X. Gong, O-GlcNAcylation regulates phosphorylation of tau: a mechanism involved in Alzheimer's disease, Proc. Natl. Acad. Sci. USA 101 (2004) 10804–10809.

[176] M.R. Reynolds, J.F. Reyes, Y. Fu, E.H. Bigio, A.L. Guillozet-Bongaarts, R.W. Berry, L.I. Binder, Tau nitration occurs at tyrosine 29 in the fibrillar lesions of Alzheimer's disease and other tauopathies, J. Neurosci. 26 (2006) 10636–10645.

[177] S.W. Min, S.H. Cho, Y. Zhou, S. Schroeder, V. Haroutunian, W.W. Seeley, E.J. Huang, Y. Shen, E. Masliah, C. Mukherjee, D. Meyers, P.A. Cole, M. Ott, L. Gan, Acetylation of tau inhibits its degradation and contributes to tauopathy, Neuron 67 (2010) 953–966.

[178] T.J. Cohen, J.L. Guo, D.E. Hurtado, L.K. Kwong, I.P. Mills, J.Q. Trojanowski, V.M. Lee, The acetylation of tau inhibits its function and promotes pathological tau aggregation, Nat. Commun. 2 (2011) 252.

[179] D.J. Irwin, T.J. Cohen, M. Grossman, S.E. Arnold, S.X. Xie, V.M. Lee, J.Q. Trojanowski, Acetylated tau, a novel pathological signature in Alzheimer's disease and other tauopathies, Brain 135 (2012) 807–818.

[180] S.A. Yuzwa, X. Shan, M.S. Macauley, T. Clark, Y. Skorobogatko, K. Vosseller, D.J. Vocadlo, Increasing O-GlcNAc slows neurodegeneration and stabilizes tau against aggregation, Nat. Chem. Biol. 8 (2012) 393–399.

[181] B. Zhang, A. Maiti, S. Shively, F. Lakhani, G. Mcdonald-Jones, J. Bruce, E.B. Lee, S.X. Xie, S. Joyce, C. Li, P.M. Toleikis, V.M. Lee, J.Q. Trojanowski, Microtubule-binding drugs offset tau sequestration by stabilizing microtubules and reversing fast axonal transport deficits in a tauopathy model, Proc. Natl. Acad. Sci. USA 102 (2005) 227–231.

[182] K.R. Brunden, B. Zhang, J. Carroll, Y. Yao, J.S. Potuzak, A.M. Hogan, M. Iba, M.J. James, S.X. Xie, C. Ballatore, A.B. Smith 3rd, V.M. Lee, J.Q. Trojanowski, Epothilone D improves microtubule density, axonal integrity, and cognition in a transgenic mouse model of tauopathy, J. Neurosci. 30 (2010) 13861–13866.

[183] D.M. Barten, P. Fanara, C. Andorfer, N. Hoque, P.Y. Wong, K.H. Husted, G.W. Cadelina, L.B. Decarr, L. Yang, V. Liu, C. Fessler, J. Protassio, T. Riff, H. Turner, C.G. Janus, S. Sankaranarayanan, C. Polson, J.E. Meredith, G. Gray, A. Hanna, R.E. Olson, S.H. Kim, G.D. Vite, F.Y. Lee, C.F. Albright, Hyperdynamic microtubules, cognitive deficits, and pathology are improved in tau transgenic mice with low doses of the microtubule-stabilizing agent BMS-241027, J. Neurosci. 32 (2012) 7137–7145.

[184] B. Zhang, J. Carroll, J.Q. Trojanowski, Y. Yao, M. Iba, J.S. Potuzak, A.M. Hogan, S.X. Xie, C. Ballatore, A.B. Smith 3rd, V.M. Lee, K.R. Brunden, The microtubule-stabilizing agent, epothilone D, reduces axonal

dysfunction, neurotoxicity, cognitive deficits, and Alzheimer-like pathology in an interventional study with aged tau transgenic mice, J. Neurosci. 32 (2012) 3601–3611.

[185] I. Gozes, I. Divinski, The femtomolar-acting NAP interacts with microtubules: novel aspects of astrocyte protection, J. Alzheimers Dis. 6 (2004) S37–S41.

[186] Y. Matsuoka, A.J. Gray, C. Hirata-Fukae, S.S. Minami, E.G. Waterhouse, M.P. Mattson, F.M. Laferla, I. Gozes, P.S. Aisen, Intranasal NAP administration reduces accumulation of amyloid peptide and tau hyper-phosphorylation in a transgenic mouse model of Alzheimer's disease at early pathological stage, J. Mol. Neurosci. 31 (2007) 165–170.

[187] Y. Matsuoka, Y. Jouroukhin, A.J. Gray, L. Ma, C. Hirata-Fukae, H.F. Li, L. Feng, L. Lecanu, B.R. Walker, E. Planel, O. Arancio, I. Gozes, P.S. Aisen, A neuronal microtubule-interacting agent, NAPVSIPQ, reduces tau pathology and enhances cognitive function in a mouse model of Alzheimer's disease, J. Pharmacol. Exp. Ther. 325 (2008) 146–153.

[188] B. Ravikumar, S. Sarkar, Z. Berger, D.C. Rubinsztein, The roles of the ubiquitin-proteasome and autoph-agy–lysosome pathways in Huntington's disease and related conditions, Clin. Neurosci. Res. 3 (2003) 141–148.

[189] N. Perez, J. Sugar, S. Charya, G. Johnson, C. Merril, L. Bierer, D. Perl, V. Haroutunian, W. Wallace, In-creased synthesis and accumulation of heat shock 70 proteins in Alzheimer's disease, Brain Res. Mol. Brain Res. 11 (1991) 249–254.

[190] J.N. Keller, K.B. Hanni, W.R. Markesbery, Impaired proteasome function in Alzheimer's disease, J. Neuro-chem. 75 (2000) 436–439.

[191] C.A. Dickey, J. Dunmore, B. Lu, J.W. Wang, W.C. Lee, A. Kamal, F. Burrows, C. Eckman, M. Hutton, L. Petrucelli, HSP induction mediates selective clearance of tau phosphorylated at proline-directed Ser/Thr sites but not KXGS (MARK) sites, FASEB J. 20 (2006) 753–755.

[192] C.A. Dickey, A. Kamal, K. Lundgren, N. Klosak, R.M. Bailey, J. Dunmore, P. Ash, S. Shoraka, J. Zlatkovic, C.B. Eckman, C. Patterson, D.W. Dickson, N.S. Nahman Jr., M. Hutton, F. Burrows, L. Petrucelli, The high-affinity HSP90-CHIP complex recognizes and selectively degrades phosphorylated tau client proteins, J. Clin. Invest. 117 (2007) 648–658.

[193] W. Luo, F. Dou, A. Rodina, S. Chip, J. Kim, Q. Zhao, K. Moulick, J. Aguirre, N. Wu, P. Greengard, G. Chiosis, Roles of heat-shock protein 90 in maintaining and facilitating the neurodegenerative phenotype in tauopathies, Proc. Natl. Acad. Sci. USA 104 (2007) 9511–9516.

[194] N. Myeku, C.L. Clelland, S. Emrani, N.V. Kukushkin, W.H. Yu, A.L. Goldberg, K.E. Duff, Tau-driven 26S proteasome impairment and cognitive dysfunction can be prevented early in disease by activating cAMP-PKA signaling, Nat. Med. 22 (2016) 46–53.

[195] A. Williams, L. Jahreiss, S. Sarkar, S. Saiki, F.M. Menzies, B. Ravikumar, D.C. Rubinsztein, Aggregate-prone proteins are cleared from the cytosol by autophagy: therapeutic implications, Curr. Top. Dev. Biol. 76 (2006) 89–101.

[196] T. Jiang, J.T. Yu, X.C. Zhu, Q.Q. Zhang, L. Cao, H.F. Wang, M.S. Tan, Q. Gao, H. Qin, Y.D. Zhang, L. Tan, Temsirolimus attenuates tauopathy in vitro and in vivo by targeting tau hyperphosphorylation and autopha-gic clearance, Neuropharmacology 85 (2014) 121–130.

[197] T. Wisniewski, A. Boutajangout, Vaccination as a therapeutic approach to Alzheimer's disease, Mt. Sinai J. Med. 77 (2010) 17–31.

[198] J.T. Pedersen, E.M. Sigurdsson, Tau immunotherapy for Alzheimer's disease, Trends Mol. Med. 21 (2015) 394–402.

[199] A.A. Asuni, A. Boutajangout, H. Scholtzova, E. Knudsen, Y.S. Li, D. Quartermain, B. Fran-gione, T. Wisniewski, E.M. Sigurdsson, Vaccination of Alzheimer's model mice with Abeta deriva-tive in alum adjuvant reduces Abeta burden without microhemorrhages, Eur J. Neurosci. 24 (2006) 2530–2542.

[200] A. Boutajangout, D. Quartermain, E.M. Sigurdsson, Immunotherapy targeting pathological tau prevents cognitive decline in a new tangle mouse model, J. Neurosci. 30 (2010) 16559–16566.

[201] L. Troquier, R. Caillierez, S. Burnouf, F.J. Fernandez-Gomez, M.E. Grosjean, N. Zommer, N. Sergeant, S. Schraen-Maschke, D. Blum, L. Buee, Targeting phospho-Ser422 by active Tau Immunotherapy in the THYTau22 mouse model: a suitable therapeutic approach, Curr. Alzheimer Res. 9 (2012) 397–405.

[202] K. Ando, A. Kabova, V. Stygelbout, K. Leroy, C. Heraud, C. Frederick, V. Suain, Z. Yilmaz, M. Authelet, R. Dedecker, M.C. Potier, C. Duyckaerts, J.P. Brion, Vaccination with Sarkosyl insoluble PHF-tau decrease neurofibrillary tangles formation in aged tau transgenic mouse model: a pilot study, J. Alzheimers Dis. 40 (Suppl. 1) (2014) S135–S145.

[203] H. Rosenmann, N. Grigoriadis, D. Karussis, M. Boimel, O. Touloumi, H. Ovadia, O. Abramsky, Tauopathy-like abnormalities and neurologic deficits in mice immunized with neuronal tau protein, Arch. Neurol. 63 (2006) 1459–1467.

[204] L. Rozenstein-Tsalkovich, N. Grigoriadis, A. Lourbopoulos, E. Nousiopoulou, I. Kassis, O. Abramsky, D. Karussis, H. Rosenmann, Repeated immunization of mice with phosphorylated-tau peptides causes neuroinflammation, Exp. Neurol. 248 (2013) 451–456.

[205] A. Boutajangout, J. Ingadottir, P. Davies, E.M. Sigurdsson, Passive immunization targeting pathological phospho-tau protein in a mouse model reduces functional decline and clears tau aggregates from the brain, J. Neurochem. 118 (2011) 658–667.

[206] X. Chai, S. Wu, T.K. Murray, R. Kinley, C.V. Cella, H. Sims, N. Buckner, J. Hanmer, P. Davies, M.J. O'Neill, M.L. Hutton, M. Citron, Passive immunization with anti-Tau antibodies in two transgenic models: reduction of Tau pathology and delay of disease progression, J. Biol. Chem. 286 (2011) 34457–34467.

[207] C. d'Abramo, C.M. Acker, H.T. Jimenez, P. Davies, Tau passive immunotherapy in mutant P301L mice: antibody affinity versus specificity, PLoS One 8 (2013) e62402.

[208] K. Yanamandra, N. Kfoury, H. Jiang, T.E. Mahan, S. Ma, S.E. Maloney, D.F. Wozniak, M.I. Diamond, D.M. Holtzman, Anti-tau antibodies that block tau aggregate seeding in vitro markedly decrease pathology and improve cognition in vivo, Neuron 80 (2013) 402–414.

[209] E.E. Congdon, J. Gu, H.B. Sait, E.M. Sigurdsson, Antibody uptake into neurons occurs primarily via clathrin-dependent Fcgamma receptor endocytosis and is a prerequisite for acute tau protein clearance, J. Biol. Chem. 288 (2013) 35452–35465.

[210] C. Theunis, N. Crespo-Biel, V. Gafner, M. Pihlgren, M.P. Lopez-Deber, P. Reis, D.T. Hickman, O. Adolfsson, N. Chuard, D.M. Ndao, P. Borghgraef, H. Devijver, F. Van Leuven, A. Pfeifer, A. Muhs, Efficacy and safety of a liposome-based vaccine against protein Tau, assessed in tau.P301L mice that model tauopathy, PLoS One 8 (2013) e72301.

[211] E. Kontsekova, N. Zilka, B. Kovacech, P. Novak, M. Novak, First-in-man tau vaccine targeting structural determinants essential for pathological tau-tau interaction reduces tau oligomerisation and neurofibrillary degeneration in an Alzheimer's disease model, Alzheimers Res. Ther. 6 (2014) 44.

[212] M. Jucker, L.C. Walker, Self-propagation of pathogenic protein aggregates in neurodegenerative diseases, Nature 501 (2013) 45–51.

[213] S.B. Prusiner, Novel proteinaceous infectious particles cause scrapie, Science 216 (1982) 136–144.

[214] N. Kfoury, B.B. Holmes, H. Jiang, D.M. Holtzman, M.I. Diamond, Trans-cellular propagation of Tau aggregation by fibrillar species, J. Biol. Chem. 287 (2012) 19440–19451.

[215] S. Calafate, A. Buist, K. Miskiewicz, V. Vijayan, G. Daneels, B. de Strooper, J. de Wit, P. Verstreken, D. Moechars, Synaptic contacts enhance cell-to-cell tau pathology propagation, Cell Rep. 11 (8.) (2015) 1176–1183.

[216] F. Clavaguera, T. Bolmont, R.A. Crowther, D. Abramowski, S. Frank, A. Probst, G. Fraser, A.K. Stalder, M. Beibel, M. Staufenbiel, M. Jucker, M. Goedert, M. Tolnay, Transmission and spreading of tauopathy in transgenic mouse brain, Nat. Cell Biol. 11 (2009) 909–913.

[217] Z. Ahmed, J. Cooper, T.K. Murray, K. Garn, E. Mcnaughton, H. Clarke, S. Parhizkar, M.A. Ward, A. Cavallini, S. Jackson, S. Bose, F. Clavaguera, M. Tolnay, I. Lavenir, M. Goedert, M.L. Hutton, M.J. O'Neill, A novel in vivo model of tau propagation with rapid and progressive neurofibrillary tangle pathology: the pattern of spread is determined by connectivity, not proximity, Acta Neuropathol. 127 (2014) 667–683.

[218] F. Clavaguera, H. Akatsu, G. Fraser, R.A. Crowther, S. Frank, J. Hench, A. Probst, D.T. Winkler, J. Reichwald, M. Staufenbiel, B. Ghetti, M. Goedert, M. Tolnay, Brain homogenates from human tauopathies induce tau inclusions in mouse brain, Proc. Natl. Acad. Sci. USA 110 (2013) 9535–9540.

[219] S. Boluda, M. Iba, B. Zhang, K.M. Raible, V.M. Lee, J.Q. Trojanowski, Differential induction and spread of tau pathology in young PS19 tau transgenic mice following intracerebral injections of pathological tau from Alzheimer's disease or corticobasal degeneration brains, Acta Neuropathol. 129 (2015) 221–237.

[220] L. Liu, V. Drouet, J.W. Wu, M.P. Witter, S.A. small, C. Clelland, K. Duff, Trans-synaptic spread of tau pathology in vivo, PLoS One 7 (2012) e31302.

[221] M. Iba, J.L. Guo, J.D. Mcbride, B. Zhang, J.Q. Trojanowski, V.M. Lee, Synthetic tau fibrils mediate transmission of neurofibrillary tangles in a transgenic mouse model of Alzheimer's-like tauopathy, J. Neurosci. 33 (2013) 1024–1037.

[222] C.A. Lasagna-Reeves, D.L. Castillo-Carranza, U. Sengupta, J. Sarmiento, J. Troncoso, G.R. Jackson, R. Kayed, Identification of oligomers at early stages of tau aggregation in Alzheimer's disease, FASEB J. 26 (2012) 1946–1959.

[223] I.C. Stancu, B. Vasconcelos, L. Ris, P. Wang, A. Villers, E. Peeraer, A. Buist, D. Terwel, P. Baatsen, T. Oyelami, N. Pierrot, C. Casteels, G. Bormans, P. Kienlen-Campard, J.N. Octave, D. Moechars, I. Dewachter, Templated misfolding of Tau by prion-like seeding along neuronal connections impairs neuronal network function and associated behavioral outcomes in Tau transgenic mice, Acta Neuropathol. 129 (2015) 875–894.

[224] D.W. Sanders, S.K. Kaufman, S.L. Devos, A.M. Sharma, H. Mirbaha, A. Li, S.J. Barker, A.C. Foley, J.R. Thorpe, L.C. Serpell, T.M. Miller, L.T. Grinberg, W.W. Seeley, M.I. Diamond, Distinct tau prion strains propagate in cells and mice and define different tauopathies, Neuron 82 (2014) 1271–1288.

[225] K. Yamada, J.R. Cirrito, F.R. Stewart, H. Jiang, M.B. Finn, B.B. Holmes, L.I. Binder, E.M. Mandelkow, M.I. Diamond, V.M. Lee, D.M. Holtzman, In vivo microdialysis reveals age-dependent decrease of brain interstitial fluid tau levels in P301S human tau transgenic mice, J. Neurosci. 31 (2011) 13110–13117.

[226] A.M. Pooler, E.C. Phillips, D.H. Lau, W. Noble, D.P. Hanger, Physiological release of endogenous tau is stimulated by neuronal activity, EMBO Rep. 14 (2013) 389–394.

[227] J. Faure, G. Lachenal, M. Court, J. Hirrlinger, C. Chatellard-Causse, B. Blot, J. Grange, G. Schoehn, Y. Goldberg, V. Boyer, F. Kirchhoff, G. Raposo, J. Garin, R. Sadoul, Exosomes are released by cultured cortical neurones, Mol. Cell. Neurosci. 31 (2006) 642–648.

[228] X. Chai, J.L. Dage, M. Citron, Constitutive secretion of tau protein by an unconventional mechanism, Neurobiol. Dis. 48 (2012) 356–366.

[229] S. Lee, W. Kim, Z. Li, G.F. Hall, Accumulation of vesicle-associated human tau in distal dendrites drives degeneration and tau secretion in an in situ cellular tauopathy model, Int. J. Alzheimers Dis. 2012 (2012) 172837.

[230] V. Plouffe, N.V. Mohamed, J. Rivest-Mcgraw, J. Bertrand, M. Lauzon, N. Leclerc, Hyperphosphorylation and cleavage at D421 enhance tau secretion, PLoS One 7 (2012) e36873.

[231] S. Saman, W. Kim, M. Raya, Y. Visnick, S. Miro, S. Saman, B. Jackson, A.C. Mckee, V.E. Alvarez, N.C. Lee, G.F. Hall, Exosome-associated tau is secreted in tauopathy models and is selectively phosphorylated in cerebrospinal fluid in early Alzheimer disease, J. Biol. Chem. 287 (2012) 3842–3849.

[232] D. Simon, E. Garcia-Garcia, A. Gomez-Ramos, J.M. Falcon-Perez, M. Diaz-Hernandez, F. Hernandez, J. Avila, Tau overexpression results in its secretion via membrane vesicles, Neurodegener. Dis. 10 (2012) 73–75.

[233] B.B. Holmes, M.I. Diamond, Prion-like properties of Tau protein: the importance of extracellular Tau as a therapeutic target, J. Biol. Chem. 289 (2014) 19855–19861.

[234] I. Santa-Maria, M. Varghese, H. Ksiezak-Reding, A. Dzhun, J. Wang, G.M. Pasinetti, Paired helical filaments from Alzheimer disease brain induce intracellular accumulation of Tau protein in aggresomes, J. Biol. Chem. 287 (2012) 20522–20533.

[235] J.W. Wu, M. Herman, L. Liu, S. Simoes, C.M. Acker, H. Figueroa, J.I. Steinberg, M. Margittai, R. Kayed, C. Zurzolo, G. Di Paolo, K.E. Duff, Small misfolded Tau species are internalized via bulk endocytosis and anterogradely and retrogradely transported in neurons, J. Biol. Chem. 288 (2013) 1856–1870.

[236] C.A. Lasagna-Reeves, D.L. Castillo-Carranza, U. Sengupta, M.J. Guerrero-Munoz, T. Kiritoshi, V. Neugebauer, G.R. Jackson, R. Kayed, Alzheimer brain-derived tau oligomers propagate pathology from endogenous tau, Sci. Rep. 2 (2012) 700.

[237] B.B. Holmes, S.L. Devos, N. Kfoury, M. Li, R. Jacks, K. Yanamandra, M.O. Ouidja, F.M. Brodsky, J. Marasa, D.P. Bagchi, P.T. Kotzbauer, T.M. Miller, D. Papy-Garcia, M.I. Diamond, Heparan sulfate proteoglycans mediate internalization and propagation of specific proteopathic seeds, Proc. Natl. Acad. Sci. USA 110 (2013) E3138–E3147.

[238] H.C. Christianson, K.J. Svensson, T.H. Van Kuppevelt, J.P. Li, M. Belting, Cancer cell exosomes depend on cell-surface heparan sulfate proteoglycans for their internalization and functional activity, Proc. Natl. Acad. Sci. USA 110 (2013) 17380–17385.

[239] A.M. Pooler, M. Polydoro, S. Wegmann, S.B. Nicholls, T.L. Spires-Jones, B.T. Hyman, Propagation of tau pathology in Alzheimer's disease: identification of novel therapeutic targets, Alzheimers Res. Ther. 5 (2013) 49.

[240] D. Morgan, D.M. Diamond, P.E. Gottschall, K.E. Ugen, C. Dickey, J. Hardy, K. Duff, P. Jantzen, G. Dicarlo, D. Wilcock, K. Connor, J. Hatcher, C. Hope, M. Gordon, G.W. Arendash, A beta peptide vaccination prevents memory loss in an animal model of Alzheimer's disease, Nature 408 (2000) 982–985.

[241] D.M. Wilcock, A. Rojiani, A. Rosenthal, G. Levkowitz, S. Subbarao, J. Alamed, D. Wilson, N. Wilson, M.J. Freeman, M.N. Gordon, D. Morgan, Passive amyloid immunotherapy clears amyloid and transiently activates microglia in a transgenic mouse model of amyloid deposition, J. Neurosci. 24 (2004) 6144–6151.

[242] A.A. Asuni, A. Boutajangout, D. Quartermain, E.M. Sigurdsson, Immunotherapy targeting pathological tau conformers in a tangle mouse model reduces brain pathology with associated functional improvements, J. Neurosci. 27 (2007) 9115–9129.

[243] M. Boimel, N. Grigoriadis, A. Lourbopoulos, E. Haber, O. Abramsky, H. Rosenmann, Efficacy and safety of immunization with phosphorylated tau against neurofibrillary tangles in mice, Exp. Neurol. 224 (2010) 472–485.

[244] M. Bi, A. Ittner, Y.D. Ke, J. Gotz, L.M. Ittner, Tau-targeted immunization impedes progression of neurofibrillary histopathology in aged P301L tau transgenic mice, PLoS One 6 (2011) e26860.

[245] C.E. Bulawa, S. Connelly, M. Devit, L. Wang, C. Weigel, J.A. Fleming, J. Packman, E.T. Powers, R.L. Wiseman, T.R. Foss, I.A. Wilson, J.W. Kelly, R. Labaudiniere, Tafamidis, a potent and selective transthyretin kinetic stabilizer that inhibits the amyloid cascade, Proc. Natl. Acad. Sci. USA 109 (2012) 9629–9634.

[246] D. Games, D. Adams, R. Alessandrini, R. Barbour, P. Berthelette, C. Blackwell, T. Carr, J. Clemens, T. Donaldson, F. Gillespie, et al. Alzheimer-type neuropathology in transgenic mice overexpressing V717F beta-amyloid precursor protein, Nature 373 (1995) 523–527.

[247] J. Gotz, F. Chen, J. van Dorpe, R.M. Nitsch, Formation of neurofibrillary tangles in P301l tau transgenic mice induced by Abeta 42 fibrils, Science 293 (2001) 1491–1495.

[248] J. Lewis, D.W. Dickson, W.L. Lin, L. Chisholm, A. Corral, G. Jones, S.H. Yen, N. Sahara, L. Skipper, D. Yager, C. Eckman, J. Hardy, M. Hutton, E. Mcgowan, Enhanced neurofibrillary degeneration in transgenic mice expressing mutant tau and APP, Science 293 (2001) 1487–1491.

[249] S. Oddo, A. Caccamo, M. Kitazawa, B.P. Tseng, F.M. Laferla, Amyloid deposition precedes tangle formation in a triple transgenic model of Alzheimer's disease, Neurobiol. Aging 24 (2003) 1063–1070.

[250] V. Echeverria, A. Ducatenzeiler, E. Dowd, J. Janne, S.M. Grant, M. Szyf, F. Wandosell, J. Avila, H. Grimm, S.B. Dunnett, T. Hartmann, L. Alhonen, A.C. Cuello, Altered mitogen-activated protein kinase signaling, tau hyperphosphorylation and mild spatial learning dysfunction in transgenic rats expressing the beta-amyloid peptide intracellularly in hippocampal and cortical neurons, Neuroscience 129 (2004) 583–592.

[251] T. Bolmont, F. Clavaguera, M. Meyer-Luehmann, M.C. Herzig, R. Radde, M. Staufenbiel, J. Lewis, M. Hutton, M. Tolnay, M. Jucker, Induction of tau pathology by intracerebral infusion of amyloid-beta -containing brain extract and by amyloid-beta deposition in APP x Tau transgenic mice, Am. J. Pathol. 171 (2007) 2012–2020.

[252] D.E. Hurtado, L. Molina-Porcel, M. Iba, A.K. Aboagye, S.M. Paul, J.Q. Trojanowski, V.M. Lee, Abeta accelerates the spatiotemporal progression of tau pathology and augments tau amyloidosis in an Alzheimer mouse model, Am. J. Pathol. 177 (2010) 1977–1988.

[253] M.A. Chabrier, M. Blurton-Jones, A.A. Agazaryan, J.L. Nerhus, H. Martinez-Coria, F.M. Laferla, Soluble abeta promotes wild-type tau pathology in vivo, J. Neurosci. 32 (2012) 17345–17350.

[254] A. Saul, F. Sprenger, T.A. Bayer, O. Wirths, Accelerated tau pathology with synaptic and neuronal loss in a novel triple transgenic mouse model of Alzheimer's disease, Neurobiol. Aging 34 (2013) 2564–2573.

[255] C. Heraud, D. Goufak, K. Ando, K. Leroy, V. Suain, Z. Yilmaz, R. De Decker, M. Authelet, V. Laporte, J.N. Octave, J.P. Brion, Increased misfolding and truncation of tau in APP/PS1/tau transgenic mice compared to mutant tau mice, Neurobiol. Dis. 62 (2014) 100–112.

[256] T. Umeda, S. Maekawa, T. Kimura, A. Takashima, T. Tomiyama, H. Mori, Neurofibrillary tangle formation by introducing wild-type human tau into APP transgenic mice, Acta Neuropathol. 127 (2014) 685–698.

[257] C. Hertel, E. Terzi, N. Hauser, R. Jakob-Rotne, J. Seelig, J.A. Kemp, Inhibition of the electrostatic interaction between beta-amyloid peptide and membranes prevents beta-amyloid-induced toxicity, Proc. Natl. Acad. Sci. USA 94 (1997) 9412–9416.

[258] H.Y. Wang, D.H. Lee, M.R. D'Andrea, P.A. Peterson, R.P. Shank, A.B. Reitz, beta-Amyloid(1-42) binds to alpha7 nicotinic acetylcholine receptor with high affinity. Implications for Alzheimer's disease pathology, J. Biol. Chem. 275 (2000) 5626–5632.

[259] S.D. Yan, A. Roher, M. Chaney, B. Zlokovic, A.M. Schmidt, D. Stern, Cellular cofactors potentiating induction of stress and cytotoxicity by amyloid beta-peptide, Biochim. Biophys. Acta 1502 (2000) 145–157.

[260] P.N. Lacor, M.C. Buniel, L. Chang, S.J. Fernandez, Y. Gong, K.L. Viola, M.P. Lambert, P.T. Velasco, E.H. Bigio, C.E. Finch, G.A. Krafft, W.L. Klein, Synaptic targeting by Alzheimer's-related amyloid beta oligomers, J. Neurosci. 24 (2004) 10191–10200.

[261] Y. Verdier, M. Zarandi, B. Penke, Amyloid beta-peptide interactions with neuronal and glial cell plasma membrane: binding sites and implications for Alzheimer's disease, J. Pept. Sci. 10 (2004) 229–248.

[262] P.N. Lacor, M.C. Buniel, P.W. Furlow, A.S. Clemente, P.T. Velasco, M. Wood, K.L. Viola, W.L. Klein, Abeta oligomer-induced aberrations in synapse composition, shape, and density provide a molecular basis for loss of connectivity in Alzheimer's disease, J. Neurosci. 27 (2007) 796–807.

[263] O. Simakova, N.J. Arispe, The cell-selective neurotoxicity of the Alzheimer's Abeta peptide is determined by surface phosphatidylserine and cytosolic ATP levels. Membrane binding is required for Abeta toxicity, J. Neurosci. 27 (2007) 13719–13729.

[264] R.M. Koffie, M. Meyer-Luehmann, T. Hashimoto, K.W. Adams, M.L. Mielke, M. Garcia-Alloza, K.D. Micheva, S.J. Smith, M.L. Kim, V.M. Lee, B.T. Hyman, T.L. Spires-Jones, Oligomeric amyloid beta associates with postsynaptic densities and correlates with excitatory synapse loss near senile plaques, Proc. Natl. Acad. Sci. USA 106 (2009) 4012–4017.

[265] J. Lauren, D.A. Gimbel, H.B. Nygaard, J.W. Gilbert, S.M. Strittmatter, Cellular prion protein mediates impairment of synaptic plasticity by amyloid-beta oligomers, Nature 457 (2009) 1128–1132.

[266] C. Balducci, M. Beeg, M. Stravalaci, A. Bastone, A. Sclip, E. Biasini, L. Tapella, L. Colombo, C. Manzoni, T. Borsello, R. Chiesa, M. Gobbi, M. Salmona, G. Forloni, Synthetic amyloid-beta oligomers impair long-term memory independently of cellular prion protein, Proc. Natl. Acad. Sci. USA 107 (2010) 2295–2300.

[267] D.A. Gimbel, H.B. Nygaard, E.E. Coffey, E.C. Gunther, J. Lauren, Z.A. Gimbel, S.M. Strittmatter, Memory impairment in transgenic Alzheimer mice requires cellular prion protein, J. Neurosci. 30 (2010) 6367–6374.

[268] M. Cisse, B. Halabisky, J. Harris, N. Devidze, D.B. Dubal, B. Sun, A. Orr, G. Lotz, D.H. Kim, P. Hamto, K. Ho, G.Q. Yu, L. Mucke, Reversing EphB2 depletion rescues cognitive functions in Alzheimer model, Nature 469 (2011) 47–52.

[269] I. Benilova, B. De Strooper, Promiscuous Alzheimer's amyloid: yet another partner, Science 341 (2013) 1354–1355.

[270] T. Kim, G.S. Vidal, M. Djurisic, C.M. William, M.E. Birnbaum, K.C. Garcia, B.T. Hyman, C.J. Shatz, Human LilrB2 is a beta-amyloid receptor and its murine homolog PirB regulates synaptic plasticity in an Alzheimer's model, Science 341 (2013) 1399–1404.

[271] Y. Li, L. Liu, S.W. Barger, W.S. Griffin, Interleukin-1 mediates pathological effects of microglia on tau phosphorylation and on synaptophysin synthesis in cortical neurons through a p38-MAPK pathway, J. Neurosci. 23 (2003) 1605–1611.

[272] E.T. Saez, M. Pehar, M.R. Vargas, L. Barbeito, R.B. Maccioni, Production of nerve growth factor by beta-amyloid-stimulated astrocytes induces p75NTR-dependent tau hyperphosphorylation in cultured hippocampal neurons, J. Neurosci. Res. 84 (2006) 1098–1106.

[273] C.J. Garwood, A.M. Pooler, J. Atherton, D.P. Hanger, W. Noble, Astrocytes are important mediators of Abeta-induced neurotoxicity and tau phosphorylation in primary culture, Cell Death Dis. 2 (2011) e167.

[274] B. Vasconcelos, I.C. Stancu, A. Buist, M. Bird, P. Wang, A. Vanoosthuyse, K. Van Kolen, A. Verheyen, P. Kienlen-Campard, J.N. Octave, P. Baatsen, D. Moechars, I. Dewachter, Heterotypic seeding of Tau fibrillization by pre-aggregated Abeta provides potent seeds for prion-like seeding and propagation of Tau-pathology in vivo, Acta Neuropathol. 131 (2016) 549–569.

[275] B.I. Giasson, M.S. Forman, M. Higuchi, L.I. Golbe, C.L. Graves, P.T. Kotzbauer, J.Q. Trojanowski, V.M. Lee, Initiation and synergistic fibrillization of tau and alpha-synuclein, Science 300 (2003) 636–640.

[276] J.L. Guo, D.J. Covell, J.P. Daniels, M. Iba, A. Stieber, B. Zhang, D.M. Riddle, L.K. Kwong, Y. Xu, J.Q. Trojanowski, V.M. Lee, Distinct alpha-synuclein strains differentially promote tau inclusions in neurons, Cell 154 (2013) 103–117.

[277] R. Morales, I. Moreno-Gonzalez, C. Soto, Cross-seeding of misfolded proteins: implications for etiology and pathogenesis of protein misfolding diseases, PLoS Pathog. 9 (2013) e1003537.

[278] M.J. Guerrero-Munoz, D.L. Castillo-Carranza, S. Krishnamurthy, A.A. Paulucci-Holthauzen, U. Sengupta, C.A. Lasagna-Reeves, Y. Ahmad, G.R. Jackson, R. Kayed, Amyloid-beta oligomers as a template for secondary amyloidosis in Alzheimer's disease, Neurobiol. Dis. 71 (2014) 14–23.

[279] G. Dziewczapolski, C.M. Glogowski, E. Masliah, S.F. Heinemann, Deletion of the alpha 7 nicotinic acetylcholine receptor gene improves cognitive deficits and synaptic pathology in a mouse model of Alzheimer's disease, J. Neurosci. 29 (2009) 8805–8815.

[280] J.J. Palop, L. Mucke, Synaptic depression and aberrant excitatory network activity in Alzheimer's disease: two faces of the same coin?, Neuromol. Med. 12 (2010) 48–55.

[281] L.M. Ittner, J. Gotz, Amyloid-beta and tau—a toxic pas de deux in Alzheimer's disease, Nat. Rev. Neurosci. 12 (2011) 65–72.

[282] R. Malinow, New developments on the role of NMDA receptors in Alzheimer's disease, Curr. Opin. Neurobiol. 22 (2012) 559–563.

[283] L. Mucke, D.J. Selkoe, Neurotoxicity of amyloid beta-protein: synaptic and network dysfunction, Cold Spring Harb. Perspect. Med. 2 (2012) a006338.

[284] E.M. Snyder, Y. Nong, C.G. Almeida, S. Paul, T. Moran, E.Y. Choi, A.C. Nairn, M.W. Salter, P.J. Lombroso, G.K. Gouras, P. Greengard, Regulation of NMDA receptor trafficking by amyloid-beta, Nat. Neurosci. 8 (2005) 1051–1058.

[285] I. Dewachter, R.K. Filipkowski, C. Priller, L. Ris, J. Neyton, S. Croes, D. Terwel, M. Gysemans, H. Devijver, P. Borghgraef, E. Godaux, L. Kaczmarek, J. Herms, F. Van Leuven, Deregulation of NMDA-receptor function and down-stream signaling in APP[V717I] transgenic mice, Neurobiol. Aging 30 (2009) 241–256.

[286] H.W. Kessels, S. Nabavi, R. Malinow, Metabotropic NMDA receptor function is required for beta-amyloid-induced synaptic depression, Proc. Natl. Acad. Sci. USA 110 (2013) 4033–4038.

[287] N.W. Hu, A.J. Nicoll, D. Zhang, A.J. Mably, T. O'Malley, S.A. Purro, C. Terry, J. Collinge, D.M. Walsh, M.J. Rowan, mGlu5 receptors and cellular prion protein mediate amyloid-beta-facilitated synaptic long-term depression in vivo, Nat. Commun. 5 (2014) 3374.

[288] M. Morris, S. Maeda, K. Vossel, L. Mucke, The many faces of tau, Neuron 70 (2011) 410–426.

[289] E.D. Roberson, K. Scearce-Levie, J.J. Palop, F. Yan, I.H. Cheng, T. Wu, H. Gerstein, G.Q. Yu, L. Mucke, Reducing endogenous tau ameliorates amyloid beta-induced deficits in an Alzheimer's disease mouse model, Science 316 (2007) 750–754.

[290] O.A. Shipton, J.R. Leitz, J. Dworzak, C.E. Acton, E.M. Tunbridge, F. Denk, H.N. Dawson, M.P. Vitek, R. Wade-Martins, O. Paulsen, M. Vargas-Caballero, Tau protein is required for amyloid beta-induced impairment of hippocampal long-term potentiation, J. Neurosci. 31 (2011) 1688–1692.

[291] C. Haass, E. Mandelkow, Fyn-tau-amyloid: a toxic triad, Cell 142 (2010) 356–358.

[292] L.M. Ittner, Y.D. Ke, F. Delerue, M. Bi, A. Gladbach, J. van Eersel, H. Wolfing, B.C. Chieng, M.J. Christie, I.A. Napier, A. Eckert, M. Staufenbiel, E. Hardeman, J. Gotz, Dendritic function of tau mediates amyloid-beta toxicity in Alzheimer's disease mouse models, Cell 142 (2010) 387–397.

[293] S. Garg, T. Timm, E.M. Mandelkow, E. Mandelkow, Y. Wang, Cleavage of Tau by calpain in Alzheimer's disease: the quest for the toxic 17 kDa fragment, Neurobiol. Aging 32 (2011) 1–14.

[294] A. Sydow, A. Van Der Jeugd, F. Zheng, T. Ahmed, D. Balschun, O. Petrova, D. Drexler, L. Zhou, G. Rune, E. Mandelkow, R. D'Hooge, C. Alzheimer, E.M. Mandelkow, Tau-induced defects in synaptic plasticity, learning, and memory are reversible in transgenic mice after switching off the toxic Tau mutant, J. Neurosci. 31 (2011) 2511–2525.

[295] S. Peineau, C. Taghibiglou, C. Bradley, T.P. Wong, L. Liu, J. Lu, E. Lo, D. Wu, E. Saule, T. Bouschet, P. Matthews, J.T. Isaac, Z.A. Bortolotto, Y.T. Wang, G.L. Collingridge, LTP inhibits LTD in the hippocampus via regulation of GSK3beta, Neuron 53 (2007) 703–717.

[296] J. Jo, D.J. Whitcomb, K.M. Olsen, T.L. Kerrigan, S.C. Lo, G. Bru-Mercier, B. Dickinson, S. Scullion, M. Sheng, G. Collingridge, K. Cho, Abeta(1-42) inhibition of LTP is mediated by a signaling pathway involving caspase-3, Akt1 and GSK-3beta, Nat. Neurosci. 14 (2011) 545–547.

[297] R.J. Perrin, A.M. Fagan, D.M. Holtzman, Multimodal techniques for diagnosis and prognosis of Alzheimer's disease, Nature 461 (2009) 916–922.

[298] A. Serrano-Pozo, M.P. Frosch, E. Masliah, B.T. Hyman, Neuropathological alterations in Alzheimer disease, Cold Spring Harb. Perspect. Med. 1 (2011) a006189.

[299] K. Bhaskar, M. Konerth, O.N. Kokiko-Cochran, A. Cardona, R.M. Ransohoff, B.T. Lamb, Regulation of tau pathology by the microglial fractalkine receptor, Neuron 68 (2010) 19–31.

[300] M. Kitazawa, D. Cheng, M.R. Tsukamoto, M.A. Koike, P.D. Wes, V. Vasilevko, D.H. Cribbs, F.M. Laferla, Blocking IL-1 signaling rescues cognition, attenuates tau pathology, and restores neuronal beta-catenin pathway function in an Alzheimer's disease model, J. Immunol. 187 (2011) 6539–6549.

[301] M. Sy, M. Kitazawa, R. Medeiros, L. Whitman, D. Cheng, T.E. Lane, F.M. Laferla, Inflammation induced by infection potentiates tau pathological features in transgenic mice, Am. J. Pathol. 178 (2011) 2811–2822.

[302] S. Ghosh, M.D. Wu, S.S. Shaftel, S. Kyrkanides, F.M. Laferla, J.A. Olschowka, M.K. O'Banion, Sustained interleukin-1beta overexpression exacerbates tau pathology despite reduced amyloid burden in an Alzheimer's mouse model, J. Neurosci. 33 (2013) 5053–5064.

[303] M. Goedert, F. Clavaguera, M. Tolnay, The propagation of prion-like protein inclusions in neurodegenerative diseases, Trends Neurosci. 33 (2010) 317–325.

[304] C. Soto, L. Estrada, J. Castilla, Amyloids, prions and the inherent infectious nature of misfolded protein aggregates, Trends Biochem. Sci. 31 (2006) 150–155.

[305] K.C. Luk, C. Song, P. O'Brien, A. Stieber, J.R. Branch, K.R. Brunden, J.Q. Trojanowski, V.M. Lee, Exogenous alpha-synuclein fibrils seed the formation of Lewy body-like intracellular inclusions in cultured cells, Proc. Natl. Acad. Sci. USA 106 (2009) 20051–20056.

[306] L.A. Volpicelli-Daley, K.C. Luk, T.P. Patel, S.A. Tanik, D.M. Riddle, A. Stieber, D.F. Meaney, J.Q. Trojanowski, V.M. Lee, Exogenous alpha-synuclein fibrils induce Lewy body pathology leading to synaptic dysfunction and neuron death, Neuron 72 (2011) 57–71.

[307] K.C. Luk, V. Kehm, J. Carroll, B. Zhang, P. O'Brien, J.Q. Trojanowski, V.M. Lee, Pathological alpha-synuclein transmission initiates Parkinson-like neurodegeneration in nontransgenic mice, Science 338 (2012) 949–953.

[308] K.C. Luk, V.M. Kehm, B. Zhang, P. O'Brien, J.Q. Trojanowski, V.M. Lee, Intracerebral inoculation of pathological alpha-synuclein initiates a rapidly progressive neurodegenerative alpha-synucleinopathy in mice, J. Exp. Med. 209 (2012) 975–986.

[309] Z. Jaunmuktane, S. Mead, M. Ellis, J.D. Wadsworth, A.J. Nicoll, J. Kenny, F. Launchbury, J. Linehan, A. Richard-Loendt, A.S. Walker, P. Rudge, J. Collinge, S. Brandner, Evidence for human transmission of amyloid-beta pathology and cerebral amyloid angiopathy, Nature 525 (2015) 247–250.

[310] J.P. Guo, T. Arai, J. Miklossy, P.L. Mcgeer, Abeta and tau form soluble complexes that may promote self aggregation of both into the insoluble forms observed in Alzheimer's disease, Proc. Natl. Acad. Sci. USA 103 (2006) 1953–1958.

[311] C.A. Lasagna-Reeves, D.L. Castillo-Carranza, M.J. Guerrero-Muoz, G.R. Jackson, R. Kayed, Preparation and characterization of neurotoxic tau oligomers, Biochemistry 49 (2010) 10039–10041.

[312] Y. Miller, B. Ma, R. Nussinov, Synergistic interactions between repeats in tau protein and Abeta amyloids may be responsible for accelerated aggregation via polymorphic states, Biochemistry 50 (2011) 5172–5181.

[313] B.T. Hyman, P. Sorger, Failure analysis of clinical trials to test the amyloid hypothesis, Ann. Neurol. 76 (2014) 159–161.

[314] E. Karran, J. Hardy, A critique of the drug discovery and phase 3 clinical programs targeting the amyloid hypothesis for Alzheimer disease, Ann. Neurol. 76 (2014) 185–205.

PARKINSON'S DISEASE

6

Mark R. Cookson
National Institute on Aging, National Institutes of Health, Bethesda, MD, United States

CHAPTER OUTLINE

Introduction 157
Current Treatment Approaches in Parkinson's Disease 158
    Pharmacological Approaches: Replacing Dopamine 158
    Nonpharmacological Approaches: Deep Brain Stimulation and Cell Replacement 160
Drugs That are Being Evaluated Clinically for Disease Modification in PD 160
Novel Targets for Disease-Modifying Therapies in PD 161
    α-Synuclein 161
    LRRK2 165
    Recessive Genes 166
Conclusions 167
Acknowledgments 167
References 168

INTRODUCTION

Initial diagnosis of Parkinson's disease (PD) usually begins when an individual presents with mild asymmetric movement problems, classically a 40 Hz rest tremor bradykinesia (slowness of movement), rigidity, and/or problems with balance and posture. However, it is recognized that even by the stage of early diagnosis, there is a constellation of additional symptoms that affect multiple parts of the nervous system, especially autonomic pathways [1]. Depression, sleep problems, constipation, and anosmia are recognized as important symptoms that precede the movement disorder in PD, perhaps by several years [2–5]. As the disease progresses, both movement and nonmovement symptoms tend to worsen and more nonmotor central nervous system (CNS) systems become involved. Particularly late in the disease, dementia is often present and becomes disabling for many people with PD [6].

This clinical progression is thought to relate to the accumulation of neuropathological events in various brain regions throughout the disease. One important marker of neuropathology in PD is the accumulation of intraneuronal inclusions called Lewy bodies and Lewy neurites in cells throughout the brain that are composed of lipids and proteins, including α-synuclein [7]. It has been suggested by Braak and coworkers that Lewy bodies form sequentially in several brain regions in a more or less stereotypical manner, starting perhaps in the peripheral nervous system before progressing through the midbrain and eventually to the cerebral cortex [8]. In at least some of the regions of the brain, where Lewy bodies are found, there is also dramatic loss of neurons, particularly of pigmented dopaminergic neurons that project from the substantia nigra pars compacta to the striatum [9]. Additionally, neuroinflammation occurs in many of the areas affects by either Lewy body and/or neuronal loss [10]. However, Lewy bodies, dopamine cell loss, and neuroinflammation can all occur in other neurodegenerative diseases and there is clinical overlap between PD and related symptomatology that can be called parkinsonism [11]. PD is therefore a progressive multibrain region disease with a distinct, but not unique, neuropathological profile.

Of the known cause(s) of disease initiation and progression, aging is the predominant required factor for the development of PD. Although there are rare early onset cases of PD that develop in the second or third decade of life [12], the overall prevalence in the population rises steadily from the age of 50 until later in life [13]. Genetics also plays a major role in the risk of PD, with both monogenic forms that run in families and risk factor genes identified in sporadic PD [14]. Importantly, the same genes have been implicated in both familial and sporadic PD, which is a form of the generalized concept that single loci can contain multiple types of genetic variants that cause disease by slightly different but related mechanisms [15]. A prominent example is the gene that encodes the Lewy body protein α-synuclein, *SNCA*, which is mutated in several families but also acts as a risk factor for sporadic PD [15]. A second, important, example is the gene *LRRK2* that encodes the multifunctional enzyme leucine-rich repeat kinase 2 [16]. Other factors, including gender and some environmental agents (principally coffee and nicotine consumption) further modify risk of disease [17].

Here, I will first discuss how what is known about the clinical and neuropathological course of PD has led to our current treatments for this disease. Then, I will cover the promise of disease-modifying therapies that come from understanding more about causation of PD.

CURRENT TREATMENT APPROACHES IN PARKINSON'S DISEASE

If one were writing strictly from the view of disease-modifying therapies in PD, the chapter would be extremely short as, put simply, there are no clinically used agents that stop progression of this disorder. However, to do so would be a disservice to the work that has gone into development of very effective symptomatic agents. Furthermore, because symptomatic treatments are available for PD, this creates an ecosystem where any potential disease-modifying agents have a high threshold to pass before being competitive. I will discuss the two major approaches, pharmacological and surgical, currently used in PD. The neurobiology underlying these approaches is outlined schematically in Fig. 6.1.

PHARMACOLOGICAL APPROACHES: REPLACING DOPAMINE

The identification of dopamine as a neurotransmitter, which was initially controversial, led to the testing of L-3,4-dihydroxyphenylalanine (L-DOPA) as a therapy for PD [18,19]. Dopamine does not cross the blood–brain barrier, but L-DOPA will and in the CNS is converted by aromatic acid decarboxylase

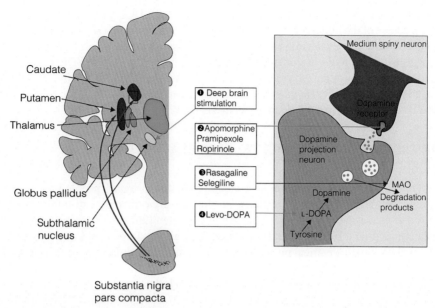

FIGURE 6.1 Cellular Neuropathology Underlies Current Approaches to Symptomatic Treatment of Parkinson's Disease (PD)

On the left, an outline of the subcortical regions of the basal ganglia that are affected in PD shows how neurons from the substantia nigra project to the caudate nucleus and putamen, collectively referred to as the corpus striatum. The boxed area on the right is a magnification of the synaptic terminal from the projection neuron where dopamine is produced and packaged into vesicles. These are released into the synaptic cleft where dopamine acts on postsynaptic receptors. Between these two schematics, the numbered and boxed compounds list some of the available drugs and treatments that are currently available to treat symptoms of PD.

into dopamine. Since L-DOPA is highly effective in restoring movement, and is safe and well tolerated for most patients, it remains the mainstay for treating PD.

Despite its widespread usage, there are some important limitations to L-DOPA as a therapy. Fundamentally, L-DOPA as a therapy is aimed at restoring the neurochemical defect that arises from loss of projection neurons in the *substantia nigra pars compacta*, that is, is based on the underlying neuropathological event in PD. This means that it does not, and cannot, suppress symptoms arising from loss of neurotransmitters other than dopamine. Due to this, L-DOPA is relatively poor at addressing many nonmotor symptoms of PD, which often require additional medications for treatment [20].

The long-term use of L-DOPA, required because of the chronic nature of PD, has further revealed some limitations to the use of this drug. L-DOPA has a relatively short half-life in plasma, which leads to fluctuations in effective drug level and is associated with wearing-off phenomena [21]. In the "off" state, a patient can remain frozen. Additionally, the requirement to increase doses of L-DOPA over time is associated with both motor and nonmotor complications, including dyskinesias and psychological disturbances, such as compulsive behaviors [22]. Alternative formulations of L-DOPA and dopamine agonists are being actively explored clinically but oral L-DOPA remains highly used for most PD patients [23].

Important for the discussion here, there is little evidence that L-DOPA treatment positively modifies the underlying course of the disease. In fact, there have been concerns that L-DOPA might induce excess toxicity in vulnerable dopamine neurons and although this issue is not fully resolved the balance of data clinically suggests that L-DOPA has no net effect on disease progression [24]. There is some evidence that rasagaline, an inhibitor of the enzyme monoamine oxidase that would normally promote the breakdown of dopamine, has some beneficial effects on disease course but the effects are relatively small and it is not clear if they represent a truly neuroprotective event [25].

NONPHARMACOLOGICAL APPROACHES: DEEP BRAIN STIMULATION AND CELL REPLACEMENT

Although L-DOPA and related drugs can be incredibly helpful for many people with PD, as the disease progresses the responsiveness of a patient to dopaminergic therapy can diminish, with longer "off" periods and increasing dyskinesia when L-DOPA is given [26]. An alternative for many patients is deep brain stimulation (DBS), where an electrode is surgically implanted in the globus pallidus or subthalamic nucleus. Small electrical pulses, somewhat akin to a pacemaker used for control of cardiac rhythmicity, are then applied to the stimulator. Although the precise physiological mechanism(s) by which DBS evokes positive responses remain uncertain [27], the result for many patients is that they can use lower dosages of L-DOPA and thereby limit side effects [28]. Like L-DOPA, DBS addresses the fundamental neuropathological event of loss of dopaminergic neurons, but at the level of the overall brain circuit.

Another nonpharmacological approach that has been explored in some depth is cell transplantation [29]. Conceptually, grafting functional dopamine neurons into the striatum would allow for L-DOPA to be released under physiological conditions and therefore improve motor function. However, mixed results from open-label trials using fetal cells have led to concerns that these transplants may not be a competitive therapy although hope remains that new types of differentiated cells, such as induced pluripotent cells might be useful clinically [29] (see also Chapter 4A).

As for L-DOPA, there are some facets of PD that are not currently resolved by DBS or by cell transplantation. Since both approaches address symptoms rather than underlying cause, progression of PD across multiple brain system will still occur. Therefore, there is still a need for approaches that will address the underlying degenerative process in PD. As discussed earlier in this chapter, such therapies do not currently exist and therefore here I will discuss drugs that are currently being evaluated then ongoing attempts to develop such treatments based on novel targets.

DRUGS THAT ARE BEING EVALUATED CLINICALLY FOR DISEASE MODIFICATION IN PD

In the past few decades, several molecules have moved forward into clinical trials aimed at testing their ability to limit disease progression in PD. Many of these have failed to show efficacy, as has been reviewed elsewhere [30]. Some of these drugs did show efficacy in preclinical models, including creatine and coenzyme Q10 that are thought to limit mitochondrial oxidative stress and were useful in blocking cell death induced by the mitochondrial toxin MPTP (1-methyl-4-phenyl-1,2,3,6-tetrahydropyridine) in animal models [31]. Antiinflammatory drugs, like pioglitazone, which again was helpful in the MPTP model [32], have not shown clinical efficacy in humans [33]. However, there are some drugs that are currently being evaluated for their potential effects on disease progression in PD that are interesting mechanistically and will be discussed here.

In my view, the key to development of disease-modifying therapy for PD is to prioritize underlying causation of the disease process over descriptive approaches based around neuropathological events. If this logic is true, then we might look at the known causes of PD, which as discussed previously include aging and genetics. Thinking about aging, it has been shown that as dopaminergic neurons mature, the complement of ion channels that they express changes. This includes the L-type calcium channel Ca(v)1.3, which is important for the pacemaking activity of dopamine neurons and is selectively expressed in mature cells that are vulnerable in PD [34]. Epidemiological studies suggest that the dihydropyridine class of L-type calcium channel blockers, which are often used to treat hypertension, diminish the risk of PD [35]. This data has led to a dose-finding trial of isradipine treatment in PD that suggested the compound was safe but did not show a significant benefit clinically, suggesting that larger follow-on studies might be required [36].

Other drugs that have epidemiological support for their utility in PD include Urate and Nicotine, both of which are being evaluated in clinical trials [30,37]. There have also been reports of positive benefit from an open-label trial of the glucagon like peptide-1 agonist exenatide, although the data are not strong enough to interpret as proving that there was neuroprotection in that study [38]. Overall, at this time, the available data shows that disease-modifying therapies are not fully proven for PD and that there is a need to continue to assess new targets for this purpose.

NOVEL TARGETS FOR DISEASE-MODIFYING THERAPIES IN PD

The lack of disease modification from current therapies has motivated many groups to look for novel targets that might influence disease course in PD. One area that has been discussed at some length is whether genetics might yield novel targets. In theory, this is an attractive idea, as we know ahead of time that modification of the gene is a strong predictor of risk of disease. Therefore, reversing the gene defect by limiting toxicity of dominant mutations or by replacing loss of function seen in recessive mutations is potentially attractive. Since none of these approaches have yet been tried in PD patients, I will focus instead on those with best preclinical data where animal models have been used to establish some efficacy against neurodegeneration. A schematic showing some of the potential targets that could be leveraged for disease modification is shown in Fig. 6.2.

α-SYNUCLEIN

If one had to prioritize a target with likely application to the broadest number of PD cases, one might, arguably, consider α-synuclein. From a genetic perspective, point mutations in the *SNCA* gene and multiplications of the *SNCA* locus are associated with a disease range from typical PD to early onset aggressive Lewy body disease [39]. Furthermore, variation around the same locus is associated with lifetime risk of sporadic disease [40]. Combining the observations that multiplication mutations and variation around *SNCA* both cause disease suggests that differences in protein expression levels are a potential mechanism

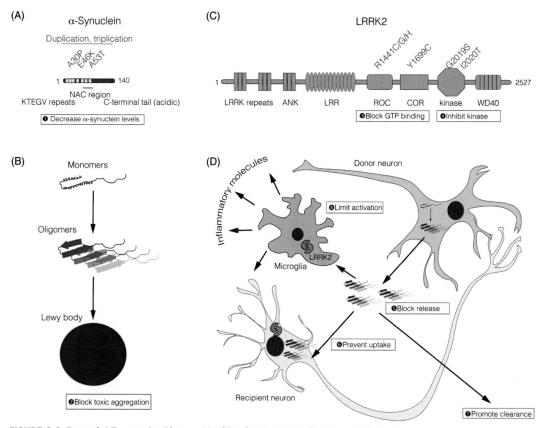

FIGURE 6.2 Potential Targets for Disease Modification From Dominantly Inherited PD

(A) Of the known genetic causes of PD, one of the most prominent is α-synuclein, where both point mutations and gene duplications are associated with inherited disease. (B) A simplified view of the hypothetical pathway by which α-synuclein aggregates to form potentially toxic oligomers and eventually, Lewy bodies. (C) Another important gene for PD, leucine-rich repeat kinase 2 (LRRK2), encodes a large, multidomain protein. Abbreviations for the domains are: *ANK*, ankyrin-like repeats; *COR*, C-terminal of ROC; *LRR*, leucine-rich repeats; *ROC*, ras of complex proteins. (D) Targets for PD based on intracellular communication. α-Synuclein is thought to be released, perhaps in an already partially aggregated form, from donor neurons *(blue)* where it can affect at least two other cell types, recipient neurons *(yellow)* and microglia *(green)*. The activation of microglia, which are LRRK2 positive (depicted here as a dimer), can promote another form of noncell autonomous communication, namely inflammation. Throughout this figure, potential ways in which these pathogenic processes might be modified to provide disease-modifying therapies are presented numbered in boxes.

for PD [15]. Also important for considering sporadic PD, and linking to one aspect of the neuropathology of this disease, α-synuclein is an important component of Lewy bodies [7]. Therefore, α-synuclein might be a disease-modifying target that could have effects on both sporadic and genetic forms of PD.

However, complicating the approach is the fact that how α-synuclein causes PD is not clear, which bears further discussion.

Is there a toxic species characterized by form or localization?

An unusual biophysical characteristic of the α-synuclein protein is that it is natively unfolded in solution but prone to aggregate into beta-sheet rich fibrillar structures [41] (see also Chapter 2). As the α-synuclein in Lewy bodies is fibrillar [42], this suggests that aggregation is associated with Lewy body formation. By extension, one might therefore consider fibrils of aggregated synuclein to therefore be toxic, and compounds that limit protein aggregation would be neuroprotective.

However, several years ago it was shown that not all mutations in *SNCA* cause α-synuclein to be more prone to form fibrils [43], work that has subsequently been confirmed using a range of different techniques [44,45]. As Lewy bodies containing fibrillar α-synuclein is, by definition, present in cells that have survived in PD there is additional reason to doubt that Lewy bodies are inherently toxic. Oligomer-prone forms of α-synuclein have been shown to induce toxicity to neurons in vivo [46,47].

This leads to the idea that antioligomeric compounds might be neuroprotective. However, designing an antioligomer compound is complicated as α-synuclein can adopt a wide range of different partially aggregated conformations [48]. The ability of α-synuclein to bind membranes provides further complexity to the range of biophysical forms available to the protein [49], as does posttranslational modification [50]. Which, if any, of these forms of α-synuclein are predominant in the cell is unclear—for example, there have been controversies over whether physiological synuclein is tetrameric [51–53]. Reagents that are able to distinguish between different oligomeric assemblies suggest that several different types of oligomers might contribute to toxicity [54]. Nonetheless, several compounds have been developed that can inhibit aggregation and limit toxicity, some of which have been tested in animal models [55,56].

Related to this idea, it might be possible to prevent toxicity by targeting downstream actions of α-synuclein oligomers. For example, oligomeric forms of α-synuclein are associated with damage to intracellular vesicular compartments, including the endoplasmic reticulum (ER). Salubrinal, which can limit ER stress, prevents the early mortality seen in mice expressing mutant α-synuclein [57].

Another concept related to a targetable form of toxic α-synuclein arises indirectly from observation of neurons transplanted into patients for symptomatic therapy, as discussed previously [29]. Lewy bodies were seen in fetal neurons grafted into the striatum of PD patients, although only in cases where those cells had survived for several years with functional benefit [58–60]. These results suggested that pathology of PD could spread from affected areas into unaffected, and in this case young, cells.

Combined with the idea that the distribution of Lewy bodies progresses throughout PD [8], these observations lead to the hypothesis that α-synuclein itself can spread between neurons in the brain. This idea was tested in a series of experiments in transgenic and nontransgenic mice, where it was shown that preformed α-synuclein fibrils can initiate toxicity and spread of pathology in vivo [61,62]. α-Synuclein fibrils may have strain-like properties in terms of the types of pathology that they can induce [63]. Both fibrils and smaller ribbon-like assemblies of α-synuclein can promote spread of inclusion pathology, although the former is associated with a greater degree of toxicity [64].

Therefore, as well as targeting a specific type of α-synuclein to limit neurodegeneration, one might also try to develop strategies to prevent the overall spread of the protein, or both. One way in which this might be done is to use antibodies against either total α-synuclein or a spreading form that transfers between cells. Experiments performed in mice suggest that intraperitoneal injection of monoclonal antibodies against α-synuclein can limit spread in the brain [65]. Two antibody-based approaches are currently in early clinical testing, mainly for safety evaluation [30].

Reducing the total amount of α-synuclein

Overall, the aforementioned results suggest that one potential antineurodegenerative approach for PD would be to limit the toxic form of α-synuclein, whether an intracellular aberrantly aggregated species or some secreted form. However, an alternative approach that would be agnostic as to the toxic form of α-synuclein would be to decrease the overall levels of the protein, thus presumably also limiting the toxic form.

One tactic for decreasing overall α-synuclein levels might be to increase the normal turnover of the protein, which predominantly occurs via lysosomal processing. For example, increasing chaperone-mediated autophagy decreases the total amount of α-synuclein expressed while limiting toxicity in a rat model [66]. Antisense gene silencing might also be tractable, as has been suggested for other genes associated with neurodegeneration [67], and gene silencing has been shown to be effective at reducing the amounts of α-synuclein protein in the primate brain [68]. Furthermore, some allele-specific constructs have been shown to affect only mutant α-synuclein in cells [69].

A reasonable question is whether reducing α-synuclein level is expected to be a safe thing to do. α-Synuclein is a very highly expressed protein in the CNS, where it has important physiological roles at the synapse [70]. Using viral vectors to produce efficient knockdown of endogenous α-synuclein promotes neurodegeneration of dopamine cells [71,72]. In contrast, partial knockdown of α-synuclein in the adult rat does not cause loss of TH-positive neurons [73] and complete knockout of the gene in mice throughout development is tolerated without obvious effects [74].

Interpretation of these results in the context of the possibility of decreasing α-synuclein for therapeutic reasons is difficult. One major intellectual concern is whether PD is associated with a gain or loss of α-synuclein function. On the one hand, the accumulation of α-synuclein protein as seen in cases with multiplication patients, which was confirmed in tissue samples several years ago [75], suggests that more α-synuclein is detrimental. Deletion or truncating mutations have not, as of the time of writing, been found associated with PD in families. Furthermore, on the other hand, the loss of cell viability after transient knockdown raises the possibility that the protein has an important normal function that cannot be removed from the mature CNS without detrimental effects. One possibility is that α-synuclein levels are explained by the Goldilocks principle—either too little or too much is detrimental depending on context. Alternatively, there could be a pool of α-synuclein that is critical for normal function that can also be depleted by increased expression. For example, it is possible that increased expression in triplication cases might cause aggregation of α-synuclein in the cytosol that recruits protein from the synaptic pool.

These considerations are somewhat speculative because we do not know the mechanisms by which overexpression or knockdown of α-synuclein cause loss of TH-positive cells, or whether they are the same. However, the possibility that lowering α-synuclein might cause some problems in the mature CNS suggests that such an approach might have a limited therapeutic window in PD. Clinical data is also clearly needed to establish if lowering α-synuclein can be successfully targeted for a chronic disease.

Overall, these considerations highlight that some form of α-synuclein might be able to be targeted to provide a disease-modifying therapy for PD, including sporadic PD, but that there is a huge amount of work to be done to generate an entity that as clinically utility. It is therefore reasonable to ask if other genes for PD might provide alternative targets for similar drugs.

LRRK2

Although α-synuclein is likely important in all cases of PD, including sporadic PD where Lewy bodies are required for pathological diagnosis, there are other genes for PD where mutations are found in numerically more cases. Of these, perhaps the most prominent is the *LRRK2* gene, which is not only mutated in a relatively large number of families [76–79] but, like *SNCA*, was also nominated as a risk gene for sporadic PD [80,81].

That LRRK2 might represent a novel target for disease-modifying therapy in PD came from an initial examination of the protein's domain structure. Mutations are clustered in three domains and affect their enzymatic function. Mutations in the ROC (ras of complex proteins)-COR (C-terminal of ROC) bidomain diminish GTPase activity of the protein [82–84] whereas mutations in the kinase domain increase activity toward a variety of model substrates [85]. The relationship between these two activities is uncertain, but it is likely that they interact to direct the overall biological output of LRRK2. It has been shown that all mutations in LRRK2 impact regulation of vesicular trafficking [86] and autophagy [87].

Several years ago, others and we asked if the enzymatic activities of the LRRK2 protein were important for the detrimental effects of mutations in this gene. Using artificial constructs engineered to lack kinase activity, first in cells [88,89] then later in animal models [90,91], multiple groups were able to establish that kinase activity of LRRk2 is required for toxicity. In part because of this information, several groups then successfully generated small molecule kinase inhibitors of LRRK2 [92–95]. Additionally, there are some compounds that have been developed that target the GTP-binding region of LRRK2 [96] with similar neuroprotective abilities. It has been proposed that kinase inhibitors, or perhaps molecules targeting different activities of LRRK2, might be useful for stopping neurodegeneration in PD, although none has been tested in the clinic at the time of writing this chapter [97].

However, there are some concerns that have been raised about the possibility of LRRK2 inhibition. The full knockout of LRRK2 in mice or rats causes the accumulation of lysosome-derived vacuoles in the collecting ducts of the kidney and type II pneumocytes in the lung [98–101]. These effects are likely related to the effects on vesicular trafficking alluded to previosuly [86,87,102]. Critically, germline knock-in of a kinase inactivating mutation in mice [103] or exposure to LRRK2 inhibitors [104,105] can also initiate some of the aforementioned kidney and lung pathologies. As for the knockdown of α-synuclein discussed previously, these results suggest that the therapeutic window for LRRK2 inhibitors might be relatively narrow. There are also some mutations that apparently have lower kinase activity than wild-type LRRK2, raising the possibility that the utility of kinase inhibitors might be mutation specific [106]. The key next steps in the assessment of LRRK2 inhibitors will, therefore, be safety assessments in people with PD.

Looking ahead, if LRRK2 inhibitors were to be shown to be relatively well tolerated, that is to say that the effects in tissues are not clinically significant, an extended consideration is which PD patients could benefit from such a therapy. Logically, people carrying mutations that have increased LRRK2 kinase activity might be the first test subjects, but after those people would there be broader applicability? Several considerations suggest that this might be the case.

One important clue comes from the association of variation around the *LRRK2* locus with sporadic PD [40]. As there is no evidence that the statistical signal observed in GWAS results from amino acid changing variants that would affect LRRK2 sequence or function, a reasonable assumption is that the effect relates to differences in expression of LRRK2, either under basal or induced conditions. The direction of effect is important here, specifically whether more LRRK2 or lower expression is associated with risk of disease. If increased LRRK2 expression or activity promotes the risk of PD, then inhibitors might reasonably be applied. And, given that the association was driven by sporadic PD cases then it would be also reasonable to consider LRRK2 inhibitors for all PD, not just inherited forms.

Another argument for the concept that LRRK2 inhibition might be helpful for sporadic PD is that there is a relationship between LRRK2 and α-synuclein. Since human LRRK2 cases do not always have α-synuclein-positive Lewy bodies [107], experimental models have been used to address the question of whether LRRK2 and α-synuclein can cooperate to induce degeneration. Overexpression of mutant LRRK2 can exacerbate the toxic effects of α-synuclein in one transgenic mouse model [108], although this has not been replicated using different promoter systems [109,110]. Perhaps more germane to the argument that LRRK2 inhibition might be helpful to prevent neurodegeneration, knockout of Lrrk2 in mice limits toxicity associated with inflammatory stimuli or α-synuclein [108,111], and these effects can be replicated using small molecule LRRK2 inhibitors [112]. Reciprocally, knockout of α-synuclein can limit the toxicity of LRRK2 mutations at least in vitro [113]. The mechanism(s) by which LRRK2 inhibition is neuroprotective in these circumstances are not clear, but may involve expression of LRRK2 in inflammatory cells, such as microglia that may respond to stimuli including release of α-synuclein from cells [114–117].

Importantly, while knockout or inhibition of LRRK2 diminishes microglial activation, dominant mutations associated with PD increase responses in these cells to stimuli [118,119]. While not completely definitive, these results suggest that mutations, which are associated with increased risk of disease, do not work by simple loss of overall LRRK2 function. By extension, this argument further supports the idea that inhibition of LRRK2 would be helpful as a disease-modifying therapy for PD.

RECESSIVE GENES

Compared to α-synuclein or LRRK2, recessive genes, such as PINK1, parkin, and DJ-1 [120] have received relatively less attention as potential targets for disease-modifying therapies. Perhaps this is because it is a little more difficult to link these genes to sporadic PD. Specifically, mutations in PINK1, parkin, or DJ-1 cause an early-onset parkinsonism that has a protracted and often benign clinical course and is not usually associated with Lewy bodies. Some have suggested that these syndromes are clinically distinct from sporadic PD and should therefore be considered separately [11]. However, there are reasons to consider that there might be interesting targets that arise from these genes.

In all cases that have been clearly documented, mutations in DJ-1, PINK1, or parkin have a clear loss of function effect. This is demonstrated by the inheritance pattern, where two alleles have to be inherited to cause disease, by the types of mutations reported, which include deletions, truncations, and destabilizing mutations [121,122], and by functional assays that show all mutations are deficient in some way or another. An example of the latter point, PINK1 and parkin are thought to participate in a cellular pathway by which damaged mitochondria are removed from the cell via a specialized form of autophagy [123]. Mutations in either PINK1 or parkin consistently fail to support this depolarization-induced mitophagy [124,125]. Thus, all mutations in these genes that have been convincingly documented are loss of function, although the precise mechanisms by which loss of function occurs may vary.

Since loss of function is associated with cell death of dopaminergic neurons in humans, and a consistent feature of the neuropathology of these cases is loss of pigmented cells in the nigra [107], it is therefore reasonable to infer that the normal function of PINK1, parkin, or DJ-1 protects neurons. By extension, increasing the same functions would be expected to have a positive, disease-modifying therapeutic effect. Along these lines, several groups have shown that overexpression of the recessive gene parkin can protect against diverse insults that include, most relevantly, the dominant PD genes α-synuclein [126,127] and LRRK2 [128,129].

Despite the persuasive argument that these pathways might be helpful in tackling neurodegeneration, there are some important questions about whether concept of increasing function of recessive genes would be practical or helpful, since overexpression of parkin in a rat model also induced dopaminergic cell death [130].

The principle difficulty is that small molecules that enhance enzyme function are not as well understood as those that limit function, that is, inhibitors. However, there are several potential ways in which PINK1 or parkin function might be increased, including mimicking the structural effects of the normal activation of parkin by inhibiting enzymes that normally antagonize parkin [131]. A smaller concern is who might benefit from such drugs, as clearly an activator of parkin would not be helpful to an individual who has parkin mutations that delete the gene and produce no protein. Due to clinical and pathological distinctions between parkin and other forms of PD [11], it is not an inherently supported assumption that drugs for one would work in the other. However, there is some evidence that pathways might be common enough to be tractable. Agents that promote mitochondrial function, specifically ursocholanic acid and related compounds, can reverse observable cellular phenotypes in fibroblasts from parkin and LRRK2 patients [132], and may be efficacious in vivo [133]. If such compounds were shown to be efficacious for multiple genetic causes of PD, and clinical data to support this needs to be generated, then it becomes reasonable to address larger populations including sporadic PD.

CONCLUSIONS

The state of the art with PD therapeutics is that most therapies that work in patients now are symptomatic and directed predominantly at the subset of movement symptoms that arise from loss of dopaminergic neurons. These are good drugs, well tolerated and safe, but they do not address the underlying disease process and therefore are unlikely to have disease-modifying effects. We therefore lack a true disease-modifying treatment in PD, which is sorely needed to address progression and, especially, late nonmotor symptoms. Although none has made it to the clinic, there are many concepts that are moving forward in the development pipeline. My argument here is that those that are most likely, but not certain, to succeed are those that are based on understanding of the underlying causes of PD and here I have argued most strongly for genetic-based treatments, although other causes should just as reasonably be considered. The potential for such drugs to have a positive impact on PD patients is high, but already there are clear areas where utility might be limited. Therefore, in the next decade or so clinical work will be especially critical in first determining which approaches are safe then which are efficacious for people living with PD.

ACKNOWLEDGMENTS

This research was supported entirely by the Intramural Research Program of the NIH, National Institute on Aging.

REFERENCES

[1] J.W. Langston, The Parkinson's complex: parkinsonism is just the tip of the iceberg, Ann. Neurol. 59 (2006) 591–596.

[2] K.L. Adams-Carr, J.P. Bestwick, S. Shribman, A. Lees, A. Schrag, A.J. Noyce, Constipation preceding Parkinson's disease: a systematic review and meta-analysis, J. Neurol. Neurosurg. Psychiatry 87 (7) (2016) 710–716.

[3] L. Ishihara, C. Brayne, A systematic review of depression and mental illness preceding Parkinson's disease, Acta Neurol. Scand. 113 (2006) 211–220.

[4] K. Suzuki, M. Miyamoto, T. Miyamoto, K. Hirata, Restless legs syndrome and leg motor restlessness in Parkinson's disease, Park. Dis. 2015 (2015) 490938.

[5] J.G. Goldman, R. Postuma, Premotor and nonmotor features of Parkinson's disease, Curr. Opin. Neurol. 27 (2014) 434–441.

[6] M. Coelho, J.J. Ferreira, Late-stage Parkinson disease, Nat. Rev. Neurol. 8 (2012) 435–442.

[7] W.S. Kim, K. Kågedal, G.M. Halliday, Alpha-synuclein biology in Lewy body diseases, Alzheimers Res. Ther. 6 (2014) 73.

[8] H. Braak, K. Del Tredici, U. Rüb, R.A.I. de Vos, E.N.H. Jansen Steur, E. Braak, Staging of brain pathology related to sporadic Parkinson's disease, Neurobiol. Aging 24 (2003) 197–211.

[9] L.V. Kalia, A.E. Lang, Parkinson's disease, Lancet Lond. Engl. 386 (2015) 896–912.

[10] E.C. Hirsch, P. Jenner, S. Przedborski, Pathogenesis of Parkinson's disease, Mov. Disord. 28 (2013) 24–30.

[11] J.W. Langston, B. Schüle, L. Rees, R.J. Nichols, C. Barlow, Multisystem Lewy body disease and the other parkinsonian disorders, Nat. Genet. 47 (2015) 1378–1384.

[12] A. Schrag, J.M. Schott, Epidemiological, clinical, and genetic characteristics of early-onset parkinsonism, Lancet Neurol. 5 (2006) 355–363.

[13] L.M.L. de Lau, M.M.B. Breteler, Epidemiology of Parkinson's disease, Lancet Neurol. 5 (2006) 525–535.

[14] R. Kumaran, M.R. Cookson, Pathways to Parkinsonism Redux: convergent pathobiological mechanisms in genetics of Parkinson's disease, Hum. Mol. Genet. 24 (2015) R32–44.

[15] A. Singleton, J. Hardy, A generalizable hypothesis for the genetic architecture of disease: pleomorphic risk loci, Hum. Mol. Genet. 20 (2011) R158–R162.

[16] M.R. Cookson, The role of leucine-rich repeat kinase 2 (LRRK2) in Parkinson's disease, Nat. Rev. Neurosci. 11 (2010) 791–797.

[17] K. Kieburtz, K.B. Wunderle, Parkinson's disease: evidence for environmental risk factors, Mov. Disord. 28 (2013) 8–13.

[18] S. Fahn, The medical treatment of Parkinson disease from James Parkinson to George Cotzias, Mov. Disord. 30 (2015) 4–18.

[19] A.J. Lees, E. Tolosa, C.W. Olanow, Four pioneers of L-dopa treatment: Arvid Carlsson, Oleh Hornykiewicz, George Cotzias, and Melvin Yahr, Mov. Disord. 30 (2015) 19–36.

[20] F. Sprenger, W. Poewe, Management of motor and non-motor symptoms in Parkinson's disease, CNS Drugs 27 (2013) 259–272.

[21] H. Reichmann, M. Emre, Optimizing levodopa therapy to treat wearing-off symptoms in Parkinson's disease: focus on levodopa/carbidopa/entacapone, Expert Rev. Neurother. 12 (2012) 119–131.

[22] M.F. Bastide, W.G. Meissner, B. Picconi, S. Fasano, P.-O. Fernagut, M. Feyder, et al. Pathophysiology of L-dopa-induced motor and non-motor complications in Parkinson's disease, Prog. Neurobiol. 132 (2015) 96–168.

[23] A.-C. Vijverman, S.H. Fox, New treatments for the motor symptoms of Parkinson's disease, Expert Rev. Clin. Pharmacol. 7 (2014) 761–777.

[24] C.W. Olanow, Levodopa: effect on cell death and the natural history of Parkinson's disease, Mov. Disord. 30 (2015) 37–44.

[25] C. Henchcliffe, W.L. Severt, Disease modification in Parkinson's disease, Drugs Aging 28 (2011) 605–615.

[26] S. Fahn, How do you treat motor complications in Parkinson's disease: Medicine, surgery, or both?, Ann. Neurol. 64 (Suppl. 2) (2008) S56–64.

[27] S. Chiken, A. Nambu, Mechanism of deep brain stimulation: inhibition, excitation, or disruption?, Neuroscientist 22 (3) (2016) 313–322.

[28] S.K. Kalia, T. Sankar, A.M. Lozano, Deep brain stimulation for Parkinson's disease and other movement disorders, Curr. Opin. Neurol. 26 (2013) 374–380.

[29] O. Lindvall, Treatment of Parkinson's disease using cell transplantation, Philos. Trans. R. Soc. Lond. B. Biol. Sci. 370 (2015) 20140370.

[30] A. Park, M. Stacy, Disease-modifying drugs in Parkinson's disease, Drugs 75 (2015) 2065–2071.

[31] L. Yang, N.Y. Calingasan, E.J. Wille, K. Cormier, K. Smith, R.J. Ferrante, et al. Combination therapy with coenzyme Q10 and creatine produces additive neuroprotective effects in models of Parkinson's and Huntington's diseases, J. Neurochem. 109 (2009) 1427–1439.

[32] C.R. Swanson, V. Joers, V. Bondarenko, K. Brunner, H.A. Simmons, T.E. Ziegler, et al. The PPAR-γ agonist pioglitazone modulates inflammation and induces neuroprotection in parkinsonian monkeys, J. Neuroinflammation. 8 (2011) 91.

[33] NINDSExploratory trials in Parkinson disease (NET-PD) FS-ZONE investigators, pioglitazone in early Parkinson's disease: a phase 2, multicentre, double-blind, randomised trial, Lancet Neurol. 14 (2015) 795–803.

[34] C.S. Chan, J.N. Guzman, E. Ilijic, J.N. Mercer, C. Rick, T. Tkatch, et al. "Rejuvenation" protects neurons in mouse models of Parkinson's disease, Nature 447 (2007) 1081–1086.

[35] B. Ritz, S.L. Rhodes, L. Qian, E. Schernhammer, J.H. Olsen, S. Friis, L-type calcium channel blockers and Parkinson disease in Denmark, Ann. Neurol. 67 (2010) 600–606.

[36] Parkinson Study GroupPhase II safety, tolerability, and dose selection study of isradipine as a potential disease-modifying intervention in early Parkinson's disease (STEADY-PD), Mov. Disord. 28 (2013) 1823–1831.

[37] M.A. Schwarzschild, A. Ascherio, M.F. Beal, M.E. Cudkowicz, G.C. Curhan, Parkinson Study Group SURE-PD Investigatorset al. Inosine to increase serum and cerebrospinal fluid urate in Parkinson disease: a randomized clinical trial, JAMA Neurol. 71 (2014) 141–150.

[38] I. Aviles-Olmos, J. Dickson, Z. Kefalopoulou, A. Djamshidian, J. Kahan, P. Ell, et al. Motor and cognitive advantages persist 12 months after exenatide exposure in Parkinson's disease, J. Park. Dis. 4 (2014) 337–344.

[39] V. Bonifati, Genetics of Parkinson's disease—state of the art, 2013, Parkinsonism Relat. Disord. 20 (Suppl. 1) (2014) S23–S28.

[40] M.A. Nalls, N. Pankratz, C.M. Lill, C.B. Do, D.G. Hernandez, M. Saad, et al. Large-scale meta-analysis of genome-wide association data identifies six new risk loci for Parkinson's disease, Nat. Genet. 46 (2014) 989–993.

[41] L. Breydo, J.W. Wu, V.N. Uversky, A-synuclein misfolding and Parkinson's disease, Biochim. Biophys. Acta 1822 (2012) 261–285.

[42] M.G. Spillantini, R.A. Crowther, R. Jakes, M. Hasegawa, M. Goedert, Alpha-synuclein in filamentous inclusions of Lewy bodies from Parkinson's disease and dementia with Lewy bodies, Proc. Natl. Acad. Sci. USA 95 (1998) 6469–6473.

[43] K.A. Conway, S.J. Lee, J.C. Rochet, T.T. Ding, R.E. Williamson, P.T. Lansbury Jr., Acceleration of oligomerization, not fibrillization, is a shared property of both alpha-synuclein mutations linked to early-onset Parkinson's disease: implications for pathogenesis and therapy, Proc. Natl. Acad. Sci. USA 97 (2000) 571–576.

[44] K. Ono, T. Ikeda, J. Takasaki, M. Yamada, Familial Parkinson disease mutations influence α-synuclein assembly, Neurobiol. Dis. 43 (2011) 715–724.

[45] L. Tosatto, M.H. Horrocks, A.J. Dear, T.P.J. Knowles, M. Dalla Serra, N. Cremades, et al. Single-molecule FRET studies on alpha-synuclein oligomerization of Parkinson's disease genetically related mutants, Sci. Rep. 5 (2015) 16696.

[46] E. Rockenstein, S. Nuber, C.R. Overk, K. Ubhi, M. Mante, C. Patrick, et al. Accumulation of oligomer-prone α-synuclein exacerbates synaptic and neuronal degeneration in vivo, Brain J. Neurol. 137 (2014) 1496–1513.

[47] B. Winner, R. Jappelli, S.K. Maji, P.A. Desplats, L. Boyer, S. Aigner, et al. In vivo demonstration that alpha-synuclein oligomers are toxic, Proc. Natl. Acad. Sci. USA 108 (2011) 4194–4199.

[48] V.N. Uversky, A protein-chameleon: conformational plasticity of alpha-synuclein, a disordered protein involved in neurodegenerative disorders, J. Biomol. Struct. Dyn. 21 (2003) 211–234.

[49] D. Snead, D. Eliezer, Alpha-synuclein function and dysfunction on cellular membranes, Exp. Neurobiol. 23 (2014) 292–313.

[50] P.J. Barrett, J. Timothy Greenamyre, Post-translational modification of α-synuclein in Parkinson's disease, Brain Res. 1628 (2015) 247–253.

[51] W. Wang, I. Perovic, J. Chittuluru, A. Kaganovich, L.T.T. Nguyen, J. Liao, et al. A soluble α-synuclein construct forms a dynamic tetramer, Proc. Natl. Acad. Sci. USA 108 (2011) 17797–17802.

[52] T. Bartels, J.G. Choi, D.J. Selkoe, α-Synuclein occurs physiologically as a helically folded tetramer that resists aggregation, Nature 477 (2011) 107–110.

[53] B. Fauvet, M.K. Mbefo, M.-B. Fares, C. Desobry, S. Michael, M.T. Ardah, et al. α-Synuclein in central nervous system and from erythrocytes, mammalian cells, and *Escherichia coli* exists predominantly as disordered monomer, J. Biol. Chem. 287 (2012) 15345–15364.

[54] W. Xin, S. Emadi, S. Williams, Q. Liu, P. Schulz, P. He, et al. Toxic oligomeric alpha-synuclein variants present in human Parkinson's disease brains are differentially generated in mammalian cell models, Biomolecules 5 (2015) 1634–1651.

[55] J. Levin, F. Schmidt, C. Boehm, C. Prix, K. Bötzel, S. Ryazanov, et al. The oligomer modulator anle138b inhibits disease progression in a Parkinson mouse model even with treatment started after disease onset, Acta Neuropathol. 127 (2014) 779–780.

[56] J. Wagner, S. Ryazanov, A. Leonov, J. Levin, S. Shi, F. Schmidt, et al. Anle138b: a novel oligomer modulator for disease-modifying therapy of neurodegenerative diseases such as prion and Parkinson's disease, Acta Neuropathol. 125 (2013) 795–813.

[57] E. Colla, P.H. Jensen, O. Pletnikova, J.C. Troncoso, C. Glabe, M.K. Lee, Accumulation of toxic α-synuclein oligomer within endoplasmic reticulum occurs in α-synucleinopathy in vivo, J. Neurosci. Off. J. Soc. Neurosci. 32 (2012) 3301–3305.

[58] J.H. Kordower, Y. Chu, R.A. Hauser, T.B. Freeman, C.W. Olanow, Lewy body-like pathology in long-term embryonic nigral transplants in Parkinson's disease, Nat. Med. 14 (2008) 504–506.

[59] J.-Y. Li, E. Englund, J.L. Holton, D. Soulet, P. Hagell, A.J. Lees, et al. Lewy bodies in grafted neurons in subjects with Parkinson's disease suggest host-to-graft disease propagation, Nat. Med. 14 (2008) 501–503.

[60] I. Mendez, A. Viñuela, A. Astradsson, K. Mukhida, P. Hallett, H. Robertson, et al. Dopamine neurons implanted into people with Parkinson's disease survive without pathology for 14 years, Nat. Med. 14 (2008) 507–509.

[61] L.A. Volpicelli-Daley, K.C. Luk, T.P. Patel, S.A. Tanik, D.M. Riddle, A. Stieber, et al. Exogenous α-synuclein fibrils induce Lewy body pathology leading to synaptic dysfunction and neuron death, Neuron 72 (2011) 57–71.

[62] K.C. Luk, V. Kehm, J. Carroll, B. Zhang, P. O'Brien, J.Q. Trojanowski, et al. Pathological α-synuclein transmission initiates Parkinson-like neurodegeneration in nontransgenic mice, Science 338 (2012) 949–953.

[63] J.L. Guo, D.J. Covell, J.P. Daniels, M. Iba, A. Stieber, B. Zhang, et al. Distinct α-synuclein strains differentially promote tau inclusions in neurons, Cell 154 (2013) 103–117.

[64] W. Peelaerts, L. Bousset, A. Van der Perren, A. Moskalyuk, R. Pulizzi, M. Giugliano, et al. α-Synuclein strains cause distinct synucleinopathies after local and systemic administration, Nature 522 (2015) 340–344.

[65] H.T. Tran, C.H.-Y. Chung, M. Iba, B. Zhang, J.Q. Trojanowski, K.C. Luk, et al. A-synuclein immunotherapy blocks uptake and templated propagation of misfolded α-synuclein and neurodegeneration, Cell Rep. 7 (2014) 2054–2065.

[66] M. Xilouri, O.R. Brekk, N. Landeck, P.M. Pitychoutis, T. Papasilekas, Z. Papadopoulou-Daifoti, et al. Boosting chaperone-mediated autophagy in vivo mitigates α-synuclein-induced neurodegeneration, Brain J. Neurol. 136 (2013) 2130–2146.

[67] T.T. Nielsen, J.E. Nielsen, Antisense gene silencing: therapy for neurodegenerative disorders?, Genes 4 (2013) 457–484.

[68] A.L. McCormack, S.K. Mak, J.M. Henderson, D. Bumcrot, M.J. Farrer, D.A. Di Monte, Alpha-synuclein suppression by targeted small interfering RNA in the primate substantia nigra, PLoS One 5 (2010) e12122.

[69] C.R. Sibley, M.J.A. Wood, Identification of allele-specific RNAi effectors targeting genetic forms of Parkinson's disease, PLoS One 6 (2011) e26194.

[70] F. Cheng, G. Vivacqua, S. Yu, The role of α-synuclein in neurotransmission and synaptic plasticity, J. Chem. Neuroanat. 42 (2011) 242–248.

[71] C.E. Khodr, M.K. Sapru, J. Pedapati, Y. Han, N.C. West, A.P. Kells, et al. An α-synuclein AAV gene silencing vector ameliorates a behavioral deficit in a rat model of Parkinson's disease, but displays toxicity in dopamine neurons, Brain Res. 1395 (2011) 94–107.

[72] O.S. Gorbatyuk, S. Li, K. Nash, M. Gorbatyuk, A.S. Lewin, L.F. Sullivan, et al. In vivo RNAi-mediated alpha-synuclein silencing induces nigrostriatal degeneration, Mol. Ther. 18 (2010) 1450–1457.

[73] A.D. Zharikov, J.R. Cannon, V. Tapias, Q. Bai, M.P. Horowitz, V. Shah, et al. shRNA targeting α-synuclein prevents neurodegeneration in a Parkinson's disease model, J. Clin. Invest. 125 (2015) 2721–2735.

[74] A. Abeliovich, Y. Schmitz, I. Fariñas, D. Choi-Lundberg, W.H. Ho, P.E. Castillo, et al. Mice lacking alpha-synuclein display functional deficits in the nigrostriatal dopamine system, Neuron 25 (2000) 239–252.

[75] D.W. Miller, S.M. Hague, J. Clarimon, M. Baptista, K. Gwinn-Hardy, M.R. Cookson, et al. Alpha-synuclein in blood and brain from familial Parkinson disease with SNCA locus triplication, Neurology 62 (2004) 1835–1838.

[76] A. Zimprich, S. Biskup, P. Leitner, P. Lichtner, M. Farrer, S. Lincoln, et al. Mutations in LRRK2 cause autosomal-dominant parkinsonism with pleomorphic pathology, Neuron 44 (2004) 601–607.

[77] C. Paisán-Ruíz, S. Jain, E.W. Evans, W.P. Gilks, J. Simón, M. van der Brug, et al. Cloning of the gene containing mutations that cause PARK8-linked Parkinson's disease, Neuron 44 (2004) 595–600.

[78] M. Funayama, K. Hasegawa, E. Ohta, N. Kawashima, M. Komiyama, H. Kowa, et al. An LRRK2 mutation as a cause for the parkinsonism in the original PARK8 family, Ann. Neurol. 57 (2005) 918–921.

[79] H.T.S. Benamer, R. de Silva, LRRK2 G2019S in the North African population: a review, Eur. Neurol. 63 (2010) 321–325.

[80] W. Satake, Y. Nakabayashi, I. Mizuta, Y. Hirota, C. Ito, M. Kubo, et al. Genome-wide association study identifies common variants at four loci as genetic risk factors for Parkinson's disease, Nat. Genet. 41 (2009) 1303–1307.

[81] J. Simón-Sánchez, C. Schulte, J.M. Bras, M. Sharma, J.R. Gibbs, D. Berg, et al. Genome-wide association study reveals genetic risk underlying Parkinson's disease, Nat. Genet. 41 (2009) 1308–1312.

[82] P.A. Lewis, E. Greggio, A. Beilina, S. Jain, A. Baker, M.R. Cookson, The R1441C mutation of LRRK2 disrupts GTP hydrolysis, Biochem. Biophys. Res. Commun. 357 (2007) 668–671.

[83] V. Daniëls, R. Vancraenenbroeck, B.M.H. Law, E. Greggio, E. Lobbestael, F. Gao, et al. Insight into the mode of action of the LRRK2 Y1699C pathogenic mutant, J. Neurochem. 116 (2011) 304–315.

[84] L. Guo, P.N. Gandhi, W. Wang, R.B. Petersen, A.L. Wilson-Delfosse, S.G. Chen, The Parkinson's disease-associated protein, leucine-rich repeat kinase 2 (LRRK2), is an authentic GTPase that stimulates kinase activity, Exp. Cell Res. 313 (2007) 3658–3670.

[85] E. Greggio, M.R. Cookson, Leucine-rich repeat kinase 2 mutations and Parkinson's disease: three questions, ASN Neuro. 1 (2009) e00002.

[86] A. Beilina, I.N. Rudenko, A. Kaganovich, L. Civiero, H. Chau, S.K. Kalia, et al. Unbiased screen for interactors of leucine-rich repeat kinase 2 supports a common pathway for sporadic and familial Parkinson disease, Proc. Natl. Acad. Sci. USA 111 (2014) 2626–2631.

[87] C. Manzoni, A. Mamais, S. Dihanich, P. McGoldrick, M.J. Devine, J. Zerle, et al. Pathogenic Parkinson's disease mutations across the functional domains of LRRK2 alter the autophagic/lysosomal response to starvation, Biochem. Biophys. Res. Commun. 441 (2013) 862–866.

[88] E. Greggio, S. Jain, A. Kingsbury, R. Bandopadhyay, P. Lewis, A. Kaganovich, et al. Kinase activity is required for the toxic effects of mutant LRRK2/dardarin, Neurobiol. Dis. 23 (2006) 329–341.

[89] W.W. Smith, Z. Pei, H. Jiang, V.L. Dawson, T.M. Dawson, C.A. Ross, Kinase activity of mutant LRRK2 mediates neuronal toxicity, Nat. Neurosci. 9 (2006) 1231–1233.

[90] B.D. Lee, J.-H. Shin, J. VanKampen, L. Petrucelli, A.B. West, H.S. Ko, et al. Inhibitors of leucine-rich repeat kinase-2 protect against models of Parkinson's disease, Nat. Med. 16 (2010) 998–1000.

[91] E. Tsika, A.P.T. Nguyen, J. Dusonchet, P. Colin, B.L. Schneider, D.J. Moore, Adenoviral-mediated expression of G2019S LRRK2 induces striatal pathology in a kinase-dependent manner in a rat model of Parkinson's disease, Neurobiol. Dis. 77 (2015) 49–61.

[92] X. Deng, N. Dzamko, A. Prescott, P. Davies, Q. Liu, Q. Yang, et al. Characterization of a selective inhibitor of the Parkinson's disease kinase LRRK2, Nat. Chem. Biol. 7 (2011) 203–205.

[93] X. Deng, H.G. Choi, S.J. Buhrlage, N.S. Gray, Leucine-rich repeat kinase 2 inhibitors: a patent review (2006–2011), Expert Opin. Ther. Pat. 22 (2012) 1415–1426.

[94] S. Göring, J.-M. Taymans, V. Baekelandt, B. Schmidt, Indolinone based LRRK2 kinase inhibitors with a key hydrogen bond, Bioorg. Med. Chem. Lett. 24 (2014) 4630–4637.

[95] Z. Sheng, S. Zhang, D. Bustos, T. Kleinheinz, C.E. Le Pichon, S.L. Dominguez, et al. Ser1292 autophosphorylation is an indicator of LRRK2 kinase activity and contributes to the cellular effects of PD mutations, Sci. Transl. Med. 4 (2012) 164ra161.

[96] T. Li, X. He, J.M. Thomas, D. Yang, S. Zhong, F. Xue, et al. A novel GTP-binding inhibitor, FX2149, attenuates LRRK2 toxicity in Parkinson's disease models, PLoS One 10 (2015) e0122461.

[97] A.B. West, Ten years and counting: moving leucine-rich repeat kinase 2 inhibitors to the clinic, Mov. Disord. 30 (2) (2015) 180–189.

[98] Y. Tong, H. Yamaguchi, E. Giaime, S. Boyle, R. Kopan, R.J. Kelleher, et al. Loss of leucine-rich repeat kinase 2 causes impairment of protein degradation pathways, accumulation of α-synuclein, and apoptotic cell death in aged mice, Proc. Natl. Acad. Sci. USA 107 (2010) 9879–9884.

[99] Y. Tong, E. Giaime, H. Yamaguchi, T. Ichimura, Y. Liu, H. Si, et al. Loss of leucine-rich repeat kinase 2 causes age-dependent bi-phasic alterations of the autophagy pathway, Mol. Neurodegener. 7 (2012) 2.

[100] M.A.S. Baptista, K.D. Dave, M.A. Frasier, T.B. Sherer, M. Greeley, M.J. Beck, et al. Loss of leucine-rich repeat kinase 2 (LRRK2) in rats leads to progressive abnormal phenotypes in peripheral organs, PLoS One 8 (2013) e80705.

[101] D. Ness, Z. Ren, S. Gardai, D. Sharpnack, V.J. Johnson, R.J. Brennan, et al. Leucine-rich repeat kinase 2 (LRRK2)-deficient rats exhibit renal tubule injury and perturbations in metabolic and immunological homeostasis, PLoS One 8 (2013) e66164.

[102] C. Manzoni, A. Mamais, S. Dihanich, R. Abeti, M.P.M. Soutar, H. Plun-Favreau, et al. Inhibition of LRRK2 kinase activity stimulates macroautophagy, Biochim. Biophys. Acta 1833 (2013) 2900–2910.

[103] M.C. Herzig, C. Kolly, E. Persohn, D. Theil, T. Schweizer, T. Hafner, et al. LRRK2 protein levels are determined by kinase function and are crucial for kidney and lung homeostasis in mice, Hum. Mol. Genet. 20 (2011) 4209–4223.

[104] M.J. Fell, C. Mirescu, K. Basu, B. Cheewatrakoolpong, D.E. DeMong, J.M. Ellis, et al. MLi-2, a potent, selective, and centrally active compound for exploring the therapeutic potential and safety of LRRK2 kinase inhibition, J. Pharmacol. Exp. Ther. 355 (2015) 397–409.

[105] R.N. Fuji, M. Flagella, M. Baca, M.A. Baptista, J. Brodbeck, B.K. Chan, et al. Effect of selective LRRK2 kinase inhibition on nonhuman primate lung, Sci. Transl. Med. 7 (2015) 273ra15.

[106] I.N. Rudenko, R. Chia, M.R. Cookson, Is inhibition of kinase activity the only therapeutic strategy for LRRK2-associated Parkinson's disease?, BMC Med. 10 (2012) 20.

[107] M.R. Cookson, J. Hardy, P.A. Lewis, Genetic neuropathology of Parkinson's disease, Int. J. Clin. Exp. Pathol. 1 (2008) 217–231.

[108] X. Lin, L. Parisiadou, X.-L. Gu, L. Wang, H. Shim, L. Sun, et al. Leucine-rich repeat kinase 2 regulates the progression of neuropathology induced by Parkinson's-disease-related mutant alpha-synuclein, Neuron 64 (2009) 807–827.

[109] J.P.L. Daher, O. Pletnikova, S. Biskup, A. Musso, S. Gellhaar, D. Galter, et al. Neurodegenerative phenotypes in an A53T α-synuclein transgenic mouse model are independent of LRRK2, Hum. Mol. Genet. 21 (2012) 2420–2431.

[110] M.C. Herzig, M. Bidinosti, T. Schweizer, T. Hafner, C. Stemmelen, A. Weiss, et al. High LRRK2 levels fail to induce or exacerbate neuronal alpha-synucleinopathy in mouse brain, PLoS One 7 (2012) e36581.

[111] J.P.L. Daher, L.A. Volpicelli-Daley, J.P. Blackburn, M.S. Moehle, A.B. West, Abrogation of α-synuclein-mediated dopaminergic neurodegeneration in LRRK2-deficient rats, Proc. Natl. Acad. Sci. USA 111 (2014) 9289–9294.

[112] J.P.L. Daher, H.A. Abdelmotilib, X. Hu, L.A. Volpicelli-Daley, M.S. Moehle, K.B. Fraser, et al. Leucine-rich repeat kinase 2 (LRRK2) pharmacological inhibition abates α-synuclein gene-induced neurodegeneration, J. Biol. Chem. 290 (2015) 19433–19444.

[113] G. Skibinski, K. Nakamura, M.R. Cookson, S. Finkbeiner, Mutant LRRK2 toxicity in neurons depends on LRRK2 levels and synuclein but not kinase activity or inclusion bodies, J. Neurosci. Off. J. Soc. Neurosci. 34 (2014) 418–433.

[114] J. Schapansky, J.D. Nardozzi, M.J. LaVoie, The complex relationships between microglia, alpha-synuclein, and LRRK2 in Parkinson's disease, Neuroscience 302 (2015) 74–88.

[115] M.S. Moehle, P.J. Webber, T. Tse, N. Sukar, D.G. Standaert, T.M. DeSilva, et al. LRRK2 inhibition attenuates microglial inflammatory responses, J. Neurosci. Off. J. Soc. Neurosci. 32 (2012) 1602–1611.

[116] B. Kim, M.-S. Yang, D. Choi, J.-H. Kim, H.-S. Kim, W. Seol, et al. Impaired inflammatory responses in murine lrrk2-knockdown brain microglia, PLoS One 7 (2012) e34693.

[117] I. Russo, G. Berti, N. Plotegher, G. Bernardo, R. Filograna, L. Bubacco, et al. Leucine-rich repeat kinase 2 positively regulates inflammation and down-regulates NF-κB p50 signaling in cultured microglia cells, J. Neuroinflammation 12 (2015) 230.

[118] I. Choi, B. Kim, J.-W. Byun, S.H. Baik, Y.H. Huh, J.-H. Kim, et al. LRRK2 G2019S mutation attenuates microglial motility by inhibiting focal adhesion kinase, Nat. Commun. 6 (2015) 8255.

[119] F. Gillardon, R. Schmid, H. Draheim, Parkinson's disease-linked leucine-rich repeat kinase 2(R1441G) mutation increases proinflammatory cytokine release from activated primary microglial cells and resultant neurotoxicity, Neuroscience 208 (2012) 41–48.

[120] V. Bonifati, Autosomal recessive parkinsonism, Parkinsonism Relat. Disord. 18 (Suppl. 1) (2012) S4–6.

[121] D.W. Miller, R. Ahmad, S. Hague, M.J. Baptista, R. Canet-Aviles, C. McLendon, et al. L166P mutant DJ-1, causative for recessive Parkinson's disease, is degraded through the ubiquitin-proteasome system, J. Biol. Chem. 278 (2003) 36588–36595.

[122] A. Beilina, M. Van Der Brug, R. Ahmad, S. Kesavapany, D.W. Miller, G.A. Petsko, et al. Mutations in PTEN-induced putative kinase 1 associated with recessive parkinsonism have differential effects on protein stability, Proc. Natl. Acad. Sci. USA 102 (2005) 5703–5708.

[123] R.J. Youle, D.P. Narendra, Mechanisms of mitophagy, Nat. Rev. Mol. Cell Biol. 12 (2011) 9–14.

[124] D.P. Narendra, S.M. Jin, A. Tanaka, D.-F. Suen, C.A. Gautier, J. Shen, et al. PINK1 is selectively stabilized on impaired mitochondria to activate Parkin, PLoS Biol. 8 (2010) e1000298.

[125] C. Vives-Bauza, C. Zhou, Y. Huang, M. Cui, R.L.A. de Vries, J. Kim, et al. PINK1-dependent recruitment of Parkin to mitochondria in mitophagy, Proc. Natl. Acad. Sci. USA 107 (2010) 378–383.

[126] L. Petrucelli, C. O'Farrell, P.J. Lockhart, M. Baptista, K. Kehoe, L. Vink, et al. Parkin protects against the toxicity associated with mutant alpha-synuclein: proteasome dysfunction selectively affects catecholaminergic neurons, Neuron 36 (2002) 1007–1019.

[127] C. Lo Bianco, B.L. Schneider, M. Bauer, A. Sajadi, A. Brice, T. Iwatsubo, et al. Lentiviral vector delivery of parkin prevents dopaminergic degeneration in an alpha-synuclein rat model of Parkinson's disease, Proc. Natl. Acad. Sci. USA 101 (2004) 17510–17515.

[128] K. Venderova, G. Kabbach, E. Abdel-Messih, Y. Zhang, R.J. Parks, Y. Imai, et al. Leucine-rich repeat kinase 2 interacts with parkin, DJ-1 and PINK-1 in a *Drosophila melanogaster* model of Parkinson's disease, Hum. Mol. Genet. 18 (2009) 4390–4404.

[129] C.-H. Ng, S.Z.S. Mok, C. Koh, X. Ouyang, M.L. Fivaz, E.-K. Tan, et al. Parkin protects against LRRK2 G2019S mutant-induced dopaminergic neurodegeneration in *Drosophila*, J. Neurosci. Off. J. Soc. Neurosci. 29 (2009) 11257–11262.

[130] A.S. Van Rompuy, E. Lobbestael, A. Van der Perren, C. Van den Haute, V. Baekelandt, Baekelandt, Long-term overexpression of human wild-type and T240R mutant Parkin in rat substantia nigra induces progressive dopaminergic neurodegeneration, J. Neuropathol. Exp. Neurol. 73 (2) (2014) 159–174.

[131] R.A. Charan, M.J. LaVoie, Pathologic and therapeutic implications for the cell biology of parkin, Mol. Cell. Neurosci. 66 (2015) 62–71.

[132] H. Mortiboys, J. Aasly, O. Bandmann, Ursocholanic acid rescues mitochondrial function in common forms of familial Parkinson's disease, Brain J. Neurol. 136 (2013) 3038–3050.

[133] H. Mortiboys, R. Furmston, G. Bronstad, J. Aasly, C. Elliott, O. Bandmann, UDCA exerts beneficial effect on mitochondrial dysfunction in LRRK2(G2019S) carriers and in vivo, Neurology 85 (2015) 846–852.

LEWY BODY DEMENTIA

Marion Delenclos, Simon Moussaud, Pamela J. McLean

Jacksonville, FL, United States

CHAPTER OUTLINE

Lewy Body Dementia 175
  Concept of Lewy Body Dementia..175
  Clinical Symptoms 177
  Genetic Association 177
  Pathophysiology 178
Management of LBD 182
  Current Symptomatic Treatment..182
  Disease-Modifying Therapy...183
Concluding Remarks 188
References 188

LEWY BODY DEMENTIA
CONCEPT OF LEWY BODY DEMENTIA

Lewy body dementias (LBD) are the second most common type of degenerative dementia in the elderly population, accounting for 15%–20% of all cases [1]. They are characterized by fluctuations in cognition (delirium), recurrent visual hallucinations, and motor features of Parkinsonism. Other significant features include sleep disorders, depression, and autonomic dysfunction [1,2]. The term "Lewy body dementia" was introduced in the last decade by physicians to describe two related disorders known as dementia with Lewy bodies (DLB) and Parkinson's disease dementia (PDD). Although the initial clinical manifestation differs in PDD and DLB, as the disorders progress the symptoms and the pathological characteristics are much more similar than they are different. As such, many researchers and clinicians think of PDD and DLB as being a continuum of similar disease processes rather than two distinct entities. The main difference lies in the timing of dementia onset in relation to motor symptoms. Indeed, the diagnosis of DLB is applied when dementia precedes or is closely followed by motor symptoms, whereas PDD diagnosis is applied when motor symptoms occur at least 1 year before dementia. This arbitrary guideline—"the 1-year rule"—was first established by an international workshop on the diagnosis of DLB in 1996 [3]. It was there that DLB was designated for the first time as a diagnostic entity in order to distinguish it from Alzheimer's disease (AD) and Parkinson's disease (PD), both of which share pathological and clinical features. Since then an LBD working group has been created to revise and refine the criteria for the clinical and pathologic diagnosis of these two types of dementia [4]. Although DLB and PPD boundary issues are still an ongoing debate, the working group concluded that clinical features justify distinguishing them as two disorders whereas a single disorder could be observed when classifying the underlying disease mechanisms [1].

DLB accounts for at least 4.2% of all dementia patients and is the second most frequent type of dementia after AD [5]. When dementia occurs in the setting of PD, it is then referred to as PDD. The prevalence of PDD is stated to be only 3.6% of all dementia cases [6], but this does not capture the fact that PDD is an age-dependent continuum of PD. In fact, up to 80% of all PD patients develop dementia during the course of their disease increasing the actual prevalence of PDD [7].

LBD can be difficult to diagnose because it can resemble and overlap with other causes of dementia like AD, but also with other parkinsonian syndromes (PD, progressive supranuclear palsy, multiple system atrophy, corticobasal syndrome), and vascular dementia. Clinically, pathologically, and biochemically LBD appears to fall somewhere in the middle of a disease spectrum ranging from AD to PD (Fig. 7.1) and as in many neurodegenerative diseases, a definitive diagnosis in only confirmed at autopsy. At the pathologic level LBD is characterized by the presence of Lewy bodies (LBs), which

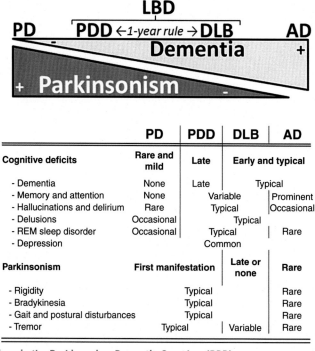

	PD	PDD	DLB	AD
Cognitive deficits	Rare and mild	Late	Early and typical	
- Dementia	None	Late	Typical	
- Memory and attention	None	Variable		Prominent
- Hallucinations and delirium	Rare	Typical		Occasional
- Delusions	Occasional	Typical		
- REM sleep disorder	Occasional	Typical		Rare
- Depression	Common			
Parkinsonism	First manifestation		Late or none	Rare
- Rigidity	Typical			Rare
- Bradykinesia	Typical			Rare
- Gait and postural disturbances	Typical			Rare
- Tremor	Typical		Variable	Rare

FIGURE 7.1 Clinical Signs in the Parkinsonism Dementia Complex *(PDD)*

The primary defining feature of PDD is dementia that develops in the setting of established Parkinson's disease *(PD)*. In PDD parkinsonism (motor deficit) develops prior to the onset of dementia. PDD can be temporally distinguished from dementia with Lewy bodies *(DLB)* by the "1-year rule"; in PDD, motor symptoms develop at least 1 year before development of dementia, while DLB is typically diagnosed when dementia occurs before or concurrently with parkinsonism (if it is present).

are eosinophilic cytoplasmic inclusions widespread in the cortex, brainstem, and limbic areas. Their distribution is slightly different than in PD where they are found mostly in the diencephalon and basal forebrain, with fewer in cortical areas. The staging of the disease is determined by the density and distribution of these inclusions and correlates with the severity of the dementia [1,8] although pathological studies report that LB densities cannot be used as criteria to distinguish cases of DLB from PDD.

CLINICAL SYMPTOMS

As with most neurodegenerative disorders, the symptoms of LBD worsen over time with cognitive and motor functions deteriorating typically over several years. The parkinsonian motor syndrome present in LBD is similar in PDD and DLB and can include slowed movement (bradykinesia), rigidity, tremor, and gait difficulties (Fig. 7.1). In terms of cognitive deficits, both disorders can involve progressive impairment of visual spatial processing, attention, executive dysfunction, with fluctuating cognition (delirium) being recognized as the most characteristic. Memory impairment may not necessarily occur in the early stages but is usually evident with progression. The cognitive features of DLB and PDD are similar and often indistinguishable from each other. However, patients with LBD generally have less severe visual and verbal memory deficits, but more marked executive dysfunction (reasoning, planning, and sequencing) than AD patients, making their cognitive profile slightly different [9,10]. Interestingly, the presence of early visuospatial deficits and development of visual hallucinations have proven to help identify patients whose clinical syndrome is LBD rather than AD [11], but hallucinations are also frequent in PD. Delusions and aggressive behavior are additional common clinical characteristics of DLB [12]. Psychiatric symptoms and autonomic dysfunction are the others component of the clinical manifestations and can include anxiety and depression. Psychotic symptoms are indistinguishable between PDD and DLB [13,14] but the frequency of occurrence is greatest in the latter. LBD patients also suffer from (REM) sleep behavior disorder, a parasomnia manifested by vivid and frightening dreams associated with simple or complex motor behavior during REM sleep [15,16]. This last feature is rather frequent in PDD and DLB and increases the diagnostic likelihood of LBD compared to AD (Fig. 7.1).

GENETIC ASSOCIATION

Although most cases of LBD are considered to be sporadic like PD and AD, family history seems to be more common in LBD than in AD, suggesting a higher contribution of genetic factors [17]. Indeed, families in which a mixed phenotype of dementia and parkinsonism is inherited in a Mendelian manner have been reported [18,19]. Also, autosomal dominant familial forms of parkinsonism due to point mutations in the alpha-synuclein (αsyn) gene (*SNCA*) (A30P, E46K, H50Q, G51D, or A53T) or the duplication or triplication of the wild-type (wt) *SNCA* gene locus, commonly present with additional atypical clinical signs, such as hallucinations, cognitive impairment, and dementia [20–23]. Similarly, autosomal dominant familial forms of dementia due to mutations in the microtubule-associated protein tau gene (*MAPT*) cause a variety of neurodegenerative phenotypes including varying extents of parkinsonism [24–26]. Interestingly, familial cases carrying mutations in the *SNCA* or *MAPT* genes present a phenotypic heterogeneity even within a single family. This variability of symptoms combining both parkinsonism and dementia strongly echoes what is seen among the LBD patients. A heterozygous mutation in the β-glucocerebrosidase gene (*GBA*) is associated with an increased risk of PD [27], but recently in a multicenter study it was shown that *GBA* mutations likely play an even larger role in the genetic etiology of DLB [28,29]. In the same study, PDD cases were also analyzed and even if the association was slightly attenuated compared to DLB, it was very similar to the one found in PD cases.

In addition to these rare autosomal dominant disease-causing mutations, genetic risk factors have been recently identified in a genome-wide association study [30]. Interestingly, several genetic loci found to increase the risk of developing AD or PD was also found in LBD cases. For instance, genetic variation around the *MAPT* gene and in particular the H1 haplotype, which is highly associated with tauopathy (i.e., AD), was linked as a risk factor in PDD cases [31]. For DLB, although no significant association of the *MAPT* locus with disease susceptibility was found in a recent genome-wide association study [30], correlation between the H1 haplotype and the degree of αsyn pathology in the brainstem was observed in a small neuropathological study [32]. In a neuropathological study, the extent of neocortical LBs and neurofibrillary pathology correlated with the *SNCA* gene single nucleotide polymorphism (SNP) rs2572324. Lastly, the APOe4 locus is identified as risk factor for AD and was predictive of PDD in a large autopsy cohort [33].

PATHOPHYSIOLOGY

Alpha-synuclein

The pathological hallmarks of LBD involve the presence of widespread LBs and Lewy neurites (LNs) that mainly contain deposits of abnormal filamentous assemblies of αsyn protein [34] (Fig. 7.2). DLB and PDD are therefore classified as alpha-synucleinopathies, meaning they are neurodegenerative conditions associated with abnormal aggregations of αsyn (see also Chapter 2). LBs are located predominantly in neurons of substantia nigra or locus coeruleus (brainstem predominant form), and/or in the cerebral cortex and limbic structure (diffuse form of DLB). Histologically they can be detected in postmortem brains by eosin staining or by antiubiquitin. Also, the gold standard marker for LBD diagnosis is now considered to be anti-αsyn immunohistochemistry. αSyn is an abundant protein in the normal brain with ambiguous function that has been implicated in synaptic transmission and induction of neurotransmitter release. Other functions including vesicular trafficking, neuroplasticity, and fatty acid binding have also been described [35–38]. αSyn is a soluble protein that may exist as a natively unfolded monomer or a metastable tetramer [39–42], but under pathological conditions is believed to misfold and dimerize, triggering an oligomerization cascade which ultimately leads to the formation of LBs and LNs (Fig. 7.3). Changes in the phosphorylation and solubility status of αSyn occur prior to LBs formation in PD and PDD [43] and are implicated in the process of αSyn aggregation. Posttranslational modifications, such as nitration, ubiquitination, or phosphorylation are prevalent and might alter the function of αSyn. In LBD brains, approximately 90% of insoluble αSyn is phosphorylated at serine residue 129 [44] suggesting its phosphorylation state has an important pathogenic role.

In the last decade evidence has accumulated which suggests that the soluble oligomers formed relatively early during the formation of fibrils are in fact the toxic species, and though the exact mechanism of toxicity remains unclear, the pathogenic process may involve multiple systems including the ubiquitin–proteasome system, the autophagy–lysosomal pathway, mitochondrial dysfunction, and the unfolded-protein response (Fig. 7.3) [45–47]. Also it is unclear whether LBs and LNs have a neuroprotective or neurotoxic function since sequestering pathological forms of αsyn into LB could be seen as an effort by the neurons to combat biological stress inside the cell, or in the contrary being seen as neurodegenerative debris. When compared to PD, αsyn pathology in PDD and DLB is often diffuse affecting the cortical and striatal area more severely. Also, the neuronal loss observed in LBD mostly

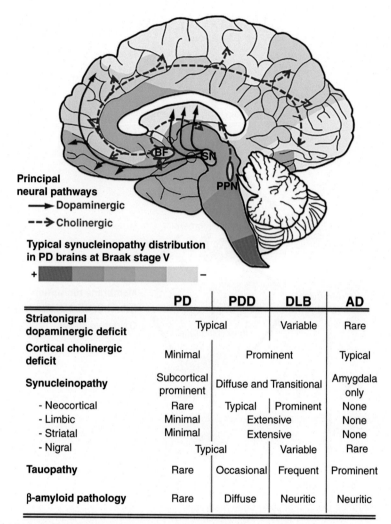

Principal
neural pathways
→ Dopaminergic
--→ Cholinergic

Typical synucleinopathy distribution
in PD brains at Braak stage V

+ ▬▬▬▬▬▬▬▬▬▬ −

	PD	PDD	DLB	AD
Striatonigral dopaminergic deficit	Typical		Variable	Rare
Cortical cholinergic deficit	Minimal	Prominent		Typical
Synucleinopathy	Subcortical prominent	Diffuse and Transitional		Amygdala only
- Neocortical	Rare	Typical	Prominent	None
- Limbic	Minimal	Extensive		None
- Striatal	Minimal	Extensive		None
- Nigral	Typical		Variable	Rare
Tauopathy	Rare	Occasional	Frequent	Prominent
β-amyloid pathology	Rare	Diffuse	Neuritic	Neuritic

FIGURE 7.2 Principal Neural Pathways and Brain Areas Affected in the Parkinsonism Dementia Complex

BF, Basal forebrain; *PPN*, pedunculopontine nucleus; *SN*, substantia nigra.

affects the dopaminergic neurons of the substantia nigra and the cholinergic neurons of the basal forebrain (Fig. 7.2)and is associated with a reduction in dopamine and acetylcholine neurotransmitters. One remarkable feature of PDD and DLB is the more profound loss of cholinergic neurons and choline acetyltransferase activity compared with both PD and AD [48]. Interestingly, even though the level and distribution of αsyn pathology are often indistinguishable between PDD and DLB, nigral neurodegeneration is sometimes minimal in DLB but it is consistent in PDD [12,49]. This finding suggests that while there are many similarities between DLB and PDD, the populations of neurons most vulnerable to neurodegeneration are different [12].

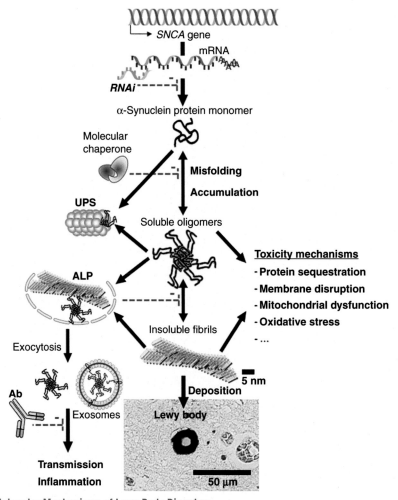

FIGURE 7.3 Molecular Mechanisms of Lewy Body Disorders

In Lewy body disorders, such as PD, PDD, and DLB, the αsyn protein abnormally misfolds and undergoes a gradual oligomerization that leads to the formation of small soluble oligomers and to large insoluble β-sheet pleated fibrils that accumulate in cells in the form of macroscopic Lewy bodies or neurites (LBs, LNs) (center panel). Although LBs and LNs disturb neuronal function by obstruction and sequestration of other proteins, the small discrete oligomers seem to be more toxic. Oligomers can catalyze their own formation by a self-templating mechanism and permeate cell membranes. These induce mitochondrial dysfunction and cytotoxicity (right panel). Cells have complex protein quality control systems to prevent such damage (left panel). Molecular chaperones can target misfolded and aggregated proteins to either properly refold them or direct them to protein catabolism pathways, such as the ubiquitin proteasome system *(UPS)* for small soluble aggregates or the autophagy/lysosome pathway for larger aggregates *(ALP)*. However, dysfunction of these systems is common in neurodegenerative disorders, making their restoration a potential therapeutic strategy *(dashed line)*. Finally, in Lewy body disorders αsyn is released by cells in an unconventional way, resulting in the presence of free soluble and exosomes-associated extracellular αsyn. This process, combined with self-templating properties of αsyn, might play a role in the disease progression and is the rational of current immunotherapeutic trials *(dashed line)*.

Tau and amyloid

Despite a central role of αsyn in LBD pathology, it is also noteworthy to mention that features of AD often coexist in DLB and PDD brains [4,50]. Pathologically, AD is associated with amyloid-beta plaques (Aβ) and neurofibrillary tangles (NFTs) composed mainly of hyperphosphorylated tau protein, distributed in the parietal, temporal, and parieto-occipital cortex (Fig. 7.2) (see also Chapter 5). Pathological studies show that the senile plaques in AD are different from those found in PDD and DLB, and are more similar to plaques found in nondemented elderly [51,52]. The deposition of Aβ plaques in DLB has been most recently shown to correlate with the amount of cortical LBs [53]. DLB is often associated with higher burdens of NFTs and Aβ than PDD, features considered as potential pathological markers for differentiating the diseases. A recent imaging study using Pittsburgh compound B (PiB) as a marker of amyloid distribution showed a significant increase in amyloid load in over 80% of DLB subjects while amyloid pathology was less frequent in PDD [54]. Along the same line, the cerebrospinal fluid (CSF) profile displayed disease-specific variations with a significantly higher level of oxidized Aβ found in DLB patients compared with PDD patients [55]. Andersson et al. [56] even showed higher levels of tau protein and a lower levels of Aβ in the CSF of DLB compared to PDD, and suggested that the combined analysis of tau and Aβ could lead to better discrimination between LBD patients than either of the measures alone. Furthermore, patients with PDD tend to have a higher Aβ cortical plaque burden and NFTs than PD patients without dementia. Thus, AD neuropathology might play an important role in the pathogenesis of dementia in LBD and more importantly might have an influence on the clinical diagnosis accuracy [57].

Chaperones proteins

Heat shock proteins (HSPs) or molecular chaperones are proteins not directly related to αsyn or amyloid plaques that may also affect LBD pathogenesis. HSPs are a class of molecular chaperones intimately involved in the folding, unfolding, and refolding of proteins which function in cohort with cochaperones. Under stress conditions the HSPs and their cochaperones are upregulated to help prevent misfolding of proteins; when this mechanism fails, however, the resultant misfolded proteins or oligomeric species may become pathogenic [58]. Given their important role in the folding of proteins and prevention of stress-induced misfolding, a key role in the pathogenesis of LBD is anticipated. In fact, LBs are immune-positive for chaperone proteins like Hsp70, Hsp40, and Hsp27. Specifically, Hsp27 and αB-crystallin, are found in LBs in DLB patients [59]. Hsp27 levels were 2.5-fold higher in DLB brains when compared to controls [60]. Also it is important to mention that several studies have found a link between HSPs expression and the proper folding of misfolded αsyn. Indeed, in a cell culture model, overexpression of Hsp70 or cochaperone HDJ-1 (HSP40 family) reduced αsyn aggregates and in in vivo experimental design crossbreeding αsyn transgenic mice with Hsp70 overexpressing mice led to a reduction of aggregation [61]. CHIP, a cochaperone Hsp70 interacting protein, immunoreactivity is also observed in LBs in postmortem brain tissue. Furthermore, overexpression of CHIP inhibits αsyn inclusion formation and reduces protein levels [62].

Neuroinflammation

Neuroinflammation is increasingly recognized as a key factor in the pathogenesis of conditions, such as AD [63,64] and PD [65,66], and is therefore most likely to be a contributor to the neurodegenerative cascade observed in LBD (see also Chapter 3). Retrospective epidemiological studies have shown that long-term treatments with nonsteroidal antiinflammatory drugs, and in particular ibuprofen, have a slight sparing effect on the risk of developing AD [67,68] or PD [69], indicating that inhibiting inflammation could be used as therapeutic strategy. However, the nonsignificant results of several clinical trials suggest that the influence of inflammation on neurodegeneration is probably more complex and subtle than first thought. In the early stages, it may be protective but it becomes increasingly detrimental and exacerbates the pathology in later stages. Indeed, it is thought that microglial activation may drive the progressive damage of dopaminergic neurons in the brain of PD patients [70]. Also in AD, Aβ fibrils and dying neurons are surrounded by activated microglia [71]. Interestingly, it was recently found that the lack of cytokine interferon-β signaling causes motor and cognitive learning impairments with formation of αsyn-containing LB, as well as nigrostriatal dopaminergic degeneration in mice [72]. Neuroinflammation in LBD has not been studied as extensively as in PD and AD, and is mostly supported by pathological and biomarker studies [73]. Imaging studies, such as positron emission tomography (PET) have shown an association between microglial activation and cognitive dysfunction in PDD patients as well as in AD [74]. Similarly in DLB, increased microglial activation in the substantia nigra, putamen, and several cortical regions was found in a PET study [75]. Interestingly, compared to AD, DLB brains showed significantly less inflammation when immunostained with an antibody against the major histocompatibility protein HLA-DR, a marker of inflammatory microglia [76]. These disease-specific patterns could be used as early biomarkers to increase the diagnosis accuracy between PD, AD, and LBD. However, future studies are required to better characterize the role of inflammation in LBD including imaging, genetic, and biomarker studies.

MANAGEMENT OF LBD

To date LBD remains incurable and only very limited treatment options exist. Over the last 15 years, DLB and PDD research has focused on clinical and pathological features in order to establish appropriate diagnostic criteria and improved patient management. However, the overlapping characteristics of LBD with others neurodegenerative disorders has hindered the discovery of new therapies and the understanding of pathological mechanisms. Currently, only symptomatic strategies are used in clinic to alleviate the neuropsychiatric manifestations and motor symptoms with more or less efficiency (see Section "Current Symptomatic Treatment"). Overall, these agents do not target the underlying pathobiology of LBD and thus do not affect the progression of disease. There is still no disease-modifying drug available but some interesting targets exist, as discussed later. Since the major component of LBs is misfolded αsyn, therapeutic interventions to modify αsyn have received the most attention in the last decade. Although targeting tau and Aβ may also be of benefit for LBD, we will focus hereafter on the current strategies and clinical trials targeting αsyn.

CURRENT SYMPTOMATIC TREATMENT

The standard approach to the relief of parkinsonism is the use of the dopamine replacement drug. L-DO-PA (levo-3,4-dihydroxyphenylalanine) is a precursor of dopamine that is able to cross the blood–brain barrier (BBB) and to increase dopamine production in the brain. Several drugs are commercialized (e.g., Levodopa, Atamet, Stalevo, Sinemet) and to date they are the most effective therapy to improve the mobility of PD patients. Other dopaminergic therapies have been marketed and include dopamine agonists, anticholinergics, selegiline, and amantadine [77]. Levodopa is also a standard dopaminergic therapy used with LBD patients, although the responsiveness appears to be different between DLB and

PDD patients [78]. Unfortunately antiparkisonian drugs have proven to have important side effects, including confusion, somnolence, and hallucinations, and psychosis and neuropsychiatric features may be triggered or even worsen [78]. Thus, they should be used with caution and at a lower dose than that used for PD therapy.

The development of cholinesterase inhibitor drugs followed the finding that cholinergic pathways are compromised not only in AD [79] but also PDD and DLB [80,81] (Fig. 7.2). Most importantly, the cholinergic deficit appears to contribute to cognitive impairment to a large extent [82,83]. Cholinesterase inhibitors are currently the only strategy used for the treatment of cognitive symptoms and hallucinations in LBD, with only modest efficacy so far. In a double-blind, placebo-controlled study, McKeith et al. [84] showed that rivastigmine, an inhibitor of acetylcholinesterase, had beneficial effects on DLB patients as they were less apathetic and anxious, and had fewer delusions and hallucinations. In another study, the use of rivastigmine was also associated with moderate improvements in cognition but this was accompanied with adverse side effects like nausea, vomiting, and tremor [85]. The use of another inhibitor, donepezil, was studied in two randomized trials and was shown to have some beneficial effect on memory but with poor tolerability [86,87]. Thus, cholinesterase inhibitors appear to offer hope for the cognitive deficits in LBD but tolerability should be carefully monitored and the potential for worsening movement symptoms, particularly tremor, should be considered [88].

Lastly, it is noteworthy to mention that medications that are often effective for psychotic symptoms, such as the atypical antipsychotics, need to be used with caution in LBD. These agents show strong side effects and worsening of motor and cognitive functioning [1,89].

DISEASE-MODIFYING THERAPY

As with most of the neurodegenerative disorders the development of drugs with disease-modifying potential is needed for the management of LBD. Symptomatic treatments are important for the quality of life of the patients but none of them has the ability to either slow or stop the degenerative process. To this end, identifying and dissecting disease-modifying pathways for LBD is a priority for the discovery of neuroprotective therapeutics and novel drug targets. Neurodegenerative diseases, such as LBD are associated with the expression of misfolded proteins that accumulate, aggregate, and can disrupt cellular protein-folding homeostasis leading to cellular dysfunction and death. In AD or PD, strategies targeting the misfolding protein process of tau, Aβ, and αsyn, respectively are the focus of intense studies in drug discovery research (see also Chapters 2, 5, and 6) and it is most likely that such strategies will be relevant to LBD as well.

Targeting αsyn protein
Clearance of αsyn
As described earlier in this chapter (i.e., Section "Pathophysiology"), αsyn accumulation and aggregation are key to the pathogenesis of DLB and PDD. αSyn is thought to play a role in different cellular pathways and several mechanisms could be involved in the toxicity process, such as mitochondrial dysfunction, impairment of autophagy–lysosome pathway (ALP), or altered inflammatory response (Fig. 7.3). Consequently, various intracellular targets for disease-modifying treatments can be anticipated. Strategies to reduce intracellular levels of αsyn by manipulating those pathways are tempting but they should be approached with caution, as manipulating such cellular machinery might lead to unforeseen side effects.

Impairment of autophagy is well recognized in neurodegeneration and particularly in AD and PD [90,91]. Recently, nilotinib entered phase I of a clinical trial [92]. Hebron et al. [93] showed that administration of nilotinib, a specific inhibitor of Abl kinase activity, promotes autophagic clearance of αsyn in transgenic mice overexpressing a mutated form of the protein. A protective effect was observed as well with significant less dopaminergic cell death [94]. In a previous study Abl knockout was able to protect against 1-methyl-4-phenyl-1,2,3,6-tetrahydropyridin (MPTP) insult in a PD animal model [95]. Overall it seems like Abl inhibition may be a potential therapeutic by promoting αsyn clearance through autophagy and/or proteasome pathways. Another potential route of αsyn degradation considered at the moment is the clearance of αsyn by the action of proteases. Indeed, neurosin, a serine protease ubiquitously present in the central nervous system has been found in LBs of PD [96] and was able to degrade αsyn in cell culture [97–99] providing a rationale for investigating its potential as a new therapy. The use of neurosin was brought to the next level recently when Spencer et al. [100] showed a reduced accumulation of αsyn accompanied by less pathology following the delivery of neurosin into the brain of a DLB mouse model.

Decreased levels of αsyn protein can be achieved by other techniques like RNA interference (RNAi). This technology allows reducing the expression of genes associated with disease. Different approaches have been taken to reduce aberrant αsyn expression in rodent models of PD, including the use of ribozymes [101], small inhibitory RNAs [102,103], short hairpin RNAs [104], and microRNAs (miR) [105]. Direct infusion or virally mediated RNAi delivery was successful in rodent and primate models to silence human αsyn. However, in most of these studies the silencing of αsyn was accompanied by neurotoxicity and neuroinflammation.

To overcome this issue several researchers attempted to embed the silencing sequence in a miR30 transcript. This was effective in PC12 cells and in a mouse model of Huntington disease [106] but unfortunately the same promising result could not be achieved in a PD rat model [107]. The inflammation was still present despite the use of miR30. Although interesting data have been generated with αsyn RNAi, there is no candidate for clinical trials at the moment. Nevertheless, further modifications of RNAi design may provide a promising outcome.

Modulating αsyn oligomerization

Recent evidence indicates that pathological oligomers rather than large fibrillary deposits of αsyn constitute the neurotoxic species [108–110]. Therefore, discovery of small molecules that reduce the burden of αsyn oligomers or enable the selective disruption or clearance of αsyn oligomers represent a viable therapeutic strategy. Innovative strategies to monitor oligomerization have been generated recently and hold great promise as powerful tools for in vitro therapeutic screening. αSyn PMCA (protein misfolding cyclic amplification) was efficient to promote the formation of aggregates with all the biochemical characteristic of those found in vivo [111]. Interestingly, this technology can be used as a high throughput screening method for the discovery of new αsyn antiaggregating compounds. Also novel stable cell lines expressing αsyn fusion proteins for either fluorescent or bioluminescent reporters were developed to enable rapid reporting of αsyn oligomerization in living cells [112]. This protein-fragment complementation assay (PCA) has been used for preliminary high-throughput profiling of modulators of αsyn aggregation and potential αsyn antiaggregating compounds, such as p38 mitogen-activated protein kinase and casein kinase 2 inhibitors have been identified [112]. Several other agents targeting αsyn aggregation have been tested in cellular or animal models with promising results [113–115]. Among them its noteworthy to mention anle138b [3-(1,3-benzodioxol-5-yl)-5-(3-bromophenyl)-1H-pyrazole], originally discovered by Wagner et al. [116] in a screen for prion protein inhibitors. Interestingly, it turned out that anle138b could inhibit αsyn oligomer formation at the same concentration range that was active in the antiprion assay. In vitro, anle138b, represents a new lead structure with several favorable features to prevent the formation of pathological aggregates of αsyn. Moreover, in three different PD mouse models, anle138b strongly inhibited oligomer accumulation and was able to reduce cell death. Lastly, anle138b had no detectable toxicity at therapeutic doses, excellent oral bioavailability, and BBB permeability.

Extracellular αsyn and it propagation

It is now well established that αsyn oligomers can be secreted and transmitted from neuron(s) to neuron(s), though a mechanism of intercellular propagation that is still poorly understood. It appears that extracellular αsyn may be the result of the passive secretion of dying neurons or an active secretion potentially by the recruitment of extracellular vesicles, such as exosomes as carriers [117–119]. Secreted αsyn is found in several biological fluids, such as CSF, human plasma, and brain interstitial fluid [120–122], and its concentration is increased under cellular stress [123]. The transfer of αsyn between cells has been proposed to be an important disease progression mechanism in PD and related synucleopathies [119,124]. A large number of studies have shown that exogenously added, recombinant αsyn can be internalized by recipient cells [125,126] and lead to toxicity. Thus, extracellular αsyn might be another target for the modulation of the disease and potentially to stop progression.

To date, the mechanism by which αsyn protein can be actively released from neurons is still controversial but the involvement of vesicles, such as exosomes has received increasing attention in synucleinopathies and neurodegeneration in general [119,127]. In fact exosomes have been termed *"The Trojan Horses of Neurodegeneration"* as they are identified as the carrier of toxic proteins from unhealthy neurons to the neighboring cells [128]. Even though exosome-associated exocytosis has been implicated in the release of misfolded αsyn, it is still not clear if vesicular-αsyn is internalized into target cells at a much higher efficiency than nonvesicular αsyn. Regardless, one could think that preventing the release of exosomes may become a novel attractive approach to stop the spreading of αsyn pathology and disease progression. The ALP is involved in the aggregation of αsyn (Fig. 7.3) but an additional role can also be attributed as it was shown to modulate αsyn release [129]. Moreover, Danzer et al. [118] showed that dysregulation in the autophagy pathway promoted exosome release of αsyn oligomers. As such, promoting autophagy in order to reduce the release of exosomes may be considered as a therapeutic strategy [130]. Additionally, the lysosomal degradation pathway may also play a role. As with autophagy, it has been found that lysosomal dysfunction increases exosome release [131]. Another interesting characteristic of these extracellular vesicles is their content. Exosomes bear a substantial amount of mRNA and microRNA, and they have capability to shuttle RNA between cells [132,133]. Exosomes could then be seen as naturally occurring RNA carriers and a source of effective delivery strategy [134]. In 2011, Alvarez-Erviti et al. [135] pioneered this approach by successfully delivering siRNA in a mouse model of AD by injection of exosomes. This method had several advantages since exosomes are stable, cross the BBB, and most importantly, avoid an immune response.

Currently, immunoregulatory therapy based on vaccination has received increasing attention and is reviewed elsewhere [136]. This can be achieved by two approaches: active immunization consisting of generating an immune response toward the immunizing agent (αsyn) or passive immunization which is achieved by the administration of antibodies against αsyn protein directly. Vaccination strategies to reduce accumulation of αsyn aggregates are very attractive; however, one has to take into account that autoimmune responses may occur. Nevertheless, Masliah et al. [137] observed promising results using an active immunization strategy in transgenic mice [137]. More recently, they demonstrated that passive immunization, where αsyn-specific antibody is injected in transgenic animals, was also protective. Antibody targeting the carboxy-terminal epitope of αsyn was able to promote clearance of aggregates in mice overexpressing αsyn, and as a result, an improvement of motor function and learning deficits was observed [138]. Later on Bae et al. [139] showed in the same animal model that immunization aids clearance of the extracellular αsyn by microglia and thereby could prevent cell-to-cell transmission. Lastly, a new generation of antibodies has been developed using small peptides, or AFFITOPEs [140] that are too short to induce a T-cell response (autoimmunity). This strategy has been successful in decreasing the level of αsyn oligomers in two different αsyn transgenic lines by modifying glial activity and cytokine profiles without activating T cells [141]. There is currently an ongoing phase I clinical trial using this antibody in early PD patients (ClinicalTrials.gov/show/NCT01568099).

Modulation of αsyn aggregation by chaperones proteins

It is now well documented that molecular chaperones may contribute to the suppression of aggregates and the accumulation of damaged proteins. In neurodegenerative diseases and in particular in synucleopathies, overexpression and pharmacological enhancement of chaperones has been shown to modify the phenotype of the disease in an array of cell and animals models (reviewed in Refs. [142–144]). The ability of chaperones to influence protein aggregation resulting in altered solubility of the mutant protein represents a promising therapeutic approach for LBD. Influencing the misfolding and subsequent aggregation of αsyn might further prevent cytotoxicity and neurodegeneration of vulnerable neurons. Lastly, αsyn has been shown to be degraded by chaperone-mediated autophagy (CMA) therefore this pathway may also be considered a major target for clearing αsyn aggregates.

Pharmacological research has focused mainly on the modulation of the level of one particular chaperone; Hsp70. Upregulating Hsp70 can be achieved either directly by overexpression or indirectly via inhibition of another chaperone, Hsp90. Indeed, inhibition of this chaperone results in the recruitment of heat shock factor 1 (HSF1) which will then upregulate protective HSPs, such as Hsp70. Hsp90-inhibiting compounds as novel and beneficial modulators of the heat shock response have been intensively studied for the treatment of PD. The antibiotic geldanamycin (GA), a naturally occurring Hsp90 inhibitor, has been well characterized in PD models. GA was found to induce Hsp70 in H4 neuroglioma cells and significantly decrease the number of cells with LB-like inclusions compared with mock-treated cells [145]; importantly, this treatment also prevented αsyn-induced toxicity. Furthermore, the antibiotic has proven to be effective in a drosophila model and in an MPTP-induced dopaminergic neurotoxicity mouse model of PD [59,146]. Unfortunately, GA does not cross the BBB and has therefore limited clinical use. Others brain permeable molecules with potent Hsp90 inhibitor activity have been developed and found to rescue αsyn-induced toxicity and to decrease oligomerization in a dose-dependent manner [147,148]. In addition, these novel Hsp90 inhibitors demonstrated excellent bioavailability and robust brain absorption in an in vivo pharmacokinetic study [147]. Unfortunately, further development of these and other Hsp90 inhibitors as PD therapeutics are limited by severe systemic toxicity associated with repeat treatments. As molecule-based therapies can be challenging and limited by BBB crossing or bioavailability, gene therapy may be a new strategy to induce chaperone overexpression. In support, preclinical studies using Hsp70 gene therapy which used recombinant adeno-associated virus (rAAV) in an MPTP mouse model of PD and a rat model was able to protect against dopaminergic cell death [149,150]. However, such overexpression of Hsp70 by genetic manipulation in human patients is not currently a feasible approach, especially due to unpredictable side effects including unexpected immune responses, such as those recently reported in clinical trials [151]. Hsp70 is the most studied molecular chaperone in regard to αsyn aggregation. However, other molecular chaperones have also been shown to act on αsyn aggregation and as such might represent alternative targets. Hsp27, Hsp40 (HDJ-1; HDJ-2), Hsp60, Hsp90, Hsp110, 14-3-3, Hsc70, and αB-crystallin were all shown to be sequestrated in LBs [152]. In addition, torsinA HDJ-1, HDJ-2, Hsp27, and to a lesser extent αB-crystallin, were all shown to reduce αsyn-induced neurotoxicity in cell culture models [60,152]. More recently, a systematic investigation of the interaction of various small HSPs (αB-crystallin, Hsp27, Hsp20, HspB8, and HspB2B3) with both wild-type and mutant αsyn showed that all small HSPs bound to the various forms of αsyn and inhibited mature fibril formation [153].

Deep brain stimulation

Deep brain stimulation (DBS) is a neurosurgical procedure that has been shown recently to be an effective form of treatment for several neurological disorders including essential tremor, dystonia, major depression, and PD [154–156]. While DBS has been traditionally used to reduce rigidity and tremor in PD patients, this nondrug-based therapy is currently considered as an alternative treatment to reduce the symptomatic and progression of cognitive dysfunction in dementia-related disorders. The technique requires the implantation of electrodes in specific regions of the brain to allow the delivery of electrical impulses in a targeted area that ultimately affects neuronal function. Although the exact mechanism of action for DBS is not fully understood, it is suggested that electrical stimulation may have an effect on neurochemical release, oscillatory activity, synaptic plasticity, and, potentially neuroprotection and neurogenesis [157].

In AD patients, several studies have reported beneficial effect of DBS in the nucleus basalis of Meynert (NBM) and in the fornix [158,159] in a recent phase I clinical study [160]. Six mild to moderate AD patients were treated with DBS in the NBM. Promising positive effects were observed on cognitive parameters including Alzheimer's disease Assessment Scale-Cognitive subscale (ADAS-cog) and Mini Mental Status Exam. The NBM is a small bundle of neurons in the basal forebrain that contains a high concentration of cholinergic neurons. It is among the structures that shrink early in AD [161] and degenerates even more in DLB and PDD, and therefore represents an interesting target for LBD patients. Recently, in a PDD study, Freund et al. [162] implanted electrodes in the NBM and observed improvement in the cognitive function of the patients. Interestingly, when the electrodes were shut off for 24 h major cognitive deterioration was detected reaching the baseline level. Later, the same group of researchers published a study using the same area of stimulation where they observed a change in apraxia in PDD patients [163]. These are so far the only published studies targeting synucleinopathy-associated dementia with DBS but more clinical studies are currently being conducted (http://clinicaltrials.gov/NCT02263937).

With regard to clinical trials, DBS seems to be a promising approach in DLB management as shown by its therapeutic effects; however, a disease-modifying profile is not clear at this point. It has been hypothesized that DBS could be neuroprotective in PD by slowing the degeneration of dopaminergic neurons in the substantia nigra [164]. A postmortem study of PD patients that received DBS showed an increased neuronal precursor cell proliferation in the subventricular zone of the lateral ventricles, the third ventricle lining, and the tissue surrounding the DBS as compared to age-matched normal control [165], although the clinical significance of these neural precursors is uncertain [157]. Extensive clinical studies show that PD symptoms continue to progress despite effective DBS therapy [166–169]. In this context, evidence that DBS influences the underlying neurodegenerative process is up to now very weak. In animal research, however, studies in both rats and monkeys have demonstrated that DBS of the subthalamic nucleus can prevent the degeneration of nigral dopamine [170–172] and hold great promise.

The use of animal models has played a substantial role in the development and refinement of DBS as a therapy in other disorders [173]. Evaluation of different target structures for DBS can be performed in rodents to exclude structures, which are less effective. Moreover, potential side effects, as well as the underlying mechanisms of DBS can be investigated. Preclinical studies will likely accelerate and validate the clinical application of DBS in LBD. While waiting for the results of ongoing clinical trials, it is worth noting that the preliminary results are promising, and undoubtedly will make a significant contribution to the growing body of research that aims to explore new applications for DBS in LBD therapy.

CONCLUDING REMARKS

LBD is one group of neurodegenerative disorders characterized as synucleinopathies due to the presence of LBs containing aggregated αsyn as the main component. Clinical criteria and assessment for an accurate diagnosis of LBD have been modified in the last decade, and are useful for the differentiation of DLB and PDD from other dementia subtypes, particularly AD, but also from parkinsonian syndromes, such as PD. There is no cure for these disorders but as in PD or AD, therapeutic strategies targeting the misfolded proteins are likely to be relevant for therapeutic interventions in LBD. Modulating αsyn toxicity by manipulating clearance mechanisms, such as CMA or by immunotherapy are promising strategies. It is important to note that as disease-modifying therapeutics are developed and become available for PD or AD, it will be crucial to try them in DLB and PDD patients as well.

REFERENCES

[1] C.F. Lippa, J.E. Duda, M. Grossman, H.I. Hurtig, D. Aarsland, B.F. Boeve, D.J. Brooks, D.W. Dickson, B. Dubois, M. Emre, S. Fahn, J.M. Farmer, D. Galasko, J.E. Galvin, C.G. Goetz, J.H. Growdon, K.A. Gwinn-Hardy, J. Hardy, P. Heutink, T. Iwatsubo, K. Kosaka, V.M. Lee, J.B. Leverenz, E. Masliah, I.G. McKeith, R.L. Nussbaum, C.W. Olanow, B.M. Ravina, A.B. Singleton, C.M. Tanner, J.Q. Trojanowski, Z.K. Wszolek, DLB and PDD boundary issues: diagnosis, treatment, molecular pathology, and biomarkers, Neurology 68 (11) (2007) 812–819.

[2] Z. Walker, K.L. Possin, B.F. Boeve, D. Aarsland, Lewy body dementias, Lancet 386 (10004) (2015) 1683–1697.

[3] I.G. McKeith, D. Galasko, K. Kosaka, E.K. Perry, D.W. Dickson, L.A. Hansen, D.P. Salmon, J. Lowe, S.S. Mirra, E.J. Byrne, G. Lennox, N.P. Quinn, J.A. Edwardson, P.G. Ince, C. Bergeron, A. Burns, B.L. Miller, S. Lovestone, D. Collerton, E.N. Jansen, C. Ballard, R.A. de Vos, G.K. Wilcock, K.A. Jellinger, R.H. Perry, Consensus guidelines for the clinical and pathologic diagnosis of dementia with Lewy bodies (DLB): report of the consortium on DLB international workshop, Neurology 47 (5) (1996) 1113–1124.

[4] I.G. McKeith, D.W. Dickson, J. Lowe, M. Emre, J.T. O'Brien, H. Feldman, J. Cummings, J.E. Duda, C. Lippa, E.K. Perry, D. Aarsland, H. Arai, C.G. Ballard, B. Boeve, D.J. Burn, D. Costa, T. Del Ser, B. Dubois, D. Galasko, S. Gauthier, C.G. Goetz, E. Gomez-Tortosa, G. Halliday, L.A. Hansen, J. Hardy, T. Iwatsubo, R.N. Kalaria, D. Kaufer, R.A. Kenny, A. Korczyn, K. Kosaka, V.M. Lee, A. Lees, I. Litvan, E. Londos, O.L. Lopez, S. Minoshima, Y. Mizuno, J.A. Molina, E.B. Mukaetova-Ladinska, F. Pasquier, R.H. Perry, J.B. Schulz, J.Q. Trojanowski, M. Yamada, Diagnosis and management of dementia with Lewy bodies: third report of the DLB Consortium, Neurology 65 (12) (2005) 1863–1872.

[5] S.A. Vann Jones, J.T. O'Brien, The prevalence and incidence of dementia with Lewy bodies: a systematic review of population and clinical studies, Psychol. Med. 44 (4) (2014) 673–683.

[6] D. Aarsland, J. Zaccai, C. Brayne, A systematic review of prevalence studies of dementia in Parkinson's disease, Mov. Disord. 20 (10) (2005) 1255–1263.

[7] M.A. Hely, W.G. Reid, M.A. Adena, G.M. Halliday, J.G. Morris, The Sydney multicenter study of Parkinson's disease: the inevitability of dementia at 20 years, Mov. Disord. 23 (6) (2008) 837–844.

[8] H.I. Hurtig, J.Q. Trojanowski, J. Galvin, D. Ewbank, M.L. Schmidt, V.M. Lee, C.M. Clark, G. Glosser, M.B. Stern, S.M. Gollomp, S.E. Arnold, Alpha-synuclein cortical Lewy bodies correlate with dementia in Parkinson's disease, Neurology 54 (10) (2000) 1916–1921.

[9] E.K. Doubleday, J.S. Snowden, A.R. Varma, D. Neary, Qualitative performance characteristics differentiate dementia with Lewy bodies and Alzheimer's disease, J. Neurol. Neurosurg. Psychiatry 72 (5) (2002) 602–607.

[10] C.G. Ballard, G. Ayre, J. O'Brien, A. Sahgal, I.G. McKeith, P.G. Ince, R.H. Perry, Simple standardised neuropsychological assessments aid in the differential diagnosis of dementia with Lewy bodies from Alzheimer's disease and vascular dementia, Dement. Geriatr. Cogn. Disord. 10 (2) (1999) 104–108.

[11] J.M. Hamilton, K.M. Landy, D.P. Salmon, L.A. Hansen, E. Masliah, D. Galasko, Early visuospatial deficits predict the occurrence of visual hallucinations in autopsy-confirmed dementia with Lewy bodies, Am. J. Geriatr. Psychiatry 20 (9) (2012) 773–781.

[12] Y. Tsuboi, D.W. Dickson, Dementia with Lewy bodies and Parkinson's disease with dementia: are they different?, Parkinsonism Relat. Disord. 11 (Suppl. 1) (2005) S47–S51.

[13] D. Aarsland, J.L. Cummings, J.P. Larsen, Neuropsychiatric differences between Parkinson's disease with dementia and Alzheimer's disease, Int. J. Geriatr. Psychiatry 16 (2) (2001) 184–191.

[14] L.A. Klatka, E.D. Louis, R.B. Schiffer, Psychiatric features in diffuse Lewy body disease: a clinicopathologic study using Alzheimer's disease and Parkinson's disease comparison groups, Neurology 47 (5) (1996) 1148–1152.

[15] B.F. Boeve, M.H. Silber, T.J. Ferman, J.A. Lucas, J.E. Parisi, Association of REM sleep behavior disorder and neurodegenerative disease may reflect an underlying synucleinopathy, Mov. Disord. 16 (4) (2001) 622–630.

[16] T.J. Ferman, B.F. Boeve, G.E. Smith, M.H. Silber, J.A. Lucas, N.R. Graff-Radford, D.W. Dickson, J.E. Parisi, R.C. Petersen, R.J. Ivnik, Dementia with Lewy bodies may present as dementia and REM sleep behavior disorder without parkinsonism or hallucinations, J. Int. Neuropsychol. Soc. 8 (7) (2002) 907–914.

[17] B.K. Woodruff, N.R. Graff-Radford, T.J. Ferman, D.W. Dickson, M.W. DeLucia, J.E. Crook, Z. Arvanitakis, S. Brassler, C. Waters, W. Barker, R. Duara, Family history of dementia is a risk factor for Lewy body disease, Neurology 66 (12) (2006) 1949–1950.

[18] V. Bogaerts, S. Engelborghs, S. Kumar-Singh, D. Goossens, B. Pickut, J. van der Zee, K. Sleegers, K. Peeters, J.J. Martin, J. Del-Favero, T. Gasser, D.W. Dickson, Z.K. Wszolek, P.P. De Deyn, J. Theuns, C. Van Broeckhoven, A novel locus for dementia with Lewy bodies: a clinically and genetically heterogeneous disorder, Brain 130 (Pt 9) (2007) 2277–2291.

[19] B. Meeus, A. Verstraeten, D. Crosiers, S. Engelborghs, M. Van den Broeck, M. Mattheijssens, K. Peeters, E. Corsmit, E. Elinck, B. Pickut, R. Vandenberghe, P. Cras, P.P. De Deyn, C. Van Broeckhoven, J. Theuns, DLB and PDD: a role for mutations in dementia and Parkinson disease genes?, Neurobiol. Aging 33 (3) (2012) 629 e5–629 e18.

[20] M.H. Polymeropoulos, C. Lavedan, E. Leroy, S.E. Ide, A. Dehejia, A. Dutra, B. Pike, H. Root, J. Rubenstein, R. Boyer, E.S. Stenroos, S. Chandrasekharappa, A. Athanassiadou, T. Papapetropoulos, W.G. Johnson, A.M. Lazzarini, R.C. Duvoisin, G. Di Iorio, L.I. Golbe, R.L. Nussbaum, Mutation in the alpha-synuclein gene identified in families with Parkinson's disease, Science 276 (5321) (1997) 2045–2047.

[21] M. Poulopoulos, O.A. Levy, R.N. Alcalay, The neuropathology of genetic Parkinson's disease, Mov. Disord. 27 (7) (2012) 831–842.

[22] A.B. Singleton, M. Farrer, J. Johnson, A. Singleton, S. Hague, J. Kachergus, M. Hulihan, T. Peuralinna, A. Dutra, R. Nussbaum, S. Lincoln, A. Crawley, M. Hanson, D. Maraganore, C. Adler, M.R. Cookson, M. Muenter, M. Baptista, D. Miller, J. Blancato, J. Hardy, K. Gwinn-Hardy, Alpha-synuclein locus triplication causes Parkinson's disease, Science 302 (5646) (2003) 841.

[23] J.J. Zarranz, J. Alegre, J.C. Gomez-Esteban, E. Lezcano, R. Ros, I. Ampuero, L. Vidal, J. Hoenicka, O. Rodriguez, B. Atares, V. Llorens, E. Gomez Tortosa, T. del Ser, D.G. Munoz, J.G. de Yebenes, The new mutation, E46K, of alpha-synuclein causes Parkinson and Lewy body dementia, Ann. Neurol. 55 (2) (2004) 164–173.

[24] M. Hutton, C.L. Lendon, P. Rizzu, M. Baker, S. Froelich, H. Houlden, S. Pickering-Brown, S. Chakraverty, A. Isaacs, A. Grover, J. Hackett, J. Adamson, S. Lincoln, D. Dickson, P. Davies, R.C. Petersen, M. Stevens, E. de Graaff, E. Wauters, J. van Baren, M. Hillebrand, M. Joosse, J.M. Kwon, P. Nowotny, L.K. Che, J. Norton, J.C. Morris, L.A. Reed, J. Trojanowski, H. Basun, L. Lannfelt, M. Neystat, S. Fahn, F. Dark, T. Tannenberg, P.R. Dodd, N. Hayward, J.B. Kwok, P.R. Schofield, A. Andreadis, J. Snowden, D. Craufurd, D. Neary, F. Owen, B.A. Oostra, J. Hardy, A. Goate, J. van Swieten, D. Mann, T. Lynch, P. Heutink, Association of missense and 5'-splice-site mutations in tau with the inherited dementia FTDP-17, Nature 393 (6686) (1998) 702–705.

[25] N. Kouri, K. Oshima, M. Takahashi, M.E. Murray, Z. Ahmed, J.E. Parisi, S.H. Yen, D.W. Dickson, Corticobasal degeneration with olivopontocerebellar atrophy and TDP-43 pathology: an unusual clinicopathologic variant of CBD, Acta. Neuropathol. 125 (5) (2013) 741–752.

[26] M. Iijima, T. Tabira, P. Poorkaj, G.D. Schellenberg, J.Q. Trojanowski, V.M. Lee, M.L. Schmidt, K. Takahashi, T. Nabika, T. Matsumoto, Y. Yamashita, S. Yoshioka, H. Ishino, A distinct familial presenile dementia with a novel missense mutation in the tau gene, Neuroreport 10 (3) (1999) 497–501.

[27] J. Neumann, J. Bras, E. Deas, S.S. O'Sullivan, L. Parkkinen, R.H. Lachmann, A. Li, J. Holton, R. Guerreiro, R. Paudel, B. Segarane, A. Singleton, A. Lees, J. Hardy, H. Houlden, T. Revesz, N.W. Wood, Glucocerebrosidase mutations in clinical and pathologically proven Parkinson's disease, Brain 132 (Pt 7) (2009) 1783–1794.

[28] D. Tsuang, J.B. Leverenz, O.L. Lopez, R.L. Hamilton, D.A. Bennett, J.A. Schneider, A.S. Buchman, E.B. Larson, P.K. Crane, J.A. Kaye, P. Kramer, R. Woltjer, W. Kukull, P.T. Nelson, G.A. Jicha, J.H. Neltner, D. Galasko, E. Masliah, J.Q. Trojanowski, G.D. Schellenberg, D. Yearout, H. Huston, A. Fritts-Penniman, I.F. Mata, J.Y. Wan, K.L. Edwards, T.J. Montine, C.P. Zabetian, GBA mutations increase risk for Lewy body disease with and without Alzheimer disease pathology, Neurology 79 (19) (2012) 1944–1950.

[29] M.A. Nalls, R. Duran, G. Lopez, M. Kurzawa-Akanbi, I.G. McKeith, P.F. Chinnery, C.M. Morris, J. Theuns, D. Crosiers, P. Cras, S. Engelborghs, P.P. De Deyn, C. Van Broeckhoven, D.M. Mann, J. Snowden, S. Pickering-Brown, N. Halliwell, Y. Davidson, L. Gibbons, J. Harris, U.M. Sheerin, J. Bras, J. Hardy, L. Clark, K. Marder, L.S. Honig, D. Berg, W. Maetzler, K. Brockmann, T. Gasser, F. Novellino, A. Quattrone, G. Annesi, E.V. De Marco, E. Rogaeva, M. Masellis, S.E. Black, J.M. Bilbao, T. Foroud, B. Ghetti, W.C. Nichols, N. Pankratz, G. Halliday, S. Lesage, S. Klebe, A. Durr, C. Duyckaerts, A. Brice, B.I. Giasson, J.Q. Trojanowski, H.I. Hurtig, N. Tayebi, C. Landazabal, M.A. Knight, M. Keller, A.B. Singleton, T.G. Wolfsberg, E. Sidransky, A multicenter study of glucocerebrosidase mutations in dementia with Lewy bodies, JAMA Neurol. 70 (6) (2013) 727–735.

[30] J. Bras, R. Guerreiro, L. Darwent, L. Parkkinen, O. Ansorge, V. Escott-Price, D.G. Hernandez, M.A. Nalls, L.N. Clark, L.S. Honig, K. Marder, W.M. Van Der Flier, A. Lemstra, P. Scheltens, E. Rogaeva, P. St George-Hyslop, E. Londos, H. Zetterberg, S. Ortega-Cubero, P. Pastor, T.J. Ferman, N.R. Graff-Radford, O.A. Ross, I. Barber, A. Braae, K. Brown, K. Morgan, W. Maetzler, D. Berg, C. Troakes, S. Al-Sarraj, T. Lashley, Y. Compta, T. Revesz, A. Lees, N. Cairns, G.M. Halliday, D. Mann, S. Pickering-Brown, D.W. Dickson, A. Singleton, J. Hardy, Genetic analysis implicates APOE, SNCA and suggests lysosomal dysfunction in the etiology of dementia with Lewy bodies, Hum. Mol. Genet. 23 (23) (2014) 6139–6146.

[31] A. Goris, C.H. Williams-Gray, G.R. Clark, T. Foltynie, S.J. Lewis, J. Brown, M. Ban, M.G. Spillantini, A. Compston, D.J. Burn, P.F. Chinnery, R.A. Barker, S.J. Sawcer, Tau and alpha-synuclein in susceptibility to, and dementia in, Parkinson's disease, Ann. Neurol. 62 (2) (2007) 145–153.

[32] M. Colom-Cadena, E. Gelpi, M.J. Marti, S. Charif, O. Dols-Icardo, R. Blesa, J. Clarimon, A. Lleo, MAPT H1 haplotype is associated with enhanced alpha-synuclein deposition in dementia with Lewy bodies, Neurobiol. Aging 34 (3) (2013) 936–942.

[33] D.J. Irwin, M.T. White, J.B. Toledo, S.X. Xie, J.L. Robinson, V. Van Deerlin, V.M. Lee, J.B. Leverenz, T.J. Montine, J.E. Duda, H.I. Hurtig, J.Q. Trojanowski, Neuropathologic substrates of Parkinson disease dementia, Ann. Neurol. 72 (4) (2012) 587–598.

[34] M.G. Spillantini, R.A. Crowther, R. Jakes, M. Hasegawa, M. Goedert, Alpha-synuclein in filamentous inclusions of Lewy bodies from Parkinson's disease and dementia with Lewy bodies, Proc. Natl. Acad. Sci. USA 95 (11) (1998) 6469–6473.

[35] A. Abeliovich, Y. Schmitz, I. Farinas, D. Choi-Lundberg, W.H. Ho, P.E. Castillo, N. Shinsky, J.M. Verdugo, M. Armanini, A. Ryan, M. Hynes, H. Phillips, D. Sulzer, A. Rosenthal, Mice lacking alpha-synuclein display functional deficits in the nigrostriatal dopamine system, Neuron 25 (1) (2000) 239–252.

[36] D.E. Cabin, K. Shimazu, D. Murphy, N.B. Cole, W. Gottschalk, K.L. McIlwain, B. Orrison, A. Chen, C.E. Ellis, R. Paylor, B. Lu, R.L. Nussbaum, Synaptic vesicle depletion correlates with attenuated synaptic responses to prolonged repetitive stimulation in mice lacking alpha-synuclein, J. Neurosci. 22 (20) (2002) 8797–8807.

[37] S. Liu, I. Ninan, I. Antonova, F. Battaglia, F. Trinchese, A. Narasanna, N. Kolodilov, W. Dauer, R.D. Hawkins, O. Arancio, Alpha-synuclein produces a long-lasting increase in neurotransmitter release, Embo. J. 23 (22) (2004) 4506–4516.

[38] L. Yavich, H. Tanila, S. Vepsalainen, P. Jakala, Role of alpha-synuclein in presynaptic dopamine recruitment, J. Neurosci. 24 (49) (2004) 11165–11170.

[39] T. Bartels, J.G. Choi, D.J. Selkoe, Alpha-synuclein occurs physiologically as a helically folded tetramer that resists aggregation, Nature 477 (7362) (2011) 107–110.

[40] J. Burre, S. Vivona, J. Diao, M. Sharma, A.T. Brunger, T.C. Sudhof, Properties of native brain alpha-synuclein, Nature 498 (7453) (2013) E4–E6 discussion E6–E7.

[41] B. Fauvet, M.K. Mbefo, M.B. Fares, C. Desobry, S. Michael, M.T. Ardah, E. Tsika, P. Coune, M. Prudent, N. Lion, D. Eliezer, D.J. Moore, B. Schneider, P. Aebischer, O.M. El-Agnaf, E. Masliah, H.A. Lashuel, Alpha-synuclein in central nervous system and from erythrocytes, mammalian cells, and Escherichia coli exists predominantly as disordered monomer, J. Biol. Chem. 287 (19) (2012) 15345–15364.

[42] D. Selkoe, U. Dettmer, E. Luth, N. Kim, A. Newman, T. Bartels, Defining the native state of alpha-synuclein, Neurodegener. Dis. 13 (2–3) (2014) 114–117.

[43] H. Fujiwara, M. Hasegawa, N. Dohmae, A. Kawashima, E. Masliah, M.S. Goldberg, J. Shen, K. Takio, T. Iwatsubo, Alpha-synuclein is phosphorylated in synucleinopathy lesions, Nat. Cell Biol. 4 (2) (2002) 160–164.

[44] J.P. Anderson, D.E. Walker, J.M. Goldstein, R. de Laat, K. Banducci, R.J. Caccavello, R. Barbour, J. Huang, K. Kling, M. Lee, L. Diep, P.S. Keim, X. Shen, T. Chataway, M.G. Schlossmacher, P. Seubert, D. Schenk, S. Sinha, W.P. Gai, T.J. Chilcote, Phosphorylation of Ser-129 is the dominant pathological modification of alpha-synuclein in familial and sporadic Lewy body disease, J. Biol. Chem. 281 (40) (2006) 29739–29752.

[45] J.T. Greenamyre, G. MacKenzie, T.I. Peng, S.E. Stephans, Mitochondrial dysfunction in Parkinson's disease, Biochem. Soc. Symp. 66 (1999) 85–97.

[46] E. Lindersson, R. Beedholm, P. Hojrup, T. Moos, W. Gai, K.B. Hendil, P.H. Jensen, Proteasomal inhibition by alpha-synuclein filaments and oligomers, J. Biol. Chem. 279 (13) (2004) 12924–12934.

[47] A.A. Cooper, A.D. Gitler, A. Cashikar, C.M. Haynes, K.J. Hill, B. Bhullar, K. Liu, K. Xu, K.E. Strathearn, F. Liu, S. Cao, K.A. Caldwell, G.A. Caldwell, G. Marsischky, R.D. Kolodner, J. Labaer, J.C. Rochet, N.M. Bonini, S. Lindquist, Alpha-synuclein blocks ER-Golgi traffic and Rab1 rescues neuron loss in Parkinson's models, Science 313 (5785) (2006) 324–328.

[48] P.T. Francis, Biochemical and pathological correlates of cognitive and behavioural change in DLB/PDD, J. Neurol. 256 (Suppl. 3) (2009) 280–285.

[49] J.E. Duda, B.I. Giasson, M.E. Mabon, V.M. Lee, J.Q. Trojanowski, Novel antibodies to synuclein show abundant striatal pathology in Lewy body diseases, Ann. Neurol. 52 (2) (2002) 205–210.

[50] Y. Compta, L. Parkkinen, S.S. O'Sullivan, J. Vandrovcova, J.L. Holton, C. Collins, T. Lashley, C. Kallis, D.R. Williams, R. de Silva, A.J. Lees, T. Revesz, Lewy- and Alzheimer-type pathologies in Parkinson's disease dementia: which is more important?, Brain 134 (Pt 5) (2011) 1493–1505.

[51] D.W. Dickson, H.A. Crystal, L.A. Mattiace, D.M. Masur, A.D. Blau, P. Davies, S.H. Yen, M.K. Aronson, Identification of normal and pathological aging in prospectively studied nondemented elderly humans, Neurobiol. Aging 13 (1) (1992) 179–189.

[52] Y. Tsuboi, H. Uchikado, D.W. Dickson, Neuropathology of Parkinson's disease dementia and dementia with Lewy bodies with reference to striatal pathology, Parkinsonism Relat. Disord. 13 (Suppl. 3) (2007) S221–S224.

[53] O. Pletnikova, N. West, M.K. Lee, G.L. Rudow, R.L. Skolasky, T.M. Dawson, L. Marsh, J.C. Troncoso, Abeta deposition is associated with enhanced cortical alpha-synuclein lesions in Lewy body diseases, Neurobiol. Aging 26 (8) (2005) 1183–1192.

[54] P. Edison, C.C. Rowe, J.O. Rinne, S. Ng, I. Ahmed, N. Kemppainen, V.L. Villemagne, G. O'Keefe, K. Nagren, K.R. Chaudhury, C.L. Masters, D.J. Brooks, Amyloid load in Parkinson's disease dementia and Lewy body dementia measured with [11C]PIB positron emission tomography, J. Neurol. Neurosurg. Psychiatry 79 (12) (2008) 1331–1338.

[55] M. Bibl, B. Mollenhauer, H. Esselmann, P. Lewczuk, H.W. Klafki, K. Sparbier, A. Smirnov, L. Cepek, C. Trenkwalder, E. Ruther, J. Kornhuber, M. Otto, J. Wiltfang, CSF amyloid-beta-peptides in Alzheimer's disease, dementia with Lewy bodies and Parkinson's disease dementia, Brain 129 (Pt 5) (2006) 1177–1187.

[56] M. Andersson, H. Zetterberg, L. Minthon, K. Blennow, E. Londos, The cognitive profile and CSF biomarkers in dementia with Lewy bodies and Parkinson's disease dementia, Int. J. Geriatr. Psychiatry 26 (1) (2011) 100–105.

[57] A.R. Merdes, L.A. Hansen, D.V. Jeste, D. Galasko, C.R. Hofstetter, G.J. Ho, L.J. Thal, J. Corey-Bloom, Influence of Alzheimer pathology on clinical diagnostic accuracy in dementia with Lewy bodies, Neurology 60 (10) (2003) 1586–1590.

[58] K.F. Winklhofer, J. Tatzelt, C. Haass, The two faces of protein misfolding: gain- and loss-of-function in neurodegenerative diseases, Embo. J. 27 (2) (2008) 336–349.

[59] P.K. Auluck, H.Y. Chan, J.Q. Trojanowski, V.M. Lee, N.M. Bonini, Chaperone suppression of alpha-synuclein toxicity in a Drosophila model for Parkinson's disease, Science 295 (5556) (2002) 865–868.

[60] T.F. Outeiro, J. Klucken, K.E. Strathearn, F. Liu, P. Nguyen, J.C. Rochet, B.T. Hyman, P.J. McLean, Small heat shock proteins protect against alpha-synuclein-induced toxicity and aggregation, Biochem. Biophys. Res. Commun. 351 (3) (2006) 631–638.

[61] J. Klucken, Y. Shin, E. Masliah, B.T. Hyman, P.J. McLean, Hsp70 reduces alpha-synuclein aggregation and toxicity, J. Biol. Chem. 279 (24) (2004) 25497–25502.

[62] Y. Shin, J. Klucken, C. Patterson, B.T. Hyman, P.J. McLean, The co-chaperone carboxyl terminus of Hsp70-interacting protein (CHIP) mediates alpha-synuclein degradation decisions between proteasomal and lysosomal pathways, J. Biol. Chem. 280 (25) (2005) 23727–23734.

[63] P.L. McGeer, E.G. McGeer, The amyloid cascade-inflammatory hypothesis of Alzheimer disease: implications for therapy, Acta. Neuropathol. 126 (4) (2013) 479–497.

[64] I. Morales, L. Guzman-Martinez, C. Cerda-Troncoso, G.A. Farias, R.B. Maccioni, Neuroinflammation in the pathogenesis of Alzheimer's disease. A rational framework for the search of novel therapeutic approaches, Front. Cell Neurosci. 8 (2014) 112.

[65] F. Blandini, Neural and immune mechanisms in the pathogenesis of Parkinson's disease, J. Neuroimmune Pharmacol. 8 (1) (2013) 189–201.

[66] E.C. Hirsch, S. Vyas, S. Hunot, Neuroinflammation in Parkinson's disease, Parkinsonism Relat. Disord. 18 (Suppl. 1) (2012) S210–S212.

[67] P.L. McGeer, J. Rogers, Anti-inflammatory agents as a therapeutic approach to Alzheimer's disease, Neurology 42 (2) (1992) 447–449.

[68] S.C. Vlad, D.R. Miller, N.W. Kowall, D.T. Felson, Protective effects of NSAIDs on the development of Alzheimer disease, Neurology 70 (19) (2008) 1672–1677.

[69] A. Samii, M. Etminan, M.O. Wiens, S. Jafari, NSAID use and the risk of Parkinson's disease: systematic review and meta-analysis of observational studies, Drugs Aging 26 (9) (2009) 769–779.

[70] P.L. McGeer, E.G. McGeer, Glial cell reactions in neurodegenerative diseases: pathophysiology and therapeutic interventions, Alzheimer Dis. Assoc. Disord. 12 (Suppl. 2) (1998) S1–S6.

[71] P. Edison, H.A. Archer, A. Gerhard, R. Hinz, N. Pavese, F.E. Turkheimer, A. Hammers, Y.F. Tai, N. Fox, A. Kennedy, M. Rossor, D.J. Brooks, Microglia, amyloid, and cognition in Alzheimer's disease: An [11C](R) PK11195-PET and [11C]PIB-PET study, Neurobiol. Dis. 32 (3) (2008) 412–419.

[72] P. Ejlerskov, J.G. Hultberg, J. Wang, R. Carlsson, M. Ambjorn, M. Kuss, Y. Liu, G. Porcu, K. Kolkova, C. Friis Rundsten, K. Ruscher, B. Pakkenberg, T. Goldmann, D. Loreth, M. Prinz, D.C. Rubinsztein, S. Issazadeh-Navikas, Lack of neuronal IFN-beta-IFNAR causes Lewy body- and Parkinson's disease-like dementia, Cell 163 (2) (2015) 324–339.

[73] A. Surendranathan, J.B. Rowe, J.T. O'Brien, Neuroinflammation in Lewy body dementia, Parkinsonism Relat. Disord. 21 (12) (2015) 1398–1406.

[74] Z. Fan, Y. Aman, I. Ahmed, G. Chetelat, B. Landeau, K. Ray Chaudhuri, D.J. Brooks, P. Edison, Influence of microglial activation on neuronal function in Alzheimer's and Parkinson's disease dementia, Alzheimers Dement. 11 (6) (2015) 608.e7–621.e7.

[75] S. Iannaccone, C. Cerami, M. Alessio, V. Garibotto, A. Panzacchi, S. Olivieri, G. Gelsomino, R.M. Moresco, D. Perani, In vivo microglia activation in very early dementia with Lewy bodies, comparison with Parkinson's disease, Parkinsonism Relat. Disord. 19 (1) (2013) 47–52.

[76] C.E. Shepherd, E. Thiel, H. McCann, A.J. Harding, G.M. Halliday, Cortical inflammation in Alzheimer disease but not dementia with Lewy bodies, Arch. Neurol. 57 (6) (2000) 817–822.

[77] A.H. Schapira, Present and future drug treatment for Parkinson's disease, J. Neurol. Neurosurg. Psychiatry 76 (11) (2005) 1472–1478.

[78] S. Molloy, I.G. McKeith, J.T. O'Brien, D.J. Burn, The role of levodopa in the management of dementia with Lewy bodies, J. Neurol. Neurosurg. Psychiatry 76 (9) (2005) 1200–1203.

[79] C. Geula, M.M. Mesulam, Cortical cholinergic fibers in aging and Alzheimer's disease: a morphometric study, Neuroscience 33 (3) (1989) 469–481.

[80] N.I. Bohnen, R.L. Albin, The cholinergic system and Parkinson disease, Behav. Brain Res. 221 (2) (2011) 564–573.

[81] I. Ziabreva, C.G. Ballard, D. Aarsland, J.P. Larsen, I.G. McKeith, R.H. Perry, E.K. Perry, Lewy body disease: thalamic cholinergic activity related to dementia and parkinsonism, Neurobiol. Aging 27 (3) (2006) 433–438.

[82] E.K. Perry, M. Curtis, D.J. Dick, J.M. Candy, J.R. Atack, C.A. Bloxham, G. Blessed, A. Fairbairn, B.E. Tomlinson, R.H. Perry, Cholinergic correlates of cognitive impairment in Parkinson's disease: comparisons with Alzheimer's disease, J. Neurol. Neurosurg. Psychiatry 48 (5) (1985) 413–421.

[83] P. Tiraboschi, L.A. Hansen, M. Alford, M.N. Sabbagh, B. Schoos, E. Masliah, L.J. Thal, J. Corey-Bloom, Cholinergic dysfunction in diseases with Lewy bodies, Neurology 54 (2) (2000) 407–411.

[84] I. McKeith, T. Del Ser, P. Spano, M. Emre, K. Wesnes, R. Anand, A. Cicin-Sain, R. Ferrara, R. Spiegel, Efficacy of rivastigmine in dementia with Lewy bodies: a randomised, double-blind, placebo-controlled international study, Lancet 356 (9247) (2000) 2031–2036.

[85] M. Emre, W. Poewe, P.P. De Deyn, P. Barone, J. Kulisevsky, E. Pourcher, T. van Laar, A. Storch, F. Micheli, D. Burn, F. Durif, R. Pahwa, F. Callegari, N. Tenenbaum, C. Strohmaier, Long-term safety of rivastigmine in Parkinson disease dementia: an open-label, randomized study, Clin. Neuropharmacol. 37 (1) (2014) 9–16.

[86] D. Aarsland, K. Laake, J.P. Larsen, C. Janvin, Donepezil for cognitive impairment in Parkinson's disease: a randomised controlled study, J. Neurol. Neurosurg. Psychiatry 72 (6) (2002) 708–712.

[87] B. Ravina, M. Putt, A. Siderowf, J.T. Farrar, M. Gillespie, A. Crawley, H.H. Fernandez, M.M. Trieschmann, S. Reichwein, T. Simuni, Donepezil for dementia in Parkinson's disease: a randomised, double blind, placebo controlled, crossover study, J. Neurol. Neurosurg. Psychiatry 76 (7) (2005) 934–939.

[88] M. Rolinski, C. Fox, I. Maidment, R. McShane, Cholinesterase inhibitors for dementia with Lewy bodies, Parkinson's disease dementia and cognitive impairment in Parkinson's disease, Cochrane Database Syst. Rev. 3 (2012) CD006504.

[89] D. Aarsland, R. Perry, J.P. Larsen, I.G. McKeith, J.T. O'Brien, E.K. Perry, D. Burn, C.G. Ballard, Neuroleptic sensitivity in Parkinson's disease and parkinsonian dementias, J. Clin. Psychiatry 66 (5) (2005) 633–637.

[90] B. Boland, A. Kumar, S. Lee, F.M. Platt, J. Wegiel, W.H. Yu, R.A. Nixon, Autophagy induction and autophagosome clearance in neurons: relationship to autophagic pathology in Alzheimer's disease, J. Neurosci. 28 (27) (2008) 6926–6937.

[91] M.A. Lynch-Day, K. Mao, K. Wang, M. Zhao, D.J. Klionsky, The role of autophagy in Parkinson's disease, Cold Spring Harb. Perspect. Med. 2 (4) (2012) a009357.

[92] A.L. Mahul-Mellier, B. Fauvet, A. Gysbers, I. Dikiy, A. Oueslati, S. Georgeon, A.J. Lamontanara, A. Bisquertt, D. Eliezer, E. Masliah, G. Halliday, O. Hantschel, H.A. Lashuel, c-Abl phosphorylates alpha-synuclein and regulates its degradation: implication for alpha-synuclein clearance and contribution to the pathogenesis of Parkinson's disease, Hum. Mol. Genet. 23 (11) (2014) 2858–2879.

[93] M.L. Hebron, I. Lonskaya, C.E. Moussa, Tyrosine kinase inhibition facilitates autophagic SNCA/alpha-synuclein clearance, Autophagy 9 (8) (2013) 1249–1250.

[94] M.L. Hebron, I. Lonskaya, C.E. Moussa, Nilotinib reverses loss of dopamine neurons and improves motor behavior via autophagic degradation of alpha-synuclein in Parkinson's disease models, Hum. Mol. Genet. 22 (16) (2013) 3315–3328.

[95] H.S. Ko, Y. Lee, J.H. Shin, S.S. Karuppagounder, B.S. Gadad, A.J. Koleske, O. Pletnikova, J.C. Troncoso, V.L. Dawson, T.M. Dawson, Phosphorylation by the c-Abl protein tyrosine kinase inhibits parkin's ubiquitination and protective function, Proc. Natl. Acad. Sci. USA 107 (38) (2010) 16691–16696.

[96] K. Ogawa, T. Yamada, Y. Tsujioka, J. Taguchi, M. Takahashi, Y. Tsuboi, Y. Fujino, M. Nakajima, T. Yamamoto, H. Akatsu, S. Mitsui, N. Yamaguchi, Localization of a novel type trypsin-like serine protease, neurosin, in brain tissues of Alzheimer's disease and Parkinson's disease, Psychiatry Clin. Neurosci. 54 (4) (2000) 419–426.

[97] H. Tatebe, Y. Watanabe, T. Kasai, T. Mizuno, M. Nakagawa, M. Tanaka, T. Tokuda, Extracellular neurosin degrades alpha-synuclein in cultured cells, Neurosci. Res. 67 (4) (2010) 341–346.

[98] T. Kasai, T. Tokuda, N. Yamaguchi, Y. Watanabe, F. Kametani, M. Nakagawa, T. Mizuno, Cleavage of normal and pathological forms of alpha-synuclein by neurosin in vitro, Neurosci. Lett. 436 (1) (2008) 52–56.

[99] A. Iwata, M. Maruyama, T. Akagi, T. Hashikawa, I. Kanazawa, S. Tsuji, N. Nukina, Alpha-synuclein degradation by serine protease neurosin: implication for pathogenesis of synucleinopathies, Hum. Mol. Genet. 12 (20) (2003) 2625–2635.

[100] B. Spencer, S. Michael, J. Shen, K. Kosberg, E. Rockenstein, C. Patrick, A. Adame, E. Masliah, Lentivirus mediated delivery of neurosin promotes clearance of wild-type alpha-synuclein and reduces the pathology in an alpha-synuclein model of LBD, Mol. Ther. 21 (1) (2013) 31–41.

[101] H. Hayashita-Kinoh, M. Yamada, T. Yokota, Y. Mizuno, H. Mochizuki, Down-regulation of alpha-synuclein expression can rescue dopaminergic cells from cell death in the substantia nigra of Parkinson's disease rat model, Biochem. Biophys. Res. Commun. 341 (4) (2006) 1088–1095.

[102] J. Lewis, H. Melrose, D. Bumcrot, A. Hope, C. Zehr, S. Lincoln, A. Braithwaite, Z. He, S. Ogholikhan, K. Hinkle, C. Kent, I. Toudjarska, K. Charisse, R. Braich, R.K. Pandey, M. Heckman, D.M. Maraganore, J. Crook, M.J. Farrer, In vivo silencing of alpha-synuclein using naked siRNA, Mol. Neurodegener. 3 (2008) 19.

[103] A.L. McCormack, S.K. Mak, J.M. Henderson, D. Bumcrot, M.J. Farrer, D.A. Di Monte, Alpha-synuclein suppression by targeted small interfering RNA in the primate substantia nigra, PLoS One 5 (8) (2010) e12122.

[104] M.K. Sapru, J.W. Yates, S. Hogan, L. Jiang, J. Halter, M.C. Bohn, Silencing of human alpha-synuclein in vitro and in rat brain using lentiviral-mediated RNAi, Exp. Neurol. 198 (2) (2006) 382–390.

[105] Y. Han, C.E. Khodr, M.K. Sapru, J. Pedapati, M.C. Bohn, A microRNA embedded AAV alpha-synuclein gene silencing vector for dopaminergic neurons, Brain Res. 1386 (2011) 15–24.

[106] J.L. McBride, R.L. Boudreau, S.Q. Harper, P.D. Staber, A.M. Monteys, I. Martins, B.L. Gilmore, H. Burstein, R.W. Peluso, B. Polisky, B.J. Carter, B.L. Davidson, Artificial miRNAs mitigate shRNA-mediated toxicity in the brain: implications for the therapeutic development of RNAi, Proc. Natl. Acad. Sci. USA 105 (15) (2008) 5868–5873.

[107] C.E. Khodr, A. Becerra, Y. Han, M.C. Bohn, Targeting alpha-synuclein with a microRNA-embedded silencing vector in the rat substantia nigra: positive and negative effects, Brain Res. 2014 (1550) 47–60.

[108] D.P. Karpinar, M.B. Balija, S. Kugler, F. Opazo, N. Rezaei-Ghaleh, N. Wender, H.Y. Kim, G. Taschenberger, B.H. Falkenburger, H. Heise, A. Kumar, D. Riedel, L. Fichtner, A. Voigt, G.H. Braus, K. Giller, S. Becker, A. Herzig, M. Baldus, H. Jackle, S. Eimer, J.B. Schulz, C. Griesinger, M. Zweckstetter, Pre-fibrillar alpha-synuclein variants with impaired beta-structure increase neurotoxicity in Parkinson's disease models, Embo. J. 28 (20) (2009) 3256–3268.

[109] K.M. Danzer, D. Haasen, A.R. Karow, S. Moussaud, M. Habeck, A. Giese, H. Kretzschmar, B. Hengerer, M. Kostka, Different species of alpha-synuclein oligomers induce calcium influx and seeding, J. Neurosci. 27 (34) (2007) 9220–9232.

[110] R. Kayed, Y. Sokolov, B. Edmonds, T.M. McIntire, S.C. Milton, J.E. Hall, C.G. Glabe, Permeabilization of lipid bilayers is a common conformation-dependent activity of soluble amyloid oligomers in protein misfolding diseases, J. Biol. Chem. 279 (45) (2004) 46363–46366.

[111] M.E. Herva, S. Zibaee, G. Fraser, R.A. Barker, M. Goedert, M.G. Spillantini, Anti-amyloid compounds inhibit alpha-synuclein aggregation induced by protein misfolding cyclic amplification (PMCA), J. Biol. Chem. 289 (17) (2014) 11897–11905.

[112] S. Moussaud, S. Malany, A. Mehta, S. Vasile, L.H. Smith, P.J. McLean, Targeting alpha-synuclein oligomers by protein-fragment complementation for drug discovery in synucleinopathies, Expert Opin. Ther. Targets 19 (5) (2015) 589–603.

[113] O.M. El-Agnaf, K.E. Paleologou, B. Greer, A.M. Abogrein, J.E. King, S.A. Salem, N.J. Fullwood, F.E. Benson, R. Hewitt, K.J. Ford, F.L. Martin, P. Harriott, M.R. Cookson, D. Allsop, A strategy for designing inhibitors of alpha-synuclein aggregation and toxicity as a novel treatment for Parkinson's disease and related disorders, Faseb. J. 18 (11) (2004) 1315–1317.

[114] M. Jiang, Y. Porat-Shliom, Z. Pei, Y. Cheng, L. Xiang, K. Sommers, Q. Li, F. Gillardon, B. Hengerer, C. Berlinicke, W.W. Smith, D.J. Zack, M.A. Poirier, C.A. Ross, W. Duan, Baicalein reduces E46K alpha-synuclein aggregation in vitro and protects cells against E46K alpha-synuclein toxicity in cell models of familiar Parkinsonism, J. Neurochem. 114 (2) (2010) 419–429.

[115] W. Zhou, K. Bercury, J. Cummiskey, N. Luong, J. Lebin, C.R. Freed, Phenylbutyrate up-regulates the DJ-1 protein and protects neurons in cell culture and in animal models of Parkinson disease, J. Biol. Chem. 286 (17) (2011) 14941–14951.

[116] J. Wagner, S. Ryazanov, A. Leonov, J. Levin, S. Shi, F. Schmidt, C. Prix, F. Pan-Montojo, U. Bertsch, G. Mitteregger-Kretzschmar, M. Geissen, M. Eiden, F. Leidel, T. Hirschberger, A.A. Deeg, J.J. Krauth, W. Zinth, P. Tavan, J. Pilger, M. Zweckstetter, T. Frank, M. Bahr, J.H. Weishaupt, M. Uhr, H. Urlaub, U. Teichmann, M. Samwer, K. Botzel, M. Groschup, H. Kretzschmar, C. Griesinger, A. Giese, Anle138b: a novel oligomer modulator for disease-modifying therapy of neurodegenerative diseases such as prion and Parkinson's disease, Acta. Neuropathol. 125 (6) (2013) 795–813.

[117] P. Desplats, H.J. Lee, E.J. Bae, C. Patrick, E. Rockenstein, L. Crews, B. Spencer, E. Masliah, S.J. Lee, Inclusion formation and neuronal cell death through neuron-to-neuron transmission of alpha-synuclein, Proc. Natl. Acad. Sci. USA 106 (31) (2009) 13010–13015.

[118] K.M. Danzer, L.R. Kranich, W.P. Ruf, O. Cagsal-Getkin, A.R. Winslow, L. Zhu, C.R. Vanderburg, P.J. McLean, Exosomal cell-to-cell transmission of alpha synuclein oligomers, Mol. Neurodegener. 7 (2012) 42.

[119] E. Emmanouilidou, K. Melachroinou, T. Roumeliotis, S.D. Garbis, M. Ntzouni, L.H. Margaritis, L. Stefanis, K. Vekrellis, Cell-produced alpha-synuclein is secreted in a calcium-dependent manner by exosomes and impacts neuronal survival, J. Neurosci. 30 (20) (2010) 6838–6851.

[120] R. Borghi, R. Marchese, A. Negro, L. Marinelli, G. Forloni, D. Zaccheo, G. Abbruzzese, M. Tabaton, Full length alpha-synuclein is present in cerebrospinal fluid from Parkinson's disease and normal subjects, Neurosci. Lett. 287 (1) (2000) 65–67.

[121] O.M. El-Agnaf, S.A. Salem, K.E. Paleologou, L.J. Cooper, N.J. Fullwood, M.J. Gibson, M.D. Curran, J.A. Court, D.M. Mann, S. Ikeda, M.R. Cookson, J. Hardy, D. Allsop, Alpha-synuclein implicated in Parkinson's disease is present in extracellular biological fluids, including human plasma, Faseb. J. 17 (13) (2003) 1945–1947.

[122] E. Emmanouilidou, D. Elenis, T. Papasilekas, G. Stranjalis, K. Gerozissis, P.C. Ioannou, K. Vekrellis, Assessment of alpha-synuclein secretion in mouse and human brain parenchyma, PLoS One 6 (7) (2011) e22225.

[123] A. Jang, H.J. Lee, J.E. Suk, J.W. Jung, K.P. Kim, S.J. Lee, Non-classical exocytosis of alpha-synuclein is sensitive to folding states and promoted under stress conditions, J. Neurochem. 113 (5) (2010) 1263–1274.

[124] O.M. El-Agnaf, S.A. Salem, K.E. Paleologou, M.D. Curran, M.J. Gibson, J.A. Court, M.G. Schlossmacher, D. Allsop, Detection of oligomeric forms of alpha-synuclein protein in human plasma as a potential biomarker for Parkinson's disease, Faseb. J. 20 (3) (2006) 419–425.

[125] (a) S.J. Lee, Origins and effects of extracellular alpha-synuclein: implications in Parkinson's disease, J. Mol. Neurosci. 34 (1) (2008) 17–22. (b) B.B. Holmes, S.L. DeVos, N. Kfoury, M. Li, R. Jacks, K. Yanamandra, M.O. Ouidja, F.M. Brodsky, J. Marasa, D.P. Bagchi, P.T. Kotzbauer, T.M. Miller, D. Papy-Garcia, M.I. Diamond, Heparan sulfate proteoglycans mediate internalization and propagation of specific proteopathic seeds, Proc. Natl. Acad. Sci. USA 110 (33) (2013) E3138–E3147.

[126] K.C. Luk, C. Song, P. O'Brien, A. Stieber, J.R. Branch, K.R. Brunden, J.Q. Trojanowski, V.M. Lee, Exogenous alpha-synuclein fibrils seed the formation of Lewy body-like intracellular inclusions in cultured cells, Proc. Natl. Acad. Sci. USA 106 (47) (2009) 20051–20056.

[127] J. De Toro, L. Herschlik, C. Waldner, C. Mongini, Emerging roles of exosomes in normal and pathological conditions: new insights for diagnosis and therapeutic applications, Front. Immunol. 6 (2015) 203.

[128] R. Ghidoni, L. Benussi, G. Binetti, Exosomes: the Trojan horses of neurodegeneration, Med. Hypotheses 70 (6) (2008) 1226–1227.

[129] A.M. Poehler, W. Xiang, P. Spitzer, V.E. May, H. Meixner, E. Rockenstein, O. Chutna, T.F. Outeiro, J. Winkler, E. Masliah, J. Klucken, Autophagy modulates SNCA/alpha-synuclein release, thereby generating a hostile microenvironment, Autophagy 10 (12) (2014) 2171–2192.

[130] F. Baixauli, C. Lopez-Otin, M. Mittelbrunn, Exosomes and autophagy: coordinated mechanisms for the maintenance of cellular fitness, Front. Immunol. 5 (2014) 403.

[131] L. Alvarez-Erviti, Y. Seow, A.H. Schapira, C. Gardiner, I.L. Sargent, M.J. Wood, J.M. Cooper, Lysosomal dysfunction increases exosome-mediated alpha-synuclein release and transmission, Neurobiol. Dis. 42 (3) (2011) 360–367.

[132] A.V. Vlassov, S. Magdaleno, R. Setterquist, R. Conrad, Exosomes: current knowledge of their composition, biological functions, and diagnostic and therapeutic potentials, Biochim. Biophys. Acta. 1820 (7) (2012) 940–948.

[133] M. Eldh, K. Ekstrom, H. Valadi, M. Sjostrand, B. Olsson, M. Jernas, J. Lotvall, Exosomes communicate protective messages during oxidative stress; possible role of exosomal shuttle RNA, PLoS One 5 (12) (2010) e15353.

[134] T.A. Shtam, R.A. Kovalev, E.Y. Varfolomeeva, E.M. Makarov, Y.V. Kil, M.V. Filatov, Exosomes are natural carriers of exogenous siRNA to human cells in vitro, Cell Commun. Signal 11 (2013) 88.

[135] L. Alvarez-Erviti, Y. Seow, H. Yin, C. Betts, S. Lakhal, M.J. Wood, Delivery of siRNA to the mouse brain by systemic injection of targeted exosomes, Nat. Biotechnol. 29 (4) (2011) 341–345.

[136] M. Romero-Ramos, M. von Euler Chelpin, V. Sanchez-Guajardo, Vaccination strategies for Parkinson disease: induction of a swift attack or raising tolerance?, Hum. Vaccin. Immunother. 10 (4) (2014) 852–867.

[137] E. Masliah, E. Rockenstein, A. Adame, M. Alford, L. Crews, M. Hashimoto, P. Seubert, M. Lee, J. Goldstein, T. Chilcote, D. Games, D. Schenk, Effects of alpha-synuclein immunization in a mouse model of Parkinson's disease, Neuron 46 (6) (2005) 857–868.

[138] E. Masliah, E. Rockenstein, M. Mante, L. Crews, B. Spencer, A. Adame, C. Patrick, M. Trejo, K. Ubhi, T.T. Rohn, S. Mueller-Steiner, P. Seubert, R. Barbour, L. McConlogue, M. Buttini, D. Games, D. Schenk, Passive immunization reduces behavioral and neuropathological deficits in an alpha-synuclein transgenic model of Lewy body disease, PLoS One 6 (4) (2011) e19338.

[139] E.J. Bae, H.J. Lee, E. Rockenstein, D.H. Ho, E.B. Park, N.Y. Yang, P. Desplats, E. Masliah, S.J. Lee, Antibody-aided clearance of extracellular alpha-synuclein prevents cell-to-cell aggregate transmission, J. Neurosci. 32 (39) (2012) 13454–13469.

[140] A. Schneeberger, M. Mandler, F. Mattner, W. Schmidt, AFFITOME(R) technology in neurodegenerative diseases: the doubling advantage, Hum. Vaccin. 6 (11) (2010) 948–952.

[141] M. Mandler, E. Valera, E. Rockenstein, H. Weninger, C. Patrick, A. Adame, R. Santic, S. Meindl, B. Vigl, O. Smrzka, A. Schneeberger, F. Mattner, E. Masliah, Next-generation active immunization approach for synucle-inopathies: implications for Parkinson's disease clinical trials, Acta. Neuropathol. 127 (6) (2014) 861–879.

[142] D.R. Jones, S. Moussaud, P. McLean, Targeting heat shock proteins to modulate alpha-synuclein toxicity, Ther. Adv. Neurol. Disord. 7 (1) (2014) 33–51.

[143] D. Ebrahimi-Fakhari, L.J. Saidi, L. Wahlster, Molecular chaperones and protein folding as therapeutic targets in Parkinson's disease and other synucleinopathies, Acta. Neuropathol. Commun. 1 (2013) 79.

[144] M. Delenclos, P.J. McLean, Molecular chaperones as potential therapeutic targets for neurological disorders, Inhibitors of Molecular Chaperones as Therapeutic Agents, The Royal Society of Chemistry, 2014, pp. 392–413 (Chapter 15) Cambridge, UK.

[145] P.J. McLean, J. Klucken, Y. Shin, B.T. Hyman, Geldanamycin induces Hsp70 and prevents alpha-synuclein aggregation and toxicity in vitro, Biochem. Biophys. Res. Commun. 321 (3) (2004) 665–669.

[146] H.Y. Shen, J.C. He, Y. Wang, Q.Y. Huang, J.F. Chen, Geldanamycin induces heat shock protein 70 and protects against MPTP-induced dopaminergic neurotoxicity in mice, J. Biol. Chem. 280 (48) (2005) 39962–39969.

[147] P. Putcha, K.M. Danzer, L.R. Kranich, A. Scott, M. Silinski, S. Mabbett, C.D. Hicks, J.M. Veal, P.M. Steed, B.T. Hyman, P.J. McLean, Brain-permeable small-molecule inhibitors of Hsp90 prevent alpha-synuclein oligomer formation and rescue alpha-synuclein-induced toxicity, J. Pharmacol. Exp. Ther. 332 (3) (2010) 849–857.

[148] N.R. McFarland, H. Dimant, L. Kibuuka, D. Ebrahimi-Fakhari, C.A. Desjardins, K.M. Danzer, M. Danzer, Z. Fan, M.A. Schwarzschild, W. Hirst, P.J. McLean, Chronic treatment with novel small molecule Hsp90 inhibitors rescues striatal dopamine levels but not alpha-synuclein-induced neuronal cell loss, PLoS One 9 (1) (2014) e86048.

[149] Z. Dong, D.P. Wolfer, H.P. Lipp, H. Bueler, Hsp70 gene transfer by adeno-associated virus inhibits MPTP-induced nigrostriatal degeneration in the mouse model of Parkinson disease, Mol. Ther. 11 (1) (2005) 80–88.

[150] A.E. Jung, H.L. Fitzsimons, R.J. Bland, M.J. During, D. Young, HSP70 and constitutively active HSF1 mediate protection against CDCrel-1-mediated toxicity, Mol. Ther. 16 (6) (2008) 1048–1055.

[151] F. Mingozzi, K.A. High, Immune responses to AAV in clinical trials, Curr. Gene Ther. 11 (4) (2011) 321–330.

[152] P.J. McLean, H. Kawamata, S. Shariff, J. Hewett, N. Sharma, K. Ueda, X.O. Breakefield, B.T. Hyman, TorsinA and heat shock proteins act as molecular chaperones: suppression of alpha-synuclein aggregation, J. Neurochem. 83 (4) (2002) 846–854.

[153] I.B. Bruinsma, K.A. Bruggink, K. Kinast, A.A. Versleijen, I.M. Segers-Nolten, V. Subramaniam, H.B. Kuiperij, W. Boelens, R.M. de Waal, M.M. Verbeek, Inhibition of alpha-synuclein aggregation by small heat shock proteins, Proteins 79 (10) (2011) 2956–2967.

[154] A.R. Rezai, A.G. Machado, M. Deogaonkar, H. Azmi, C. Kubu, N.M. Boulis, Surgery for movement disorders, Neurosurgery 62 (Suppl. 2) (2008) 809–838 discussion 838–839.

[155] K.A. Follett, F.M. Weaver, M. Stern, K. Hur, C.L. Harris, P. Luo, W.J. Marks Jr., J. Rothlind, O. Sagher, C. Moy, R. Pahwa, K. Burchiel, P. Hogarth, E.C. Lai, J.E. Duda, K. Holloway, A. Samii, S. Horn, J.M. Bronstein, G. Stoner, P.A. Starr, R. Simpson, G. Baltuch, A. De Salles, G.D. Huang, D.J. Reda, Pallidal versus subtha-lamic deep-brain stimulation for Parkinson's disease, N. Engl. J. Med. 362 (22) (2010) 2077–2091.

[156] D.M. Blumberger, B.H. Mulsant, Z.J. Daskalakis, What is the role of brain stimulation therapies in the treatment of depression?, Curr. Psychiatry Rep. 15 (7) (2013) 368.

[157] T.M. Herrington, J.J. Cheng, E.N. Eskandar, Mechanisms of deep brain stimulation, J. Neurophysiol. 115 (1) (2016) 19–38.

[158] A.W. Laxton, D.F. Tang-Wai, M.P. McAndrews, D. Zumsteg, R. Wennberg, R. Keren, J. Wherrett, G. Naglie, C. Hamani, G.S. Smith, A.M. Lozano, A phase I trial of deep brain stimulation of memory circuits in Alzheimer's disease, Ann. Neurol. 68 (4) (2010) 521–534.

[159] T. Sankar, M.M. Chakravarty, A. Bescos, M. Lara, T. Obuchi, A.W. Laxton, M.P. McAndrews, D.F. Tang-Wai, C.I. Workman, G.S. Smith, A.M. Lozano, Deep brain stimulation influences brain structure in Alzheimer's disease, Brain Stimul. 8 (3) (2015) 645–654.

[160] J. Kuhn, K. Hardenacke, D. Lenartz, T. Gruendler, M. Ullsperger, C. Bartsch, J.K. Mai, K. Zilles, A. Bauer, A. Matusch, R.J. Schulz, M. Noreik, C.P. Buhrle, D. Maintz, C. Woopen, P. Haussermann, M. Hellmich, J. Klosterkotter, J. Wiltfang, M. Maarouf, H.J. Freund, V. Sturm, Deep brain stimulation of the nucleus basalis of Meynert in Alzheimer's dementia, Mol. Psychiatry 20 (3) (2015) 353–360.

[161] M. Grothe, H. Heinsen, S. Teipel, Longitudinal measures of cholinergic forebrain atrophy in the transition from healthy aging to Alzheimer's disease, Neurobiol. Aging 34 (4) (2013) 1210–1220.

[162] H.J. Freund, J. Kuhn, D. Lenartz, J.K. Mai, T. Schnell, J. Klosterkoetter, V. Sturm, Cognitive functions in a patient with Parkinson-dementia syndrome undergoing deep brain stimulation, Arch. Neurol. 66 (6) (2009) 781–785.

[163] T.T. Barnikol, N.B. Pawelczyk, U.B. Barnikol, J. Kuhn, D. Lenartz, V. Sturm, P.A. Tass, H.J. Freund, Changes in apraxia after deep brain stimulation of the nucleus basalis Meynert in a patient with Parkinson dementia syndrome, Mov. Disord. 25 (10) (2010) 1519–1520.

[164] P.D. Charles, C.E. Gill, T.L. Davis, P.E. Konrad, A.L. Benabid, Is deep brain stimulation neuroprotective if applied early in the course of PD?, Nat. Clin. Pract. Neurol. 4 (8) (2008) 424–426.

[165] V. Vedam-Mai, B. Gardner, M.S. Okun, F.A. Siebzehnrubl, M. Kam, P. Aponso, D.A. Steindler, A.T. Yachnis, D. Neal, B.U. Oliver, S.J. Rath, R.L. Faull, B.A. Reynolds, M.A. Curtis, Increased precursor cell proliferation after deep brain stimulation for Parkinson's disease: a human study, PLoS One 9 (3) (2014) e88770.

[166] B. Lilleeng, M. Gjerstad, R. Baardsen, I. Dalen, J.P. Larsen, Motor symptoms after deep brain stimulation of the subthalamic nucleus, Acta. Neurol. Scand. 131 (5) (2015) 298–304.

[167] A. Fasano, L.M. Romito, A. Daniele, C. Piano, M. Zinno, A.R. Bentivoglio, A. Albanese, Motor and cognitive outcome in patients with Parkinson's disease 8 years after subthalamic implants, Brain 133 (9) (2010) 2664–2676.

[168] R. Hilker, A.T. Portman, J. Voges, M.J. Staal, L. Burghaus, T. van Laar, A. Koulousakis, R.P. Maguire, J. Pruim, B.M. de Jong, K. Herholz, V. Sturm, W.D. Heiss, K.L. Leenders, Disease progression continues in patients with advanced Parkinson's disease and effective subthalamic nucleus stimulation, J. Neurol. Neurosurg. Psychiatry 76 (9) (2005) 1217–1221.

[169] A. Merola, M. Zibetti, S. Angrisano, L. Rizzi, V. Ricchi, C.A. Artusi, M. Lanotte, M.G. Rizzone, L. Lopiano, Parkinson's disease progression at 30 years: a study of subthalamic deep brain-stimulated patients, Brain 134 (Pt 7) (2011) 2074–2084.

[170] D. Harnack, W. Meissner, J.A. Jira, C. Winter, R. Morgenstern, A. Kupsch, Placebo-controlled chronic high-frequency stimulation of the subthalamic nucleus preserves dopaminergic nigral neurons in a rat model of progressive Parkinsonism, Exp. Neurol. 210 (1) (2008) 257–260.

[171] S. Maesawa, Y. Kaneoke, Y. Kajita, N. Usui, N. Misawa, A. Nakayama, J. Yoshida, Long-term stimulation of the subthalamic nucleus in hemiparkinsonian rats: neuroprotection of dopaminergic neurons, J. Neurosurg. 100 (4) (2004) 679–687.

[172] Y. Temel, V. Visser-Vandewalle, S. Kaplan, R. Kozan, M.A. Daemen, A. Blokland, C. Schmitz, H.W. Steinbusch, Protection of nigral cell death by bilateral subthalamic nucleus stimulation, Brain Res. 1120 (1) (2006) 100–115.

[173] S. Hescham, L.W. Lim, A. Jahanshahi, A. Blokland, Y. Temel, Deep brain stimulation in dementia-related disorders, Neurosci. Biobehav. Rev. 37 (10 Pt 2) (2013) 2666–2675.

FRONTOTEMPORAL DEMENTIA

8

Eline Wauters*,**, Kristel Sleegers*,**, Marc Cruts*,**, Christine Van Broeckhoven*,**

*VIB Center for Molecular Neurology, University of Antwerp, Antwerp, Belgium
**Institute Born–Bunge, University of Antwerp, Antwerp, Belgium

CHAPTER OUTLINE

Overview of Frontotemporal Dementia 199
   Clinical Presentations 200
   Pathological Heterogeneity 201
   Genetics of FTD 202
   FTD: A Heterogeneous Disorder 205
Disease Management 206
   Pharmacologic Symptomatic Treatment 206
   Nonpharmacologic Management 209
Novel Possibilities in Disease-Modifying Drug Development 209
   Tau 209
   Granulin 214
   C9orf72 216
   TDP-43 219
   Cellular Waste Clearance Systems 220
   Heterogeneity = Opportunity 222
Discussion: Essentials for the Development of a Disease-Modifying Therapy 224
   Gaps in our Understanding of FTD 224
   Toward Early and Targeted Intervention 226
Conclusions 227
References 227

OVERVIEW OF FRONTOTEMPORAL DEMENTIA

Frontotemporal dementia (FTD) is the second most common neurodegenerative dementia at young age affecting up to 20% of the patients younger than 65 years [1,2]. FTD refers to a group of neurodegenerative disorders that are clinically characterized by progressive changes in behavior, personality, and/or language difficulties. The etiology of FTD has a strong genetic component, and molecular genetic studies have led to the identification of causal mutations in six genes. The pathological hallmark of these diseases is neuronal loss in the frontal and temporal lobes of the brain, or frontotemporal lobar degeneration (FTLD) [3]. FTLD is a proteinopathy and encompasses several microscopic subtypes. The prevalence of FTD is age dependent, but the overall prevalence has been estimated to be between 2 and 35 per 100,000 individuals [4]. Reported estimates of the incidence of FTD vary between 1.3 and 4.1, but can go up to 16.7 in the age group older than 65 years [4]. Large-scale epidemiological studies are, however, lacking and these numbers are likely to be underestimated, as a subset of patients is misdiagnosed and/or referred to psychiatric services.

CLINICAL PRESENTATIONS

The clinical picture of FTD consists of a spectrum of overlapping syndromes. Based on the predominant presenting features of either behavioral and personality changes or language disturbances, three clinical subtypes are defined: the behavioral variant of FTD (bvFTD) and two forms of primary progressive aphasia (PPA); the nonfluent and the semantic variant.

About half of the FTD patients present with bvFTD, characterized by changes in behavior and personality, such as disinhibition, apathy, loss of empathy, hyperorality, and stereotypic behavior. Most patients lack disease insight. Executive functioning is impaired, while memory and visuospatial functions are usually spared [5–7].

When patients show language impairment, a diagnosis of PPA is made, which is characterized by, for example, difficulties in word finding, word usage, and language comprehension or articulatory difficulties. In PPA patients, episodic memory and other cognitive domains are relatively preserved. Based on the type of impairment, three subtypes of PPA are described. The nonfluent variant of PPA is characterized by effortful speech, agrammatism, and apraxia of speech, while the comprehension of language is relatively preserved. Patients suffering from semantic variant PPA present with loss of word and object knowledge, while speech fluency and grammar are spared. In addition to the nonfluent and the semantic variant, a third variant of PPA was defined, namely logopenic variant PPA. This subtype is mostly associated with a neuropathological diagnosis of Alzheimer disease (AD) [8,9].

Concomitant motor neuron disease (MND) has been described in up to 40% of FTD patients, with up to 14% of patients fulfilling the diagnostic criteria of amyotrophic lateral sclerosis (ALS) [10,11]. Motor neuron dysfunction has been observed most frequently in patients with bvFTD, although MND can occur in PPA patients as well [12–15]. The motor symptoms can precede, follow, or coincide with the development of FTD. Conversely, in a large study of ALS patients, around 50% showed some degree of cognitive impairment, with 15% of ALS patients meeting the diagnostic criteria of FTD [15]. Individuals in whom both diseases occur are referred to as FTD–ALS patients.

In addition to MND, parkinsonian features are often present in FTD patients and the symptoms may resemble progressive supranuclear palsy (PSP) or corticobasal syndrome (CBS) [16]. Most PSP patients show abnormalities in cognition and behavior that overlap with bvFTD and are further affected by postural instability and falls, supranuclear ophthalmoparesis, and motor symptoms [17,18]. Also CBS presents as a movement disorder, combined with cortical dysfunction and cognitive impairment, among other disease features [19]. FTD also shows overlap with argyrophilic grain disease (AGD) and inclusion body myopathy with Paget disease of bone and frontotemporal dementia (IBMPFD). Patients who suffer from AGD present with cognitive decline and dementia [20]. IBMPFD is a disease that is characterized by muscle weakness (myopathy), which can cooccur with FTD and Paget disease of bone (PDB) [21]. PDB disrupts the body's normal bone recycling process with bone pain and/or deformities as the major symptoms.

The onset age of FTD is highly variable, ranging from around 20 to 80 years. In approximately 70% of the patients, the disease onset occurs before the age of 65 years, with an average onset age of about 58 years [22–24]. The median survival from the onset of symptoms varies from 6 to 12 years, independent of factors such as onset age, gender, and education level. Several studies reported a longer survival of patients suffering from semantic variant PPA compared to bvFTD, and also the underlying pathological subtype was associated with survival from symptom onset [22,25,26]. Concomitant ALS was found to be associated with a reduced survival, with FTD–ALS patients showing a median survival of 2–3 years from the onset of symptoms [22,27]. Patients who suffer from FTD eventually may have problems with swallowing, chewing, and bladder and/or bowel control. Death is often caused by the consequences of these difficulties, mostly by infections of the lungs due to aspiration pneumonia or infections of the skin or urinary tract. Although pneumonia and sudden unexplained deaths were found to be particularly frequent [28], causes of death of FTD patients are very diverse.

PATHOLOGICAL HETEROGENEITY

FTLD patients typically show atrophy of the frontal and/or temporal lobes of the brain and the occurrence of microscopic features, such as neuronal loss, microvacuolization, astrocytosis, and white-matter myelin loss. FTLD is a proteinopathy characterized by the presence of abnormal, ubiquitinated neuronal and glial protein inclusions. Based on the nature of these inclusions FTLD is classified into pathological subtypes: FTLD–tau, FTLD–TAR DNA-binding protein 43 (FTLD–TDP), FTLD–fused in sarcoma (FTLD–FUS), and FTLD–ubiquitin proteasome system (FTLD–UPS) [29–31].

FTLD–tau pathology is characterized by deposits of misfolded, hyperphosphorylated, and insoluble tau protein within neuronal and glial cells. FTLD–tau comprises the pathology associated with mutations in the *MAPT* gene, as well as cases of Pick's disease, corticobasal degeneration, PSP, and AGD. Six protein isoforms are formed by alternative splicing of microtubule-associated protein tau (*MAPT*). Three isoforms contain three microtubule-binding (MTB) domains (3R) and three contain four MTB domains (4R). In the normal brain, all isoforms are present with equal levels of 3R and 4R isoforms. In the different tauopathies, a preferential accumulation of either 3R or 4R tau is detected. 3R tau is characteristic of Pick's disease, whereas 4R tau predominates in corticobasal degeneration, PSP, and AGD. For forms of FTLD–tau related to mutations in *MAPT*, the predominating tau isoforms can be 3R, 4R, or a combination thereof, depending on the mutation [32,33].

FTLD–TDP is the most frequent FTLD subtype and is characterized by neuronal and glial inclusions which are immunoreactive for the TAR DNA-binding protein 43 (TDP-43). This protein is localized in the nucleus under normal conditions, whereas in disease it is aberrantly localized to the cytoplasm. TDP-43 proteinopathy is characterized by pathological TDP-43 modifications, such as aggregation, proteolytic cleavage, hyperphosphorylation, and ubiquitination. FTLD–TDP is further classified into four pathological subtypes based on the morphology, distribution, density, and intracellular localization of the inclusions. Type A is defined by numerous short dystrophic neurites (DN) and crescentic or oval neuronal cytoplasmic inclusions (NCI), which are primarily located in the neocortical layer 2. Furthermore, moderate numbers of lentiform neuronal intranuclear inclusions can be present. Type B shows moderate numbers of NCIs throughout all the cortical layers and very few DNs, whereas type C has a predominance of elongated DNs in the upper cortical layers and a very few NCIs. Also in Type D, few NCIs are observed, but numerous short DNs and lentiform neuronal intranuclear inclusions are found in the entire cortical thickness [34]. TDP pathology is not only observed in FTD patients, but also in patients suffering from MND with or without FTD [29,35].

A subset of FTLD–TDP patients present with inclusions that are TDP-43 negative, in addition to FTLD–TDP pathology. The pathology is a direct consequence of a specific disease-causing mutation, that is, a G_4C_2 repeat expansion in the *C9orf72* gene, which will be discussed in more detail in the section "Genetics of FTD." Translation of the expanded repeat in all six reading frames results in the formation of aggregating dipeptide repeat (DPR) proteins [36,37]. It was suggested that the nomenclature used to describe the pathology of FTLD would be revised to include the DPR pathology [36].

Most tau and TDP-43-negative and ubiquitin-positive cases show immunoreactivity for the FUS protein, classified as FTLD–FUS. In FTLD–UPS cases, the pathological inclusions show no immunoreactivity against TDP-43, tau, and FUS, but are ubiquitin/p62 positive. Either the main pathological constituent of these inclusions still needs to be discovered or they are comprised of a heterogeneous group of proteins [30].

GENETICS OF FTD

Causal genes

About 25%–50% of FTD patients have a family history of FTD, often showing an autosomal dominant inheritance pattern [38,39]. This indicates a significant genetic contribution to disease etiology. Causal mutations have been identified in six genes, that is, granulin (*GRN*), chromosome 9 open reading frame 72 (*C9orf72*), *MAPT*, TANK-binding kinase 1 (*TBK1*), charged multivesicular body protein 2B (*CHMP2B*), and valosin-containing protein (*VCP*). Patients who carry a mutation in the ALS genes encoding TDP-43 (*TARDBP*) or fused in sarcoma (*FUS*) can present clinically with FTD in rare cases. Mutation frequencies vary between studies and among different populations. In general mutations in *C9orf72*, *GRN*, *MAPT*, and *TBK1* are the most common, while mutations in *CHMP2B* and *VCP* account for less than 1% of FTD patients [4]. Together, mutations in known genes explain at least 17% of familial patients [40]. For example, in a Belgian FTD cohort, mutations in known genes account for around 15% of FTD patients and 30% of familial FTD patients [41].

Linkage analysis in families affected by FTD with parkinsonism and characterized by inclusions containing tau protein pointed toward the chromosomal region 17q21–22 [42]. Sequencing analysis revealed pathological mutations in the *MAPT* gene, segregating with a dominant pattern of inheritance [43–45]. Mutations in *MAPT* lead to heterogeneous FTD phenotypes, namely bvFTD, the nonfluent variant and the semantic variant of PPA [46,47]. The normal function of the tau protein is to promote microtubule (MT) assembly and stabilization and the regulation of MT dynamics. This activity is regulated further by tau phosphorylation. Nonphosphorylated sites lead to a stronger binding, while phosphorylation decreases MT binding, which makes the MTs less stable. Most mutations reduce the ability of tau to promote MT polymerization, leading to instability, disruption of the cytoskeleton, negative effects on axonal transport, etc. and/or increase its ability to aggregate into fibrils. Certain mutations, however, have the opposite effect and increase the polymerization of MTs and/or decrease the propensity to aggregate [47].

A *MAPT* mutation could not be identified in all families in which the disease was linked to 17q21. The presence of tau-negative inclusions also suggested that mutations in another gene were causing the disease. Sequence analysis of candidate genes in the same genetic locus led to the identification of causal mutations in the *GRN* gene [48,49]. It encodes a growth factor consisting of a signal sequence and 7.5 tandem repeats of a 10 to 12 cysteine–containing motif, the GRN domains. The signal peptide is cleaved off resulting in mature GRN, which is subsequently glycosylated and secreted. Proteolytic cleavage of this protein results in the generation of GRN peptides or granulins. The biological functions of GRN and of GRN peptides are not fully understood, but they are believed to be implicated in inflammation, wound repair, and tumorigenesis (reviewed in Refs. [50,51]). In the central nervous system, studies suggest an involvement in neuronal survival and neurite outgrowth [52]. Complete deficiency of GRN is a known cause of neuronal ceroid lipofuscinosis, a lysosomal storage disorder. In FTD, the majority of mutations are heterozygous loss-of-function mutations, leading to the degradation of the mutant transcript and haploinsufficiency. Correspondingly, GRN levels in blood or cerebrospinal fluid (CSF) are reduced in these mutation carriers [53,54]. Pathological missense mutations are known as well, affecting, for example, GRN secretion and degradation, proper protein folding, subcellular localization, and proteolytic processing [50,51,55,56]. *GRN* mutation carriers can display a wide range of clinical presentations even within one family, including bvFTD, the nonfluent variant of PPA, AD, and Parkinson disease (PD) [57–59]. MND is rare, while parkinsonian symptoms are frequent [58,60]. FTD due to *GRN* mutations is characterized by a FTLD–TDP type A proteinopathy [61–63].

The strong link between FTD and ALS, exemplified by the observed cooccurrence in families and in individual patients and the presence of neuronal TDP-43 inclusions in both patient groups, was further reinforced with the identification of a disease locus on chromosome 9 for autosomal dominant FTD and ALS [64–70]. Furthermore, genome-wide association studies (GWAS) in ALS provided evidence for a risk factor located in the same region [71–73]. The identification of a causal expansion of a noncoding G_4C_2 hexanucleotide repeat in the *C9orf72* gene in this locus was a major leap forward in FTD and ALS research [74–76]. The size of the repeat ranges from 2 to over 4000 repeat units [76–85], with normal repeat sizes ranging from 2 to 24 units [86,87]. Various cutoffs between normal and pathogenic alleles have been defined [74,75]; however, the shortest pathogenic expansion described thus far counts 47 repeat units [85]. The function of C9orf72 is still unclear, but the protein might be involved in autophagy and endosomal trafficking [88]. Several hypotheses have been suggested to explain the pathogenic effect of the repeat expansion, namely haploinsufficiency, RNA toxicity, and DPR toxicity [36,37]. The *C9orf72* G_4C_2 repeat expansion is the most important cause of FTD, FTD–ALS, and ALS. Patients with the expansion mutation mostly display FTLD–TDP pathology and the presence of DPR aggregates.

Until recently, *C9orf72* was the only known common disease gene for the FTD–ALS phenotype. This has changed with the recent identification of loss-of-function mutations in *TBK1* as another important cause of FTD, FTD–ALS, and ALS [41,89–91]. TBK1 is a serine/threonine kinase that is known to phosphorylate proteins involved in the innate immune system [92], autophagy [93,94], and cell proliferation [95]. Optineurin (OPTN) and p62 (SQSTM1), both known to be involved in ALS and/or FTD, are substrates of TBK1 [96–100]. All *TBK1* mutation carriers reported until now present pathologically with FTLD–TDP type A or B [90,91,101].

CHMP2B, *VCP*, *TARDBP*, and *FUS* are less frequently mutated genes in FTD. *CHMP2B* encodes the multivesicular body protein 2B, which is a component of the endosomal sorting complexes required for transport (ESCRT)-III machinery. Patients with a *CHMP2B* mutation present mostly with bvFTD and FTLD–UPS pathology. Mutations in the *VCP* gene are found in patients suffering from FTD, ALS, and IBMPFD. *VCP* mutations in FTD patients are most frequently reported in bvFTD and semantic variant PPA patients, who show FTLD–TDP type D pathology [102]. TDP-43 and FUS are

RNA-binding ribonucleoproteins, which are involved in multiple levels of RNA processing, including splicing, RNA transport, and transcription [103]. Mutations in *TARDBP* and *FUS* are typically found in patients suffering from ALS, although they are also a rare cause of FTD [104–106].

SQSTM1 and *TREM2* mutations are an uncommon cause of FTD (-like) disease. Mutations in the *SQSTM1* gene, encoding the p62 protein, were found in patients suffering from PDB, ALS, and FTD [99,100,107,108]. The p62 protein is an autophagic receptor that recognizes and enables the removal of ubiquitinated proteins. It is present in neuronal inclusions in the brain of patients suffering from various disorders, such as AD, PD, and FTD. In FTD, p62 colocalizes with TDP-43 and FUS in the brain. Furthermore, DPR inclusions also show immunoreactivity against p62. Homozygous mutations in *TREM2*, encoding the triggering receptor expressed on myeloid cells 2, were identified in consanguineous families as a cause of bvFTD. TREM2 is a receptor of the innate immune system and triggers the activation of immune responses in macrophages and dendritic cells [109].

Disease risk

In addition to the causal genes, several susceptibility factors have been identified in FTD. A GWAS conducted on FTLD–TDP patients and control individuals implicated variations in *TMEM106B* as a strong risk factor for FTLD–TDP [110]. An association between a *TMEM106B* genetic risk variation and plasma GRN levels [111,112] or onset age [111] was reported in patients carrying a mutation in the *GRN* gene, although this was not seen in all studies [113]. The effect of *TMEM106B* variations on disease penetrance, risk, onset, and survival was also investigated in patients carrying a *C9orf72* G_4C_2 repeat expansion. Here as well an association between *TMEM106B* variations and disease risk was found [114,115]. Unexpectedly, this was associated with a later disease onset and a longer survival [115]. In addition to variations in *TMEM106B*, GWAS identified risk variations in the locus containing the *RAB38* (member RAS oncogene family) and *CTSC* (cathepsin C) genes for bvFTD and the human leukocyte antigen (HLA) locus for the FTD spectrum [116].

Also variations within the causal genes *MAPT*, *GRN*, *SQSTM1*, and *TREM2* are known to modify disease risk. The genomic region of the *MAPT* gene contains two extended haplotypes, H1 and H2. The H1 haplotype has been associated with an increased risk for PSP, CBS, PD, as well as FTD [117–123]. Conversely, in some studies, the H2 haplotype appeared to be associated with familial FTD [124] or with an earlier onset of FTD [125]. The common *GRN* variant rs5848, which is located in the 3′-untranslated region, was identified as a risk factor for disease in a series of pathologically confirmed FTLD–TDP patients without *GRN* mutations [126]. The variation is located within a binding site for miR-659. This miRNA binds more efficiently to the high-risk T-allele of rs5848, which results in a translational inhibition of GRN. In contrast, analysis of the variation in three independent European clinical FTD cohorts and matched controls did not demonstrate an association of rs5848 with FTD in any of the individual cohorts, nor when the data were combined [127]. Recently, a significant association between rs5848 and survival after onset was found in carriers of *C9orf72* G_4C_2 repeat expansions [128]. Rare mutations in *SQSTM1* which cluster in the ubiquitin-associated domain of the p62 protein, were associated with FTD, doubling the risk of disease development [107]. Several studies also reported an association between variations in *TREM2* and FTD [129–132], while others did not [133,134].

Studies into the nongenetic risk factors for FTD are limited. One retrospective case–control study indicated a higher risk for FTD in individuals suffering from head trauma [135], while a small-scale case–control study of cardiovascular risk factors identified diabetes mellitus as a risk factor for FTD [136].

FTD: A HETEROGENEOUS DISORDER

FTD is a genetically, pathologically, and phenotypically heterogeneous group of disorders. Correlations between the clinical phenotype and the underlying proteinopathy were found, but a strict one-to-one relationship is lacking (illustrated in Fig. 8.1). Furthermore, a substantial overlap exists between FTD and related neurodegenerative diseases. Multiple genetic and neuropathological factors are involved in

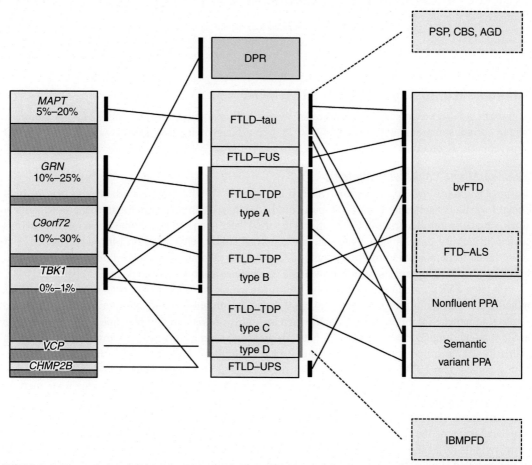

FIGURE 8.1 Illustration of the Genetic, Pathological, and Clinical Correlations in Frontotemporal Dementia (FTD)

For each causal gene, the percentage indicates its mutation frequency in FTD cases overall [40]. *AGD,* argyrophilic grain disease; *ALS,* amyotrophic lateral sclerosis; *bvFTD,* behavioral variant of FTD; *CBS,* corticobasal syndrome; *CHMP2B,* charged multivesicular body protein 2B; *DPR,* dipeptide repeat pathology; *C9orf72,* chromosome 9 open reading frame 72; *FTLD,* frontotemporal lobar degeneration; *FTLD-FUS,* FTLD-fused in sarcoma; *FTLD-TDP,* FTLD-TAR DNA-binding protein 43; *FTLD-UPS,* FTLD-ubiquitin proteasome system; *GRN,* granulin; *IBMPFD,* inclusion body myopathy with Paget disease of bone and frontotemporal dementia; *MAPT,* microtubule-associated protein tau; *PPA,* primary progressive aphasia; *PSP,* progressive supranuclear palsy; *TBK1,* TANK-binding kinase 1; *VCP,* valosin containing protein.

various neurodegenerative diseases. *GRN* mutations are, for example, also present in patients suffering from AD and PD (for an example, see Ref. [59]), *TBK1* mutations and the G_4C_2 repeat expansion in *C9orf72* are present in patients suffering from ALS, FTD, and FTD–ALS (for examples, see Refs. [74,101]), and TDP-43 pathology is identified in FTD, ALS, FTD–ALS, AD, and PD (for examples, see Refs. [59,137,138]). Variability in onset age and disease duration and the presence of reduced or age-dependent penetrance adds further to the observed heterogeneity. This complex picture of overlapping genetics, pathological features, and phenotypes poses a challenge to the development and the evaluation of therapeutics.

DISEASE MANAGEMENT
PHARMACOLOGIC SYMPTOMATIC TREATMENT

At present, no disease-modifying therapies are available for patients suffering from FTD. The rationale behind current symptomatic treatment options is their efficacy in treating patients who suffer from AD or psychiatric disorders.

Behavioral and cognitive symptoms

Selective serotonin reuptake inhibitors (SSRIs) are typically used as antidepressants and may be beneficial for patients suffering from FTD, particularly for those with behavioral disturbances. Relatively small, often uncontrolled trials in FTD patients reported these agents to ameliorate behavioral symptoms such as irritability, disinhibition, agitation, and abnormal eating behavior [139–146], whereas one brief study failed to demonstrate improvements [147]. A recent randomized, double-blinded, placebo-controlled study of citalopram showed improvement in response inhibition systems in FTD patients [148].

Also non-SSRI antidepressants may be useful for managing some behavioral symptoms, such as trazodone, a serotonin receptor antagonist and reuptake inhibitor. Venlafaxine, a noradrenaline and serotonin reuptake inhibitor, is considered when apathy is prominent, and bupropion, a noradrenaline and dopamine reuptake inhibitor, when parkinsonism is present [146,149]. Methylphenidate, which primarily acts as a noradrenaline–dopamine reuptake inhibitor, showed to ameliorate abnormal risk-taking behavior in FTD [150]. A case report demonstrated significant behavioral improvement and partial normalization of the cortical activity in the frontotemporal areas of the left cortex upon methylphenidate treatment [151].

A metaanalysis of the effects of SSRIs and trazodone on behavioral symptoms in bvFTD suggested a reduced score on the Neuropsychiatric inventory (NPI), a measure of behavioral impairments [152]. However, this result should be interpreted with caution, given the absence of a placebo control in many of the included studies, the small sample sizes, and publication bias. Furthermore, it might be that the NPI is not applicable in FTD, as longitudinal improvement on the NPI in FTD patients may reflect increasing apathy and inertia due to disease progression, which decreases the impact of disruptive behaviors [149,153].

In patients with severe agitation, aggression, or delusions that are refractory to SSRIs, treatment with *atypical antipsychotic agents* may be considered. The motivation to use these agents in FTD is currently supported by a limited number of studies only. A trial with olanzapine showed improvements in delusions, NPI scores, and caregiver stress, and small case reports indicated benefits of the use of aripiprazole [154–156]. However no differences in NPI were seen in a study where quetiapine was administered to eight bvFTD patients [157]. Caution is warranted when using both typical and atypical antipsychotics, as FTD patients were reported to be sensitive to extrapyramidal adverse effects [158,159]. Furthermore, antipsychotics are associated with an increased mortality risk in elderly patients suffering from dementia [160]. The increased mortality and side effects have to be discussed with the patient and family and weighed against the potential benefits to quality of life. Often the use of atypical antipsychotics is only temporarily and can be reduced as patients become more apathetic with disease progression [149,161].

Cholinesterase inhibitors and *N*-methyl-D-aspartic acid (NMDA) receptor antagonists are drugs that are commonly used in the symptomatic treatment of AD. In FTD no cholinergic deficits are known and studies with *acetylcholinesterase inhibitors* in FTD showed mixed results. In one study some bvFTD patients that were given either donepezil or rivastigmine exhibited advances in the Mini-Mental State Examination (MMSE), a clock-drawing test, and SPECT scans [162]. Another study of rivastigmine in patients with bvFTD demonstrated improvements in behavioral, depressive symptoms via the NPI score, a stabilization of executive functioning, as well as a reduction in caregiver stress [163]. In contrast a more recent study with galantamine reported no significant cognitive or behavioral differences [164]. Donepezil even exacerbated behavioral symptoms in patients suffering from bvFTD [165]. After cessation, the patients returned to baseline levels of disinhibited and compulsive behavior. This was also seen in another donepezil discontinuation study, where an increase in NPI score was reported [166]. Also the effects of memantine, an *NMDA receptor antagonist*, are inconsistent. In three patients suffering from bvFTD improved NPI scores were seen after treatment with memantine, encouraging further investigation [167]. Two uncontrolled trials in bvFTD patients showed different results: in one study no significant differences in NPI scores were identified [168], while another trial showed a transient improvement in the NPI score [169]. However, two double-blind, placebo-controlled trials of memantine for FTD failed to demonstrate significant differences in the NPI score and in the clinical global impression of change score [170,171]. Taken together, current evidence suggests cholinesterase inhibitors and NMDA receptor antagonists are not effective treatments for FTD [172].

A clinical trial of AVP-786 for symptoms of agitation in AD patients revealed a significant reduction of the NPI agitation/aggression scores in comparison to placebo. AVP-786 is a combination of dextromethorphan hydrobromide and quinidine. Dextromethorphan has NMDA receptor antagonist activity and functions as a serotonin and norepinephrine reuptake inhibitor, among other activities (clinicaltrials.gov identifier NCT01584440) [173]. A phase 2 trial for the treatment of disinhibition syndrome in patients suffering from neurodegenerative disorders, including FTD, PSP, and CBS has been initiated (clinicaltrials.gov identifier NCT02534038).

Another phase 2 clinical trial investigates the effects of tolcapone in patients suffering from FTD (clinicaltrials.gov identifier NCT00604591). Tolcapone is an *inhibitor of catechol-O-methyl transferase* (COMT) and is used as a treatment for PD. It increases the availability of dopamine in the brain, a neurotransmitter that may be lowered in FTD, although not all studies find evidence of dopaminergic dysfunction (neurotransmitter deficits in FTD reviewed in Ref. [152]). Two case reports describe a beneficial effect of *monoamine oxidase* (MAO) *inhibitors* in FTD patients. In one study with selegiline, a MAO-B inhibitor, patients showed improved NPI scores and a renewed interest for different aspects of life [174]. Treatment with moclobemide, a MAO-A inhibitor, resulted in improvements mainly in the domains of behavior, affect, and speech [175]. MAO inhibitors are also commonly used as a treatment for PD, as an effective adjuvant to L-dopa. Selegiline also showed promising results in AD trials initially [176,177]. On the other hand, a recent phase 2 "MAyflOwer RoAD" study with sembragiline, a MAO-B inhibitor, in moderate AD patients had no impact on its primary endpoint, the AD assessment scale–cognitive behavior subscale (ADAS–cog) (clinicaltrials.gov identifier NCT01677754, compound RO4602522). Conversely, one of the secondary endpoints suggested an effect on behavioral symptoms [178]. Administration of *dextroamphetamine*, a nonspecific monoaminergic agonist, to eight bvFTD patients decreased the total NPI score, but did not affect the Repeatable battery for the assessment of neuropsychological status (RBANS) [157].

The neuropeptide *oxytocin* is thought to mediate social behavior. When administered to healthy individuals, it improves empathy [179], trust among humans [180], and the ability to accurately identify positive emotional facial expressions [181]. In patients with FTD, a single dose of intranasal oxytocin was associated with a transient improvement in neuropsychiatric behaviors (clinicaltrials.gov identifier NCT01002300) [182]. A randomized, double-blind, placebo-controlled study on the safety and tolerability of intranasally administered oxytocin in FTD patients indicated no significant adverse events or changes in the overall NPI, although convergent changes in subscales of the NPI and other behavioral scales could be identified (clinicaltrials.gov identifier NCT01386333) [183]. A study to evaluate the effects of a one-time dose of oxytocin on emotions and neural activity in bvFTD patients has been initiated (clinicaltrials.gov identifier NCT01937013).

Drugs that are commonly used as *anticonvulsants* and have *mood-stabilizing effects*, such as carbamazepine may be helpful in managing some of the extreme behavioral symptoms [184]. Another example is topiramate, which showed to reduce hyperorality in FTD patients in several case reports [185–188]. However, the use and efficacy of these agents in FTD has not been studied thoroughly. In fact, some of these agents also demonstrate biological disease-modifying effects that will be discussed in the section "Novel Possibilities in Disease-Modifying Drug Development."

Aphasia

Bromocriptine has been used in patients suffering from aphasia. A study in PPA patients demonstrated only a limited effect and did not appear to alter the overall course of the disease [189]. Galantamine treatment of patients with PPA demonstrated, on the other hand, a stable language function in the treated PPA group compared to the placebo group [164]. Memantine revealed different outcomes in patients suffering from bvFTD, the nonfluent variant of PPA, and the semantic variant of PPA. The bvFTD and semantic variant PPA patients deteriorated on cognitive and behavioral outcome measures and remained stable on the unified Parkinson's disease rating scale (UPDRS), while the opposite was seen in patients suffering from nonfluent variant PPA [169]. Another trial of memantine in PPA patients could not identify significant differences between the placebo and treated groups. However a trend was observed for a smaller degree of decline on the Western Aphasia Battery quotient in the treated group compared to the placebo group [190].

Motor symptom management

Riluzole is an agent that is currently in use for the treatment of ALS. It increases survival with approximately 2–3 months in these patients. At present, no evidence exists to support the efficacy of riluzole in FTD–ALS. Nevertheless, given the pathological and genetic overlap between the diseases, one may suggest that such therapy is an option in FTD–ALS cases [191]. Riluzole treatment is often started in patients suffering from FTD–ALS and is generally well tolerated [192]. Currently a phase 2 clinical study is initiated to investigate the effects of riluzole in mild AD (clinicaltrials.gov identifier NCT01703117);

however, a trial in FTD (–ALS) is lacking. A phase 3 study in PSP and multiple system atrophy showed no effect on survival or progression rate (clinicaltrials.gov identifier NCT00211224) [193].

FTD patients suffering from parkinsonism are minimally responsive to *dopamine replacement*, such as levodopa or carbidopa [161]. These agents are used to manage motor symptoms in PD. Benefits have, however, been observed in some FTD cases [194].

NONPHARMACOLOGIC MANAGEMENT

Current disease management depends heavily on nonpharmacologic management, certainly considering the limited pharmacologic treatment options, which are modestly effective at best. Nonpharmacologic management includes psychosocial support and education of caregivers and family, as well as environmental, behavioral, and physical interventions designed to minimize the occurrence and consequences of undesired behaviors. Also physical, occupational, and speech/swallow and language therapy, home safety evaluations, and the implementation of augmentative communication devices are helpful. Support groups are valuable to provide information, education, and support to patients and caregivers [149]. For a comprehensive discussion on nonpharmacologic management, we refer to Ref. [195].

NOVEL POSSIBILITIES IN DISEASE-MODIFYING DRUG DEVELOPMENT

Studies into the genetic component of FTD have enriched our understanding of disease pathogenesis and provided novel targets for therapeutic intervention. These advances enable a shift of focus from symptomatic to targeted modifying treatments. Many novel approaches for disease therapies focus on a specific group of FTD patients, for example, patients characterized by a causal mutation in a particular gene or sharing a common pathological signature like FTLD–TDP.

TAU

One of the key questions concerning tau-related FTD and other tauopathies, such as PSP and AD, is the identity of the actual toxic form of tau. Under physiological conditions tau exists in a dynamic equilibrium between a MT-bound form and the phosphorylated form, which dissociates from the MTs. When tau becomes hyperphosphorylated it can oligomerize and subsequently aggregate into fibrils. Originally, these fibrils were deemed the most neurotoxic species, while other findings suggest that tau oligomers are the most deleterious. This is, for example, seconded by studies that observe a better correlation between the accumulation of oligomeric tau and neurodegeneration and behavioral deficits in tau transgenic mouse models [196,197]. In fact, lowering the expression of the tau transgene in a mouse line with inducible mutant tau, following the formation of neurofibrillary tangles recovered the cognitive function, although tangle formation continued [198]. Taken together, these and other studies [199–203] weigh against tau tangles as the sole toxic tau species. Either way, *preventing the aggregation of tau* could, thus, be a promising therapeutic strategy (Fig. 8.2, 1). Screening assays have identified various compounds that are able to inhibit tau self-assembly, such as phenylthiazolyl hydrazides, rhodanines, anthraquinones, *N*-phenylamines, and thiacarbocyanines [204–210]. One of the aggregation inhibitors is methylene blue [211], a compound that is used for the treatment of malaria

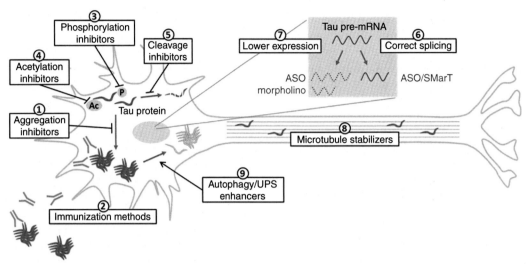

FIGURE 8.2 Therapeutic Strategies Under Investigation for Tauopathies, Including FTLD–Tau, Alzheimer Disease, and progressive supranuclear palsy

These comprise inhibition of tau aggregation *(1)*, immunotherapies *(2)*, inhibition of tau phosphorylation using either kinase inhibitors or phosphatase activators *(3)*, inhibition of tau acetylation *(4)*, inhibition of tau cleavage *(5)*, interference with the splicing machinery to correct the 3R/4R ratio *(6)*, lowering the expression of tau *(7)*, microtubule stabilization *(8)*, and enhancing tau clearance *(9)*. See text for details. ASO, Antisense oligonucleotide; SMarT, spliceosome-mediated RNA transsplicing; *UPS*, ubiquitine proteasome system.

among other conditions. After promising phase 2 studies in AD (Rember, TRx0014, clinicaltrials.gov identifier NCT00515333), TauRx Therapeutics (Singapore) generated a second-generation aggregation inhibitor, which is now in phase 3 clinical trials for bvFTD and AD [LMTX, TRx0237, clinicaltrials. gov identifiers NCT02245568 (AD and bvFTD), NCT01626378 (bvFTD), NCT01689246 (mild to moderate AD), NCT01689233 (mild AD)]. Recently the diphenyl-pyrazole anle138b reduced phosphorylated tau and tau aggregation, prolonged survival, and improved cognition in transgenic mice overexpressing mutant P301S human tau [212].

Tau pathology occurs intracellularly; however, it is believed that the pathology can spread from neuron to neuron via an extracellular stage. Several mechanisms are proposed to explain this phenomenon. It is possible that aggregated tau species are released upon neuronal cell death and taken up via endocytosis by other neurons. Potential alternative routes are exocytotic release of tau, as well as constitutive secretion of nonmembrane-enveloped tau by living cells [213–215]. Tunneling nanotubes have been identified as structures involved in cell–cell communication that are used by prion proteins to spread between cells. Whether this route is also used for tau spreading, is not clear yet [216,217].

A promising approach for the prevention and treatment of tauopathies involves the use of the immune system (Fig. 8.2, 2). Both active and passive *immunization methods* have been employed in several studies, which showed beneficial effects in transgenic mouse models of tauopathy. One possible mechanism of action is that when tau is spreading from neuron to neuron, antibodies interact with it before it can be taken up by neighboring neurons. Antibodies might promote the uptake of tau aggregates into microglia, which might thus clear antibody-bound tau from the extracellular space. Another plausible mechanism could be the neuronal entry of antibodies, binding to tau, and subsequent clearance, presumably via the endosomal–autophagy–lysosomal pathway [218–221].

Several tau vaccines have already entered clinical trials in AD or PSP. Based on structural determinants of tau protein which are important for pathological tau–tau interactions, an active tau peptide vaccine was designed, which generates antibodies that can discriminate between pathological and physiological tau [222]. This AADvac-1 vaccine (Axon Neuroscience SE, Larnaca, Cyprus) demonstrated a reduction in tau oligomerization, pathological tau phosphorylation, and neurofibrillary pathology and an improvement in sensorimotor function in transgenic rats expressing human truncated tau protein in the brain and spinal cord [223]. A phase 1 clinical trial in patients with mild to moderate AD showed it was safe and well tolerated and induced increasing titers of antibody with repeat injections. The ADAS–cog scores remained stable over 6 months (clinicaltrials.gov identifier NCT01850238 and NCT02031198) [224]. A phase 2 study in patients suffering from AD has been initiated (clinicaltrials.gov identifier NCT02579252). A phase 1 clinical trial of BMS-986168 (Bristol–Myers Squibb), an antibody against extracellular, N-terminally fragmented tau species, has been initiated for PSP patients (clinicaltrials.gov identifier NCT02460094) and healthy individuals (clinicaltrials.gov identifier NCT02294851) [225]. Similarly, PSP patients are being recruited for a phase 1 study of C_2N-8E12, an antibody that recognizes an aggregated extracellular form of tau (C_2N Diagnostics, LLC, St. Louis, Missouri, USA; clinicaltrials.gov identifier NCT02494024). In tau transgenic mice, it reduced hyperphosphorylated, aggregated, and insoluble tau, decreased brain atrophy and microglial activation, ameliorated motor function and cognitive deficits, and blocked tau seeding activity [226–228]. A phase 1 clinical trial in patients suffering from mild to moderate AD that started in 2013 investigates the effects of a liposome-based vaccine ACI-35 (AC Immune SA, Lausanne, Switzerland; Janssen, Beerse, Belgium). This vaccine is designed to elicit an immune response to a pathological tau conformer by incorporating short phosphorylated peptides into liposomes to mimic a pathological tau epitope [229]. Recently, a virus-like particle–based vaccine, targeting pathological phosphorylated tau, was described. Here, tau peptides are packed within viral capsids to elicit an immune response. In tau transgenic mice, this resulted in a decrease of phosphorylated tau and neurofibrillary tangles. The mice also showed reduced levels of inflammation, decreased apoptotic neurons, and improved learning and memory performance [230]. Tau immunotherapies are reviewed in Refs. [219,220].

Tau undergoes several modifications that may potentiate toxicity. In addition to the aforementioned phosphorylation, tau modifications include O-linked glycosylation, ubiquitination, nitration, acetylation, and glycation.

Research mainly concentrated on two tau kinases: glycogen synthase kinase-3 beta (GSK3β) and cyclin-dependent kinase-5 (Cdk5). To *inhibit tau phosphorylation* (Fig. 8.2, 3), several studies focused on agents that inhibit tau kinases, either by selectively inhibiting GSK3 or Cdk5 or by inhibiting a larger group of kinases. A relatively nonspecific kinase inhibitor, SRN-003-556, prevented okadaic acid–induced tau hyperphosphorylation in hippocampal slices, and delayed the onset of motor deficits and reduced the amount of insoluble aggregated tau in tau transgenic mice, however, without decreasing neurofibrillary tangles [231]. Tideglusib, a more selective GSK3 inhibitor, also showed promising results in model systems [232,233] and was recently tested in patients suffering from AD and PSP. An initial study in 30 mild to moderate AD patients indicated tideglusib was well tolerated and suggested positive trends in cognitive performance (clinicaltrials.gov identifier NCT00948259) [234]. However, a subsequent trial in over 300 AD patients produced no clinical benefits (clinicaltrials.gov identifier

NCT01350362) [235]. This was also the conclusion after the trial in PSP patients [236]; although, here the progression of brain atrophy was reduced in patients who were treated with tideglusib [237] (clinicaltrials.gov identifier NCT01049399). MMBO, a compound with a good selectivity for GSK3 [238], inhibited tau phosphorylation in primary neural cells and in the mouse brain. Furthermore, MMBO treatment in an AD mouse model resulted in decreased tau phosphorylation and improved memory [239]. Studies in tau transgenic mice and AD *APP/PS1* mouse models suggested lithium and valproate, two kinase inhibitors that are commonly used mood-stabilizing drugs, to be promising therapeutics [240–247]. Conversely, other studies demonstrated reversible brain atrophy and cognitive impairment associated with valproate treatment [248–251]. Lithium strongly inhibits GSK3β, as well as a number of other kinases, though with slightly lower potency [252]. Valproate is believed to reduce the activity of GSK3β and Cdk5 [245]. In contrast, it was also reported that valproate does not inhibit GSK3-mediated tau phosphorylation [253,254]. The results of clinical trials of lithium in AD patients or patients with mild cognitive impairment suggest mixed outcomes so far: some studies observe an improvement in cognitive measures after treatment [255–259], while others do not [260,261]. A study in PSP and CBS patients was terminated due to poor tolerability of the drug (clinicaltrials.gov identifier NCT00703677). A phase 2 trial for the treatment of agitation/aggression with or without psychosis in AD patients (clinicaltrials.gov identifier NCT02129348) and a trial in individuals with mild cognitive impairment (clinicaltrials.gov identifier NCT02601859) have been initiated. Also for valproate the results in clinical trials for AD are diverse, with several studies observing improvements in behavior [262–265], while valproate did not prove to be beneficial [266] and even aggravated agitation and aggression or caused a greater decline in the hippocampal and brain volumes in other studies [267–269].

While valproate and lithium inhibit tau phosphorylation, other substances promote its dephosphorylation, for example by enhancing the activity of protein phosphatase 2A (PP2A). Metformin, a drug approved for use in diabetes, showed to reduce tau phosphorylation through an increased PP2A activity in mice [270]. Also sodium selenate had this effect on PP2A activity and tau phosphorylation in mouse models of tauopathy. Further it improved memory functions and motor performance, abrogated the formation of neurofibrillary tangles, and prevented neurodegeneration [271,272].

Acetylation at lysines often competes with ubiquitination at these sites. Therefore, acetylation can prevent the degradation of misfolded tau by the UPS. Thus, *reducing tau acetylation* and, thereby, promoting ubiquitination and proteasomal degradation of misfolded tau is another therapeutic strategy (Fig. 8.2, 4) [215,273]. One of the main enzymes that acetylates tau is the acetyltransferase p300. Inhibition of p300 by salsalate lowered levels of total tau and tau acetylated at K174, rescued tau-induced memory deficits, and prevented hippocampal atrophy in the P301S tauopathy mouse model [274]. A phase 1 clinical trial to evaluate salsalate in PSP patients has been initiated (clinicaltrials.gov identifier NCT02422485).

Tau is *cleaved* by various proteolytic enzymes, such as caspases, calpains, thrombin, cathepsins, and puromycin-sensitive aminopeptidase (reviewed in Ref. [275]). The generated fragments would have different propensities toward aggregation and induction of neurotoxicity. Inhibition of proteolytic enzymes has proven to be beneficial in vitro and in vivo (Fig. 8.2, 5). For example, treatment of tau P301L mice with calpastatin, a calpain inhibitor, prevented tauopathy and loss of motor axons, delayed disease onset, and extended survival to the normal life span [276]. Also the small-molecule calpain inhibitor A-705253 blocked temperature-induced tau phosphorylation in hippocampal slices, and administration to a triple-transgenic mouse model of AD reduced cognitive impairment, synaptic dysfunction, and tau hyperphosphorylation, among other beneficial effects [277,278].

As mentioned in the section "Overview of Frontotemporal Dementia," six isoforms of tau are expressed that differ in the number of MTB domains. About half of the FTD-related *MAPT* mutations affect splicing of exon 10, thereby disturbing the balance between 4R (+exon 10) and 3R (−exon 10) isoforms, predominantly toward a 4R preponderance [47,215]. This affects the binding affinity to MTs, as 3R tau isoforms bind less tightly than 4R tau isoforms to MTs [279]. To *correct the 3R/4R ratio* (Fig. 8.2, 6), a strategy using antisense oligonucleotides (ASO) was proposed. ASOs are short, synthetic, modified nucleic acids that bind RNA and modulate its function. Depending on the chemistry and target site, ASO binding can modify mRNA maturation (prevent 5′-cap formation, modulate splicing, and the location of polyadenylation), activate RNA degradation via RNase H, sterically block the binding of RNA-binding proteins (RBPs), such as ribosomal subunits, and affect translation in this way, or bind to miRNA or antisense transcripts and, thereby, prevent them from inhibiting their mRNA targets (reviewed in Refs. [280–282]). To modify alternative splicing of tau, two types of ASOs were investigated: 2′-O-methyl–modified ASOs that are complementary to either the 3′- or the 5′-splice site of exon 10, and a bipartite ASO that interacts simultaneously with both flanking regions of the splice regulatory hairpin structure, which forms at the boundary between exon 10 and intron 10. Transfection of cell lines with these ASOs resulted in a significant reduction of endogenous 4R tau [283–285].

In addition to the use of ASOs, reprogramming of tau alternative splicing by spliceosome-mediated RNA transsplicing (SMarT) has been suggested. With this technology, RNA transsplicing between an endogenous target pre-mRNA and a delivered RNA molecule is mediated by the spliceosome. The applicability of this technique in tau-related FTD was demonstrated in cellular systems, as well as in a mouse model of tau missplicing. These experiments show that exon 10 inclusion can be reduced, as well as increased by transsplicing, resulting in a conversion of 3R into 4R or vice versa, depending on the delivered molecule [286–288]. Another area of investigation is the identification of small molecules or compounds that affect tau splicing [289–294].

Interestingly, miR-132 has been identified to be downregulated in PSP, a 4R tau disorder. miR-132 levels are inversely correlated with expression of PTBP2, a protein that participates in the regulation of tau isoform abundance [295]. miR-132/212 knockout mice display memory problems and increased tau expression, phosphorylation, and aggregation [296,297].

The use of ASOs in tau-related FTD was also proposed as a means to *lower the expression of total tau* (Fig. 8.2, 7). ASO-based reduction of endogenous tau in adult mice resulted in no significant deviations from baseline levels in any sensory, motor, or cognitive behavior task [298]. Furthermore, repressible tauopathy mouse models showed, depending on the model, recovery of memory function and synaptic plasticity, stabilization of neuron numbers, and reversal of neurofibrillary tangles [198,299,300]. ASO-based reduction of tau expression enables the translation of these findings to human therapeutics. ASO treatment in a tau transgenic mouse model lowered human tau mRNA and protein levels and hyperphosphorylated tau deposition. In aged mice, the emerged tau pathology was even reversed and hippocampal volume and neuronal loss restored [301a,302]. A study with morpholino oligonucleotides inducing exon skipping demonstrated reduced *MAPT* mRNA and tau protein levels in human neuroblastoma cell lines and in mice transgenic for the human *MAPT* gene [303]. miR-based therapies might also be a possibility to lower *MAPT* expression. miR-219, for example, was identified to bind to the 3′-untranslated region of *MAPT* mRNA and to repress tau synthesis [304]. A drug screen aimed at detecting compounds that are able to reduce tau protein levels also identified aggregation inhibitors, such as methylene blue, as well as antiproliferatives, antibiotics, receptor antagonists, steroids, and MT regulators, such as paclitaxel [305].

The rationale behind the use of *MT stabilizers* as a therapy is that hyperphosphorylated and aggregated tau cannot bind and stabilize the MTs. To preserve MT organization and normal axonal transport, MT stabilizers have been investigated (Fig. 8.2, 8). Paclitaxel could bind to MTs and improve axonal transport and motor impairments in tau transgenic mice [306]. However, it does not effectively penetrate the blood -brain barrier (BBB) [307]. TPI-287 is a derivative of the taxane drugs used in cancer therapy, such as paclitaxel. Unlike most taxanes, TPI-287 crosses the BBB [308]. A study of TPI-287 in CBS and PSP (clinicaltrials.gov identifier NCT02133846) and in AD (clinicaltrials.gov identifier NCT01966666) has been initiated. Davunetide (NAPVSIPQ, NAP), a growth factor–derived octapeptide, also protected memory, increased soluble tau, and reduced tau hyperphosphorylation in a tauopathy mouse model [309], and also rescued neuronal dysfunction in a *Drosophila* model of tauopathy [310]. In a triple-transgenic AD mouse model, davunetide reduced amyloid accumulation and tau hyperphosphorylation and improved performance in the Morris water maze [311–313]. A phase 2 trial was performed in individuals with mild cognitive impairment (clinicaltrials.gov identifier NCT00422981) and a phase 2/3 trial in patients suffering from PSP (clinicaltrials.gov identifier NCT01110720). This study failed to show efficacy of davunetide [314].

GRANULIN

As discussed in the section "Overview of Frontotemporal Dementia," loss-of-function mutations in *GRN* cause FTD through a mechanism of haploinsufficiency. An obvious therapeutic choice would be to counteract GRN deficiency by stimulating GRN expression or by inhibiting its breakdown, for example, through inhibition of sortilin 1 (SORT1)-mediated endocytosis.

Upregulating GRN levels could be achieved using gene therapy, administering recombinant protein [315], or via strategies to boost endogenous GRN expression (Fig. 8.3). Several strategies have focused

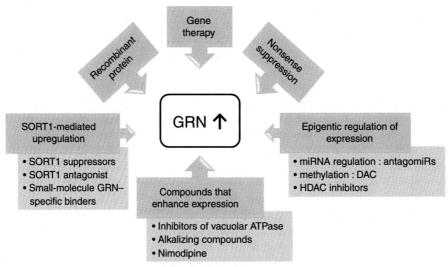

FIGURE 8.3 Therapeutic Strategies to Enhance GRN Expression

See text for details. *DAC*, 5-Aza-2′-deoxycytidine; *HDAC*, histone deacetylase; *SORT1*, sortilin.

on the administration of compounds to upregulate GRN levels. An example is the study by Capell et al. [316], which demonstrates that *inhibitors of vacuolar ATPase*, as well as *alkalizing compounds*, including three drugs used for other indications, are able to restore GRN levels in organotypic cortical slice cultures from GRN$^{+/-}$ mice and in lymphoblast cell lines derived from FTD patients with a *GRN* mutation (FTD–GRN patients) [316]. Interestingly, vacuolar ATPase inhibitors also cause increased TMEM106B levels in a cellular system [317]. Also *nimodipine*, a calcium channel blocker that is approved by the Food and Drug Administration (FDA), increases GRN levels [318]. A phase 1 clinical trial is ongoing in FTD–GRN patients (clinicaltrials.gov identifier NCT01835665). A drug screen with 1200 FDA-approved drugs identified a *histone deacetylase* (HDAC) *inhibitor*, suberoylanilide hydroxamic acid (SAHA), as an activator of GRN expression [319]. SAHA is a small-molecule enhancer of progranulin transcription and increased GRN mRNA levels in patient-derived lymphoblast and fibroblast cell lines [319]. FORUM Pharmaceuticals developed a series of HDAC inhibitors and demonstrated that compound FRM-0334 induces a fivefold increase in GRN mRNA in cultured primary rat neurons and in patient-derived lymphoblast cell lines. Currently they are performing a phase 2 clinical trial of FRM-0334 in FTD–GRN patients (clinicaltrials.gov identifier NCT02149160). FORUM Pharmaceuticals also reported a screen in a mouse microglial cell line to identify modulators of GRN expression. They identified cytoskeletal regulators, lysosomal alkalizing agents, cholesterol synthesis inhibitors, kinase inhibitors, and modulators of other signaling pathways that were able to modulate GRN levels [301b,c].

Also other epigenetic mechanisms seem to be involved in the regulation of GRN expression. Posttranscriptional modification of GRN expression by *miRNAs* [126,320,321], as well as significant alterations in miRNA expression in brain of FTD–GRN patients have been identified [322,323]. Although replication of these miRNA targets is recommended, miRNA-based therapeutics could be an option. Anti-miRs or antagomiRs can be designed to bind to the active miRNA strands of miRNAs that target *GRN*. A general concern of using antagomiR as therapeutic agents is their potential nonspecific binding to other RNAs, which could result in side effects. A possibility to circumvent this issue is the use of ASOs that block the mRNA target site. This target-specific site blocker prevents miRNAs from gaining access to the site. Also *methylation* of the *GRN* promoter region was shown to be inversely correlated with GRN expression in lymphoblast cell lines. Studies on brain material of FTLD–TDP patients revealed evidence of hypermethylation of the *GRN* promoter and an upregulation of DNA methyltransferase 3a (DNMT3a) compared to control individuals. Treatment of cells with a DNA-demethylating drug, 5-aza-2′-deoxycytidine (DAC), led to reexpression of GRN in lymphoblast cell lines. This indicates that targeting the DNA methylation machinery might be a potential strategy to elevate GRN levels [324].

Pathogenic mutations in *GRN* are predominantly null mutations, which cause loss of expression via nonsense-mediated mRNA decay. A potential therapeutic approach to boost GRN expression in patients with a nonsense mutation is based on *nonsense suppression* to restore the production of full-length protein. In vitro and in vivo studies with compounds, such as ataluren (also known as Translarna or PTC124, PTC Therapeutics; South Plainfield, New Jersey, USA) showed ribosomal read-through activity of premature termination codons in multiple disease models, including the neurological disorder infantile neuronal ceroid lipofuscinosis [325,326]. Ataluren is in development for the treatment of Duchenne muscular dystrophy and cystic fibrosis caused by nonsense mutations. It is currently licensed in Europe for the treatment of Duchenne muscular dystrophy caused by nonsense mutations in ambulatory patients aged 5 years and older. In the United States, it is only available through clinical trials. Based on these encouraging results, clinical trials with read-through compounds for patients who carry a *GRN* nonsense mutation would be interesting and are envisioned (Herz of the Consortium for Frontotemporal Research in Ref. [192]).

In addition to these strategies, which are focused on altering GRN expression via an effect on GRN transcription or translation, other approaches involving SORT1 are investigated. Sortilin, a membrane receptor, was identified as a high-affinity neuronal receptor of GRN [327,328]. SORT1 normally endocytoses GRN and directs it to the endolysosomal pathway for degradation. It is possible to inhibit this process to maintain GRN levels via either the development of suppressors of SORT1 expression, SORT1 antagonists, or small-molecule GRN-specific binders. In a study by Lee et al. [329] these three strategies were investigated. Both in mammalian cell lines, as well as in induced pluripotent stem cell (iPSC)-derived neurons and lymphoblast cell lines of FTD–GRN patients, a small molecule termed *MPEP* (1-[2-(2-tert-butyl-5-methylphenoxy)-ethyl]-3-methylpiperidine) decreases SORT1 levels and increases extracellular GRN levels. Further, high-affinity SORT1 ligands, such as *neurotensin*, prevent access of extracellular GRN to the SORT1 binding sites and inhibit GRN endocytosis in this way. On the other hand, Lee et al. [329] discuss that the use of SORT1 antagonists can potentially trigger off-target effects and cause clinical side effects. Therefore, they hypothesize that modulating the GRN–SORT1 interaction via GRN-specific interactors may be a more promising approach. They identified a small molecule, *BVFP* (4-[2-(3-bromophenyl)vinyl]-6-(trifluoromethyl)-2(1H)-pyrimidinone), which targets a motif in GRN that is essential for the interaction with SORT1. This reduces the amount of GRN to be captured by SORT1 [329].

The question remains if inhibition of GRN endocytosis by blocking SORT1 has deleterious effects on other cell functions, for example, in lysosomes, as discussed by Lee et al. [329]. In this regard, it is of interest that sortilin and prosaposin (PSAP) were recently described as two independent and complementary pathways for GRN lysosomal targeting in both biosynthetic and endocytic pathways [330].

C9ORF72

The *C9orf72* G_4C_2 repeat is located in the first intron after the noncoding exon 1 of transcript variants 1 and 3 and upstream of transcript variant 2. Repeat expansions might influence *gene expression* by affecting the binding of regulatory elements, by altering the DNA methylation, a distance effect, or a combination [331]. Several lines of evidence support a loss-of-function mechanism, such as allele-specific reduced *C9orf72* mRNA and protein expression [74,76,84,332–334] and the presence of a (motor) phenotype in some *C9orf72* knockdown/out models [335–337]. Furthermore, reporter gene studies with *C9orf72* promoter constructs containing a variable number of repeat units indicate an inverse correlation between repeat size and gene expression [85,87]. In patients carrying a G_4C_2 expansion, hypermethylation of the G_4C_2 repeat, and the flanking CpG island [85,338–344], as well as histone trimethylation [333] was observed, suggesting a loss-of-function mechanism through transcriptional silencing of the promoter.

On the other hand, the absence of other causal loss-of-function mutations in the *C9orf72* gene [87,345], the fact that a homozygous repeat expansion does not seem to lead to a more severe or different clinical phenotype [332], as well as the absence of a motor phenotype in some knockdown/out models [346,347] does not favor loss of function as the sole central mechanism. Therefore, therapeutic strategies have not focused on increasing the expression of *C9orf72*.

One of the gain-of-toxic-function mechanisms to explain how G_4C_2 repeat expansions in *C9orf72* could cause disease is the abnormal bidirectional transcription of expanded repeat sequences and the formation of *RNA foci*. These are aggregates of repeat-containing sense and antisense RNA transcripts, which fold into hairpin and sense G-quadruplex structures and sequestered RBPs. RNA foci are present in the brain and spinal cord of patients who carry the *C9orf72* hexanucleotide expansion and have been identified in patient tissues and iPSC-derived neurons (iPSNs) [76,334,346,348–353]. The RNA foci burden in the frontal cortex showed a significant inverse correlation with onset age [350] and repeat length [354]. However, at this moment it is still unclear whether nuclear RNA foci induce neurotoxicity. The sequestered RBPs were shown to be involved in RNA splicing, editing, nuclear export, and nucleolar function. Examples of identified RBPs are hnRNPA1, hnRNP H, SF2, and Pur-alpha [334,348,353,355–357]. Loss of function of these proteins could affect downstream cellular processes. Overexpression of Pur-alpha mitigated repeat-induced toxicity in mouse neuronal cells and *Drosophila* [355].

Another gain-of-function mechanism is the *DPR protein toxicity* due to repeat-associated non-ATG-initiated (RAN) translation, where polymers of dipeptides are produced by translation in all reading frames. These DPR proteins are prone to aggregation and accumulate in the brain and, to a lesser extent, in spinal cord of patients carrying a *C9orf72* G_4C_2 repeat expansion [36,37,351,352,358,359]. Experimental work in cultured cells, primary neurons, and *Drosophila* supports a toxic effect of DPRs [352,354,360–366]. In fact, some studies find evidence that arginine-rich DPRs are the most toxic DPR species [354,361,364–366].

Several studies show a lack of spatial correlation between the DPR deposits and the degree of neurodegeneration [367,368]. In contrast, the distribution of TDP-43 pathology correlates well with areas of degeneration and the clinical phenotype [29,368]. Therefore, it was hypothesized that DPR pathology is a stressor that triggers a cascade of biological events, which eventually promote FTLD–TDP pathology and, as a consequence of this, neurodegeneration [369]. This was also supported by the observation of aggregates with a DPR core surrounded by aggregated TDP-43, suggesting that DPR aggregation may precede TDP-43 accumulation [36,368]. Some patients with a *C9orf72* expansion mutation even lack the TDP-43 pathology [74,370].

To target gain-of-function mechanisms *ASOs* are an exciting possibility (Fig. 8.4, 1). Thus far, three studies investigated the effects of ASO treatment in fibroblasts and iPSC-derived (motor) neurons from patients carrying a *C9orf72* hexanucleotide repeat expansion [334,346,348] (reviewed in Ref. [371]). Two main types of ASOs were tested: an ASO that blocked binding of RBPs to the hexanucleotide repeat expansion and ASOs that induce RNase H–mediated RNA degradation. ASOs targeting the region downstream of the repeat expansion showed a more prominent downregulation, as this blocks transcription of all three transcript variants. All studies show that ASOs are able to reduce RNA foci, regardless of the effect on *C9orf72* RNA levels. Furthermore, Donnelly et al. [334] demonstrated that ASO treatment normalized the dysregulated gene expression of several genes which are differentially expressed between control iPSNs and *C9orf72* hexanucleotide expansion carrier–derived iPSNs. They noted an increased susceptibility to glutamate-mediated excitotoxicity in *C9orf72* iPSNs, a phenotype that was rescued upon ASO treatment. Poly-GP DPRs were still present after ASO treatment. Whether this suggests that the detected DPRs are not major contributors to neurotoxicity in this model or that other RAN peptides contribute to the observed phenotypes is not sure [334]. Sareen et al. [348] studied the effect of ASOs in iPSC-derived motor neurons and found ASO-mediated reversal of gene expression alterations, which were observed in patient cells compared to control cells. On the other hand, in the study by Lagier-Tourenne et al. [346], the RNA signature in *C9orf72* patient fibroblasts could not

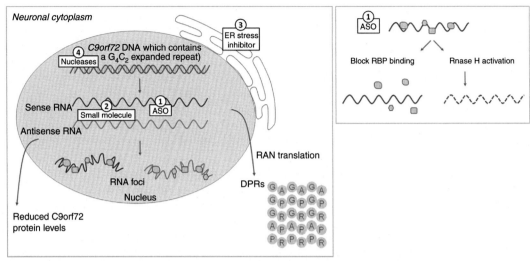

FIGURE 8.4 Disease Mechanisms Involved in *C9orf72*-Related Neurodegenerative Disease and Possible Therapeutic Strategies

Expansion of the *C9orf72* G_4C_2 repeat has different consequences: reduced expression of the *C9orf72* gene and sense and antisense transcription and repeat-associated non-ATG translation *(RAN)* of the expanded repeat resulting in nuclear RNA foci and aggregated dipeptide repeat *(DPR)* neuropathology. RNA-binding proteins *(RBPs)* bind to the repeat RNA. Main therapies proposed thus far are antisense oligonucleotide *(ASOs)* that either block RBP binding or activate RNase H to degrade the RNA *(1)*, small molecules that bind the repeat RNA *(2)*, and inhibitors of endoplasmic reticulum *(ER)* stress *(3)*. A more speculative method might be the use of nucleases to contract repeats *(4)*.

be corrected by ASO treatment. They argue that this might be due to the unaltered accumulation of antisense RNA foci that potentially disrupts the function of proteins binding to the RNA. Recently, duplex RNAs that recognize both sense and antisense repeat transcripts have been engineered. Recognition inhibits the formation of foci formed by G_4C_2 and C_4G_2 RNA [372].

Small molecules that bind to the G_4C_2 RNA repeat represent another promising area of therapeutic interventions (Fig. 8.4, 2). One of the identified compounds was able to inhibit RAN translation and foci formation in iPSNs from individuals carrying a *C9orf72* repeat expansion and did not affect *C9orf72* mRNA levels. However as the expansion is bidirectionally transcribed, therapeutics may have to target both G_4C_2, as well as C_4G_2 repeats [373]. 5,10,15,20-Tetra(*N*-methyl-4-pyridyl)porphyrin (TMPyP4) was found to bind to the G_4C_2 RNA repeat and distort the G-quadruplex structure, diminishing its interactions with the RBPs hnRNPA1 and SF2. This strategy may prove valuable in the case hnRNPA1 or SF2 is involved in a toxic RNA pathway [374]. At this moment, the precise role of RNA structure in disease pathogenesis is not yet known. It was hypothesized that RNA–RNA interactions mediated through G-quadruplexes might help explain the transcript aggregation and RNA foci formation [375]. If the formation of G-quadruplexes is involved, perturbation of this structure by small-molecule ligands might also be useful to further elucidate the disease mechanisms [374].

In in vitro models of DPR toxicity, evidence of endoplasmic reticulum (ER) stress was recognized. ER stress inhibitors salubrinal and TUDCA, protected cells against poly-GA-induced toxicity, which suggests that small molecules that target the ER may be promising therapeutics (Fig. 8.4, 3) [360].

Genetic screens in yeast and *Drosophila* identified modifiers of *C9orf72*-repeat associated toxicity, with a striking enrichment of genes involved in nucleocytoplasmic transport. These studies give insight into the mechanisms of *C9orf72*-related disease and point to modulation of nucleocytoplasmic shuttling as a potential novel drug target [365,366,376,377]. Treatment of *Drosophila* expressing a G_4C_2 hexanucleotide expansion with TMPyP4 rescued nuclear import in a dose-dependent manner. The same was observed after treatment with KPT-276, an exportin 1 inhibitor. Furthermore, ASO treatment of *Drosophila* or *C9orf72* iPSNs mitigated the nucleocytoplasmic transport deficits. G_4C_2-mediated neurodegeneration in the eye of *Drosophila* was suppressed by ASO, TMPyP4, or KPT-276 treatment [377].

The described ASOs and compounds do not tackle the loss of *C9orf72* and some of them even reduce *C9orf72* RNA levels. It remains to be elucidated whether it is pivotal for an effective therapy to address this as well. As discussed in the section "Overview of Frontotemporal Dementia," the function of the *C9orf72* gene is currently unclear. If its expression levels are crucial for normal cell function, it could be preferable to use therapies that do not alter the expression of the nonexpanded *C9orf72* transcript in order not to interfere with the other functions of the translated nonexpanded protein. It was, therefore, suggested that therapies would focus on posttranscriptional degradation of the repeat-expanded RNA or promoting transcriptional silencing of mutant *C9orf72*. Therapies that aim to increase the gene expression may even be detrimental if they boost the expression of the mutant allele [343,344,371,378].

A speculative approach to tackle repeat expansion diseases is the possibility of shortening repeats. A recent review highlights the early experiments and possibilities using meganucleases, zinc-finger nucleases, transcription activator–like effector nucleases (TALENs), and CRISPR–Cas nucleases to contract trinucleotide repeats (Fig. 8.4, 4) [379].

TDP-43

TDP-43 is a conserved nuclear RNA-binding protein, which functions in transcription and splicing regulation. Aggregates of TDP-43 are present in various neurodegenerative disorders, including FTD. Only few FTD patients carry a causal mutation in the *TARDBP* gene, meaning that in most patients the wild-type protein is aggregated and located in the cytosol. However, the pathogenicity of these aggregates and its protein modifications, including hyperphosphorylation, ubiquitination, and cleavage into C-terminal fragments is currently poorly understood and controversial (for examples, see Refs. [380–382]). One possible disease mechanism might be the loss of the normal nuclear functions of TDP-43 due to nuclear depletion and cytosolic accumulation. On the other hand, aggregated TDP-43 might be toxic for the cell.

In agreement with the results in tau-related FTD, methylene blue also *decreased TDP-43 aggregation* in cellular models [383]. Furthermore, it protected against TDP-43 toxicity in *Caenorhabditis elegans* and zebrafish models expressing mutant human TDP-43 [384]. Also dimebon, a nonselective antihistaminergic compound, reduced TDP-43 aggregation in cells [383]. Dimebon has been tested in patients suffering from AD; however, the phase 3 clinical trial did not show any significant effect in two coprimary and several secondary outcome measures [385]. A drug screen in iPSC-derived motor and cortical neurons from an ALS patient which displayed TDP-43 aggregates, identified four classes of compounds that reduced the percentage of cells with TDP-43 aggregates, namely cyclin-dependent kinase inhibitors, c-Jun N-terminal kinase inhibitors, triptolide, and FDA-approved cardiac glycosides [386]. Another compound, LDN-0130436, was found to inhibit TDP-43 aggregation via a cell-based compound screening and validation in a *C. elegans* transgenic model of TDP-43 toxicity [387].

In addition to these examples of compound screenings aimed at identifying drugs that modify TDP-43 aggregation, screens have been performed to identify *modifiers of TDP-43 toxicity* or pathways controlling TDP-43 *nucleocytoplasmic shuttling*. Examples are the identification of ataxin-2 (ATXN2), debranching RNA lariats 1 (DBR1), and RAD-23 as modifiers of TDP-43 toxicity. Loss of function suppressed TDP-43 toxicity and, therefore, they represent potential therapeutic targets for small-molecule inhibitors or RNA interference [388–390]. Another example is a genome-wide RNAi screen that identified ITPR1, encoding an intracellular receptor for inositol-1,4,5-trisphosphate, as one of the factors involved in nucleocytoplasmic transport of TDP-43. Silencing of ITPR1 increased the cytosolic localization of TDP-43 [391].

A genome-wide genetic screen in yeast identified human upframeshift protein 1 (UPF1) as a modifier of FUS toxicity [392]. In primary rat cortical neurons transfected with mutant TDP-43 and FUS coexpression with UPF1 diminished the risk of cell death by 20%–40% [393]. Adeno-associated virus–based delivery of wild-type human TDP-43 to rats induces limb paralysis. Simultaneous delivery of UPF1 improved motor function by protecting forelimb function [394].

Modifications of TDP-43, such as phosphorylation and acetylation, might also attenuate toxicity. Modifying TDP-43 phosphorylation might be an attractive intervention and studies are performed in this direction [395–397]. For example, inhibitors of the protein kinase CK-1 have been identified that prevented TDP-43 phosphorylation in cell cultures and increased the life span of *Drosophila* by reducing TDP-43 neurotoxicity [397]. However, conflicting data arguing that phosphorylation would mitigate toxicity exist as well [398,399]. TDP-43 acetylation would impair binding to target mRNAs and enhance the formation of aggregates. It was, therefore, predicted that approaches to reduce levels of acetylated TDP-43 by, for example, pharmacologically targeting putative TDP-43 acetyltransferases might be beneficial [400].

TDP-43 is implicated in RNA processing, including miRNA biogenesis [401–404]. Studies of *TARDBP* mutation carriers or patients with FTLD–TDP pathology identified deregulated miRNAs [323,405,406], including the miR-132 cluster, which may target *TMEM106B* mRNA species [323]. Although more research into these miRNAs and their role in disease pathogenesis is required, these findings are first steps into a *miR-based therapy*, be it miRNA-replacement therapy or the use of antagomiRs. Interestingly, miR-132 was also found to be differentially expressed in the brains of AD patients [407] and its loss was associated with tau exon 10 inclusion in PSP [295].

CELLULAR WASTE CLEARANCE SYSTEMS

Many neurodegenerative diseases, including FTD, are characterized by the accumulation of misfolded and ubiquitinated proteins into neuronal aggregates. This suggests defects in the protein quality control systems and clearance machinery. Due to the defects in the pathways of protein degradation, cells might be unable to clear even a normal load of aberrant proteins. Furthermore, the demand on the degradation mechanisms might be too high due to the accumulation of aggregated, and possibly toxic, proteins in many neurodegenerative disorders. Consequently, strategies to enhance the cellular waste clearance systems, be it the chaperone network, the UPS, and/or the autophagy–lysosome pathway might be beneficial to clear aggregated tau (Fig. 8.2, 9) or TDP-43, for example.

One of the systems to clear misfolded proteins is the proteasome, which degrades proteins that are modified by ubiquitination. A small-molecule inhibitor of the deubiquitinase USP14, promotes TDP-43 clearance in cultured cells [408], while silencing of another deubiquitinase UBPY exacerbates the toxicity of TDP-43 in *Drosophila* [409]. USP14 would trim ubiquitin chains and, thereby, inhibit proteasomal degradation. Enhancement of proteasome activity through the inhibition of USP14 also proved to be beneficial for the degradation of tau protein in cells [408].

Protein kinase A (PKA) phosphorylation of the proteasome can boost proteolytic activity [410–412]. Agents that increase cAMP levels and, consequently, PKA activity enhanced the degradation of tau and TDP-43 in cells [413]. Moreover, administration of the cAMP-enhancing agent rolipram to double-transgenic rTg4510 mice expressing human 4R tau and the *MAPT* P301L mutation reduced the levels of total and phosphorylated tau, enhanced proteasome activity, and improved cognition in early-stage tauopathy [414]. Rolipram is, however, no longer in clinical use due to adverse effects. Therefore, other drugs in its class are under investigation [415].

Heat shock proteins are involved in the prevention of tau aggregation or assist in tau degradation [416]. Several inhibitors of these molecular chaperones have proven to be promising targets for tau therapies (reviewed in Ref. [417]). An example are inhibitors of Hsp90, such as the antibiotic geldanamycin [416,418,419]. Many of the Hsp90 inhibitors induce Hsp70 expression, but also Hsp70-specific drugs have been studied and seem beneficial [420–422]. Incubation of cells with an Hsp90 inhibitor also reduced full-length and cleaved TDP-43 levels [423].

Potentiated Hsp104 variants eliminated cytoplasmic TDP-43 aggregates and restored TDP-43 to the nucleus in yeast. Hsp104 is an AAA+ ATPase and protein disaggregase from yeast. Whether it will be feasible to introduce Hsp104 as a therapeutic is, however, not clear [424].

In FTD patients, mutations are identified in various autophagy-related genes, and autophagy can be disturbed at several stages of the pathway (role of autophagy in neurodegenerative disease is reviewed in Refs. [425,426]). OPTN and p62, two proteins associated with FTD (−ALS), are autophagic receptors which are involved in selective autophagy. They bind to ubiquitinated cargo molecules via their ubiquitin-associated domain and to LC3 through their LC3 interaction region motif [427,428]. LC3 is localized to the membrane of autophagosomes. With the identification of *TBK1* mutations in ALS and FTD, the link between autophagy and FTD (−ALS) became even more prominent. TBK1 phosphorylates OPTN [429] and p62 [93], and it also phosphorylates and interacts with VPS37C, a component of the ESCRT-I [430]. *CHMP2B* encodes a component of the ESCRT-III complex. The ESCRT complexes are involved in the formation of multivesicular bodies. These are late endosomes, which contain internal vesicles through the inward budding of vesicles into the lumen. Multivesicular bodies can fuse with autophagosomes prior to lysosomal fusion to form amphisomes (an intermediate between an autophagosome and autolysosome), or fuse with lysosomes to degrade their contents. VCP is implicated in the regulation of endolysosomal sorting of endocytosed ubiquitinated cargoes, and mutations in the gene lead to the accumulation of immature autophagosomes [431–433]. The C9orf72 protein might be involved in endosomal trafficking [88] and is linked with autophagy via coexpression studies [434].

The mTOR inhibitor *rapamycin*, as well as more soluble analogs, such as temsirolimus, induce autophagy and enhance the clearance of various aggregation-prone proteins involved in neurodegeneration in cellular and/or animal models, including tau and TDP-43. For example, in P301S mutant tau transgenic mice, rapamycin treatment significantly reduced tau pathology and tau hyperphosphorylation [435], while in a mouse model expressing 4R tau with the P301L mutation, a reduction of tau-induced neuronal loss was observed among other beneficial effects [436]. Treatment of P301S mice with temsirolimus, furthermore, rescued spatial learning and memory impairments [437]. Recently, transduction of P301L-mutated tau into iPSC-derived neural progenitor cells and differentiation into cortical neurons triggered tau aggregation and hyperphosphorylation. Administration of rapamycin and trehalose reduced tau aggregation in this model [438]. In a mouse model with TDP-43 proteinopathy, rapamycin treatment rescued the learning/memory impairment, ameliorated motor neuron function, and increased the clearance of TDP inclusions [439]. On the other hand, rapamycin exacerbated the pathology and weakness in a VCP mouse model [440].

mTOR signaling has various autophagy-independent functions, which causes adverse effects in patients who are treated with rapamycin. Therefore, it might be more promising to identify and use mTOR-independent upregulators of autophagy, and screening studies have been undertaken in this direction (for examples, see Refs. [441,442]).

The mood-stabilizers *lithium*, *valproate*, and *carbamazepine* exert an effect on autophagy as well. Lithium enhances autophagy by inhibiting inositol monophosphatase and the phosphoinositol cycle, while valproate and carbamazepine inhibit inositol synthesis [443]. The effects of lithium and valproate in disease models and clinical trials with these compounds were discussed earlier in this chapter. Carbamazepine has been shown to be protective in an *APP/PS1* transgenic mouse model of AD [444] and rescued motor dysfunction, reduced cytosolic TDP-43 inclusions, and enhanced neuronal survival in a transgenic TDP-43 mouse model [439]. Furthermore as described in the section "Disease Management", extreme behavioral symptoms in FTD patients may be ameliorated by carbamazepine treatment [184].

Another mTOR-independent autophagy inducer is *trehalose*. Treatment of cells with this FDA-approved dissaccharide resulted in decreased levels of TDP-43 protein and its pathogenic truncated form, while in a neuronal cell model of tauopathy tau aggregation was inhibited [445,446]. Blockage of the UPS in neuroblastoma cells caused increased levels of proteins involved in neurodegenerative diseases, such as phosphorylated tau, an effect that was reverted by trehalose-mediated stimulation of autophagy [447]. Mice overexpressing human mutant tau and a parkin deletion presented with an amelioration of tau pathology after trehalose treatment. Also in a tau transgenic mouse model, a reduction in tau aggregates was observed [448,449]. Trehalose might also be beneficial in related neurodegenerative diseases, such as ALS. In *SOD1* mouse models for ALS, trehalose prolonged the life span and reduced motor neuron loss in the spinal cord, among other beneficial effects [450,451].

Transcription factor EB (TFEB) is a master regulator of the autophagy–lysosome pathway (reviewed in Ref. [452]). Upregulation of TFEB rescued neurodegenerative effects resulting from huntingtin toxicity [453] and reduced the accumulation of aggregated α-synuclein by inducing autophagic clearance [454]. Pharmacological activation of TFEB was obtained using 2-hydroxypropyl-β-cyclodextrin [455]. Given these encouraging results in other proteinopathies, TFEB activation might be an interesting approach for tau- or TDP-43-related FTD.

HETEROGENEITY = OPPORTUNITY

In the section "Overview of Frontotemporal Dementia," the heterogeneity of FTD was described as a problematic factor in the treatment of patients and the design and execution of clinical trials. However, one might also recognize this as an opportunity. The observed heterogeneity can serve as a basis to identify factors modifying onset age, disease duration, and clinical phenotype. As an example, Fig. 8.5A illustrates the wide onset age ranges per causal FTD gene.

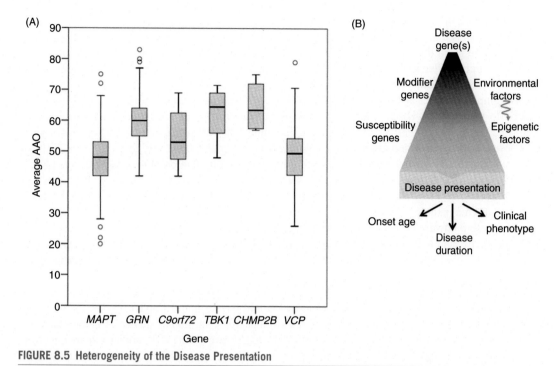

FIGURE 8.5 Heterogeneity of the Disease Presentation

(A) Box plot showing disease onset age distributions per gene. Family-based average onset ages of established pathogenic variants were used. *Boxes* represent the interquartile onset age distribution, *horizontal lines* indicate medians, *whiskers* show standard deviations, and *circles* indicate outliers. (B) Various factors affecting the expression of one genetic mutation, leading to diverse disease phenotypes. *AAO*, age at onset.

One approach to identify such factors in patients is by performing GWAS using genotype data. These studies enable the identification of loci that are associated with, for example, onset age or disease phenotype. This has already been done in, for instance, PD and Huntington's disease, where variations at several genomic loci were identified to modify the onset of disease [456,457]. Likewise, genome or exome data sets of large patient cohorts could be used to identify genetic modifiers, as illustrated by the successful application of this strategy to identify risk genes and pathways in ALS [89]. A family-based approach is possible as well, as heterogeneity is even observed within families in which one causal mutation cosegregates with disease [458]. In large families striking observations such as a high variability in onset age or the occurrence of different clinical phenotypes, have been made. For AD, modifier genes of age at onset have been identified in families segregating a causal presenilin 1 (*PSEN1*) mutation [459]. An example in FTD is the large Flemish *GRN* founder family in which a causal *GRN* mutation segregates with disease [48,53,59,113,460]. In this family the onset age ranges from 45 to 80 years and a quantitative trait locus for onset age was identified [301d,461]. Variability of the clinical phenotype is observed frequently in families segregating the G_4C_2 repeat expansion in *C9orf72*, with the manifestation of FTD, FTD–ALS, and ALS in one family (for an example, see Ref. [74]). Furthermore, in families segregating a causal *GRN* mutation, the observed phenotypes are variable as well, ranging from bvFTD to PPA, AD, and PD (for examples, see Refs. [59,462]). These observations point toward the existence of modifiers affecting the age at onset and clinical presentation of the disease. We have to keep in mind that it is possible that in certain pedigrees, (a part of) the observed clinical heterogeneity might be due to the presence of double-mutation carriers. With the advances of sequencing technologies, the identification of double-mutation carriers has increased. Patients carrying mutations in, for example, *C9orf72* and *GRN*, *MAPT*, *TARDBP*, or *TBK1* have been reported [101,463–466]. In addition to these patient-based studies, (genome-wide) modifier screens in model organisms are an often used approach to identify modifier genes. Examples are the earlier mentioned screens in yeast and *Drosophila* that identified modifiers of *C9orf72*-repeat associated toxicity [365,366,376,377]. Alternatively, the genetic variation that is present in model organisms can point toward modifiers [467].

Modifiers of onset age are targets that hold promise for disease-delaying therapies, as they alter biological processes in a way that delays or accelerates FTD pathogenesis. The identification of the modifier genes and the pathways by which the modification occurs will allow the development of therapies that alter these pathways. It is likely that multiple factors shape the disease phenotype, including genetic, epigenetic, and environmental modifiers (Fig. 8.5B). Although the effect of one single modifier, thus, could be rather limited, it is possible that pharmaceuticals that target the involved pathway have a stronger impact. Furthermore, by targeting the pathways in which the modifier acts, these therapies could have a broader applicability than only in those patients who carry a modifying allele [468].

The development of a disease-delaying therapy would have a major impact on both patients and their relatives and our society as a whole, and lead to a decrease in the prevalence of dementia [469], a longer healthy life for the patients, a reduction of the healthcare costs [470], and an expansion of the economically active population. Modeling of the effect on dementia prevalence indicated that an onset delay of 2 years will reduce the prevalence in 2040 with 16%, while delaying dementia onset by 5 years will reduce the prevalence by even 37% [469].

Disease onset and duration, as well as clinical phenotype have been the subject of genetic studies and several variations have demonstrated to modify disease in patients suffering from FTD and related neurodegenerative diseases (for examples, see Refs. [114,128,471]). The promise of therapeutically targeting modifier genes in neurodegenerative diseases was recently exemplified by the suggestion to target ataxin-2 (*ATXN2*) in ALS [472].

DISCUSSION: ESSENTIALS FOR THE DEVELOPMENT OF A DISEASE-MODIFYING THERAPY

GAPS IN OUR UNDERSTANDING OF FTD

Over the past few years, our knowledge about the molecular biology, genetics, and neuropathology of FTD has advanced remarkably. Yet a lot of questions are unanswered, such as the cause of disease in the around 60% genetically unexplained familial FTD patients, the disease mechanisms beyond the genetic defects, the risk factors, and disease modifiers. We do not know the influence of epigenetic and environmental factors on disease, and are searching for the reason for the selective vulnerability of the frontal and temporal lobes of the brain. Nonetheless, it is possible that we do not need to solve all these questions to find a disease-modifying therapy for a specific genetic subgroup of patients or FTD in general.

However, also some key questions related to the promising therapies discussed in this chapter are still left unanswered. A possible therapeutic strategy for GRN-related FTD could be the compensation of the loss of GRN by enhancing *GRN* expression or replacement therapy and investigations are being undertaken in that direction. The question remains whether it is necessary to upregulate full-length GRN or whether it is sufficient to enhance the expression of certain granulins to modify the disease? If the latter would be the case, we might speculate that *GRN* could be a good target for exon-skipping therapies. Granulins are encoded by two nonidentical exons [473]. Patients with causative frameshift mutations might be treated with ASOs, which mediate skipping of exons encoding the mutant GRN peptide and restore the reading frame. A phase 3 clinical trial of an ASO-based exon-skipping therapy for Duchenne muscular dystrophy is currently ongoing (Eteplirsen; Sarepta Therapeutics, Cambridge, Massachusetts, USA; clinicaltrials.gov identifier NCT02255552).

As both under- and overexpression of GRN are associated with disease (neurodegeneration and tumor growth, respectively), the upregulation of GRN might be difficult to control within the bounds of nonpathological levels. Conversely, inhibition of GRN expression was proposed as a target to combat cancers (for an example, see Ref. [474]). Also TBK1 inhibitors are considered as a promising therapy for inflammation and cancer [475]. Given the identification of loss-of-function mutations in both genes in FTD, FTD–ALS, and/or ALS, it might be important to keep the levels of GRN and TBK1 between certain well-tolerated boundaries.

For *C9orf72*-related FTD (–ALS), different mechanisms have been postulated by which the hexanucleotide repeat expansion could cause disease: reduced expression of the *C9orf72* gene and sense and antisense transcription and RAN translation of the expanded repeat, resulting in nuclear RNA foci and aggregated DPR neuropathology. However, the exact role of each of these mechanisms in the disease process remains unclear. Possibly a combination of proposed mechanisms contributes to the disease, while some observations might represent benign side effects or protective mechanisms. In the light of gain-of-toxic-function mechanisms, *C9orf72* hypermethylation can be regarded as a rescue mechanism to prevent the formation of RNA foci and DPR protein aggregates and might, therefore, be neuroprotective [343,344,476,477]. Further research into these mechanisms will be instrumental to pinpoint the optimal therapeutic approach. Recently a short–*C9orf72* repeat expansion of 66 units in a mouse model caused the appearance of RNA foci, TDP-43 and DPR pathology, neuronal loss, and behavioral deficits [478]. This and other *C9orf72* animal models, as well as patient derived induced (motor) neurons could prove valuable to further decipher the pathomechanisms and to test potential therapeutics, such as ASOs and small molecules.

The observation that the distribution of TDP-43 pathology correlates well with areas of neurodegeneration, in contrast to DPR pathology, argues that targeting more downstream mechanisms, such as inhibiting TDP-43 aggregation and toxicity, would also be a possible therapeutic intervention [369].

In tau-related FTD gain of toxic function and loss of functional tau are both described as disease mechanisms. Loss of function due to tau disengagement from MTs, MT destabilization, and axonal transport defects is targeted by therapeutics such as kinase inhibitors and MT stabilizers. However, a reduction of total tau levels is also considered as a therapeutic strategy. Reports on *MAPT* knockout mice suggest that the mice are mostly normal, although changes in axonal structure, MT organization, muscle strength, and other defects were reported, including the occurrence of mild parkinsonism characteristics later in life [479–485]. It is possible that during development, expression changes of other MT-associated proteins occurred in these mice to compensate for the loss of tau. The simultaneous consideration of tau loss of function as a disease mechanism and loss of tau as a therapy seems rather counterintuitive. Further elucidation of the role of tau in physiological conditions and in disease state is necessary.

An analogous duality can be appreciated for TDP-43 proteinopathy. Whether the nuclear loss of function and/or the cytosolic gain of (toxic?) aggregated TDP-43 is the culprit, is currently unknown.

A consideration for all proposed therapies and targets is the BBB permeability. This needs to be addressed either by using drugs crossing through the BBB or by applying techniques that bypass the BBB, such as intranasal drug delivery [486].

TOWARD EARLY AND TARGETED INTERVENTION

When disease-modifying therapies become available and a patient is diagnosed with FTD, the question is whether the neuronal loss has progressed too far for therapies to be effective? In fact, the disease process in FTD is ongoing for years before the first clinical symptoms become apparent, as seen in other neurodegenerative diseases, such as AD. Studies performed in presymptomatic carriers of a causal *GRN*, *MAPT*, or *C9orf72* mutation indicate that by the time the mutation carrier expresses clinical symptoms, irreversible changes are likely to have already been set in motion [487]. These preclinical changes in the brain demonstrate that accurate diagnosis is important, but it implies that we might even have to move to prediction and prevention. It suggests that it would be advisable to intervene early in the disease process, before the clinical presentation, with a disease-modifying therapy. This is exemplified by the existence of modifier variations that can postpone the onset of disease in patients with a causal mutation. This shows how—in this case a genetic—early intervention can modify disease.

To accomplish early detection and intervention, we need biomarkers to monitor disease onset, progression in the presymptomatic period, and staging of the disease process. Current diagnostic criteria for FTD incorporate clinical features, neuroimaging, neuropathology, and genetic testing. Distinct criteria are put forward for bvFTD and for PPA [5–7,9]. Differential diagnosis is often not straightforward and FTD is often initially misdiagnosed as a psychiatric problem, AD, or PD. Discriminating early stages of disease from normal aging by subtle deviations from baseline measures could point toward individuals who would benefit from presymptomatic therapies. There are, however, many considerations: who should be screened and when, how will these results be communicated, and what would be the ideal timing for therapeutic intervention to be effective? Furthermore, when individuals enter a trial when they are still presymptomatic, some people might develop symptoms 10 years later, while others might develop disease symptoms within a year. Therefore, it might be necessary to monitor the changes within participants to account for this type of heterogeneity. Determining which individuals to include is easier in the familial cases where a known genetic defect is detectable. Furthermore, for AD where large-scale GWAS advanced the knowledge of risk variations remarkably, genetic risk scores can be calculated [488]. In this way individuals at high risk who might benefit from certain therapeutics can be identified.

In the Alzheimer field, efforts for earlier intervention are exemplified by the planned clinical trial of an Aβ vaccine and a BACE inhibitor on healthy individuals homozygous for the *APOE ε4* allele (clinicaltrials.gov identifier NCT02565511), who have an increased risk to develop the disease.

The combination of genetic, CSF, blood, and imaging markers would not only be helpful in the identification of individuals at risk and for diagnostic purposes, but also to reduce the heterogeneity in clinical trials. Certain ongoing clinical trials already focus on a specific subgroup of FTD patients to perform targeted trials in more homogenous patient populations, such as patients with the same underlying genetic defect. However, apart from genetics, current biomarkers are poor predictors of underlying pathology. To be able to better define homogeneous patient groups, further research into the genetics of FTD and other biomarkers, such as neuroimaging and CSF or blood biomarkers is necessary. For novel disease-modifying therapies to be evaluated and applied in the clinic, we are also in need of progression biomarkers, to evaluate the efficiency of novel drugs over time. For example, progranulin levels in CSF might be a potential biomarker to monitor the effects of therapies [489], in addition to cognitive tests and neuroimaging. Recently, detection of poly(GP) DPRs in CSF was suggested as a potential biomarker [373].

CONCLUSIONS

FTD is a devastating disease for which currently only symptomatic treatments are available. With the increased understanding of the genetics and pathomechanisms involved in FTD, the focus shifts toward targeted disease-modifying therapies. Promising therapies are on their way, some still in the first stages of preclinical research, while others are ready to be tested in clinical trials. These advances, together with diagnostic progress, will be of paramount importance to be able to provide a disease-delaying or disease-modifying therapy to patients suffering from FTD in the future.

REFERENCES

[1] J.S. Snowden, D. Bathgate, A. Varma, A. Blackshaw, Z.C. Gibbons, D. Neary, Distinct behavioural profiles in frontotemporal dementia and semantic dementia, J. Neurol. Neurosurg. Psychiatry 70 (2001) 323–332.

[2] N.J. Cairns, M. Neumann, E.H. Bigio, et al. TDP-43 in familial and sporadic frontotemporal lobar degeneration with ubiquitin inclusions, Am. J. Pathol. 171 (2007) 227–240.

[3] D. Neary, J. Snowden, D. Mann, Frontotemporal dementia, Lancet Neurol. 4 (2005) 771–780.

[4] C.U. Onyike, J. Diehl-Schmid, The epidemiology of frontotemporal dementia, Int. Rev. Psychiatry 25 (2013) 130–137.

[5] K. Rascovsky, J.R. Hodges, D. Knopman, et al. Sensitivity of revised diagnostic criteria for the behavioural variant of frontotemporal dementia, Brain 134 (2011) 2456–2477.

[6] K. Rascovsky, J.R. Hodges, C.M. Kipps, et al. Diagnostic criteria for the behavioral variant of frontotemporal dementia (bvFTD): current limitations and future directions, Alzheimer Dis. Assoc. Disord. 21 (2007) S14–S18.

[7] K. Rascovsky, M. Grossman, Clinical diagnostic criteria and classification controversies in frontotemporal lobar degeneration, Int. Rev. Psychiatry 25 (2013) 145–158.

[8] G.D. Rabinovici, W.J. Jagust, A.J. Furst, et al. Abeta amyloid and glucose metabolism in three variants of primary progressive aphasia, Ann. Neurol. 64 (2008) 388–401.

[9] M.L. Gorno-Tempini, A.E. Hillis, S. Weintraub, et al. Classification of primary progressive aphasia and its variants, Neurology 76 (2011) 1006–1014.

[10] C. Lomen-Hoerth, T. Anderson, B. Miller, The overlap of amyotrophic lateral sclerosis and frontotemporal dementia, Neurology 59 (2002) 1077–1079.

[11] J.R. Burrell, M.C. Kiernan, S. Vucic, J.R. Hodges, Motor neuron dysfunction in frontotemporal dementia, Brain 134 (2011) 2582–2594.

[12] C. Lomen-Hoerth, J. Murphy, S. Langmore, J.H. Kramer, R.K. Olney, B. Miller, Are amyotrophic lateral sclerosis patients cognitively normal?, Neurology 60 (2003) 1094–1097.

[13] R.J. Caselli, A.J. Windebank, R.C. Petersen, et al. Rapidly progressive aphasic dementia and motor neuron disease, Ann. Neurol. 33 (1993) 200–207.

[14] Y. Mitsuyama, Presenile dementia with motor neuron disease in Japan: clinico-pathological review of 26 cases, J. Neurol. Neurosurg. Psychiatry 47 (1984) 953–959.

[15] G.M. Ringholz, S.H. Appel, M. Bradshaw, N.A. Cooke, D.M. Mosnik, P.E. Schulz, Prevalence and patterns of cognitive impairment in sporadic ALS, Neurology 65 (2005) 586–590.

[16] J. Siuda, S. Fujioka, Z.K. Wszolek, Parkinsonian syndrome in familial frontotemporal dementia, Parkinsonism Relat. Disord. 20 (2014) 957–964.

[17] K.L. Donker, A.J. Boon, W. Kamphorst, R. Ravid, H.J. Duivenvoorden, J.C. Van Swieten, Frontal presentation in progressive supranuclear palsy, Neurology 69 (2007) 723–729.

[18] U. Nath, Y. Ben-Shlomo, R.G. Thomson, A.J. Lees, D.J. Burn, Clinical features and natural history of progressive supranuclear palsy: a clinical cohort study, Neurology 60 (2003) 910–916.

[19] A. Kertesz, P. McMonagle, Behavior and cognition in corticobasal degeneration and progressive supranuclear palsy, J. Neurol. Sci. 289 (2010) 138–143.

[20] I. Ferrer, G. Santpere, F.W. van Leeuwen, Argyrophilic grain disease, Brain 131 (2008) 1416–1432.

[21] V. Kimonis, S. Donkervoort, G. Watts, Inclusion body myopathy with Paget disease of bone and/or frontotemporal dementia, (2011). Available from: https://www.ncbi.nlm.nih.gov/books/NBK1476/

[22] J.R. Hodges, R. Davies, J. Xuereb, J. Kril, G. Halliday, Survival in frontotemporal dementia, Neurology 61 (2003) 349–354.

[23] S.M. Rosso, K.L. Donker, T. Baks, et al. Frontotemporal dementia in The Netherlands: patient characteristics and prevalence estimates from a population-based study, Brain 126 (2003) 2016–2022.

[24] J.K. Johnson, J. Diehl, M.F. Mendez, et al. Frontotemporal lobar degeneration: demographic characteristics of 353 patients, Arch. Neurol. 62 (2005) 925–930.

[25] S. Nunnemann, D. Last, T. Schuster, H. Forstl, A. Kurz, J. Diehl-Schmid, Survival in a German population with frontotemporal lobar degeneration, Neuroepidemiology 37 (2011) 160–165.

[26] E.D. Roberson, J.H. Hesse, K.D. Rose, et al. Frontotemporal dementia progresses to death faster than Alzheimer disease, Neurology 65 (2005) 719–725.

[27] R.K. Olney, J. Murphy, D. Forshew, et al. The effects of executive and behavioral dysfunction on the course of ALS, Neurology 65 (2005) 1774–1777.

[28] J. Diehl-Schmid, C. Pohl, R. Perneczky, J. Hartmann, H. Forstl, A. Kurz, Initial symptoms, survival and causes of death in 115 patients with frontotemporal lobar degeneration, Fortschr. Neurol. Psychiatr. 75 (2007) 708–713.

[29] M. Neumann, D.M. Sampathu, L.K. Kwong, et al. Ubiquitinated TDP-43 in frontotemporal lobar degeneration and amyotrophic lateral sclerosis, Science 314 (2006) 130–133.

[30] I.R. Mackenzie, M. Neumann, E.H. Bigio, et al. Nomenclature and nosology for neuropathologic subtypes of frontotemporal lobar degeneration: an update, Acta Neuropathol. 119 (2010) 1–4.

[31] M. Neumann, R. Rademakers, S. Roeber, M. Baker, H.A. Kretzschmar, I.R. Mackenzie, A new subtype of frontotemporal lobar degeneration with FUS pathology, Brain 132 (2009) 2922–2931.

[32] T. Lashley, J.D. Rohrer, S. Mead, T. Revesz, Review: an update on clinical, genetic and pathological aspects of frontotemporal lobar degenerations, Neuropathol. Appl. Neurobiol. 41 (2015) 858–881.

[33] D.W. Dickson, N. Kouri, M.E. Murray, K.A. Josephs, Neuropathology of frontotemporal lobar degeneration-tau (FTLD-tau), J. Mol. Neurosci. 45 (2011) 384–389.

[34] I.R. Mackenzie, M. Neumann, A. Baborie, et al. A harmonized classification system for FTLD-TDP pathology, Acta Neuropathol. 122 (2011) 111–113.

[35] Deleted in review.

[36] K. Mori, S.M. Weng, T. Arzberger, et al. The C9orf72 GGGGCC repeat is translated into aggregating dipeptide-repeat proteins in FTLD/ALS, Science 339 (2013) 1335–1338.

[37] P.E. Ash, K.F. Bieniek, T.F. Gendron, et al. Unconventional translation of C9ORF72 GGGGCC expansion generates insoluble polypeptides specific to c9FTD/ALS, Neuron 77 (2013) 639–646.

[38] T. Bird, D. Knopman, J. VanSwieten, et al. Epidemiology and genetics of frontotemporal dementia/Pick's disease, Ann. Neurol. 54 (Suppl. 5) (2003) S29–S31.

[39] J.S. Goldman, J.M. Farmer, E.M. Wood, et al. Comparison of family histories in FTLD subtypes and related tauopathies, Neurology 65 (2005) 1817–1819.

[40] A. Sieben, L.T. Van, S. Engelborghs, et al. The genetics and neuropathology of frontotemporal lobar degeneration, Acta Neuropathol. 124 (2012) 353–372.

[41] I. Gijselinck, M.S. Van, J. van der Zee, et al. Loss of TBK1 is a frequent cause of frontotemporal dementia in a Belgian cohort, Neurology 85 (2015) 2116–2125.

[42] N.L. Foster, K. Wilhelmsen, A.A. Sima, M.Z. Jones, C.J. D'Amato, S. Gilman, Frontotemporal dementia and parkinsonism linked to chromosome 17: a consensus. Conference Participants, Ann. Neurol. 41 (1997) 706–715.

[43] M. Hutton, C.L. Lendon, P. Rizzu, et al. Association of missense and 5'-splice-site mutations in tau with the inherited dementia FTDP-17, Nature 393 (1998) 702–705.

[44] P. Poorkaj, T.D. Bird, E. Wijsman, et al. Tau is a candidate gene for chromosome 17 frontotemporal dementia, Ann. Neurol. 43 (1998) 815–825.

[45] M.G. Spillantini, J.R. Murrell, M. Goedert, M.R. Farlow, A. Klug, B. Ghetti, Mutation in the tau gene in familial multiple system tauopathy with presenile dementia, Proc. Natl. Acad. Sci. USA 95 (1998) 7737–7741.

[46] B. Ghetti, A.L. Oblak, B.F. Boeve, K.A. Johnson, B.C. Dickerson, M. Goedert, Invited review: frontotemporal dementia caused by microtubule-associated protein tau gene (MAPT) mutations: a chameleon for neuropathology and neuroimaging, Neuropathol. Appl. Neurobiol. 41 (2015) 24–46.

[47] G. Rossi, F. Tagliavini, Frontotemporal lobar degeneration: old knowledge and new insight into the pathogenetic mechanisms of tau mutations, Front. Aging Neurosci. 7 (2015) 192.

[48] M. Cruts, I. Gijselinck, J. van der Zee, et al. Null mutations in progranulin cause ubiquitin-positive frontotemporal dementia linked to chromosome 17q21, Nature 442 (2006) 920–924.

[49] M. Baker, I.R. Mackenzie, S.M. Pickering-Brown, et al. Mutations in progranulin cause tau-negative frontotemporal dementia linked to chromosome 17, Nature 442 (2006) 916–919.

[50] G. Kleinberger, A. Capell, C. Haass, B.C. Van, Mechanisms of granulin deficiency: lessons from cellular and animal models, Mol. Neurobiol. 47 (2013) 337–360.

[51] L. De Muynck, D.P. Van, Cellular effects of progranulin in health and disease, J. Mol. Neurosci. 45 (2011) 549–560.

[52] P. Van Damme, H.A. Van, D. Lambrechts, et al. Progranulin functions as a neurotrophic factor to regulate neurite outgrowth and enhance neuronal survival, J. Cell. Biol. 181 (2008) 37–41.

[53] K. Sleegers, N. Brouwers, D.P. Van, et al. Serum biomarker for progranulin-associated frontotemporal lobar degeneration, Ann. Neurol. 65 (2009) 603–609.

[54] R. Ghidoni, L. Benussi, M. Glionna, M. Franzoni, G. Binetti, Low plasma progranulin levels predict progranulin mutations in frontotemporal lobar degeneration, Neurology 71 (2008) 1235–1239.

[55] S.S. Shankaran, A. Capell, A.T. Hruscha, et al. Missense mutations in the progranulin gene linked to frontotemporal lobar degeneration with ubiquitin-immunoreactive inclusions reduce progranulin production and secretion, J. Biol. Chem. 283 (2008) 1744–1753.

[56] J. Wang, D.P. Van, C. Cruchaga, et al. Pathogenic cysteine mutations affect progranulin function and production of mature granulins, J. Neurochem. 112 (2010) 1305–1315.

[57] R. Rademakers, M. Baker, J. Gass, et al. Phenotypic variability associated with progranulin haploinsufficiency in patients with the common 1477C-->T (Arg493X) mutation: an international initiative, Lancet Neurol. 6 (2007) 857–868.

[58] J.C. Van Swieten, P. Heutink, Mutations in progranulin (GRN) within the spectrum of clinical and pathological phenotypes of frontotemporal dementia, Lancet Neurol. 7 (2008) 965–974.

[59] N. Brouwers, K. Nuytemans, J. van der Zee, et al. Alzheimer and Parkinson diagnoses in progranulin null mutation carriers in an extended founder family, Arch. Neurol. 64 (2007) 1436–1446.

[60] A.S. Chen-Plotkin, M. Martinez-Lage, P.M. Sleiman, et al. Genetic and clinical features of progranulin-associated frontotemporal lobar degeneration, Arch. Neurol. 68 (2011) 488–497.

[61] K.A. Josephs, Z. Ahmed, O. Katsuse, et al. Neuropathologic features of frontotemporal lobar degeneration with ubiquitin-positive inclusions with progranulin gene (PGRN) mutations, J. Neuropathol. Exp. Neurol. 66 (2007) 142–151.

[62] I.R. Mackenzie, M. Baker, S. Pickering-Brown, et al. The neuropathology of frontotemporal lobar degeneration caused by mutations in the progranulin gene, Brain 129 (2006) 3081–3090.

[63] I.R. Mackenzie, The neuropathology and clinical phenotype of FTD with progranulin mutations, Acta Neuropathol. 114 (2007) 49–54.

[64] A.L. Boxer, I.R. Mackenzie, B.F. Boeve, et al. Clinical, neuroimaging and neuropathological features of a new chromosome 9p-linked FTD-ALS family, J. Neurol. Neurosurg. Psychiatry 82 (2011) 196–203.

[65] I. Gijselinck, S. Engelborghs, G. Maes, et al. Identification of 2 Loci at chromosomes 9 and 14 in a multiplex family with frontotemporal lobar degeneration and amyotrophic lateral sclerosis, Arch. Neurol. 67 (2010) 606–616.

[66] I. Le Ber, A. Camuzat, E. Berger, et al. Chromosome 9p-linked families with frontotemporal dementia associated with motor neuron disease, Neurology 72 (2009) 1669–1676.

[67] A.A. Luty, J.B. Kwok, E.M. Thompson, et al. Pedigree with frontotemporal lobar degeneration--motor neuron disease and Tar DNA binding protein-43 positive neuropathology: genetic linkage to chromosome 9, BMC Neurol. 8 (2008) 32.

[68] M. Morita, A. Al-Chalabi, P.M. Andersen, et al. A locus on chromosome 9p confers susceptibility to ALS and frontotemporal dementia, Neurology 66 (2006) 839–844.

[69] P.N. Valdmanis, N. Dupre, J.P. Bouchard, et al. Three families with amyotrophic lateral sclerosis and frontotemporal dementia with evidence of linkage to chromosome 9p, Arch. Neurol. 64 (2007) 240–245.

[70] C. Vance, A. Al-Chalabi, D. Ruddy, et al. Familial amyotrophic lateral sclerosis with frontotemporal dementia is linked to a locus on chromosome 9p13.2-21.3, Brain 129 (2006) 868–876.

[71] H. Laaksovirta, T. Peuralinna, J.C. Schymick, et al. Chromosome 9p21 in amyotrophic lateral sclerosis in Finland: a genome-wide association study, Lancet Neurol. 9 (2010) 978–985.

[72] A. Shatunov, K. Mok, S. Newhouse, et al. Chromosome 9p21 in sporadic amyotrophic lateral sclerosis in the UK and seven other countries: a genome-wide association study, Lancet Neurol. 9 (2010) 986–994.

[73] M.A. van Es, J.H. Veldink, C.G. Saris, et al. Genome-wide association study identifies 19p13.3 (UNC13A) and 9p21.2 as susceptibility loci for sporadic amyotrophic lateral sclerosis, Nat. Genet. 41 (2009) 1083–1087.

[74] I. Gijselinck, T. Van Langenhove, J. van der Zee, et al. A C9orf72 promoter repeat expansion in a Flanders-Belgian cohort with disorders of the frontotemporal lobar degeneration-amyotrophic lateral sclerosis spectrum: a gene identification study, Lancet Neurol. 11 (2012) 54–65.

[75] A.E. Renton, E. Majounie, A. Waite, et al. A hexanucleotide repeat expansion in C9ORF72 is the cause of chromosome 9p21-linked ALS-FTD, Neuron 72 (2011) 257–268.

[76] M. Dejesus-Hernandez, I.R. Mackenzie, B.F. Boeve, et al. Expanded GGGGCC hexanucleotide repeat in noncoding region of C9ORF72 causes chromosome 9p-linked FTD and ALS, Neuron 72 (2011) 245–256.

[77] J. Beck, M. Poulter, D. Hensman, et al. Large C9orf72 hexanucleotide repeat expansions are seen in multiple neurodegenerative syndromes and are more frequent than expected in the UK population, Am. J. Hum. Genet. 92 (2013) 345–353.

[78] C. Dobson-Stone, M. Hallupp, C.T. Loy, et al. C9ORF72 repeat expansion in Australian and Spanish frontotemporal dementia patients, PLoS One 8 (2013) e56899.

[79] H. Ishiura, Y. Takahashi, J. Mitsui, et al. C9ORF72 repeat expansion in amyotrophic lateral sclerosis in the Kii peninsula of Japan, Arch. Neurol. 69 (2012) 1154–1158.

[80] V.L. Buchman, J. Cooper-Knock, N. Connor-Robson, et al. Simultaneous and independent detection of C9ORF72 alleles with low and high number of GGGGCC repeats using an optimised protocol of Southern blot hybridisation, Mol. Neurodegener. 8 (2013) 12.

[81] M. van Blitterswijk, M. Dejesus-Hernandez, E. Niemantsverdriet, et al. Association between repeat sizes and clinical and pathological characteristics in carriers of C9ORF72 repeat expansions (Xpansize-72): a cross-sectional cohort study, Lancet Neurol. 12 (2013) 978–988.

[82] O. Dols-Icardo, A. Garcia-Redondo, R. Rojas-Garcia, et al. Characterization of the repeat expansion size in C9orf72 in amyotrophic lateral sclerosis and frontotemporal dementia, Hum. Mol. Genet. 23 (2013) 749–754.

[83] A. Hubers, N. Marroquin, B. Schmoll, et al. Polymerase chain reaction and Southern blot-based analysis of the C9orf72 hexanucleotide repeat in different motor neuron diseases, Neurobiol. Aging 35 (2014) 1214–1216.

[84] A.J. Waite, D. Baumer, S. East, et al. Reduced C9orf72 protein levels in frontal cortex of amyotrophic lateral sclerosis and frontotemporal degeneration brain with the C9ORF72 hexanucleotide repeat expansion, Neurobiol. Aging 35 (2014) 1779.

[85] I. Gijselinck, S. Van Mossevelde, J. van der Zee, et al. The C9orf72 repeat size correlates with onset age of disease, DNA methylation and transcriptional downregulation of the promoter, Mol. Psychiatry 21 (2016) 1112–1124.

[86] E. Majounie, A.E. Renton, K. Mok, et al. Frequency of the C9orf72 hexanucleotide repeat expansion in patients with amyotrophic lateral sclerosis and frontotemporal dementia: a cross-sectional study, Lancet Neurol. 11 (2012) 323–330.

[87] J. van der Zee, I. Gijselinck, L. Dillen, et al. A pan-European study of the C9orf72 repeat associated with FTLD: geographic prevalence, genomic instability, and intermediate repeats, Hum. Mutat. 34 (2013) 363–373.

[88] M.A. Farg, V. Sundaramoorthy, J.M. Sultana, et al. C9ORF72, implicated in amytrophic lateral sclerosis and frontotemporal dementia, regulates endosomal trafficking, Hum. Mol. Genet. 23 (2014) 3579–3595.

[89] E.T. Cirulli, B.N. Lasseigne, S. Petrovski, et al. Exome sequencing in amyotrophic lateral sclerosis identifies risk genes and pathways, Science 347 (2015) 1436–1441.

[90] A. Freischmidt, T. Wieland, B. Richter, et al. Haploinsufficiency of TBK1 causes familial ALS and frontotemporal dementia, Nat. Neurosci. 18 (2015) 631–636.

[91] C. Pottier, K.F. Bieniek, N. Finch, et al. Whole-genome sequencing reveals important role for TBK1 and OPTN mutations in frontotemporal lobar degeneration without motor neuron disease, Acta Neuropathol. 130 (2015) 77–92.

[92] J.F. Clement, S. Meloche, M.J. Servant, The IKK-related kinases: from innate immunity to oncogenesis, Cell Res. 18 (2008) 889–899.

[93] M. Pilli, J. Arko-Mensah, M. Ponpuak, et al. TBK-1 promotes autophagy-mediated antimicrobial defense by controlling autophagosome maturation, Immunity 37 (2012) 223–234.

[94] T.L. Thurston, G. Ryzhakov, S. Bloor, M.N. von, F. Randow, The TBK1 adaptor and autophagy receptor NDP52 restricts the proliferation of ubiquitin-coated bacteria, Nat. Immunol. 10 (2009) 1215–1221.

[95] D.A. Barbie, P. Tamayo, J.S. Boehm, et al. Systematic RNA interference reveals that oncogenic KRAS-driven cancers require TBK1, Nature 462 (2009) 108–112.

[96] H. Weidberg, Z. Elazar, TBK1 mediates crosstalk between the innate immune response and autophagy, Sci. Signal 4 (2011) e39.

[97] C.E. Gleason, A. Ordureau, R. Gourlay, J.S. Arthur, P. Cohen, Polyubiquitin binding to optineurin is required for optimal activation of TANK-binding kinase 1 and production of interferon beta, J. Biol. Chem. 286 (2011) 35663–35674.

[98] H. Maruyama, H. Morino, H. Ito, et al. Mutations of optineurin in amyotrophic lateral sclerosis, Nature 465 (2010) 223–226.

[99] E. Rubino, I. Rainero, A. Chio, et al. SQSTM1 mutations in frontotemporal lobar degeneration and amyotrophic lateral sclerosis, Neurology 79 (2012) 1556–1562.

[100] F. Fecto, J. Yan, S.P. Vemula, et al. SQSTM1 mutations in familial and sporadic amyotrophic lateral sclerosis, Arch. Neurol. 68 (2011) 1440–1446.

[101] I. Gijselinck, S. Van Mossevelde, J. van der Zee, et al. Loss of TBK1 is a frequent cause of frontotemporal dementia in a Belgian cohort, Neurology 85 (2015) 2116–2125.

[102] C.C. Weihl, Valosin containing protein associated fronto-temporal lobar degeneration: clinical presentation, pathologic features and pathogenesis, Curr. Alzheimer Res. 8 (2011) 252–260.

[103] C. Lagier-Tourenne, M. Polymenidou, D.W. Cleveland, TDP-43 and FUS/TLS: emerging roles in RNA processing and neurodegeneration, Hum. Mol. Genet. 19 (2010) R46–R64.

[104] B. Borroni, S. Archetti, B.R. Del, et al. TARDBP mutations in frontotemporal lobar degeneration: frequency, clinical features, and disease course, Rejuvenation Res. 13 (2010) 509–517.

[105] I.R. Mackenzie, R. Rademakers, M. Neumann, TDP-43 and FUS in amyotrophic lateral sclerosis and frontotemporal dementia, Lancet Neurol. 9 (2010) 995–1007.

[106] T. Van Langenhove, J. van der Zee, K. Sleegers, et al. Genetic contribution of FUS to frontotemporal lobar degeneration, Neurology 74 (2010) 366–371.

[107] J. van der Zee, T. Van Langenhove, G.G. Kovacs, et al. Rare mutations in SQSTM1 modify susceptibility to frontotemporal lobar degeneration, Acta Neuropathol. 128 (2014) 397–410.

[108] I. Le Ber, A. Camuzat, R. Guerreiro, et al. SQSTM1 mutations in French patients with frontotemporal dementia or frontotemporal dementia with amyotrophic lateral sclerosis, JAMA Neurol. 70 (2013) 1403–1410.

[109] R.J. Guerreiro, E. Lohmann, J.M. Bras, et al. Using exome sequencing to reveal mutations in TREM2 presenting as a frontotemporal dementia-like syndrome without bone involvement, JAMA Neurol. 70 (2013) 78–84.

[110] V.M. Van Deerlin, P.M. Sleiman, M. Martinez-Lage, et al. Common variants at 7p21 are associated with frontotemporal lobar degeneration with TDP-43 inclusions, Nat. Genet. 42 (2010) 234–239.

[111] C. Cruchaga, C. Graff, H.H. Chiang, et al. Association of TMEM106B gene polymorphism with age at onset in granulin mutation carriers and plasma granulin protein levels, Arch Neurol. 68 (2011) 581–586.

[112] N. Finch, M.M. Carrasquillo, M. Baker, et al. TMEM106B regulates progranulin levels and the penetrance of FTLD in GRN mutation carriers, Neurology 76 (2011) 467–474.

[113] J. van der Zee, L.T. Van, G. Kleinberger, et al. TMEM106B is associated with frontotemporal lobar degeneration in a clinically diagnosed patient cohort, Brain 134 (2011) 808–815.

[114] M. van Blitterswijk, B. Mullen, A.M. Nicholson, et al. TMEM106B protects C9ORF72 expansion carriers against frontotemporal dementia, Acta Neuropathol. 127 (2014) 397–406.

[115] M.D. Gallagher, E. Suh, M. Grossman, et al. TMEM106B is a genetic modifier of frontotemporal lobar degeneration with C9orf72 hexanucleotide repeat expansions, Acta Neuropathol. 127 (2014) 407–418.

[116] R. Ferrari, D.G. Hernandez, M.A. Nalls, et al. Frontotemporal dementia and its subtypes: a genome-wide association study, Lancet Neurol. 13 (2014) 686–699.

[117] E. Di Maria, M. Tabaton, T. Vigo, et al. Corticobasal degeneration shares a common genetic background with progressive supranuclear palsy, Ann. Neurol. 47 (2000) 374–377.

[118] H. Houlden, M. Baker, H.R. Morris, et al. Corticobasal degeneration and progressive supranuclear palsy share a common tau haplotype, Neurology 56 (2001) 1702–1706.

[119] M. Baker, I. Litvan, H. Houlden, et al. Association of an extended haplotype in the tau gene with progressive supranuclear palsy, Hum. Mol. Genet. 8 (1999) 711–715.

[120] R. de Silva, M. Weiler, H.R. Morris, E.R. Martin, N.W. Wood, A.J. Lees, Strong association of a novel Tau promoter haplotype in progressive supranuclear palsy, Neurosci. Lett. 311 (2001) 145–148.

[121] P. Verpillat, A. Camuzat, D. Hannequin, et al. Association between the extended tau haplotype and frontotemporal dementia, Arch. Neurol. 59 (2002) 935–939.

[122] A. Hughes, D. Mann, S. Pickering-Brown, Tau haplotype frequency in frontotemporal lobar degeneration and amyotrophic lateral sclerosis, Exp. Neurol. 181 (2003) 12–16.

[123] D.G. Healy, P.M. Abou-Sleiman, A.J. Lees, et al. Tau gene and Parkinson's disease: a case-control study and meta-analysis, J. Neurol. Neurosurg. Psychiatry 75 (2004) 962–965.

[124] R. Ghidoni, S. Signorini, L. Barbiero, et al. The H2 MAPT haplotype is associated with familial frontotemporal dementia, Neurobiol. Dis. 22 (2006) 357–362.

[125] B. Borroni, D. Yancopoulou, M. Tsutsui, et al. Association between tau H2 haplotype and age at onset in frontotemporal dementia, Arch. Neurol. 62 (2005) 1419–1422.

[126] R. Rademakers, J.L. Eriksen, M. Baker, et al. Common variation in the miR-659 binding-site of GRN is a major risk factor for TDP43-positive frontotemporal dementia, Hum. Mol. Genet. 17 (2008) 3631–3642.

[127] S. Rollinson, J.D. Rohrer, J. van der Zee, et al. No association of PGRN 3′UTR rs5848 in frontotemporal lobar degeneration, Neurobiol. Aging 32 (2011) 754–755.

[128] M. van Blitterswijk, B. Mullen, A. Wojtas, et al. Genetic modifiers in carriers of repeat expansions in the C9ORF72 gene, Mol. Neurodegener. 9 (2014) 38.

[129] M. Thelen, C. Razquin, I. Hernandez, et al. Investigation of the role of rare TREM2 variants in frontotemporal dementia subtypes, Neurobiol. Aging 35 (2014) 2657–2659.

[130] B. Borroni, F. Ferrari, D. Galimberti, et al. Heterozygous TREM2 mutations in frontotemporal dementia, Neurobiol. Aging 35 (2014) 934.e7–934.e10.

[131] E. Cuyvers, K. Bettens, S. Philtjens, et al. Investigating the role of rare heterozygous TREM2 variants in Alzheimer's disease and frontotemporal dementia, Neurobiol. Aging 35 (2014) 726–729.

[132] S. Rayaprolu, B. Mullen, M. Baker, et al. TREM2 in neurodegeneration: evidence for association of the p.R47H variant with frontotemporal dementia and Parkinson's disease, Mol. Neurodegener. 8 (2013) 19.

[133] C.M. Lill, A. Rengmark, L. Pihlstrom, et al. The role of TREM2 R47H as a risk factor for Alzheimer's disease, frontotemporal lobar degeneration, amyotrophic lateral sclerosis, and Parkinson's disease, Alzheimers Dement. 11 (2015) 1407–1416.

[134] C.F. Slattery, J.A. Beck, L. Harper, et al. R47H TREM2 variant increases risk of typical early-onset Alzheimer's disease but not of prion or frontotemporal dementia, Alzheimers Dement. 10 (2014) 602–608.

[135] S.M. Rosso, E.J. Landweer, M. Houterman, K.L. Donker, C.M. van Duijn, J.C. Van Swieten, Medical and environmental risk factors for sporadic frontotemporal dementia: a retrospective case-control study, J. Neurol. Neurosurg. Psychiatry 74 (2003) 1574–1576.

[136] A. Golimstok, N. Campora, J.I. Rojas, et al. Cardiovascular risk factors and frontotemporal dementia: a case-control study, Transl. Neurodegener. 3 (2014) 13.

[137] T. Arai, I.R. Mackenzie, M. Hasegawa, et al. Phosphorylated TDP-43 in Alzheimer's disease and dementia with Lewy bodies, Acta Neuropathol. 117 (2009) 125–136.

[138] K.A. Jellinger, Recent advances in our understanding of neurodegeneration, J. Neural. Transm. 116 (2009) 1111–1162.

[139] N. Herrmann, S.E. Black, T. Chow, J. Cappell, D.F. Tang-Wai, K.L. Lanctot, Serotonergic function and treatment of behavioral and psychological symptoms of frontotemporal dementia, Am. J. Geriatr. Psychiatry 20 (2012) 789–797.

[140] M. Ikeda, K. Shigenobu, R. Fukuhara, et al. Efficacy of fluvoxamine as a treatment for behavioral symptoms in frontotemporal lobar degeneration patients, Dement. Geriatr. Cogn. Disord. 17 (2004) 117–121.

[141] J.R. Swartz, B.L. Miller, I.M. Lesser, A.L. Darby, Frontotemporal dementia: treatment response to serotonin selective reuptake inhibitors, J. Clin. Psychiatry 58 (1997) 212–216.

[142] M.F. Mendez, J.S. Shapira, B.L. Miller, Stereotypical movements and frontotemporal dementia, Mov. Disord. 20 (2005) 742–745.

[143] J.M. Anneser, R.J. Jox, G.D. Borasio, Inappropriate sexual behaviour in a case of ALS and FTD: successful treatment with sertraline, Amyotroph. Lateral Scler. 8 (2007) 189–190.

[144] C.I. Prodan, M. Monnot, E.D. Ross, Behavioural abnormalities associated with rapid deterioration of language functions in semantic dementia respond to sertraline, J. Neurol. Neurosurg. Psychiatry 80 (2009) 1416–1417.

[145] R. Moretti, P. Torre, R.M. Antonello, G. Cazzato, A. Bava, Frontotemporal dementia: paroxetine as a possible treatment of behavior symptoms. A randomized, controlled, open 14-month study, Eur. Neurol. 49 (2003) 13–19.

[146] F. Lebert, W. Stekke, C. Hasenbroekx, F. Pasquier, Frontotemporal dementia: a randomised, controlled trial with trazodone, Dement. Geriatr. Cogn. Disord. 17 (2004) 355–359.

[147] J.B. Deakin, S. Rahman, P.J. Nestor, J.R. Hodges, B.J. Sahakian, Paroxetine does not improve symptoms and impairs cognition in frontotemporal dementia: a double-blind randomized controlled trial, Psychopharmacology 172 (2004) 400–408.

[148] L.E. Hughes, T. Rittman, R. Regenthal, T.W. Robbins, J.B. Rowe, Improving response inhibition systems in frontotemporal dementia with citalopram, Brain 138 (2015) 1961–1975.

[149] G.D. Rabinovici, B.L. Miller, Frontotemporal lobar degeneration: epidemiology, pathophysiology, diagnosis and management, CNS Drugs 24 (2010) 375–398.

[150] S. Rahman, T.W. Robbins, J.R. Hodges, et al. Methylphenidate ('Ritalin') can ameliorate abnormal risk-taking behavior in the frontal variant of frontotemporal dementia, Neuropsychopharmacology 31 (2006) 651–658.

[151] H.W. Goforth, L. Konopka, M. Primeau, et al. Quantitative electroencephalography in frontotemporal dementia with methylphenidate response: a case study, Clin. EEG Neurosci. 35 (2004) 108–111.

[152] E.D. Huey, K.T. Putnam, J. Grafman, A systematic review of neurotransmitter deficits and treatments in frontotemporal dementia, Neurology 66 (2006) 17–22.

[153] D.S. Knopman, J.H. Kramer, B.F. Boeve, et al. Development of methodology for conducting clinical trials in frontotemporal lobar degeneration, Brain 131 (2008) 2957–2968.

[154] R. Moretti, P. Torre, R.M. Antonello, G. Cazzato, S. Griggio, A. Bava, Olanzapine as a treatment of neuropsychiatric disorders of Alzheimer's disease and other dementias: a 24-month follow-up of 68 patients, Am. J. Alzheimers Dis. Other Demen. 18 (2003) 205–214.

[155] A. Fellgiebel, M.J. Muller, C. Hiemke, P. Bartenstein, M. Schreckenberger, Clinical improvement in a case of frontotemporal dementia under aripiprazole treatment corresponds to partial recovery of disturbed frontal glucose metabolism, World J. Biol. Psychiatry 8 (2007) 123–126.

[156] R.R. Reeves, C.L. Perry, Aripiprazole for sexually inappropriate vocalizations in frontotemporal dementia, J. Clin. Psychopharmacol. 33 (2013) 145–146.

[157] E.D. Huey, C. Garcia, E.M. Wassermann, M.C. Tierney, J. Grafman, Stimulant treatment of frontotemporal dementia in 8 patients, J. Clin. Psychiatry 69 (2008) 1981–1982.

[158] Y.A. Pijnenburg, E.L. Sampson, R.J. Harvey, N.C. Fox, M.N. Rossor, Vulnerability to neuroleptic side effects in frontotemporal lobar degeneration, Int. J. Geriatr. Psychiatry 18 (2003) 67–72.

[159] K. Czarnecki, N. Kumar, K.A. Josephs, Parkinsonism and tardive antecollis in frontotemporal dementia—increased sensitivity to newer antipsychotics?, Eur. J. Neurol. 15 (2008) 199–201.

[160] L.S. Schneider, K.S. Dagerman, P. Insel, Risk of death with atypical antipsychotic drug treatment for dementia: meta-analysis of randomized placebo-controlled trials, JAMA 294 (2005) 1934–1943.

[161] R.M. Tsai, A.L. Boxer, Treatment of frontotemporal dementia, Curr. Treat. Options Neurol. 16 (2014) 319.

[162] Y. Lampl, M. Sadeh, M. Lorberboym, Efficacy of acetylcholinesterase inhibitors in frontotemporal dementia, Ann. Pharmacother. 38 (2004) 1967–1968.

[163] R. Moretti, P. Torre, R.M. Antonello, T. Cattaruzza, G. Cazzato, A. Bava, Rivastigmine in frontotemporal dementia: an open-label study, Drugs Aging 21 (2004) 931–937.

[164] A. Kertesz, D. Morlog, M. Light, et al. Galantamine in frontotemporal dementia and primary progressive aphasia, Dement. Geriatr. Cogn. Disord. 25 (2008) 178–185.

[165] M.F. Mendez, J.S. Shapira, A. McMurtray, E. Licht, Preliminary findings: behavioral worsening on donepezil in patients with frontotemporal dementia, Am. J. Geriatr. Psychiatry 15 (2007) 84–87.

[166] T. Kimura, J. Takatsu, Pilot study of pharmacological treatment for frontotemporal dementia: risk of donepezil treatment for behavioral and psychological symptoms, Geriatr. Gerontol. Int. 13 (2013) 506–507.

[167] M.M. Swanberg, Memantine for behavioral disturbances in frontotemporal dementia: a case series, Alzheimer Dis. Assoc. Disord. 21 (2007) 164–166.

[168] J. Diehl-Schmid, H. Forstl, R. Perneczky, C. Pohl, A. Kurz, A 6-month, open-label study of memantine in patients with frontotemporal dementia, Int. J. Geriatr. Psychiatry 23 (2008) 754–759.

[169] A.L. Boxer, A.M. Lipton, K. Womack, et al. An open-label study of memantine treatment in 3 subtypes of frontotemporal lobar degeneration, Alzheimer Dis. Assoc. Disord. 23 (2009) 211–217.

[170] A.L. Boxer, D.S. Knopman, D.I. Kaufer, et al. Memantine in patients with frontotemporal lobar degeneration: a multicentre, randomised, double-blind, placebo-controlled trial, Lancet Neurol. 12 (2013) 149–156.

[171] M. Vercelletto, C. Boutoleau-Bretonniere, C. Volteau, et al. Memantine in behavioral variant frontotemporal dementia: negative results, J. Alzheimers Dis. 23 (2011) 749–759.

[172] G.A. Kerchner, M.C. Tartaglia, A. Boxer, Abhorring the vacuum: use of Alzheimer's disease medications in frontotemporal dementia, Expert Rev. Neurother. 11 (2011) 709–717.

[173] J.L. Cummings, C.G. Lyketsos, E.R. Peskind, et al. Effect of dextromethorphan-quinidine on agitation in patients with Alzheimer disease dementia: a randomized clinical trial, JAMA 314 (2015) 1242–1254.

[174] R. Moretti, P. Torre, R.M. Antonello, G. Cazzato, A. Bava, Effects of selegiline on fronto-temporal dementia: a neuropsychological evaluation, Int. J. Geriatr. Psychiatry 17 (2002) 391–392.

[175] G. Adler, M. Teufel, L.M. Drach, Pharmacological treatment of frontotemporal dementia: treatment response to the MAO-A inhibitor moclobemide, Int. J. Geriatr. Psychiatry 18 (2003) 653–655.

[176] I. Alafuzoff, S. Helisalmi, E.H. Heinonen, et al. Selegiline treatment and the extent of degenerative changes in brain tissue of patients with Alzheimer's disease, Eur. J. Clin. Pharmacol. 55 (2000) 815–819.

[177] T. Thomas, Monoamine oxidase-B inhibitors in the treatment of Alzheimer's disease, Neurobiol. Aging 21 (2000) 343–348.

[178] S. Nave, R.S. Doody, M. Boada Rovira, et al. Sembragiline in moderate Alzheimer's disease dementia: results of a phase 2 trial (Mayflower road), J. Prev. Alz. Dis. 2 (2015) 269–396.

[179] R. Hurlemann, A. Patin, O.A. Onur, et al. Oxytocin enhances amygdala-dependent, socially reinforced learning and emotional empathy in humans, J. Neurosci. 30 (2010) 4999–5007.

[180] M. Kosfeld, M. Heinrichs, P.J. Zak, U. Fischbacher, E. Fehr, Oxytocin increases trust in humans, Nature 435 (2005) 673–676.

[181] A.A. Marsh, H.H. Yu, D.S. Pine, R.J. Blair, Oxytocin improves specific recognition of positive facial expressions, Psychopharmacology 209 (2010) 225–232.

[182] S. Jesso, D. Morlog, S. Ross, et al. The effects of oxytocin on social cognition and behaviour in frontotemporal dementia, Brain 134 (2011) 2493–2501.

[183] E.C. Finger, J. MacKinley, M. Blair, et al. Oxytocin for frontotemporal dementia: a randomized dose-finding study of safety and tolerability, Neurology 84 (2015) 174–181.

[184] C.E. Poetter, J.T. Stewart, Treatment of indiscriminate, inappropriate sexual behavior in frontotemporal dementia with carbamazepine, J. Clin. Psychopharmacol. 32 (2012) 137–138.

[185] M. Cruz, V. Marinho, L.F. Fontenelle, E. Engelhardt, J. Laks, Topiramate may modulate alcohol abuse but not other compulsive behaviors in frontotemporal dementia: case report, Cogn. Behav. Neurol. 21 (2008) 104–106.

[186] P.J. Nestor, Reversal of abnormal eating and drinking behaviour in a frontotemporal lobar degeneration patient using low-dose topiramate, J. Neurol. Neurosurg. Psychiatry 83 (2012) 349–350.

[187] C. Singam, M. Walterfang, R. Mocellin, A. Evans, D. Velakoulis, Topiramate for abnormal eating behaviour in frontotemporal dementia, Behav. Neurol. 27 (2013) 285–286.

[188] S. Shinagawa, N. Tsuno, K. Nakayama, Managing abnormal eating behaviours in frontotemporal lobar degeneration patients with topiramate, Psychogeriatrics 13 (2013) 58–61.

[189] D.A. Reed, N.A. Johnson, C. Thompson, S. Weintraub, M.M. Mesulam, A clinical trial of bromocriptine for treatment of primary progressive aphasia, Ann. Neurol. 56 (2004) 750.

[190] N.A. Johnson, A. Rademaker, S. Weintraub, D. Gitelman, C. Wienecke, M. Mesulam, Pilot trial of memantine in primary progressive aphasia, Alzheimer Dis. Assoc. Disord. 24 (2010) 308.

[191] G.A. Jicha, P.T. Nelson, Management of frontotemporal dementia: targeting symptom management in such a heterogeneous disease requires a wide range of therapeutic options, Neurodegener. Dis. Manag. 1 (2011) 141–156.

[192] K.A. Vossel, B.L. Miller, New approaches to the treatment of frontotemporal lobar degeneration, Curr. Opin. Neurol. 21 (2008) 708–716.

[193] G. Bensimon, A. Ludolph, Y. Agid, M. Vidailhet, C. Payan, P.N. Leigh, Riluzole treatment, survival and diagnostic criteria in Parkinson plus disorders: the NNIPPS study, Brain 132 (2009) 156–171.

[194] T.W. Chow, M.F. Mendez, Goals in symptomatic pharmacologic management of frontotemporal lobar degeneration, Am. J. Alzheimers Dis. Other Demen. 17 (2002) 267–272.

[195] S. Shinagawa, S. Nakajima, E. Plitman, et al. Non-pharmacological management for patients with frontotemporal dementia: a systematic review, J. Alzheimers Dis. 45 (2015) 283–293.

[196] Z. Berger, H. Roder, A. Hanna, et al. Accumulation of pathological tau species and memory loss in a conditional model of tauopathy, J. Neurosci. 27 (2007) 3650–3662.

[197] T.L. Spires, J.D. Orne, K. SantaCruz, et al. Region-specific dissociation of neuronal loss and neurofibrillary pathology in a mouse model of tauopathy, Am. J. Pathol. 168 (2006) 1598–1607.

[198] K. SantaCruz, J. Lewis, T. Spires, et al. Tau suppression in a neurodegenerative mouse model improves memory function, Science 309 (2005) 476–481.

[199] C. Andorfer, C.M. Acker, Y. Kress, P.R. Hof, K. Duff, P. Davies, Cell-cycle reentry and cell death in transgenic mice expressing nonmutant human tau isoforms, J. Neurosci. 25 (2005) 5446–5454.

[200] M. Polydoro, C.M. Acker, K. Duff, P.E. Castillo, P. Davies, Age-dependent impairment of cognitive and synaptic function in the htau mouse model of tau pathology, J. Neurosci. 29 (2009) 10741–10749.

[201] Y. Yoshiyama, M. Higuchi, B. Zhang, et al. Synapse loss and microglial activation precede tangles in a P301S tauopathy mouse model, Neuron 53 (2007) 337–351.

[202] C.W. Wittmann, M.F. Wszolek, J.M. Shulman, et al. Tauopathy in *Drosophila*: neurodegeneration without neurofibrillary tangles, Science 293 (2001) 711–714.

[203] B.C. Kraemer, B. Zhang, J.B. Leverenz, J.H. Thomas, J.Q. Trojanowski, G.D. Schellenberg, Neurodegeneration and defective neurotransmission in a *Caenorhabditis elegans* model of tauopathy, Proc. Natl. Acad. Sci. USA 100 (2003) 9980–9985.

[204] A. Crowe, C. Ballatore, E. Hyde, J.Q. Trojanowski, V.M. Lee, High throughput screening for small molecule inhibitors of heparin-induced tau fibril formation, Biochem. Biophys. Res. Commun. 358 (2007) 1–6.

[205] M. Pickhardt, Z. Gazova, B.M. von, et al. Anthraquinones inhibit tau aggregation and dissolve Alzheimer's paired helical filaments in vitro and in cells, J. Biol. Chem. 280 (2005) 3628–3635.

[206] M. Pickhardt, G. Larbig, I. Khlistunova, et al. Phenylthiazolyl-hydrazide and its derivatives are potent inhibitors of tau aggregation and toxicity in vitro and in cells, Biochemistry 46 (2007) 10016–10023.

[207] M. Pickhardt, J. Biernat, I. Khlistunova, et al. N-phenylamine derivatives as aggregation inhibitors in cell models of tauopathy, Curr. Alzheimer Res. 4 (2007) 397–402.

[208] B. Bulic, M. Pickhardt, I. Khlistunova, et al. Rhodanine-based tau aggregation inhibitors in cell models of tauopathy, Angew. Chem. Int. Ed. Engl. 46 (2007) 9215–9219.

[209] C. Chirita, M. Necula, J. Kuret, Ligand-dependent inhibition and reversal of tau filament formation, Biochemistry 43 (2004) 2879–2887.

[210] M. Pickhardt, T. Neumann, D. Schwizer, et al. Identification of small molecule inhibitors of tau aggregation by targeting monomeric tau as a potential therapeutic approach for tauopathies, Curr. Alzheimer Res. 12 (2015) 814–828.

[211] C.M. Wischik, P.C. Edwards, R.Y. Lai, M. Roth, C.R. Harrington, Selective inhibition of Alzheimer disease-like tau aggregation by phenothiazines, Proc. Natl. Acad. Sci. USA 93 (1996) 11213–11218.

[212] J. Wagner, S. Krauss, S. Shi, et al. Reducing tau aggregates with anle138b delays disease progression in a mouse model of tauopathies, Acta Neuropathol. 130 (2015) 619–631.

[213] X. Chai, J.L. Dage, M. Citron, Constitutive secretion of tau protein by an unconventional mechanism, Neurobiol. Dis. 48 (2012) 356–366.

[214] J.E. Gerson, R. Kayed, Formation and propagation of tau oligomeric seeds, Front. Neurol. 4 (2013) 93.

[215] M.S. Wolfe, The role of tau in neurodegenerative diseases and its potential as a therapeutic target, Scientifica 2012 (2012) 796024.

[216] X. Sun, Y. Wang, J. Zhang, et al. Tunneling-nanotube direction determination in neurons and astrocytes, Cell Death. Dis. 3 (2012) e438.

[217] H.H. Gerdes, A. Rustom, X. Wang, Tunneling nanotubes, an emerging intercellular communication route in development, Mech. Dev. 130 (2013) 381–387.

[218] K.E. Funk, H. Mirbaha, H. Jiang, D.M. Holtzman, M.I. Diamond, Distinct therapeutic mechanisms of tau antibodies: promoting microglial clearance versus blocking neuronal uptake, J. Biol. Chem. 290 (2015) 21652–21662.

[219] J.T. Pedersen, E.M. Sigurdsson, Tau immunotherapy for Alzheimer's disease, Trends Mol. Med. 21 (2015) 394–402.

[220] E.M. Sigurdsson, Tau immunotherapy, Neurodegener. Dis. 16 (2016) 34–38.

[221] T.E. Golde, J. Lewis, N.R. McFarland, Anti-tau antibodies: hitting the target, Neuron 80 (2013) 254–256.

[222] E. Kontsekova, N. Zilka, B. Kovacech, R. Skrabana, M. Novak, Identification of structural determinants on tau protein essential for its pathological function: novel therapeutic target for tau immunotherapy in Alzheimer's disease, Alzheimers Res. Ther. 6 (2014) 45.

[223] E. Kontsekova, N. Zilka, B. Kovacech, P. Novak, M. Novak, First-in-man tau vaccine targeting structural determinants essential for pathological tau-tau interaction reduces tau oligomerisation and neurofibrillary degeneration in an Alzheimer's disease model, Alzheimers Res. Ther. 6 (2014) 44.

[224] Alzforum, AADvac-1, (2016).

[225] Alzforum, BMS-986168, (2016).

[226] N. Kfoury, B.B. Holmes, H. Jiang, D.M. Holtzman, M.I. Diamond, Trans-cellular propagation of Tau aggregation by fibrillar species, J. Biol. Chem. 287 (2012) 19440–19451.

[227] K. Yanamandra, N. Kfoury, H. Jiang, et al. Anti-tau antibodies that block tau aggregate seeding in vitro markedly decrease pathology and improve cognition in vivo, Neuron 80 (2013) 402–414.

[228] K. Yanamandra, H. Jiang, T.E. Mahan, et al. Anti-tau antibody reduces insoluble tau and decreases brain atrophy, Ann. Clin. Transl. Neurol. 2 (2015) 278–288.

[229] C. Theunis, N. Crespo-Biel, V. Gafner, et al. Efficacy and safety of a liposome-based vaccine against protein Tau, assessed in tau. P301L mice that model tauopathy, PLoS One 8 (2013) e72301.

[230] T. Fagan, Preclinical research offers new angles on immunotherapy, Alzforum, (2016).

[231] S. Le Corre, H.W. Klafki, N. Plesnila, et al. An inhibitor of tau hyperphosphorylation prevents severe motor impairments in tau transgenic mice, Proc. Natl. Acad. Sci. USA 103 (2006) 9673–9678.

[232] J.A. Morales-Garcia, R. Luna-Medina, S. Alonso-Gil, et al. Glycogen synthase kinase 3 inhibition promotes adult hippocampal neurogenesis in vitro and in vivo, ACS Chem. Neurosci. 3 (2012) 963–971.

[233] L. Sereno, M. Coma, M. Rodriguez, et al. A novel GSK-3beta inhibitor reduces Alzheimer's pathology and rescues neuronal loss in vivo, Neurobiol. Dis. 35 (2009) 359–367.

[234] T. del Ser, K.C. Steinwachs, H.J. Gertz, et al. Treatment of Alzheimer's disease with the GSK-3 inhibitor tideglusib: a pilot study, J. Alzheimers Dis. 33 (2013) 205–215.

[235] S. Lovestone, M. Boada, B. Dubois, et al. A phase II trial of tideglusib in Alzheimer's disease, J. Alzheimers Dis. 45 (2015) 75–88.

[236] E. Tolosa, I. Litvan, G.U. Hoglinger, et al. A phase 2 trial of the GSK-3 inhibitor tideglusib in progressive supranuclear palsy, Mov. Disord. 29 (2014) 470–478.

[237] G.U. Hoglinger, H.J. Huppertz, S. Wagenpfeil, et al. Tideglusib reduces progression of brain atrophy in progressive supranuclear palsy in a randomized trial, Mov. Disord. 29 (2014) 479–487.

[238] M. Saitoh, J. Kunitomo, E. Kimura, et al. 2-{3-[4-(Alkylsulfinyl)phenyl]-1-benzofuran-5-yl}-5-methyl-1,3,4-oxadiazole derivatives as novel inhibitors of glycogen synthase kinase-3beta with good brain permeability, J. Med. Chem. 52 (2009) 6270–6286.

[239] T. Onishi, H. Iwashita, Y. Uno, et al. A novel glycogen synthase kinase-3 inhibitor 2-methyl-5-(3-{4-[(S)-methylsulfinyl]phenyl}-1-benzofuran-5-yl)-1,3,4-oxadiazole decreases tau phosphorylation and ameliorates cognitive deficits in a transgenic model of Alzheimer's disease, J. Neurochem. 119 (2011) 1330–1340.

[240] W. Noble, E. Planel, C. Zehr, et al. Inhibition of glycogen synthase kinase-3 by lithium correlates with reduced tauopathy and degeneration in vivo, Proc. Natl. Acad. Sci. USA 102 (2005) 6990–6995.

[241] T. Engel, P. Goni-Oliver, J.J. Lucas, J. Avila, F. Hernandez, Chronic lithium administration to FTDP-17 tau and GSK-3beta overexpressing mice prevents tau hyperphosphorylation and neurofibrillary tangle formation, but pre-formed neurofibrillary tangles do not revert, J. Neurochem. 99 (2006) 1445–1455.

[242] K. Leroy, K. Ando, C. Heraud, et al. Lithium treatment arrests the development of neurofibrillary tangles in mutant tau transgenic mice with advanced neurofibrillary pathology, J. Alzheimers Dis. 19 (2010) 705–719.

[243] M.A. Nunes, N.M. Schowe, K.C. Monteiro-Silva, et al. Chronic microdose lithium treatment prevented memory loss and neurohistopathological changes in a transgenic mouse model of Alzheimer's disease, PLoS One 10 (2015) e0142267.

[244] M. Perez, F. Hernandez, F. Lim, J. Diaz-Nido, J. Avila, Chronic lithium treatment decreases mutant tau protein aggregation in a transgenic mouse model, J. Alzheimers Dis. 5 (2003) 301–308.

[245] J.P. Hu, J.W. Xie, C.Y. Wang, et al. Valproate reduces tau phosphorylation via cyclin-dependent kinase 5 and glycogen synthase kinase 3 signaling pathways, Brain Res. Bull. 85 (2011) 194–200.

[246] Z.M. Long, L. Zhao, R. Jiang, et al. Valproic acid modifies synaptic structure and accelerates neurite outgrowth via the glycogen synthase kinase-3beta signaling pathway in an Alzheimer's disease model, CNS Neurosci. Ther. 21 (2015) 887–897.

[247] H. Qing, G. He, P.T. Ly, et al. Valproic acid inhibits Abeta production, neuritic plaque formation, and behavioral deficits in Alzheimer's disease mouse models, J. Exp. Med. 205 (2008) 2781–2789.

[248] R. Guerrini, A. Belmonte, R. Canapicchi, C. Casalini, E. Perucca, Reversible pseudoatrophy of the brain and mental deterioration associated with valproate treatment, Epilepsia 39 (1998) 27–32.

[249] C.A. Galimberti, M. Diegoli, I. Sartori, et al. Brain pseudoatrophy and mental regression on valproate and a mitochondrial DNA mutation, Neurology 67 (2006) 1715–1717.

[250] O. Papazian, E. Canizales, I. Alfonso, R. Archila, M. Duchowny, J. Aicardi, Reversible dementia and apparent brain atrophy during valproate therapy, Ann. Neurol. 38 (1995) 687–691.

[251] R. Straussberg, S. Kivity, R. Weitz, L. Harel, N. Gadoth, Reversible cortical atrophy and cognitive decline induced by valproic acid, Eur J. Paediatr. Neurol. 2 (1998) 213–218.

[252] J. Bain, L. Plater, M. Elliott, et al. The selectivity of protein kinase inhibitors: a further update, Biochem. J. 408 (2007) 297–315.

[253] C.J. Phiel, F. Zhang, E.Y. Huang, M.G. Guenther, M.A. Lazar, P.S. Klein, Histone deacetylase is a direct target of valproic acid, a potent anticonvulsant, mood stabilizer, and teratogen, J. Biol. Chem. 276 (2001) 36734–36741.

[254] N. Jin, A.D. Kovacs, Z. Sui, S. Dewhurst, S.B. Maggirwar, Opposite effects of lithium and valproic acid on trophic factor deprivation-induced glycogen synthase kinase-3 activation, c-Jun expression and neuronal cell death, Neuropharmacology 48 (2005) 576–583.

[255] M.A. Nunes, T.A. Viel, H.S. Buck, Microdose lithium treatment stabilized cognitive impairment in patients with Alzheimer's disease, Curr. Alzheimer Res. 10 (2013) 104–107.

[256] P.V. Nunes, O.V. Forlenza, W.F. Gattaz, Lithium and risk for Alzheimer's disease in elderly patients with bipolar disorder, Br. J. Psychiatry 190 (2007) 359–360.

[257] T. Leyhe, G.W. Eschweiler, E. Stransky, et al. Increase of BDNF serum concentration in lithium treated patients with early Alzheimer's disease, J. Alzheimers Dis. 16 (2009) 649–656.

[258] O.V. Forlenza, B.S. Diniz, M. Radanovic, F.S. Santos, L.L. Talib, W.F. Gattaz, Disease-modifying properties of long-term lithium treatment for amnestic mild cognitive impairment: randomised controlled trial, Br. J. Psychiatry 198 (2011) 351–356.

[259] S. Matsunaga, T. Kishi, P. Annas, H. Basun, H. Hampel, N. Iwata, Lithium as a treatment for Alzheimer's disease: a systematic review and meta-analysis, J. Alzheimers Dis. 48 (2015) 403–410.

[260] H. Hampel, M. Ewers, K. Burger, et al. Lithium trial in Alzheimer's disease: a randomized, single-blind, placebo-controlled, multicenter 10-week study, J. Clin. Psychiatry 70 (2009) 922–931.

[261] S.D. Brinkman, N. Pomara, N. Barnett, R. Block, E.F. Domino, S. Gershon, Lithium-induced increases in red blood cell choline and memory performance in Alzheimer-type dementia, Biol. Psychiatry 19 (1984) 157–164.

[262] A.P. Porsteinsson, P.N. Tariot, L.J. Jakimovich, et al. Valproate therapy for agitation in dementia: open-label extension of a double-blind trial, Am. J. Geriatr. Psychiatry 11 (2003) 434–440.

[263] A.P. Porsteinsson, P.N. Tariot, R. Erb, S. Gaile, An open trial of valproate for agitation in geriatric neuropsychiatric disorders, Am. J. Geriatr. Psychiatry 5 (1997) 344–351.

[264] A.P. Porsteinsson, P.N. Tariot, R. Erb, et al. Placebo-controlled study of divalproex sodium for agitation in dementia, Am. J. Geriatr. Psychiatry 9 (2001) 58–66.

[265] C. Dolder, J. McKinsey, Low-dose divalproex in agitated patients with Alzheimer's disease, J. Psychiatr. Pract. 16 (2010) 63–67.

[266] P.N. Tariot, R. Raman, L. Jakimovich, et al. Divalproex sodium in nursing home residents with possible or probable Alzheimer Disease complicated by agitation: a randomized, controlled trial, Am. J. Geriatr. Psychiatry 13 (2005) 942–949.

[267] A.S. Fleisher, D. Truran, J.T. Mai, et al. Chronic divalproex sodium use and brain atrophy in Alzheimer disease, Neurology 77 (2011) 1263–1271.

[268] N. Herrmann, K.L. Lanctot, L.S. Rothenburg, G. Eryavec, A placebo-controlled trial of valproate for agitation and aggression in Alzheimer's disease, Dement. Geriatr. Cogn. Disord. 23 (2007) 116–119.

[269] P.N. Tariot, L.S. Schneider, J. Cummings, et al. Chronic divalproex sodium to attenuate agitation and clinical progression of Alzheimer disease, Arch. Gen. Psychiatry 68 (2011) 853–861.

[270] E. Kickstein, S. Krauss, P. Thornhill, et al. Biguanide metformin acts on tau phosphorylation via mTOR/protein phosphatase 2A (PP2A) signaling, Proc. Natl. Acad. Sci. USA 107 (2010) 21830–21835.

[271] J. van Eersel, Y.D. Ke, X. Liu, et al. Sodium selenate mitigates tau pathology, neurodegeneration, and functional deficits in Alzheimer's disease models, Proc. Natl. Acad. Sci. USA 107 (2010) 13888–13893.

[272] N.M. Corcoran, D. Martin, B. Hutter-Paier, et al. Sodium selenate specifically activates PP2A phosphatase, dephosphorylates tau and reverses memory deficits in an Alzheimer's disease model, J. Clin. Neurosci. 17 (2010) 1025–1033.

[273] S.W. Min, S.H. Cho, Y. Zhou, et al. Acetylation of tau inhibits its degradation and contributes to tauopathy, Neuron 67 (2010) 953–966.

[274] S.W. Min, X. Chen, T.E. Tracy, et al. Critical role of acetylation in tau-mediated neurodegeneration and cognitive deficits, Nat. Med. 21 (2015) 1154–1162.

[275] D.P. Hanger, S. Wray, Tau cleavage and tau aggregation in neurodegenerative disease, Biochem. Soc. Trans. 38 (2010) 1016–1020.

[276] M.V. Rao, M.K. McBrayer, J. Campbell, et al. Specific calpain inhibition by calpastatin prevents tauopathy and neurodegeneration and restores normal lifespan in tau P301L mice, J. Neurosci. 34 (2014) 9222–9234.

[277] R. Medeiros, M. Kitazawa, M.A. Chabrier, et al. Calpain inhibitor A-705253 mitigates Alzheimer's disease-like pathology and cognitive decline in aged 3xTgAD mice, Am. J. Pathol. 181 (2012) 616–625.

[278] A.L. Nikkel, B. Martino, S. Markosyan, et al. The novel calpain inhibitor A-705253 prevents stress-induced tau hyperphosphorylation in vitro and in vivo, Neuropharmacology 63 (2012) 606–612.

[279] K.A. Butner, M.W. Kirschner, Tau protein binds to microtubules through a flexible array of distributed weak sites, J. Cell Biol. 115 (1991) 717–730.

[280] S.L. DeVos, T.M. Miller, Antisense oligonucleotides: treating neurodegeneration at the level of RNA, Neurotherapeutics 10 (2013) 486–497.

[281] A.L. Southwell, N.H. Skotte, C.F. Bennett, M.R. Hayden, Antisense oligonucleotide therapeutics for inherited neurodegenerative diseases, Trends Mol. Med. 18 (2012) 634–643.

[282] M.M. Evers, L.J. Toonen, W.M. van Roon-Mom, Antisense oligonucleotides in therapy for neurodegenerative disorders, Adv. Drug Deliv. Rev. 87 (2015) 90–103.

[283] C.P. Donahue, C. Muratore, J.Y. Wu, K.S. Kosik, M.S. Wolfe, Stabilization of the tau exon 10 stem loop alters pre-mRNA splicing, J. Biol. Chem. 281 (2006) 23302–23306.

[284] E. Peacey, L. Rodriguez, Y. Liu, M.S. Wolfe, Targeting a pre-mRNA structure with bipartite antisense molecules modulates tau alternative splicing, Nucleic Acids Res. 40 (2012) 9836–9849.

[285] B. Kalbfuss, S.A. Mabon, T. Misteli, Correction of alternative splicing of tau in frontotemporal dementia and parkinsonism linked to chromosome 17, J. Biol. Chem. 276 (2001) 42986–42993.

[286] M.E. Avale, T. Rodriguez-Martin, J.M. Gallo, Trans-splicing correction of tau isoform imbalance in a mouse model of tau mis-splicing, Hum. Mol. Genet. 22 (2013) 2603–2611.

[287] T. Rodriguez-Martin, K. Anthony, M.A. Garcia-Blanco, S.G. Mansfield, B.H. Anderton, J.M. Gallo, Correction of tau mis-splicing caused by FTDP-17 MAPT mutations by spliceosome-mediated RNA trans-splicing, Hum. Mol. Genet. 18 (2009) 3266–3273.

[288] T. Rodriguez-Martin, M.A. Garcia-Blanco, S.G. Mansfield, et al. Reprogramming of tau alternative splicing by spliceosome-mediated RNA trans-splicing: implications for tauopathies, Proc. Natl. Acad. Sci. USA 102 (2005) 15659–15664.

[289] J. Zhou, Q. Yu, T. Zou, Alternative splicing of exon 10 in the tau gene as a target for treatment of tauopathies, BMC Neurosci. 9 (Suppl. 2) (2008) S10.

[290] L. Varani, M.G. Spillantini, M. Goedert, G. Varani, Structural basis for recognition of the RNA major groove in the tau exon 10 splicing regulatory element by aminoglycoside antibiotics, Nucleic Acids Res. 28 (2000) 710–719.

[291] S. Zheng, Y. Chen, C.P. Donahue, M.S. Wolfe, G. Varani, Structural basis for stabilization of the tau pre-mRNA splicing regulatory element by novantrone (mitoxantrone), Chem. Biol. 16 (2009) 557–566.

[292] C.P. Donahue, J. Ni, E. Rozners, M.A. Glicksman, M.S. Wolfe, Identification of tau stem loop RNA stabilizers, J. Biomol. Screen. 12 (2007) 789–799.

[293] Y. Liu, E. Peacey, J. Dickson, et al. Mitoxantrone analogues as ligands for a stem-loop structure of tau pre-mRNA, J. Med. Chem. 52 (2009) 6523–6526.

[294] P. Lopez-Senin, I. Gomez-Pinto, A. Grandas, V. Marchan, Identification of ligands for the Tau exon 10 splicing regulatory element RNA by using dynamic combinatorial chemistry, Chemistry 17 (2011) 1946–1953.

[295] P.Y. Smith, C. Delay, J. Girard, et al. MicroRNA-132 loss is associated with tau exon 10 inclusion in progressive supranuclear palsy, Hum. Mol. Genet. 20 (2011) 4016–4024.

[296] J. Hernandez-Rapp, P.Y. Smith, M. Filali, et al. Memory formation and retention are affected in adult miR-132/212 knockout mice, Behav. Brain Res. 287 (2015) 15–26.

[297] P.Y. Smith, J. Hernandez-Rapp, F. Jolivette, et al. miR-132/212 deficiency impairs tau metabolism and promotes pathological aggregation in vivo, Hum. Mol. Genet. 24 (2015) 6721–6735.

[298] S.L. DeVos, D.K. Goncharoff, G. Chen, et al. Antisense reduction of tau in adult mice protects against seizures, J. Neurosci. 33 (2013) 12887–12897.

[299] A. Sydow, A. Van der Jeugd, F. Zheng, et al. Tau-induced defects in synaptic plasticity, learning, and memory are reversible in transgenic mice after switching off the toxic Tau mutant, J. Neurosci. 31 (2011) 2511–2525.

[300] M. Polydoro, C.A. de, M. Suarez-Calvet, et al. Reversal of neurofibrillary tangles and tau-associated phenotype in the rTgTauEC model of early Alzheimer's disease, J. Neurosci. 33 (2013) 13300–13311.

[301a] T.M. Miller, Antisense oligonucleotide therapy for FTD, 9th International Conference on Frontotemporal Dementias, Am. J. Neurodegener. Dis. 3 (2014).

[301b] J.-F. Blain, F. Albayya, Z. Tu, H. Hodgdon, D. Costa, D. Flood, G. Koenig, H. Patzke, FORUM Pharmaceuticals, The clinical stage HDAC inhibitor FRM-0334 induces progranulin in rodent brain and in FTLD-GRN patient-derived lymphoblasts, 9th International Conference on Frontotemporal Dementias, Am. J. Neurodegener. Dis. 3 (2014) 53.

[301c] K. Larson, A. Cook, J. Soper, L. Herl Martens, F. Albayya, V. Mahadomrongkul, D. Burnett, G. Koenig, M. Townsend, H. Patzke, FORUM Pharmaceuticals, A phenotypic screen of a mouse microglial cell line

reveals novel mechanisms to modulate levels of progranulin, 9th International Conference on Frontotemporal Dementias, Am. J. Neurodegener. Dis. 3 (2014) 155.

[301d] E. Wauters, I. Gijselinck, T. Van Langenhove, S. Engelborghs, M. Vandenbulcke, M. Mattheijssens, K. Peeters, J.-J. Martin, P. Cras, P. Santens, R. Vandenberghe, P.P. De Deyn, J. van der Zee, K. Sleegers, C. Van Broeckhoven, M. Cruts, An integrative approach to identify onset age modifier genes in a large founder GRN FTLD family, 9th International Conference on Frontotemporal Dementias, Am. J. Neurodegener. Dis. 3 (2014) 271.

[302] S.L. DeVos, Antisense reduction of the protein tau attenuates neuronal hyperexcitability and permits clearance of intraneuronal tau accumulations in vivo, (2014).

[303] R. Sud, E.T. Geller, G.D. Schellenberg, Antisense-mediated exon skipping decreases tau protein expression: a potential therapy for tauopathies, Mol. Ther. Nucleic Acids 3 (2014) e180.

[304] I. Santa-Maria, M.E. Alaniz, N. Renwick, et al. Dysregulation of microRNA-219 promotes neurodegeneration through post-transcriptional regulation of tau, J. Clin. Invest. 125 (2015) 681–686.

[305] C.A. Dickey, P. Ash, N. Klosak, et al. Pharmacologic reductions of total tau levels; implications for the role of microtubule dynamics in regulating tau expression, Mol. Neurodegener. 1 (2006) 6.

[306] B. Zhang, A. Maiti, S. Shively, et al. Microtubule-binding drugs offset tau sequestration by stabilizing microtubules and reversing fast axonal transport deficits in a tauopathy model, Proc. Natl. Acad. Sci. USA 102 (2005) 227–231.

[307] S. Fellner, B. Bauer, D.S. Miller, et al. Transport of paclitaxel (Taxol) across the blood-brain barrier in vitro and in vivo, J. Clin. Invest. 110 (2002) 1309–1318.

[308] Alzforum, TPI 287, (2016).

[309] N. Shiryaev, Y. Jouroukhin, E. Giladi, et al. NAP protects memory, increases soluble tau and reduces tau hyperphosphorylation in a tauopathy model, Neurobiol. Dis. 34 (2009) 381–388.

[310] S. Quraishe, C.M. Cowan, A. Mudher, NAP (davunetide) rescues neuronal dysfunction in a *Drosophila* model of tauopathy, Mol. Psychiatry 18 (2013) 834–842.

[311] I. Vulih-Shultzman, A. Pinhasov, S. Mandel, et al. Activity-dependent neuroprotective protein snippet NAP reduces tau hyperphosphorylation and enhances learning in a novel transgenic mouse model, J. Pharmacol. Exp. Ther. 323 (2007) 438–449.

[312] Y. Matsuoka, Y. Jouroukhin, A.J. Gray, et al. A neuronal microtubule-interacting agent, NAPVSIPQ, reduces tau pathology and enhances cognitive function in a mouse model of Alzheimer's disease, J. Pharmacol. Exp. Ther. 325 (2008) 146–153.

[313] Y. Matsuoka, A.J. Gray, C. Hirata-Fukae, et al. Intranasal NAP administration reduces accumulation of amyloid peptide and tau hyperphosphorylation in a transgenic mouse model of Alzheimer's disease at early pathological stage, J. Mol. Neurosci. 31 (2007) 165–170.

[314] M. Bowman Rogers, Tau-targeting drug davunetide washes out in phase 3 trials, Alzforum, (2016).

[315] S. Herdewyn, M.L. De, L. Van Den Bosch, W. Robberecht, D.P. Van, Progranulin does not affect motor neuron degeneration in mutant SOD1 mice and rats, Neurobiol. Aging 34 (2013) 2302–2303.

[316] A. Capell, S. Liebscher, K. Fellerer, et al. Rescue of progranulin deficiency associated with frontotemporal lobar degeneration by alkalizing reagents and inhibition of vacuolar ATPase, J. Neurosci. 31 (2011) 1885–1894.

[317] C.M. Lang, K. Fellerer, B.M. Schwenk, et al. Membrane orientation and subcellular localization of transmembrane protein 106B (TMEM106B), a major risk factor for frontotemporal lobar degeneration, J. Biol. Chem. 287 (2012) 19355–19365.

[318] J.D. Rohrer, J.D. Warren, N.C. Fox, M.N. Rossor, Presymptomatic studies in genetic frontotemporal dementia, Rev. Neurol. 169 (2013) 820–824.

[319] B. Cenik, C.F. Sephton, C.M. Dewey, et al. Suberoylanilide hydroxamic acid (vorinostat) up-regulates progranulin transcription: rational therapeutic approach to frontotemporal dementia, J. Biol. Chem. 286 (2011) 16101–16108.

[320] W.X. Wang, B.R. Wilfred, S.K. Madathil, et al. miR-107 regulates granulin/progranulin with implications for traumatic brain injury and neurodegenerative disease, Am. J. Pathol. 177 (2010) 334–345.

[321] J. Jiao, L.D. Herl, R.V. Farese, F.B. Gao, MicroRNA-29b regulates the expression level of human progranulin, a secreted glycoprotein implicated in frontotemporal dementia, PLoS One 5 (2010) e10551.

[322] J. Kocerha, N. Kouri, M. Baker, et al. Altered microRNA expression in frontotemporal lobar degeneration with TDP-43 pathology caused by progranulin mutations, BMC Genomics 12 (2011) 527.

[323] A.S. Chen-Plotkin, T.L. Unger, M.D. Gallagher, et al. TMEM106B, the risk gene for frontotemporal dementia, is regulated by the microRNA-132/212 cluster and affects progranulin pathways, J. Neurosci. 32 (2012) 11213–11227.

[324] J. Banzhaf-Strathmann, R. Claus, O. Mucke, et al. Promoter DNA methylation regulates progranulin expression and is altered in FTLD, Acta Neuropathol. Commun. 1 (2013) 16.

[325] S.W. Peltz, M. Morsy, E.M. Welch, A. Jacobson, Ataluren as an agent for therapeutic nonsense suppression, Annu. Rev. Med. 64 (2013) 407–425.

[326] C. Sarkar, Z. Zhang, A.B. Mukherjee, Stop codon read-through with PTC124 induces palmitoyl-protein thioesterase-1 activity, reduces thioester load and suppresses apoptosis in cultured cells from INCL patients, Mol. Genet. Metab. 104 (2011) 338–345.

[327] T. Braulke, J.S. Bonifacino, Sorting of lysosomal proteins, Biochim. Biophys. Acta 1793 (2009) 605–614.

[328] F. Hu, T. Padukkavidana, C.B. Vaegter, et al. Sortilin-mediated endocytosis determines levels of the frontotemporal dementia protein, progranulin, Neuron 68 (2010) 654–667.

[329] W.C. Lee, S. Almeida, M. Prudencio, et al. Targeted manipulation of the sortilin-progranulin axis rescues progranulin haploinsufficiency, Hum. Mol. Genet. 23 (2014) 1467–1478.

[330] X. Zhou, L. Sun, O.F. Bastos de, et al. Prosaposin facilitates sortilin-independent lysosomal trafficking of progranulin, J. Cell Biol. 210 (2015) 991–1002.

[331] M. Cruts, I. Gijselinck, T. Van Langenhove, J. van der Zee, C. Van Broeckhoven, Current insights into the C9orf72 repeat expansion diseases of the FTLD/ALS spectrum, Trends Neurosci. 36 (2013) 450–459.

[332] P. Fratta, M. Poulter, T. Lashley, et al. Homozygosity for the C9orf72 GGGGCC repeat expansion in frontotemporal dementia, Acta Neuropathol. 126 (2013) 401–409.

[333] V.V. Belzil, P.O. Bauer, M. Prudencio, et al. Reduced C9orf72 gene expression in c9FTD/ALS is caused by histone trimethylation, an epigenetic event detectable in blood, Acta Neuropathol. 126 (2013) 895–905.

[334] C.J. Donnelly, P.W. Zhang, J.T. Pham, et al. RNA toxicity from the ALS/FTD C9ORF72 expansion is mitigated by antisense intervention, Neuron 80 (2013) 415–428.

[335] M. Therrien, G.A. Rouleau, P.A. Dion, J.A. Parker, Deletion of C9ORF72 results in motor neuron degeneration and stress sensitivity in *C. elegans*, PLoS One 8 (2013) e83450.

[336] S. Ciura, S. Lattante, I. Le Ber, et al. Loss of function of C9orf72 causes motor deficits in a zebrafish model of amyotrophic lateral sclerosis, Ann Neurol 74 (2013) 180–187.

[337] N. Suzuki, A.M. Maroof, F.T. Merkle, et al. The mouse C9ORF72 ortholog is enriched in neurons known to degenerate in ALS and FTD, Nat. Neurosci. 16 (2013) 1725–1727.

[338] V.V. Belzil, P.O. Bauer, T.F. Gendron, M.E. Murray, D. Dickson, L. Petrucelli, Characterization of DNA hypermethylation in the cerebellum of c9FTD/ALS patients, Brain Res. 1584 (2014) 15–21.

[339] C.T. McMillan, J. Russ, E.M. Wood, et al. C9orf72 promoter hypermethylation is neuroprotective: neuroimaging and neuropathologic evidence, Neurology 84 (2015) 1622–1630.

[340] Z. Xi, L. Zinman, D. Moreno, et al. Hypermethylation of the CpG island near the G4C2 repeat in ALS with a C9orf72 expansion, Am. J. Hum. Genet. 92 (2013) 981–989.

[341] Z. Xi, I. Rainero, E. Rubino, et al. Hypermethylation of the CpG-island near the C9orf72 G(4)C(2)-repeat expansion in FTLD patients, Hum. Mol. Genet. 23 (2014) 5630–5637.

[342] Z. Xi, M. Zhang, A.C. Bruni, et al. The C9orf72 repeat expansion itself is methylated in ALS and FTLD patients, Acta Neuropathol. 129 (2015) 715–727.

[343] E.Y. Liu, J. Russ, K. Wu, et al. C9orf72 hypermethylation protects against repeat expansion-associated pathology in ALS/FTD, Acta Neuropathol. 128 (2014) 525–541.

[344] J. Russ, E.Y. Liu, K. Wu, et al. Hypermethylation of repeat expanded C9orf72 is a clinical and molecular disease modifier, Acta Neuropathol. 129 (2015) 39–52.

[345] M.B. Harms, J. Cady, C. Zaidman, et al. Lack of C9ORF72 coding mutations supports a gain of function for repeat expansions in amyotrophic lateral sclerosis, Neurobiol. Aging 34 (2013) 2234–2239.

[346] C. Lagier-Tourenne, M. Baughn, F. Rigo, et al. Targeted degradation of sense and antisense C9orf72 RNA foci as therapy for ALS and frontotemporal degeneration, Proc. Natl. Acad. Sci. USA 110 (2013) E4530–E4539.

[347] M. Koppers, A.M. Blokhuis, H.J. Westeneng, et al. C9orf72 ablation in mice does not cause motor neuron degeneration or motor deficits, Ann. Neurol. 78 (2015) 426–438.

[348] D. Sareen, J.G. O'Rourke, P. Meera, et al. Targeting RNA foci in iPSC-derived motor neurons from ALS patients with a C9ORF72 repeat expansion, Sci. Transl. Med. 5 (2013) 208ra149.

[349] S. Almeida, E. Gascon, H. Tran, et al. Modeling key pathological features of frontotemporal dementia with C9ORF72 repeat expansion in iPSC-derived human neurons, Acta Neuropathol. 126 (2013) 385–399.

[350] S. Mizielinska, T. Lashley, F.E. Norona, et al. C9orf72 frontotemporal lobar degeneration is characterised by frequent neuronal sense and antisense RNA foci, Acta Neuropathol. 126 (2013) 845–857.

[351] T.F. Gendron, K.F. Bieniek, Y.J. Zhang, et al. Antisense transcripts of the expanded C9ORF72 hexanucleotide repeat form nuclear RNA foci and undergo repeat-associated non-ATG translation in c9FTD/ALS, Acta Neuropathol. 126 (2013) 829–844.

[352] T. Zu, Y. Liu, M. Banez-Coronel, et al. RAN proteins and RNA foci from antisense transcripts in C9ORF72 ALS and frontotemporal dementia, Proc. Natl. Acad. Sci. USA 110 (2013) E4968–E4977.

[353] Y.B. Lee, H.J. Chen, J.N. Peres, et al. Hexanucleotide repeats in ALS/FTD form length-dependent RNA foci, sequester RNA binding proteins, and are neurotoxic, Cell Rep. 5 (2013) 1178–1186.

[354] S. Mizielinska, S. Gronke, T. Niccoli, et al. C9orf72 repeat expansions cause neurodegeneration in *Drosophila* through arginine-rich proteins, Science 345 (2014) 1192–1194.

[355] Z. Xu, M. Poidevin, X. Li, et al. Expanded GGGGCC repeat RNA associated with amyotrophic lateral sclerosis and frontotemporal dementia causes neurodegeneration, Proc. Natl. Acad. Sci. USA 110 (2013) 7778–7783.

[356] K. Mori, S. Lammich, I.R. Mackenzie, et al. hnRNP A3 binds to GGGGCC repeats and is a constituent of p62-positive/TDP43-negative inclusions in the hippocampus of patients with C9orf72 mutations, Acta Neuropathol. 125 (2013) 413–423.

[357] A.R. Haeusler, C.J. Donnelly, G. Periz, et al. C9orf72 nucleotide repeat structures initiate molecular cascades of disease, Nature 507 (2014) 195–200.

[358] J. Gomez-Deza, Y.B. Lee, C. Troakes, et al. Dipeptide repeat protein inclusions are rare in the spinal cord and almost absent from motor neurons in C9ORF72 mutant amyotrophic lateral sclerosis and are unlikely to cause their degeneration, Acta Neuropathol. Commun. 3 (2015) 38.

[359] K. Mori, T. Arzberger, F.A. Grasser, et al. Bidirectional transcripts of the expanded C9orf72 hexanucleotide repeat are translated into aggregating dipeptide repeat proteins, Acta Neuropathol. 126 (2013) 881–893.

[360] Y.J. Zhang, K. Jansen-West, Y.F. Xu, et al. Aggregation-prone c9FTD/ALS poly(GA) RAN-translated proteins cause neurotoxicity by inducing ER stress, Acta Neuropathol. 128 (2014) 505–524.

[361] I. Kwon, S. Xiang, M. Kato, et al. Poly-dipeptides encoded by the C9orf72 repeats bind nucleoli, impede RNA biogenesis, and kill cells, Science 345 (2014) 1139–1145.

[362] S. May, D. Hornburg, M.H. Schludi, et al. C9orf72 FTLD/ALS-associated Gly-Ala dipeptide repeat proteins cause neuronal toxicity and Unc119 sequestration, Acta Neuropathol. 128 (2014) 485–503.

[363] M. Yamakawa, D. Ito, T. Honda, et al. Characterization of the dipeptide repeat protein in the molecular pathogenesis of c9FTD/ALS, Hum. Mol. Genet. 24 (2015) 1630–1645.

[364] X. Wen, W. Tan, T. Westergard, et al. Antisense proline-arginine RAN dipeptides linked to C9ORF72-ALS/FTD form toxic nuclear aggregates that initiate in vitro and in vivo neuronal death, Neuron 84 (2014) 1213–1225.

[365] S. Boeynaems, E. Bogaert, E. Michiels, et al. *Drosophila* screen connects nuclear transport genes to DPR pathology in c9ALS/FTD, Sci. Rep. 6 (2016) 20877.

[366] A. Jovicic, J. Mertens, S. Boeynaems, et al. Modifiers of C9orf72 dipeptide repeat toxicity connect nucleocytoplasmic transport defects to FTD/ALS, Nat. Neurosci. 18 (2015) 1226–1229.

[367] Y. Davidson, A.C. Robinson, X. Liu, et al. Neurodegeneration in frontotemporal lobar degeneration and motor neurone disease associated with expansions in C9orf72 is linked to TDP-43 pathology and not associated with aggregated forms of dipeptide repeat proteins, Neuropathol. Appl. Neurobiol. 42 (2016) 242–254.

[368] I.R. Mackenzie, T. Arzberger, E. Kremmer, et al. Dipeptide repeat protein pathology in C9ORF72 mutation cases: clinico-pathological correlations, Acta Neuropathol. 126 (2013) 859–879.

[369] D. Edbauer, C. Haass, An amyloid-like cascade hypothesis for C9orf72 ALS/FTD, Curr. Opin. Neurobiol. 36 (2015) 99–106.

[370] J.S. Snowden, S. Rollinson, J.C. Thompson, et al. Distinct clinical and pathological characteristics of frontotemporal dementia associated with C9ORF72 mutations, Brain 135 (2012) 693–708.

[371] G. Riboldi, C. Zanetta, M. Ranieri, et al. Antisense oligonucleotide therapy for the treatment of C9ORF72 ALS/FTD diseases, Mol. Neurobiol. 50 (2014) 721–732.

[372] J. Hu, J. Liu, L. Li, K.T. Gagnon, D.R. Corey, Engineering duplex rnas for challenging targets: recognition of GGGGCC/CCCCGG repeats at the ALS/FTD C9orf72 locus, Chem. Biol. 22 (2015) 1505–1511.

[373] Z. Su, Y. Zhang, T.F. Gendron, et al. Discovery of a biomarker and lead small molecules to target r(GGGGCC)-associated defects in c9FTD/ALS, Neuron 83 (2014) 1043–1050.

[374] B. Zamiri, K. Reddy, R.B. Macgregor Jr., C.E. Pearson, TMPyP4 porphyrin distorts RNA G-quadruplex structures of the disease-associated r(GGGGCC)n repeat of the C9orf72 gene and blocks interaction of RNA-binding proteins, J. Biol. Chem. 289 (2014) 4653–4659.

[375] K. Reddy, B. Zamiri, S.Y. Stanley, R.B. Macgregor Jr., C.E. Pearson, The disease-associated r(GGGGCC)n repeat from the C9orf72 gene forms tract length-dependent uni- and multimolecular RNA G-quadruplex structures, J. Biol. Chem. 288 (2013) 9860–9866.

[376] B.D. Freibaum, Y. Lu, R. Lopez-Gonzalez, et al. GGGGCC repeat expansion in C9orf72 compromises nucleocytoplasmic transport, Nature 525 (2015) 129–133.

[377] K. Zhang, C.J. Donnelly, A.R. Haeusler, et al. The C9orf72 repeat expansion disrupts nucleocytoplasmic transport, Nature 525 (2015) 56–61.

[378] M. van Blitterswijk, T.F. Gendron, M.C. Baker, et al. Novel clinical associations with specific C9ORF72 transcripts in patients with repeat expansions in C9ORF72, Acta Neuropathol. 130 (2015) 863–876.

[379] G.F. Richard, Shortening trinucleotide repeats using highly specific endonucleases: a possible approach to gene therapy?, Trends Genet. 31 (2015) 177–186.

[380] E.S. Arnold, S.C. Ling, S.C. Huelga, et al. ALS-linked TDP-43 mutations produce aberrant RNA splicing and adult-onset motor neuron disease without aggregation or loss of nuclear TDP-43, Proc. Natl. Acad. Sci. UA 110 (2013) E736–E745.

[381] E.B. Lee, V.M. Lee, J.Q. Trojanowski, Gains or losses: molecular mechanisms of TDP43-mediated neurodegeneration, Nat. Rev. Neurosci. 13 (2012) 38–50.

[382] R. Liu, G. Yang, T. Nonaka, T. Arai, W. Jia, M.S. Cynader, Reducing TDP-43 aggregation does not prevent its cytotoxicity, Acta Neuropathol. Commun. 1 (2013) 49.

[383] M. Yamashita, T. Nonaka, T. Arai, et al. Methylene blue and dimebon inhibit aggregation of TDP-43 in cellular models, FEBS Lett. 583 (2009) 2419–2424.

[384] A. Vaccaro, S.A. Patten, S. Ciura, et al. Methylene blue protects against TDP-43 and FUS neuronal toxicity in *C. elegans* and *D. rerio*, PLoS One 7 (2012) e42117.

[385] T. Fagan, Dimebon disappoints in phase 3 trial, Alzforum, (2010).

[386] M.F. Burkhardt, F.J. Martinez, S. Wright, et al. A cellular model for sporadic ALS using patient-derived induced pluripotent stem cells, Mol. Cell. Neurosci. 56 (2013) 355–364.

[387] J.D. Boyd, J.P. Lee-Armandt, M.S. Feiler, et al. A high-content screen identifies novel compounds that inhibit stress-induced TDP-43 cellular aggregation and associated cytotoxicity, J. Biomol. Screen. 19 (2014) 44–56.

[388] M. Armakola, M.J. Higgins, M.D. Figley, et al. Inhibition of RNA lariat debranching enzyme suppresses TDP-43 toxicity in ALS disease models, Nat. Genet. 44 (2012) 1302–1309.

[389] A.C. Elden, H.J. Kim, M.P. Hart, et al. Ataxin-2 intermediate-length polyglutamine expansions are associated with increased risk for ALS, Nature 466 (2010) 1069–1075.

[390] A.M. Jablonski, T. Lamitina, N.F. Liachko, et al. Loss of RAD-23 protects against models of motor neuron disease by enhancing mutant protein clearance, J. Neurosci. 35 (2015) 14286–14306.

[391] S.H. Kim, L. Zhan, K.A. Hanson, R.S. Tibbetts, High-content RNAi screening identifies the type 1 inositol triphosphate receptor as a modifier of TDP-43 localization and neurotoxicity, Hum. Mol. Genet. 21 (2012) 4845–4856.

[392] S. Ju, D.F. Tardiff, H. Han, et al. A yeast model of FUS/TLS-dependent cytotoxicity, PLoS Biol. 9 (2011) e1001052.

[393] S.J. Barmada, S. Ju, A. Arjun, et al. Amelioration of toxicity in neuronal models of amyotrophic lateral sclerosis by hUPF1, Proc. Natl. Acad. Sci. USA 112 (2015) 7821–7826.

[394] K.L. Jackson, R.D. Dayton, E.A. Orchard, et al. Preservation of forelimb function by UPF1 gene therapy in a rat model of TDP-43-induced motor paralysis, Gene Ther. 22 (2015) 20–28.

[395] N.F. Liachko, P.J. McMillan, C.R. Guthrie, T.D. Bird, J.B. Leverenz, B.C. Kraemer, CDC7 inhibition blocks pathological TDP-43 phosphorylation and neurodegeneration, Ann. Neurol. 74 (2013) 39–52.

[396] N.F. Liachko, P.J. McMillan, T.J. Strovas, et al. The tau tubulin kinases TTBK1/2 promote accumulation of pathological TDP-43, PLoS Genet. 10 (2014) e1004803.

[397] I.G. Salado, M. Redondo, M.L. Bello, et al. Protein kinase CK-1 inhibitors as new potential drugs for amyotrophic lateral sclerosis, J. Med. Chem. 57 (2014) 2755–2772.

[398] O.A. Brady, P. Meng, Y. Zheng, Y. Mao, F. Hu, Regulation of TDP-43 aggregation by phosphorylation and p62/SQSTM1, J. Neurochem. 116 (2011) 248–259.

[399] H.Y. Li, P.A. Yeh, H.C. Chiu, C.Y. Tang, B.P. Tu, Hyperphosphorylation as a defense mechanism to reduce TDP-43 aggregation, PLoS One 6 (2011) e23075.

[400] T.J. Cohen, A.W. Hwang, C.R. Restrepo, C.X. Yuan, J.Q. Trojanowski, V.M. Lee, An acetylation switch controls TDP-43 function and aggregation propensity, Nat. Commun. 6 (2015) 5845.

[401] Y. Kawahara, A. Mieda-Sato, TDP-43 promotes microRNA biogenesis as a component of the Drosha and Dicer complexes, Proc. Natl. Acad. Sci. USA 109 (2012) 3347–3352.

[402] E. Buratti, C.L. De, C. Stuani, M. Romano, M. Baralle, F. Baralle, Nuclear factor TDP-43 can affect selected microRNA levels, FEBS J. 277 (2010) 2268–2281.

[403] Z. Fan, X. Chen, R. Chen, Transcriptome-wide analysis of TDP-43 binding small RNAs identifies miR-NID1 (miR-8485), a novel miRNA that represses NRXN1 expression, Genomics 103 (2014) 76–82.

[404] I.N. King, V. Yartseva, D. Salas, et al. The RNA-binding protein TDP-43 selectively disrupts microRNA-1/206 incorporation into the RNA-induced silencing complex, J. Biol. Chem. 289 (2014) 14263–14271.

[405] J. Shugart, Do microRNAs cause mayhem across frontotemporal dementia spectrum? Alzforum, (2014).

[406] Z. Zhang, S. Almeida, Y. Lu, et al. Downregulation of microRNA-9 in iPSC-derived neurons of FTD/ALS patients with TDP-43 mutations, PLoS One 8 (2013) e76055.

[407] S.S. Hebert, W.X. Wang, Q. Zhu, P.T. Nelson, A study of small RNAs from cerebral neocortex of pathology-verified Alzheimer's disease, dementia with lewy bodies, hippocampal sclerosis, frontotemporal lobar dementia, and non-demented human controls, J. Alzheimers Dis. 35 (2013) 335–348.

[408] B.H. Lee, M.J. Lee, S. Park, et al. Enhancement of proteasome activity by a small-molecule inhibitor of USP14, Nature 467 (2010) 179–184.

[409] F. Hans, F.C. Fiesel, J.C. Strong, et al. UBE2E ubiquitin-conjugating enzymes and ubiquitin isopeptidase Y regulate TDP-43 protein ubiquitination, J. Biol. Chem. 289 (2014) 19164–19179.

[410] M. Asai, O. Tsukamoto, T. Minamino, et al. PKA rapidly enhances proteasome assembly and activity in in vivo canine hearts, J. Mol. Cell. Cardiol. 46 (2009) 452–462.

[411] N. Myeku, H. Wang, M.E. Figueiredo-Pereira, cAMP stimulates the ubiquitin/proteasome pathway in rat spinal cord neurons, Neurosci. Lett. 527 (2012) 126–131.

[412] F. Zhang, Y. Hu, P. Huang, C.A. Toleman, A.J. Paterson, J.E. Kudlow, Proteasome function is regulated by cyclic AMP-dependent protein kinase through phosphorylation of Rpt6, J. Biol. Chem. 282 (2007) 22460–22471.

[413] S. Lokireddy, N.V. Kukushkin, A.L. Goldberg, cAMP-induced phosphorylation of 26S proteasomes on Rpn6/PSMD11 enhances their activity and the degradation of misfolded proteins, Proc. Natl. Acad. Sci. USA 112 (2015) E7176–E7185.

[414] N. Myeku, C.L. Clelland, S. Emrani, et al. Tau-driven 26S proteasome impairment and cognitive dysfunction can be prevented early in disease by activating cAMP-PKA signaling, Nat. Med. 22 (2016) 46–53.

[415] M. Bowman Rogers, Protecting proteasomes from toxic tau keeps mice sharp, Alzforum, (2015).

[416] L. Petrucelli, D. Dickson, K. Kehoe, et al. CHIP and Hsp70 regulate tau ubiquitination, degradation and aggregation, Hum. Mol. Genet. 13 (2004) 703–714.

[417] L.J. Blair, B. Zhang, C.A. Dickey, Potential synergy between tau aggregation inhibitors and tau chaperone modulators, Alzheimers Res. Ther. 5 (2013) 41.

[418] F. Dou, L.D. Yuan, J.J. Zhu, Heat shock protein 90 indirectly regulates ERK activity by affecting Raf protein metabolism, Acta Biochim. Biophys. Sin. 37 (2005) 501–505.

[419] A. Opattova, P. Filipcik, M. Cente, M. Novak, Intracellular degradation of misfolded tau protein induced by geldanamycin is associated with activation of proteasome, J. Alzheimers Dis. 33 (2013) 339–348.

[420] J. Abisambra, U.K. Jinwal, Y. Miyata, et al. Allosteric heat shock protein 70 inhibitors rapidly rescue synaptic plasticity deficits by reducing aberrant tau, Biol. Psychiatry 74 (2013) 367–374.

[421] U.K. Jinwal, Y. Miyata, J. Koren III, et al. Chemical manipulation of hsp70 ATPase activity regulates tau stability, J. Neurosci. 29 (2009) 12079–12088.

[422] Y. Miyata, X. Li, H.F. Lee, et al. Synthesis and initial evaluation of YM-08, a blood-brain barrier permeable derivative of the heat shock protein 70 (Hsp70) inhibitor MKT-077, which reduces tau levels, ACS Chem. Neurosci. 4 (2013) 930–939.

[423] U.K. Jinwal, J.F. Abisambra, J. Zhang, et al. Cdc37/Hsp90 protein complex disruption triggers an autophagic clearance cascade for TDP-43 protein, J. Biol. Chem. 287 (2012) 24814–24820.

[424] M.E. Jackrel, M.E. DeSantis, B.A. Martinez, et al. Potentiated Hsp104 variants antagonize diverse proteotoxic misfolding events, Cell 156 (2014) 170–182.

[425] R.A. Nixon, The role of autophagy in neurodegenerative disease, Nat. Med. 19 (2013) 983–997.

[426] F.M. Menzies, A. Fleming, D.C. Rubinsztein, Compromised autophagy and neurodegenerative diseases, Nat. Rev. Neurosci. 16 (2015) 345–357.

[427] S. Pankiv, T.H. Clausen, T. Lamark, et al. p62/SQSTM1 binds directly to Atg8/LC3 to facilitate degradation of ubiquitinated protein aggregates by autophagy, J. Biol. Chem. 282 (2007) 24131–24145.

[428] P. Wild, H. Farhan, D.G. McEwan, et al. Phosphorylation of the autophagy receptor optineurin restricts *Salmonella* growth, Science 333 (2011) 228–233.

[429] S. Morton, L. Hesson, M. Peggie, P. Cohen, Enhanced binding of TBK1 by an optineurin mutant that causes a familial form of primary open angle glaucoma, FEBS Lett. 582 (2008) 997–1002.

[430] Q. Da, X. Yang, Y. Xu, G. Gao, G. Cheng, H. Tang, TANK-binding kinase 1 attenuates PTAP-dependent retroviral budding through targeting endosomal sorting complex required for transport-I, J. Immunol. 186 (2011) 3023–3030.

[431] J.S. Ju, R.A. Fuentealba, S.E. Miller, et al. Valosin-containing protein (VCP) is required for autophagy and is disrupted in VCP disease, J. Cell Biol. 187 (2009) 875–888.

[432] E. Tresse, F.A. Salomons, J. Vesa, et al. VCP/p97 is essential for maturation of ubiquitin-containing autophagosomes and this function is impaired by mutations that cause IBMPFD, Autophagy 6 (2010) 217–227.

[433] D. Ritz, M. Vuk, P. Kirchner, et al. Endolysosomal sorting of ubiquitylated caveolin-1 is regulated by VCP and UBXD1 and impaired by VCP disease mutations, Nat. Cell Biol. 13 (2011) 1116–1123.

[434] C. Behrends, M.E. Sowa, S.P. Gygi, J.W. Harper, Network organization of the human autophagy system, Nature 466 (2010) 68–76.

[435] S. Ozcelik, G. Fraser, P. Castets, et al. Rapamycin attenuates the progression of tau pathology in P301S tau transgenic mice, PLoS One 8 (2013) e62459.

[436] R. Siman, R. Cocca, Y. Dong, The mTOR inhibitor rapamycin mitigates perforant pathway neurodegeneration and synapse loss in a mouse model of early-stage Alzheimer-type tauopathy, PLoS One 10 (2015) e0142340.

[437] T. Jiang, J.T. Yu, X.C. Zhu, et al. Temsirolimus attenuates tauopathy in vitro and in vivo by targeting tau hyperphosphorylation and autophagic clearance, Neuropharmacology 85 (2014) 121–130.

[438] A. Verheyen, A. Diels, J. Dijkmans, et al. Using human iPSC-derived neurons to model tau aggregation, PLoS One 10 (2015) e0146127.

[439] I.F. Wang, B.S. Guo, Y.C. Liu, et al. Autophagy activators rescue and alleviate pathogenesis of a mouse model with proteinopathies of the TAR DNA-binding protein 43, Proc. Natl. Acad. Sci. USA 109 (2012) 15024–15029.

[440] J.K. Ching, C.C. Weihl, Rapamycin-induced autophagy aggravates pathology and weakness in a mouse model of VCP-associated myopathy, Autophagy 9 (2013) 799–800.

[441] A. Williams, S. Sarkar, P. Cuddon, et al. Novel targets for Huntington's disease in an mTOR-independent autophagy pathway, Nat. Chem. Biol. 4 (2008) 295–305.

[442] L. Zhang, J. Yu, H. Pan, et al. Small molecule regulators of autophagy identified by an image-based high-throughput screen, Proc. Natl. Acad. Sci. USA 104 (2007) 19023–19028.

[443] R.L. Vidal, S. Matus, L. Bargsted, C. Hetz, Targeting autophagy in neurodegenerative diseases, Trends Pharmacol. Sci. 35 (2014) 583–591.

[444] L. Li, S. Zhang, X. Zhang, et al. Autophagy enhancer carbamazepine alleviates memory deficits and cerebral amyloid-beta pathology in a mouse model of Alzheimer's disease, Curr. Alzheimer Res. 10 (2013) 433–441.

[445] U. Kruger, Y. Wang, S. Kumar, E.M. Mandelkow, Autophagic degradation of tau in primary neurons and its enhancement by trehalose, Neurobiol. Aging 33 (2012) 2291–2305.

[446] X. Wang, H. Fan, Z. Ying, B. Li, H. Wang, G. Wang, Degradation of TDP-43 and its pathogenic form by autophagy and the ubiquitin-proteasome system, Neurosci. Lett. 469 (2010) 112–116.

[447] M.J. Casarejos, R.M. Solano, A. Gomez, J. Perucho, J.G. de Yebenes, M.A. Mena, The accumulation of neurotoxic proteins, induced by proteasome inhibition, is reverted by trehalose, an enhancer of autophagy, in human neuroblastoma cells, Neurochem. Int. 58 (2011) 512–520.

[448] J.A. Rodriguez-Navarro, L. Rodriguez, M.J. Casarejos, et al. Trehalose ameliorates dopaminergic and tau pathology in parkin deleted/tau overexpressing mice through autophagy activation, Neurobiol. Dis. 39 (2010) 423–438.

[449] V. Schaeffer, I. Lavenir, S. Ozcelik, M. Tolnay, D.T. Winkler, M. Goedert, Stimulation of autophagy reduces neurodegeneration in a mouse model of human tauopathy, Brain 135 (2012) 2169–2177.

[450] K. Castillo, M. Nassif, V. Valenzuela, et al. Trehalose delays the progression of amyotrophic lateral sclerosis by enhancing autophagy in motoneurons, Autophagy 9 (2013) 1308–1320.

[451] X. Zhang, S. Chen, L. Song, et al. MTOR-independent, autophagic enhancer trehalose prolongs motor neuron survival and ameliorates the autophagic flux defect in a mouse model of amyotrophic lateral sclerosis, Autophagy 10 (2014) 588–602.

[452] C. Settembre, D.L. Medina, TFEB and the CLEAR network, Methods Cell Biol. 126 (2015) 45–62.

[453] T. Tsunemi, T.D. Ashe, B.E. Morrison, et al. PGC-1alpha rescues Huntington's disease proteotoxicity by preventing oxidative stress and promoting TFEB function, Sci. Transl. Med. 4 (2012) 142ra97.

[454] K. Kilpatrick, Y. Zeng, T. Hancock, L. Segatori, Genetic and chemical activation of TFEB mediates clearance of aggregated alpha-synuclein, PLoS One 10 (2015) e0120819.

[455] W. Song, F. Wang, P. Lotfi, M. Sardiello, L. Segatori, 2-Hydroxypropyl-beta-cyclodextrin promotes transcription factor EB-mediated activation of autophagy: implications for therapy, J. Biol. Chem. 289 (2014) 10211–10222.

[456] J.C. Latourelle, N. Pankratz, A. Dumitriu, et al. Genomewide association study for onset age in Parkinson disease, BMC Med. Genet. 10 (2009) 98.

[457] Genetic Modifiers of Huntington's Disease (GeM-HD) ConsortiumIdentification of Genetic factors that modify clinical onset of Huntington's disease, Cell 162 (2015) 516–526.

[458] A. Benussi, A. Padovani, B. Borroni, Phenotypic heterogeneity of monogenic frontotemporal dementia, Front. Aging Neurosci. 7 (2015) 171.

[459] J.H. Lee, R. Cheng, B. Vardarajan, et al. Genetic modifiers of age at onset in carriers of the G206A mutation in PSEN1 with familial Alzheimer disease among Caribbean Hispanics, JAMA Neurol. 72 (2015) 1043–1051.

[460] Deleted in review.

[461] J. Shugart, Stream of genetics pushes FTD research forward, Alzforum, (2014).

[462] C. Bonvicini, E. Milanesi, A. Pilotto, et al. Understanding phenotype variability in frontotemporal lobar degeneration due to granulin mutation, Neurobiol. Aging 35 (2014) 1206–1211.

[463] A. Chio, G. Restagno, M. Brunetti, et al. ALS/FTD phenotype in two Sardinian families carrying both C9ORF72 and TARDBP mutations, J. Neurol. Neurosurg. Psychiatry 83 (2012) 730–733.

[464] P. Origone, J. Accardo, S. Verdiani, et al. Neuroimaging features in C9orf72 and TARDBP double mutation with FTD phenotype, Neurocase 21 (2015) 529–534.

[465] M. van Blitterswijk, M.C. Baker, M. Dejesus-Hernandez, et al. C9ORF72 repeat expansions in cases with previously identified pathogenic mutations, Neurology 81 (2013) 1332–1341.

[466] T. Lashley, J.D. Rohrer, C. Mahoney, et al. A pathogenic progranulin mutation and C9orf72 repeat expansion in a family with frontotemporal dementia, Neuropathol. Appl. Neurobiol. 40 (2014) 502–513.

[467] T. Fagan, Can common genetic variation in mice nail genes of aging, Alzheimer's? Alzforum, (2015).

[468] J.F. Gusella, M.E. Macdonald, J.M. Lee, Genetic modifiers of Huntington's disease, Mov. Disord. 29 (2014) 1359–1365.

[469] V. Vickland, G. McDonnell, J. Werner, B. Draper, L.F. Low, H. Brodaty, A computer model of dementia prevalence in Australia: foreseeing outcomes of delaying dementia onset, slowing disease progression, and eradicating dementia types, Dement. Geriatr. Cogn. Disord. 29 (2010) 123–130.

[470] Alzheimer's Association, Changing the trajectory of Alzheimer's disease: how a treatment by 2025 saves lives and dollars, (2015).

[471] M. van Blitterswijk, B. Mullen, M.G. Heckman, et al. Ataxin-2 as potential disease modifier in C9ORF72 expansion carriers, Neurobiol. Aging 35 (2014) 2421–2427.

[472] D.M. van den Heuvel, O. Harschnitz, L.H. van den Berg, R.J. Pasterkamp, Taking a risk: a therapeutic focus on ataxin-2 in amyotrophic lateral sclerosis?, Trends Mol. Med. 20 (2014) 25–35.

[473] R.G. Palfree, H.P. Bennett, A. Bateman, The evolution of the secreted regulatory protein progranulin, PLoS One 10 (2015) e0133749.

[474] S. Demorrow, Progranulin: a novel regulator of gastrointestinal cancer progression, Transl. Gastrointest. Cancer 2 (2013) 145–151.

[475] T. Yu, Y. Yang, Yin dQ, et al. TBK1 inhibitors: a review of patent literature (2011–2014), Expert Opin. Ther. Pat. 25 (2015) 1385–1396.

[476] C.T. McMillan, J. Russ, E.M. Wood, et al. C9orf72 promoter hypermethylation is neuroprotective: neuroimaging and neuropathologic evidence, Neurology 129 (2015) 39–52.

[477] P.O. Bauer, Methylation of C9orf72 expansion reduces RNA foci formation and dipeptide-repeat proteins expression in cells, Neurosci. Lett. 612 (2016) 204–209.

[478] J. Chew, T.F. Gendron, M. Prudencio, et al. Neurodegeneration. C9ORF72 repeat expansions in mice cause TDP-43 pathology, neuronal loss, and behavioral deficits, Science 348 (2015) 1151–1154.

[479] H.N. Dawson, A. Ferreira, M.V. Eyster, N. Ghoshal, L.I. Binder, M.P. Vitek, Inhibition of neuronal maturation in primary hippocampal neurons from tau deficient mice, J. Cell Sci. 114 (2001) 1179–1187.

[480] A. Harada, K. Oguchi, S. Okabe, et al. Altered microtubule organization in small-calibre axons of mice lacking tau protein, Nature 369 (1994) 488–491.

[481] S. Ikegami, A. Harada, N. Hirokawa, Muscle weakness, hyperactivity, and impairment in fear conditioning in tau-deficient mice, Neurosci. Lett. 279 (2000) 129–132.

[482] P. Lei, S. Ayton, D.I. Finkelstein, et al. Tau deficiency induces parkinsonism with dementia by impairing APP-mediated iron export, Nat. Med. 18 (2012) 291–295.

[483] Z. Li, A.M. Hall, M. Kelinske, E.D. Roberson, Seizure resistance without parkinsonism in aged mice after tau reduction, Neurobiol. Aging 35 (2014) 2617–2624.

[484] M. Morris, P. Hamto, A. Adame, N. Devidze, E. Masliah, L. Mucke, Age-appropriate cognition and subtle dopamine-independent motor deficits in aged tau knockout mice, Neurobiol. Aging 34 (2013) 1523–1529.

[485] Y. Takei, J. Teng, A. Harada, N. Hirokawa, Defects in axonal elongation and neuronal migration in mice with disrupted tau and map1b genes, J. Cell Biol. 150 (2000) 989–1000.

[486] L. Crawford, J. Rosch, D. Putnam, Concepts, technologies, and practices for drug delivery past the blood-brain barrier to the central nervous system, J. Control. Release. 240 (2016) 251–266.

[487] J.D. Rohrer, J.M. Nicholas, D.M. Cash, et al. Presymptomatic cognitive and neuroanatomical changes in genetic frontotemporal dementia in the Genetic Frontotemporal dementia Initiative (GENFI) study: a cross-sectional analysis, Lancet Neurol. 14 (2015) 253–262.

[488] K. Sleegers, K. Bettens, R.A. De, et al. A 22-single nucleotide polymorphism Alzheimer's disease risk score correlates with family history, onset age, and cerebrospinal fluid Abeta42, Alzheimers Dement. 11 (2015) 1452–1460.

[489] E. Feneberg, P. Steinacker, A.E. Volk, et al. Progranulin as a candidate biomarker for therapeutic trial in patients with ALS and FTLD, J. Neural Transm. 123 (2016) 289–296.

FROM HUNTINGTIN GENE TO HUNTINGTON'S DISEASE-ALTERING STRATEGIES

Nicole Déglon

Lausanne University Hospital (CHUV), Lausanne, Switzerland
Neuroscience Research Center (CRN), Lausanne University Hospital (CHUV), Lausanne, Switzerland

CHAPTER OUTLINE

Huntington's Disease..251
Huntingtin Gene and Transcripts 253
Huntingtin Protein...257
HD Pathogenic Mechanisms 258
Molecular Strategies for HD 259
ASO...260
   RNA Interference 261
Genome Editing..262
Conclusions and Perspectives 263
Acknowledgments...264
References...264

HUNTINGTON'S DISEASE

Huntington's disease (HD) (MIM 143100) was first described by George Huntington in 1872 as a neurological disorder inherited in a dominant manner and characterized by involuntary movements in adult life [1]. Typical clinical features include involuntary dancing movements (chorea) evolving progressively into slow movements (bradykinesia) with rigidity and dystonia, cognitive impairment with dementia, and neuropsychiatric deficits including depression and apathy, leading to death after 15–20 years. More than a century passed before the underlying genetic mutation in HD was identified. In 1993, the HD mutation was found to be a CAG repeat expansion in exon 1 of the huntingtin (*HTT*) gene [2] (Fig. 9.1). The diagnosis is, in most cases, based on genetic testing for the presence of the mutation as well as a motor examination using the unified Huntington's disease rating scale (UHDRS) [3]. The trinucleotide repeat encodes an expanded polyglutamine (polyQ) stretch close to the N-terminus of the HTT protein. There is a negative correlation between the length of the CAG expansion and age at onset of HD symptoms [4–6]. However, the expanded polyglutamine tract only partially accounts for the age of onset [7]. Environmental factors, CAG repeat somatic mosaicism, and genetic risk factors may contribute to the disease [8–13]. Evidence of genetic modifiers was suggested by the HD-MAPS

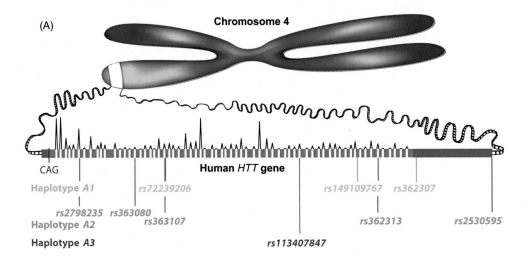

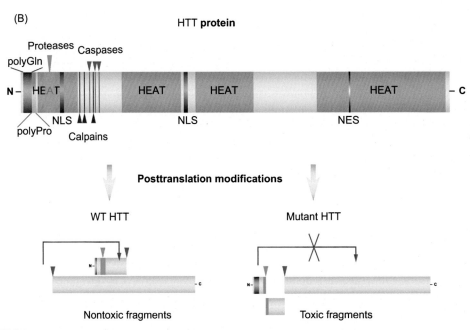

FIGURE 9.1

(A) The human *HTT* gene is located on the short (p) arm of chromosome 4 at position 16.3. The gene spans 180 kb, contains 67 exons with the first exon containing the CAG expansion. The analysis of all common target alleles across the *HTT* gene revealed that mutant *HTT* is enriched in A1, A2, and A3 haplotypes. A panel of single nucleotide polymorphisms (SNPs) uniquely defines these haplotypes and provides selective targets for mutant *HTT* silencing. (B) Schematic diagram of the huntingtin protein. PolyGln and polyPro designate the polyglutamine and polyproline tracts, respectively. The brown rectangles indicate the main groups of HEAT repeats. The nuclear export signals *(NES)* and nuclear localization *(NLS)* are indicated. The blue triangles signal the presence of caspase cleavage sites while the red triangles indicate the calpain cleavage sites. In healthy conditions, HTT proteolysis is limited and generates N-terminal fragments that interact with the C-terminal fragment. In Huntington's disease (HD), the fragmentation of mutant HTT is increased and produces toxic N-terminal fragments that no longer interact with the C-terminal fragment. El-Daher et al. demonstrated that this nonpolyQ C-terminal fragment has also deleterious effects in neurons.

(Modifiers of Age at onset in Pairs of Sibs) study [14], but genome-wide association (GWA) analysis on large cohorts of patients was necessary to identify loci that modify the age at onset of HD [7,15].

In contrast to the ubiquitous expression of the *HTT* gene, HD pathology is characterized by the specific vulnerability of neuronal populations in the striatum and to a lesser extent in the cortex, as well as other structures in the CNS and periphery [16,17]. In the striatum, degeneration of medium-spiny GABAergic neurons progresses along a caudal-rostral and dorsal-ventral axis toward the putamen [18]. In the cerebral cortex, neuronal death also occurs in frontal, parietal, and temporal regions [19,20]. The CAG-Age Product (CAP) score is used as a predictor of clinical progression and brain pathology (calculated as follows: age of the patient × CAG length − L = constant near the threshold of CAG repeat expansions) [21].

Patient care currently focuses on the management of symptoms and improvement of the quality of life [22–25]. Most of the experimental treatments that have been evaluated in clinical trials during the last 20 years have failed to demonstrate efficacy [26–28]. However, the mutation that causes HD provides a strong and rational target for intervention, the *HTT* gene and its products. A prerequisite for the development of HTT-lowering strategies is a solid understanding of the pathological and genetic mechanisms of HD. Large studies of HD patients and fundamental analyses of mutant HTT toxicity have provided new insights on the natural history, and RNA- and protein-mediated mechanisms operating in HD, offering new opportunities for therapeutic interventions [5,29–35]. In this review, we describe the structure of the *HTT* gene, its expression, and the biological features relevant for HTT-lowering strategies. Based on this knowledge, we then review strategies to reduce mutant *HTT* expression and current preclinical/clinical developments.

HUNTINGTIN GENE AND TRANSCRIPTS

The prevalence rate for HD in western populations is estimated to be between 7 and 13 individuals per 100,000, with an incidence of 4–7 new cases per million per year [36]. HD is present on all continents, but European ancestry probably accounts for the spread of HD mutations and their worldwide distribution [37–39]. Approximately 50% of Europeans with HD share a common ancestor [37,40–43]. This ancestor phenomenon was particularly well studied in the Maracaibo region, Venezuela, which has a very high HD prevalence due to a single ancestor, and in British Columbia, Canada [9,44]. Haplotype analysis identified a subset of single nucleotide polymorphisms (SNPs) that is highly associated with the disease chromosome [37,40–43]. Taking advantage of the 1000 genome project and HD patient cohorts, Kay et al. analyzed the haplotypes covering the entire *HTT* locus [45]. This study revealed that (1) recombination within the *HTT* gene is rare, (2) specific SNPs are present in HD and control chromosomes, and (3) the most common HD haplotypes (A1, A2, and A3) account for the majority of HD chromosomes (Fig. 9.1).

The human *HTT* gene is a large locus spanning 180 kb and consisting of 67 exons located on chromosome 4p16.3 [2,46,47] (Fig. 9.1). An open reading frame (ORF) encoding a 21-amino acid peptide is present within the 5′ untranslated region (5′ UTR) of the *HTT* gene [48]. *HTT* messenger RNA (mRNA) expression is negatively influenced by this ORF, probably due to reduced access of ribosomes to the downstream *HTT* initiation sites [48]. Lin et al. 1995 reported the presence of two putative *HTT* transcription start sites (TSS: −135 and −145 relative to the adenosine nucleotide of the ATG codon) [49]. However, the initiation of transcription of the *HTT* gene is probably more heterogeneous with multiple sites spanning nucleotides −145 to −129 [48]. Partial functional analysis of the HD promoter showed the presence of a GC rich region and binding sites for regulatory elements (Sp1, AP2, p53, and NF-κB) within a fragment extending from position −324 to +20, but the absence of TATA or CCAAT regulatory elements [49–54]. Two potential transcription factors and nuclear shuttle proteins, HDBP1 and HDPB2, bind to a 7 bp (GCCGGCG) triplicate sequence (−213 to −174), and are essential for HTT expression [55]. The first HD transgenic mice (R6/2 mice) show that a 1.9 kb genomic fragment, including approximately 1 kb of sequence upstream of the initiator ATG codon, is sufficient to control *HTT* expression [56]. Becanovic et al. demonstrated that genetic variation in the *HTT* promoter influences the transcriptional rate of *HTT*. They identified a SNP (rs13102260: G > A) that modulates the binding of NF-κB and alters the expression of *HTT*. The A variant delays HD pathology when associated with the disease allele and induces an earlier age of onset when present on the wild-type (WT) allele [57].

The *HTT* gene is expressed from two alternatively polyadenylated forms [58]. The 13.7 kb transcript is expressed predominantly in adult and fetal brain, whereas the 10.3 kb transcript is more widely expressed. In 2013, Sathasivam described the CAG repeat length-dependent aberrant splicing of exon 1 coding for a pathogenic and truncated HTT protein [59]. Additional alternatively spliced transcripts have been reported in mouse and human *HTT* [60,61]. However, these transcripts represent a small proportion of all *HTT* mRNAs and are present in both WT and HD samples (Fig. 9.2).

Two *HTT* antisense transcripts (*HTTAS_*v1: exon 1 containing the CTG repeat and exon 3; *HTTAS_*v2: exons 2 and 3), which are 5′ capped and poly (A) tailed, have also been identified [62] (Fig. 9.3). *HTTAS_*v1 is expressed at low levels in multiple tissue types and throughout the brain. The level of expression of *HTTAS_*v1 controls the expression of the sense *HTT* mRNA in a repeat length- and dicer-dependent manner [62]. *HTTAS_*v1 is detected in control brains, but not in postmortem HD samples. It was suggested that *HTTAS* negatively regulates *HTT* transcript expression and that the loss of *HTTAS* in HD increases mutant *HTT* mRNA levels. However, the loss of *HTTAS* is balanced by the direct positive effect of the repeat expansion on *HTT* expression thus the global impact on *HTT* expression in HD and control brains is almost neutral [62]. The analysis of postmortem tissues supports this conclusion and shows that the abundance of HTT mRNA and protein is not significantly altered by the disease process [63–67].

HTT is a ubiquitous protein [66,68–71] but the pattern of expression of WT and mutant HTT mRNA/protein is still only partially characterized and the results are sometimes inconsistent. *HTT* mRNA expression is low during early development and high in adult CNS neurons and testis [68–70,72]. Quantitative in situ hybridization studies have shown that human *HTT* mRNA is expressed in all brain regions. The highest levels are seen in cerebellum, hippocampus, cerebral cortex, substantia nigra pars compacta, and pontine nuclei [63]. The level of expression in the striatum is intermediate and that of the globus pallidus is low. *HTT* mRNA is expressed predominantly in neurons and at a low but appreciable level in glial cells [63,73]. All human neurons of the striatum are labeled by in situ hybridization [63], whereas only 65% of rat medium-sized striatal neurons (calbindin-labeled neurons localized to the matrix compartment) contain detectable *HTT* mRNA [74,75]. Human *HTT* mRNA expression is low in undifferentiated hESCs and human fibroblasts, higher in differentiated neurons [2.8-fold higher than neural stem cells (NSCs)], but still much lower than in human striatum and cortical samples (16- and 39-fold higher than NSCs) [76]. No difference between WT and mutant *HTT* RNA expression was observed in patient-derived lymphoblasts [77]. However, mutant *HTT* mRNA is more abundant than WT *HTT* mRNA in the cortex and striatum of 75% of HD patients based on allele-

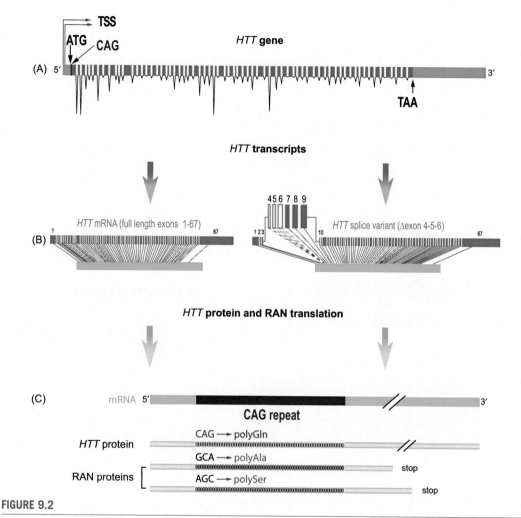

FIGURE 9.2

(A) Representation of the human *HTT* gene and the corresponding transcription start sites *(TSS; arrows)*. The *HTT* start codon (ATG) and CAG expansion in exon 1 are indicated as well as the stop codon (TAA) in exon 67. (B) Multiple splice variants have been identified in human brain samples. The canonical *HTT* transcript is shown on the left, and one of the novel variant (HTTΔex4,5,6) is shown on the right. (C) Schematic diagram showing the *HTT* mRNA and the production of polyGln containing HTT protein (ATG initiated). In addition, non-ATG initiated polyAla and polySer RAN proteins are generated. This noncanonical type of protein translation is length and hairpin dependent.

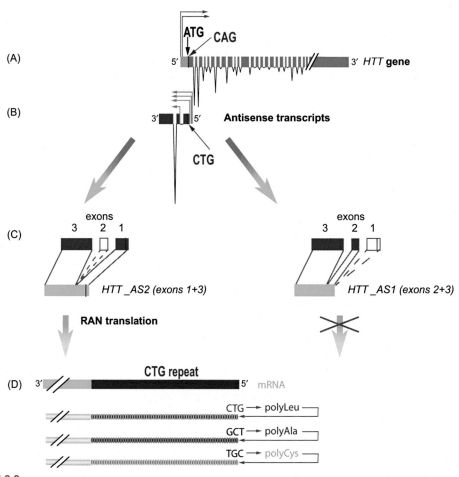

FIGURE 9.3

(A) Representation of the 5′ end of the human HTT gene. The *HTT* start codon (ATG) and CAG expansion in exon 1 are indicated. (B) The antisense transcripts, the corresponding transcription start sites (TSS; *arrows*) and the CTG repeats are indicated. (C) The antisense transcript HTT_AS1 contains exons 1 and 3 and HTT_AS2 contains exons 2 and 3. Exon 3 contains an alternative polyA+ signal. (D) Antisense RAN proteins are produced in all three reading frames and these RAN proteins (polyLeu, polyAla, polyCys) are toxic to cells.

specific quantitative PCR [78]. This difference is more pronounced in the early stage of the pathology (grade 1–2). No significant differences between two *HTT* alleles were measured in HD or control cerebellum. Lower expression of mutant than WT *HTT* mRNA was detected in a recent study performed on grade 3 and 4 samples [79].

In summary, although *HTT* is widely expressed, its level of expression varies greatly between organs, areas of the brain and cell populations. Its expression pattern is further complicated by alternative splicing, polyadenylation, and the presence of antisense transcripts. Ultimately, the pathological process appears to have only a limited impact on the production of both WT and mutant HTT.

HUNTINGTIN PROTEIN

HTT is a highly conserved protein of 348 kD containing 3144 amino acids. It is expressed in all vertebrates and has no sequence homology with other proteins [80] (Fig. 9.1). The HTT protein contains 16 elongation factor 3, protein phosphatase 2A, and TOR 1 (HEAT) repeat domains organized in 4 clusters. Soluble HTT has a half-life of approximately 24 h [81]. A crystal structure is not available due to its size, but HTT is predicted to form an elongated, superhelical solenoid, and to have a flexible structure [82]. The polyQ tract present at the N-terminus forms a polar zipper that mediates binding with various factors [83,84]. The HD mutation induces a conformational change and accumulation of misfolded HTT protein [85]. Several studies have questioned the role of these neuritic and nuclear aggregates in neuronal dysfunction and degeneration [86–89], although these aggregates constitute a hallmark of the pathology. HTT may traffic between the nucleus and cytoplasm. This is suggested by the presence of a nuclear export signal (NES) sequence and nuclear localization signal (NLS), as well as the interaction of the first 18 amino acids with the nuclear pore protein TPR [90,91]. HTT is also subjected to various posttranslational modifications, which regulate its intracellular localization, stability, and function [80]. The protease cleavage sites (caspases, calpain, and aspartyl proteases) generate N-terminal fragments that contain the polyQ stretch; the toxicity of these fragments is well established [92–100]. Recently, Saudou and coworkers have shown that the C-terminal fragments generated by these cleavages are also toxic due to dilation of the endoplasmic reticulum (ER) and increased ER stress [101] (Fig. 9.1).

The WT HTT is predominantly found in the cytoplasm and colocalizes with many organelles, including the ER, golgi complex, endosomes, and mitochondria [69,102–104]. HTT-interacting proteins have been analyzed to provide insights into its function. This has revealed a network of interactions with a large number of effector proteins that control various physiological processes including clathrin-mediated endocytosis and vesicle transport, cell signaling, morphogenesis, autophagy, and transcriptional regulation [69,105–122]. Recent reviews are providing comprehensive information on current knowledge of HTT biology [35,123–125].

Variable levels of HTT have been observed in human and rodent medium-sized striatal neurons by immunolabeling using various antibodies [69,71,75,102,126]. HTT is rarely detected in somatostatinergic striatal interneurons. A small fraction of parvalbuminergic interneurons is positive for HTT; the progressive depletion of these neurons in advanced HD has been postulated to be associated with dystonia in HD [127]. HTT protein and mRNA are very abundant in large cholinergic interneurons, which are spared in HD [75,126,128–131]. Striatonigral neurons containing both substance P and D1a dopamine receptors express a higher level of HTT protein than striatopallidal neurons [132], a result supported by single-cell RT-qPCR analysis [75]. HTT is present in the globus pallidus, the entopeduncular nucleus, the basal nucleus of Meynert, and the ventral pallidum. Finally, HTT is enriched in cortical pyramidal neurons in layers III and V that project to the striatal neurons [75]. In the striatum, enkephalin (enk) expressing neurons projecting to the external pallidum are affected in early stages of HD, whereas neurons containing substance P (SP) and projecting to the internal pallidum and the substantia nigra degenerate in advanced stages of the pathology [133–139].

In the cortex, large pyramidal projection neurons in layers V, VI and to a lesser extent, layer III, are preferentially lost in HD [140,141].

HD PATHOGENIC MECHANISMS

The mechanisms leading to HD are not fully understood, but HTT is an organelle-associated scaffolding protein playing a role in various intracellular pathways. A large number of studies have demonstrated the role of WT HTT in vesicle/RNA transport, cytoskeletal organization, postsynaptic signaling, transcription, proteasomal degradation, Ca^{2+} anomalies, autophagy, antiapoptotic processes, and bioenergetic defects [35,123–125]. Differential roles of HTT in early development, adulthood, cell-type specific functions, and age-dependent mechanisms are emerging but we are still lacking a clear picture of the contribution of these different pathways in HD and how this ultimately lead to selective neuronal death in the brain [34,142–144]. Excellent reviews are covering this topic, which is beyond the scope of the present chapter [35,123–125].

The autosomal dominant inheritance of polyglutamine expansion in HD can result in a gain-of-function mutation or a more complex gain of function with a partial loss of function. The toxic gain-of-function is supported by the fact that HD homozygotes are indistinguishable from their heterozygous siblings while they do not express WT *HTT* [145–148]. Individuals with a chromosomal translocation that inactivates one *HTT* allele do not exhibit HD symptoms [81,149] A 50% reduction of WT *HTT* expression in mice does not affect the survival of neurons [150–152]. The WT *HTT* gene was inactivated in mice to further demonstrate gain- versus loss-of-function in HD [150,153,154]. Knock-out of the Huntington's disease homolog (Hdh) gene in the mouse results in early embryonic lethality. However, the analysis of chimeras reveals the role of HTT protein in extraembryonic membrane function, presumably in the vesicular transport of nutrients [155]. Perinatal lethality is observed in transgenic animals with a reduced level of mutant HTT (Q50 expressed below 50% of normal levels) and no WT HTT [156]. Conditional gene inactivation leading to an 84% reduction in HTT protein triggers degeneration of striatal and cortical neurons [157]. Human mutant *HTT* compensates for the loss of murine *HTT* and rescues the embryonic lethal phenotype [158]. Recently, Wang and collaborators show that ablation of WT HTT, in adult but not in young animals, is nondeleterious [159,160]. These data support the toxic gain-of-function model. However, several studies demonstrate a partial failure of mutant HTT to exhibit functional activities of the WT protein suggesting a partial loss of function in HD pathology [161]. Jacobsen et al. analyzed the transcriptomic profiles from WT, null and knock-in (different CAG length) mouse embryonic stem cells to discriminate between a simple gain of function or a more complex gain of function coupled to a loss-of-function mechanism (e.g., dominant negative-graded loss of function) [162]. If CAG expansion contributes to HTT loss of function, then an overlap between the null and knock-in profiles would be expected, whereas this would not be the case for a simple gain of function. The comparison of the *HTT*-null and ES with CAG expansions of various length (HdhQ20/7, HdhQ50/7, HdhQ91/7, HdhQ111/7) revealed virtually no overlap. Altogether, these studies support the hypothesis that a gain-of-function mechanism plays a major role in HD transcriptional dysregulation.

The pathogenesis of HD is commonly attributed to the mutant HTT protein. However, the presence of long hairpin-forming repeats in *HTT* mRNA induces sense and antisense repeat-associated non-ATG (RAN) translation (Figs. 9.1 and 9.2). RAN translation in HD brain leads to the synthesis and accumulation of homopolymeric polyglutamine, polyalanine, polyserine, polyleucine, and polycysteine proteins [163]. RAN translation is polyglutamine length dependent, and RAN proteins are toxic to neural cells [164,165]. The CAG expansion also induces a translational frameshift that generates hybrid PolyQ/A proteins due to the depletion of charged glutaminyl-transfer RNA (tRNA)$^{Gln-CUG}$ that pairs with the CAG codon [166]. The frequency of this translational frameshift is higher in structures with low concentrations of transfer (tRNA)$^{Gln-CUG}$, such as the striatum [166]. The contribution of this phenomenon to the pathological process is a matter of debate.

Beside HTT protein toxicity, growing evidence supports the importance of RNA toxicity [167]. Mutant *HTT* RNA sequesters the muscleblind-like 1 (MBNL1) splicing factor and induces misregulation of alternative splicing [168]. The aberrant splicing of *HTT* exon 1 is triggered by the increased binding of the alternative splicing factor SRSF6 to expanded CAG repeats [59]. Krol et al. showed that dicer cleaves transcripts containing long hairpin structures [169]. In dicer-deficient cells, the level of mutant HD transcripts is approximately 50% higher than in cells containing normal dicer activity, whereas the level of normal *HTT* mRNA is unchanged [169]. The interaction of mutant CAG repeats with nucleolin induces stress [170]. Perturbation of the small nuclear RNA U2-associated factor U2AF65/NXF1 complex that interacts with transcripts containing expanded CAG results in the accumulation of mutant *HTT* mRNA in the nucleus [170]. Finally, Woerner et al. show that cytoplasmic mutant Htt but not nuclear aggregates affect the transport of RNA and induce the accumulation of nuclear mRNA in cells [171]. Altogether, these data suggest that both RNA and protein toxicity play a role in the HD phenotype.

The mechanisms that contribute to HD are partially explained. It is not clear whether the *HTT* mutation itself may induce neuronal dysfunction and cell death (cell autonomous) or whether cell–cell communication is involved (noncell autonomous). Both cell- and noncell-autonomous mechanisms may contribute to behavioral deficits and neuronal degeneration whereas data clearly demonstrate that aggregates are formed in a cell-autonomous manner [172]. These studies have also highlighted the contribution of astrocytes, microglial cells [173–176], and brain circuitry in the pathology of the disease [177,178]. Two recent publications show that HTT protein transfer occurs in neuronal networks and in grafted patients [179,180]. The relative contribution of these different mechanisms and in particular HTT protein transfer to HD progression has not been established.

MOLECULAR STRATEGIES FOR HD

Experimental approaches based on drug, cell, or gene therapy are currently in development and are progressively reaching the clinic [181]. Several preclinical or clinical studies with drugs to counteract cellular perturbations and/or directly target polyglutamine protein expression, cleavage, or conformations have been initiated [182,183]. Strategies that directly interfere with the activity of the neurotoxic gene are particularly appealing for autosomal dominant neurological disorders [184–188]: the concept of antisense oligonucleotides was first described in the 1970s [189], RNA interference (RNAi) was discovered in plants in the 1990s [190], and the powerful CRISPR/Cas9 system for human genome editing was first reported a few years ago [191]. The most advanced HTT-lowering therapeutics target the *HTT* mRNA. In 2015, the first clinical trial using an antisense oligonucleotide (ASO) that targets the human *HTT* was initiated in Europe and Canada (Isis Pharmaceuticals and Roche; http://en.hdbuzz.net/182).

ASO

The use of ASOs is the most established approach for RNA-based therapies. It is based on the use of single-stranded 20–25 base pair (bp) long DNA molecules that hybridize with complementary mRNA via Watson–Crick base pairing [192,193]. ASO–mRNA heteroduplex formation in the nucleus inhibits gene expression by various mechanisms. The most prominent is the activation of RNase H activity and degradation of the corresponding mRNA [192], and accounts for the majority of RNA-based drugs in development. Algorithms are available to identify target sites with appropriate nucleotide composition, binding energy, and secondary structure of the mRNA [194,195]. RNase H1 and RNase H2 are present in human cells. RNase H1 is the enzyme responsible for ASO-mediated cleavage [196]. Initial *HTT* mRNA targeting studies demonstrated a decrease in the level of endogenous *HTT* mRNA and a reduction of misfolded HTT protein accumulation [197–199]. Stanek et al. showed that ICV delivery of ASOs directed against mutant *HTT* in YAC128 mice reduced mutant HTT levels, corrected transcriptional alterations, and improved behavioral deficits [200]. Second generation ASOs, including 2′-O-methoxyethylribose (MOE) gapmer oligonucleotides, have an increased affinity for RNA and are highly resistant to nuclease degradation, thus improving both their pharmacokinetic properties and pharmacological activity. Strong preclinical evidence supports the use of these MOE gapmer ASOs in HD [201]. Continuous infusion of HuASO (a sequence targeting exon 36 of human *HTT*; 10–50 µg/day) for 2 weeks in the right ventricle of HD transgenic mice induced a dose-dependent and selective reduction of *HTT* mRNA for up to 12 weeks. Suppression of human mutant *HTT* mRNA in these HD transgenic mice reduced the disease phenotype. A 4-week treatment of symptomatic R6/2 HD mice increased survival and reduced brain atrophy without modifying the number of HTT aggregates. Early treatment of YAC128 and BACHD mice significantly improved motor coordination on a rotarod compared to untreated animals. Transient suppression (4 months) of mutant *HTT* mRNA was sufficient to obtain a long-term benefit (9 months) [201]. The authors also showed that an ASO targeting both human and murine *HTT* (MoHuASO targeting a sequence in exon 30 of *HTT* reduces human and mouse mRNA to 31% ± 4% and 17% ± 4% of untreated levels, respectively) in BACHD mice was tolerated and preserved therapeutic benefits. Finally, the scale-up of the delivery was assessed with the intrathecal administration of ASO in primates (MkHuASO targeting a sequence in exon 37 of *HTT*). Regions close to the ventricles had high levels of ASO. *HTT* mRNA was reduced in the cortex, whereas deeper brain structures, such as the caudate were unaffected [201].

In the phase I clinical trial initiated in 2015 (http://www.prnewswire.com/news-releases/isis-pharmaceuticals-initiates-clinical-study-of-isis-htt-rx-in-patients-with-huntingtons-disease-300115250.html), the ASO-HTTRx will be administered to the cerebrospinal fluid (CSF) of a small number of HD patients to induce the degradation of both WT and mutant *HTT* mRNA. They will be treated with various doses of ASO-HTTRx in order to identify potential adverse effects and the optimal dose. A placebo group (25% of the patients) is included. ASO-HTTRx will be administered once a month, as it should have an effect after 4–6 weeks with activity lasting up to 4 months. Clinical rating scales will be used to monitor HD symptoms, but the primary outcome measures will be the safety and tolerability of ASO-HTTRx. This first clinical trial will investigate several issues of HTT-lowering strategies. It should provide information on whether preclinical rodent results translate to HD patients and whether both the mutant and WT *HTT* alleles can be targeted or if allele-specific approaches should be considered.

SNPs in the *HTT* gene have been described, and genotype and allele frequencies have been reported [202–204]. The recent description of the most frequent HD haplotypes will facilitate the selection and validation of potential allele-specific sequences. Screening of a few SNPs per patient should yield at least one disease allele-linked SNP for selective targeting [45]. The proof-of-concept of allele-specific degradation of mutant *HTT* mRNA has already been obtained with an ASO that selectively targets single SNPs in *HTT* exons or introns [45,205]. Chemical modifications have been incorporated based on structure-activity design to increase selectivity. A screening pipeline has been developed using primary neurons that express human WT and mutant *HTT* with or without endogenous murine *HTT* [206–208]. These new ASOs limit RNase H cleavage of the nontargeted transcript and should provide a panel of allele-specific ASOs for the treatment of most HD patients.

RNA INTERFERENCE

RNAi, based on naturally occurring and conserved molecular machinery, inhibits expression of a target gene [209]. This cellular machinery can be hijacked to induce specific gene expression silencing. The exogenous and artificial RNAi "moieties" that are applied to mammalian cells include synthetic siRNAs, short-hairpin RNA (shRNA) and artificial miRNAs. Artificial miRNAs act as siRNA shuttles, also called miR-embedded siRNA, to promote gene silencing in a manner similar to naturally occurring miRNAs [210–214]. Vectorized shRNA and miR-embedded siRNA ensure continuous and long-term expression in the CNS. The double-stranded precursor RNA is first processed in the nucleus by the endoribonuclease enzyme Drosha and transported to the cytoplasm by the nuclear exportin-5. The cytoplasmic endoribonuclease enzyme Dicer cleaves the molecules into small interfering RNAs (siRNAs), which are 19–25 nt long double-stranded RNA. The guide strand associates with the Ago2 protein to form the RNA-induced silencing complex (RISC) and the passenger strand is degraded. The complementarity of the guide strand RNA with its target mRNA induces the cleavage and elimination of the corresponding transcript and, as a consequence, its protein product [215]. Over 30 siRNA-based therapeutics have reached the clinic [216–218]. Within the field of polyglutamine diseases, RNAi was first tested in SBMA [219], SCA3/MJD [220], and HD [221–226]. Today over 15 studies have been conducted [186,227–236].

Long-term delivery of siRNA to the CNS remains a challenge for the application of RNAi. The negative charge of siRNAs as well as their size makes it difficult for them to cross the cell membrane. Various delivery strategies include nanoparticles, cationic lipids, cholesterol, ultrasound, and viral vectors [226,228,237,238]. The siRNAs must also cross the blood–brain barrier; a rabies glycoprotein peptide [239] and gold nanoparticles [240] have been proposed. Gene transfer with viral vectors is a promising strategy to ensure continuous and long-term expression of siRNAs in the CNS. The use of viral-mediated delivery of shRNAs and miR-embedded siRNA has led to reduced *HTT* transcript levels, decreased formation of inclusions, and improvements in behavioral tests in HD animal models [223,225,231,232]. RNAi treatment was mostly initiated when the animals were still presymptomatic. Thus, limited conclusions can be drawn on the reversal of HD neuropathology. But data from conditional HD mice show that inhibiting mutant human *HTT* expression in symptomatic animals reversed the neuropathology [241]. Two rodent studies showed that RNAi treatment could block and partially reverse the accumulation of HTT inclusions, and improve striatal dysfunction [227,235]. These results suggest that siRNA therapy should be effective in early symptomatic patients.

Most of these proof of concept studies have been performed with siRNA that specifically targeted human *HTT* sequences, leaving the endogenous rodent WT *HTT* mRNA intact. These siRNAs would degrade both WT and mutant *HTT* transcripts if applied to humans. Several studies have shown that, similar to what has been obtained with ASO, the cosilencing of WT and mutant *HTT* provides therapeutic efficacy and is well tolerated in rodent models of HD [233,235]. The long-term silencing of both mutant and WT *HTT* has no impact on the survival of GABAergic neurons up to 9 months post-treatment and is associated with a dramatic reduction of HD neuropathology coupled with an improvement of behavioral deficits. Silencing of endogenous WT *HTT* in monkeys is also well tolerated [242–244]. Altogether, these results suggest that long-term partial reduction of WT *HTT* expression in adult brain may be possible and that residual levels of *HTT* transcripts are sufficient to maintain biological functions.

Allele-specific therapeutic strategies have been developed as an alternative strategy to preserve WT *HTT* expression and functions. Most of these rely on the presence of SNPs to discriminate WT and mutant *HTT* alleles. Several studies, including one publication from our group, have successfully used this approach for allele-specific silencing of mutant *HTT* [202,204,236,245,246]. These studies have shown that preferential silencing of mutant *HTT* is feasible, but additional development is required to improve the selectivity of some candidates. Regardless of the strategy, long-term treatment/delivery to the CNS will be required to degrade *HTT* mRNA and maintain a therapeutic benefit for 15–20 years.

GENOME EDITING

The recent development of gene editing in mammals provides new avenues to translate human genetics into therapeutic tools by broadening our ability to perform gene disruption or gene correction strategies. Examples of therapeutic editing with the permanent correction of disease-causing mutations are growing [247–252]. Mutations responsible for monogenic diseases are promising targets for these approaches. The first protein-guided genome engineering systems are based on meganucleases, zinc finger nucleases (ZFN), or transcription-activator like effector nucleases (TALEN) [253]. Customized DNA-binding domains that recognize specific target DNA sequences are fused to nucleases or repressors to induce gene inactivation or transcriptional repression. Garriga-Canut et al. designed artificial ZFN chains designed to bind to longer CAG repeats more efficiently than shorter ones. They demonstrated that viral delivery by intrastriatal injection of R6/2 mice with an adeno-associated virus (AAV) efficiently repressed (up to 60%) mutant *HTT* expression in the brain, reduced HTT aggregates, and alleviated behavioral deficits based on rotarod and clasping analyses [254].

A new genome editing approach based on clustered, regularly interspaced, short palindromic repeat (CRISPR)-associated protein Cas9/Cpf1 systems has greatly facilitated genome interrogation [255]. The ease of design, high efficiency, and relatively low cost of CRISPR/Cas9 has contributed to the rapid implementation of this technology in the scientific community. The CRISPR system contains two components: (1) an artificial single guide RNA (sgRNA) of approximately 20 nucleotides, which recognizes the target sequence by Watson–Crick base pairing, fused at the 3′ end to the tracrRNA that binds to Cas9 and (2) the Cas9 nuclease that cleaves the DNA following sgRNA hybridization with the complementary strand. Any sequence of 23 nucleotides, including the 3′ NGG/NAG protospacer adjacent motif, can be a CRISPR target sequence. This provides a large number of potential targets for any gene and greatly facilitates the development of DNA editing strategies compared to other systems. Mutant cells lines and mice bearing multiple modified alleles have been generated using this technology [256] and large collections of sgRNAs have been produced to interrogate gene function (gene disruption) on a genome-wide scale in human cell lines [257–259]. These technologies represent a paradigm shift and may offer the first opportunity to permanently correct genetic defects. Allele-specific (SNP) and multiplexed gene editing have been achieved [260–262]. The first proof-of-principle of in vivo gene repair for a genetic disease was obtained using the CRISPR system in the liver of mice carrying the mutation that causes type I hereditary tyrosinemia [250]. In vivo gene editing of the Mecp2 locus in the CNS has also been obtained [263]. In this study, ~65% of the neurons transduced with the viral vectors (GFP-KASH labeling and isolation of nuclei) were genetically modified. This level of efficiency confirms the interest of the approach for HD.

Our group recently incorporated genes coding for the CRISPR system into viral vectors for delivery to the brain and optimized the approach using fluorescent reporter proteins to facilitate the monitoring of gene targeting. Up to 50% gene disruption was achieved in neurons and astrocytes with a corresponding loss of green fluorescence. Viral-mediated delivery of the CRISPR system was associated with high efficiency of gene editing and up to 97% loss of reporter gene expression in the striatum of adult mice with no notable toxicity [264,265]. Based on these data, we targeted the human *HTT* gene and induced gene disruption to prevent HTT protein expression. Experiments in cultured cells and in the striatum of mice expressing the human mutant *HTT* gene reveal extremely efficient gene disruption and an up to 90% loss of HTT protein. *HTT* disruption also leads to a strong reduction of misfolded HTT protein. These results are an important step toward the use of the CRISPR system for *HTT* gene disruption, allele-specific *HTT* inactivation, or gene repair.

CONCLUSIONS AND PERSPECTIVES

Tremendous progress and promising preclinical proof-of-concept have been obtained for HTT-lowering therapeutics. However, appropriate clinical trial design, efficient delivery systems, and biosafety issues are major challenges for the development of molecular therapies for a rare and chronic disease. Conventional clinical trials usually involve large patient cohorts and a relatively long follow-up. The identification and validation of unbiased and sensitive clinical parameters to detect subtle changes over short periods of time are needed. They must reflect disease severity and be suitable for measuring the efficacy of HTT-lowering treatment. Some of these issues are currently being addressed in large longitudinal observational studies (PREDICT-HD, TRACK-HD, COHORT, REGISTRY, Enroll-HD) and ongoing clinical trials [30,31,266,267]. The availability of surrogate markers [268–271] will be essential and include: MRI imaging for brain atrophy measurement [272,273], diffusion tensor imaging or PET to measure global brain connectivity, [274–276] and biochemical measures to detect HTT protein [277]. These will be necessary to design appropriately powered clinical trials to evaluate therapeutic strategies in premanifest patients, and to demonstrate meaningful therapeutic benefit with objective and quantitative measures.

The second challenge for the clinical implementation of HTT-lowering strategies is a safe and long-term delivery system. Intrathecal (ASO) and intraventricular or intraparenchymal (viral vectors) administration have been successfully applied in CNS clinical trials. Patients with a familial form of amyotrophic lateral sclerosis (SOD1 mutation) were injected in their CSF with ASO [278]. This phase I clinical trial demonstrates the safety of the approach. The SOD1 protein level in the CSF was unchanged, but postmortem analysis of the spinal cord of one patient revealed the presence of the ASO and SOD1 protein levels at the low end of the normal range [278]. The administration of ASO to correct an SMN2 splicing error has recently been approved for clinical trial in humans

[279]. Stereotactic injection of AAV-2 based vectors has been used in clinical trials in patients suffering from Parkinson's, Alzheimer's, Batten or Canavan's disease [280–285]. AAVs are considered to be safe viral gene delivery systems, because WT AAVs are not associated with any known human or animal diseases and recombinant vectors persist predominantly as episomes without known cytotoxicity. Lentiviral vectors (LV) are extremely efficient at transducing dividing and quiescent cells, resulting in high level and stable transgene expression with no cytotoxicity. Their relatively simple production has made them highly attractive in research programs and has led to their widespread use. One potential limitation of LV is the risk of insertional mutagenesis. However, recent studies suggest that LV has a low genotoxicity profile compared to the murine retrovirus MLV and targeting postmitotic cells in the CNS further reduces the risk [286–289]. Several clinical trials with LVs [290] are ongoing [291,292] including in the CNS for the treatment of adrenoleukodystrophy [293] and Parkinson's disease [294,295].

Oligonucleotide-based approaches are maturing into potentially powerful therapeutic strategies for HD. Several issues still need to be addressed, mostly concerning the long-term specificity and biosafety of these approaches. More work is needed to support allele- and nonallele-specific HTT targeting. The development and validation of highly specific and efficient candidates with minimal potential off-target effects is crucial to maximize therapeutic benefit and reduce potential cytotoxicity. This will be critical for future clinical trials, which may be performed in early stage or premanifest HD patients.

ACKNOWLEDGMENTS

We would like to thank Dr. Christian wider for critical reading of the manuscript and Maria Rey for the production of the figures.

REFERENCES

[1] G. Huntington, On chorea, Med. Surg. Rep. 26 (1872) 317–332.

[2] The, et al. A novel gene containing a trinucleotide repeat that is expanded and unstable on Huntington's disease chromosome, Cell 72 (1993) 971–983.

[3] Huntington, Study & GroupUnified Huntington's disease rating scale: reliability and consistency, Mov. Disord. 11 (1996) 136–142.

[4] J.B. Penney Jr., J.P. Vonsattel, M.E. MacDonald, J.F. Gusella, R.H. Myers, CAG repeat number governs the development rate of pathology in Huntington's disease, Ann. Neurol. 41 (1997) 689–692.

[5] S. Finkbeiner, Huntington's disease, Cold Spring Harb. Perspect. Biol. 3 (2011) a007476.

[6] G.P. Bates, et al. Natural history of clinical Huntington disease, Nat. Rev. Dis. Prim. 1 (2015) 1–21.

[7] J.F. Gusella, M.E. MacDonald, J.M. Lee, Genetic modifiers of Huntington's disease, Mov. Disord. 29 (2014) 1359–1365.

[8] V.C. Wheeler, et al. Length-dependent gametic CAG repeat instability in the Huntington's disease knock-in mouse, Hum. Mol. Genet. 8 (1999) 115–122.

[9] TUS-VCR Project, et al. Venezuelan kindreds reveal that genetic and environmental factors modulate Huntington's disease age of onset, Proc. Natl. Acad. Sci. USA 101 (2004) 3498–3503.

[10] A.G. Mason, et al. Expression levels of DNA replication and repair genes predict regional somatic repeat instability in the brain but are not altered by polyglutamine disease protein expression or age, Hum. Mol. Genet. 23 (2014) 1606–1618.

[11] S. Metzger, et al. The S18Y polymorphism in the UCHL1 gene is a genetic modifier in Huntington's disease, Neurogenetics 7 (2006) 27–30.

[12] W. Zeng, et al. Genetic analysis of the GRIK2 modifier effect in Huntington's disease, BMC Neurosci. 7 (2006) 62.

[13] L. Arning, et al. NR2A and NR2B receptor gene variations modify age at onset in Huntington disease in a sex-specific manner, Hum. Genet. 122 (2007) 175–182.

[14] J.L. Li, et al. A genome scan for modifiers of age at onset in Huntington disease: the HD MAPS study: interaction of normal and expanded CAG repeat sizes influences age at onset of Huntington disease, Am. J. Hum. Genet. 119A (2003) 279–282.

[15] Genetic Modifiers of Huntington's Disease, CIdentification of genetic factors that modify clinical onset of Huntington's disease, Cell 162 (2015) 516–526.

[16] I. Han, Y. You, J.H. Kordower, S.T. Brady, G.A. Morfini, Differential vulnerability of neurons in Huntington's disease: the role of cell type-specific features, J. Neurochem. 113 (2010) 1073–1091.

[17] J.M. van der Burg, M. Bjorkqvist, P. Brundin, Beyond the brain: widespread pathology in Huntington's disease, Lancet Neurol. 8 (2009) 765–774.

[18] J.P. Vonsattel, M. DiFiglia, Huntington disease, J. Neuropath. Exp. Neurol. 57 (1998) 369–384.

[19] D.M. Mann, R. Oliver, J.S. Snowden, The topographic distribution of brain atrophy in Huntington's disease and progressive supranuclear palsy, Acta. Neuropathol. 85 (1993) 553–559.

[20] H. Heinsen, et al. Cortical and striatal neurone number in Huntington's disease, Acta. Neuropathol. 88 (1994) 320–333.

[21] Y. Zhang, et al. Indexing disease progression at study entry with individuals at-risk for Huntington disease, Am. J. Med. Genet. B Neuropsychiatr. Genet. 156B (2011) 751–763.

[22] M.B. Hocaoglu, E.A. Gaffan, A.K. Ho, The Huntington's disease health-related quality of life questionnaire (HDQoL): a disease-specific measure of health-related quality of life, Clin. Genet. 81 (2012) 117–122.

[23] E. Clay, et al. Validation of the first quality-of-life measurement for patients with Huntington's disease: the Huntington quality of life instrument, Int. Clin. Psychopharmacol. 27 (2012) 208–214.

[24] J. Read, et al. Quality of life in Huntington's disease: a comparative study investigating the impact for those with pre-manifest and early manifest disease, and their partners, J Huntingtons Dis. 2 (2013) 159–175.

[25] D. Zielonka, M. Mielcarek, G.B. Landwehrmeyer, Update on Huntington's disease: advances in care and emerging therapeutic options, Parkinsonism Relat. Disord. 21 (2015) 169–178.

[26] T.A. Mestre, J.J. Ferreira, An evidence-based approach in the treatment of Huntington's disease, Parkinsonism Relat. Disord. 18 (2012) 316–320.

[27] C. Sampaio, B. Borowsky, R. Reilmann, Clinical trials in Huntington's disease: interventions in early clinical development and newer methodological approaches, Mov. Disord. 29 (2014) 1419–1428.

[28] A. Kumar, et al. Huntington's disease: an update of therapeutic strategies, Gene 556 (2015) 91–97.

[29] E.R. Dorsey, et al. Natural history of Huntington disease, JAMA Neurol. 70 (2013) 1520–1530.

[30] J.S. Paulsen, et al. Clinical and biomarker changes in premanifest Huntington disease show trial feasibility: a decade of the PREDICT-HD study, Front. Aging Neurosci. 6 (2014) 78.

[31] S.J. Tabrizi, et al. Predictors of phenotypic progression and disease onset in premanifest and early-stage Huntington's disease in the TRACK-HD study: analysis of 36-month observational data, Lancet Neurol. 12 (2013) 637–649.

[32] A. Rosenblatt, et al. Predictors of neuropathological severity in 100 patients with Huntington's disease, Ann. Neurol. 54 (2003) 488–493.

[33] R.A. Barker, S.L. Mason, Neurodegenerative disease: mapping the natural history of Huntington disease, Nat. Rev. Neurol. 10 (2014) 12–13.

[34] J.J. Weber, A.S. Sowa, T. Binder, J. Hubener, From pathways to targets: understanding the mechanisms behind polyglutamine disease, BioMed Res. Int. 2014 (2014) 701758.

[35] C.A. Ross, et al. Huntington disease: natural history, biomarkers and prospects for therapeutics, Nat. Rev. Neurol. 10 (2014) 204–216.

[36] T. Pringsheim, et al. The incidence and prevalence of Huntington's disease: a systematic review and meta-analysis, Mov. Disord. 27 (2012) 1083–1091.

[37] S.C. Warby, et al. HTT haplotypes contribute to differences in Huntington disease prevalence between Europe and East Asia, Eur. J. Hum. Genet. 19 (2011) 561–566.

[38] M.R. Hayden, H.C. Hopkins, M. Macrea, P.H. Beighton, The origin of Huntington's chorea in the Afrikaner population of South Africa, S. Afr. Med. J. 58 (1980) 197–200.

[39] E. Almqvist, et al. Ancestral differences in the distribution of the delta 2642 glutamic acid polymorphism is associated with varying CAG repeat lengths on normal chromosomes: insights into the genetic evolution of Huntington disease, Hum. Mol. Genet. 4 (1995) 207–214.

[40] J. Scholefield, J. Greenberg, A common SNP haplotype provides molecular proof of a founder effect of Huntington disease linking two South African populations, Eur. J. Hum. Genet. 15 (2007) 590–595.

[41] D. Falush, Haplotype background, repeat length evolution, and Huntington's disease, Am. J. Hum. Genet. 85 (2009) 939–942.

[42] S.C. Warby, et al. CAG expansion in the Huntington disease gene is associated with a specific and targetable predisposing haplogroup, Am. J. Hum. Genet. 84 (2009) 351–366.

[43] J.M. Lee, et al. Common SNP-based haplotype analysis of the 4p16.3 Huntington disease gene region, Am. J. Hum. Genet. 90 (2012) 434–444.

[44] E.R. Fisher, M.R. Hayden, Multisource ascertainment of Huntington disease in Canada: prevalence and population at risk, Mov. Disord. 29 (2014) 105–114.

[45] C. Kay, et al. Huntingtin haplotypes provide prioritized target panels for allele specific silencing in Huntington disease patients of European ancestry, Mol. Ther. 23 (2015) 1759–1771.

[46] B. Lin, et al. Sequence of the murine Huntington disease gene: evidence for conservation, alternate splicing and polymorphism in a triplet (CCG) repeat [corrected], Hum. Mol. Genet. 3 (1994) 85–92.

[47] I. Schmitt, et al. Expression of the Huntington disease gene in rodents: cloning the rat homologue and evidence for downregulation in non-neuronal tissues during development, Hum. Mol. Genet. 4 (1995) 1173–1182.

[48] J. Lee, et al. An upstream open reading frame impedes translation of the huntingtin gene, Nucleic Acids Res. 30 (2002) 5110–5119.

[49] B. Lin, et al. Structural analysis of the 5' region of mouse and human Huntington disease genes reveals conservation of putative promoter region and di- and trinucleotide polymorphisms, Genomics 25 (1995) 707–715.

[50] R. Coles, R. Caswell, D.C. Rubinsztein, Functional analysis of the Huntington's disease (HD) gene promoter, Hum. Mol. Genet. 7 (1998) 791–800.

[51] C. Holzmann, et al. Isolation and characterization of the rat huntingtin promoter, Biochem. J. 336 (Pt 1) (1998) 227–234.

[52] C. Holzmann, T. Schmidt, G. Thiel, J.T. Epplen, O. Riess, Functional characterization of the human Huntington's disease gene promoter, Brain Res. Mol. Brain Res. 92 (2001) 85–97.

[53] Z. Feng, et al. p53 tumor suppressor protein regulates the levels of huntingtin gene expression, Oncogene 25 (2006) 1–7.

[54] R. Wang, et al. Sp1 regulates human huntingtin gene expression, J. Mol. Neurosci. 47 (2012) 311–321.

[55] K. Tanaka, J. Shouguchi-Miyata, N. Miyamoto, J.E. Ikeda, Novel nuclear shuttle proteins, HDBP1 and HDBP2, bind to neuronal cell-specific cis-regulatory element in the promoter for the human Huntington's disease gene, J. Biol. Chem. 279 (2004) 7275–7286.

[56] L. Mangiarini, et al. Exon 1 of the HD gene with an expanded CAG repeat is sufficient to cause a progressive neurological phenotype in transgenic mice, Cell 87 (1996) 493–506.

[57] K. Becanovic, et al. A SNP in the HTT promoter alters NF-kappaB binding and is a bidirectional genetic modifier of Huntington disease, Nat. Neurosci. 18 (2015) 807–816.

[58] B. Lin, et al. Differential 3' polyadenylation of the Huntington disease gene results in two mRNA species with variable tissue expression, Hum. Mol. Genet. 2 (1993) 1541–1545.

[59] K. Sathasivam, et al. Aberrant splicing of HTT generates the pathogenic exon 1 protein in Huntington disease, Proc. Natl. Acad. Sci. USA 110 (2013) 2366–2370.

[60] A.C. Hughes, et al. Identification of novel alternative splicing events in the huntingtin gene and assessment of the functional consequences using structural protein homology modelling, J. Mol. Biol. 426 (2014) 1428–1438.

[61] M. Mort, et al. Huntingtin exists as multiple splice forms in human rain, J. Huntington Dis. 4 (2015) 161–171.

[62] D.W. Chung, D.D. Rudnicki, L. Yu, R.L. Margolis, A natural antisense transcript at the Huntington's disease repeat locus regulates HTT expression, Hum. Mol. Genet. 20 (2011) 3467–3477.

[63] G.B. Landwehrmeyer, et al. Huntington's disease gene: regional and cellular expression in brain of normal and affected individuals, Ann. Neurol. 37 (1995) 218–230.

[64] G. Schilling, et al. Expression of the Huntington's disease (IT15) protein product in HD patients, Hum. Mol. Genet. 4 (1995) 1365–1371.

[65] Y. Trottier, et al. Cellular localization of the Huntington's disease protein and discrimination of the normal and mutated form, Nat. Genet. 10 (1995) 104–110.

[66] I. Gourfinkel-An, et al. Differential distribution of the normal and mutated forms of huntingtin in the human brain, Ann. Neurol. 42 (1997) 712–719.

[67] E. Sapp, et al. Huntingtin localization in brains of normal and Huntington's disease patients, Ann. Neurol. 42 (1997) 604–612.

[68] S.H. Li, et al. Huntington's disease gene (IT15) is widely expressed in human and rat tissues, Neuron 11 (1993) 985–993.

[69] M. DiFiglia, et al. Huntingtin is a cytoplasmic protein associated with vesicles in human and rat brain neurons, Neuron 14 (1995) 1075–1081.

[70] A.H. Sharp, et al. Widespread expression of Huntington's disease gene (IT15) protein product, Neuron 14 (1995) 1065–1074.

[71] P.G. Bhide, et al. Expression of normal and mutant huntingtin in the developing brain, J. Neurosci. 16 (1996) 5523–5535.

[72] T.V. Strong, et al. Widespread expression of the human and rat Huntington's disease gene in brain and non-neural tissues, Nat. Genet. 5 (1993) 259–265.

[73] J.Y. Shin, et al. Expression of mutant huntingtin in glial cells contributes to neuronal excitotoxicity, J. Cell Biol. 171 (2005) 1001–1012.

[74] C.M. Kosinski, et al. Huntingtin immunoreactivity in the rat neostriatum: differential accumulation in projection and interneurons, Exp. Neurol. 144 (1997) 239–247.

[75] F.R. Fusco, et al. Cellular localization of huntingtin in striatal and cortical neurons in rats: lack of correlation with neuronal vulnerability in Huntington's disease, J. Neurosci. 19 (1999) 1189–1202.

[76] M. Feyeux, et al. Early transcriptional changes linked to naturally occurring Huntington's disease mutations in neural derivatives of human embryonic stem cells, Hum. Mol. Genet. 21 (2012) 3883–3895.

[77] J.M. Lee, et al. Dominant effects of the Huntington's disease HTT CAG repeat length are captured in gene-expression data sets by a continuous analysis mathematical modeling strategy, Hum. Mol. Genet. 22 (2013) 3227–3238.

[78] W. Liu, et al. Increased steady-state mutant huntingtin mRNA in Huntington's disease brain, J. Huntingtons Dis. 2 (2013) 491–500.

[79] M.M. Evers, et al. Making (anti-) sense out of huntingtin levels in Huntington disease, Mol. Neurodegener. 10 (2015) 21.

[80] C. Zuccato, M. Valenza, E. Cattaneo, Molecular mechanisms and potential therapeutical targets in Huntington's disease, Physiol. Rev. 90 (2010) 905–981.

[81] F. Persichetti, et al. Differential expression of normal and mutant Huntington's disease gene alleles, Neurobiol. Dis. 3 (1996) 183–190.

[82] W. Li, L.C. Serpell, W.J. Carter, D.C. Rubinsztein, J.A. Huntington, Expression and characterization of full-length human huntingtin, an elongated HEAT repeat protein, J. Biol. Chem. 281 (2006) 15916–15922.

[83] M.F. Perutz, T. Johnson, M. Suzuki, J.T. Finch, Glutamine repeats as polar zippers: their possible role in inherited neurodegenerative diseases, Proc. Natl. Acad. Sci. USA 91 (1994) 5355–5358.

[84] P. Harjes, E.E. Wanker, The hunt for huntingtin function: interaction partners tell many different stories, Trends Biochem. Sci. 28 (2003) 425–433.

[85] M. DiFiglia, et al. Aggregation of huntingtin in neuronal intranuclear inclusions and dystrophic neurites in brain, Science 277 (1997) 1990–1993.

[86] F. Saudou, S. Finkbeiner, D. Devys, M.E. Greenberg, Huntingtin acts in the nucleus to induce apoptosis but death does not correlate with the formation of intranuclear inclusions, Cell 95 (1998) 55–66.

[87] M. Kim, et al. Mutant huntingtin expression in clonal striatal cells: dissociation of inclusion formation and neuronal survival by caspase inhibition, J. Neurosci. 19 (1999) 964–973.

[88] M. Arrasate, S. Mitra, E.S. Schweitzer, M.R. Segal, S. Finkbeiner, Inclusion body formation reduces levels of mutant huntingtin and the risk of neuronal death, Nature 431 (2004) 805–8108.

[89] J. Miller, et al. Quantitative relationships between huntingtin levels, polyglutamine length, inclusion body formation, and neuronal death provide novel insight into Huntington's disease molecular pathogenesis, J. Neurosci. 30 (2010) 10541–10550.

[90] J. Cornett, et al. Polyglutamine expansion of huntingtin impairs its nuclear export, Nat. Genet. 37 (2005) 198–204.

[91] R.S. Atwal, et al. Huntingtin has a membrane association signal that can modulate huntingtin aggregation, nuclear entry and toxicity, Hum. Mol. Genet. 16 (2007) 2600–2615.

[92] Y.P. Goldberg, et al. Cleavage of huntingtin by apopain, a proapoptotic cysteine protease, is modulated by the polyglutamine tract, Nat. Genet. 13 (1996) 442–449.

[93] C.L. Wellington, et al. Inhibiting caspase cleavage of huntingtin reduces toxicity and aggregate formation in neuronal and nonneuronal cells, J. Biol. Chem. 275 (2000) 19831–19838.

[94] Y.J. Kim, et al. Caspase 3-cleaved N-terminal fragments of wild-type and mutant huntingtin are present in normal and Huntington's disease brains, associate with membranes, and undergo calpain-dependent proteolysis, Proc. Natl. Acad. Sci. USA 98 (2001) 12784–12789.

[95] L.M. Mende-Mueller, T. Toneff, S.R. Hwang, M.F. Chesselet, V.Y. Hook, Tissue-specific proteolysis of Huntingtin (htt) in human brain: evidence of enhanced levels of N- and C-terminal htt fragments in Huntington's disease striatum, J. Neurosci. 21 (2001) 1830–1837.

[96] C.L. Wellington, et al. Caspase cleavage of mutant huntingtin precedes neurodegeneration in Huntington's disease, J. Neurosci. 22 (2002) 7862–7872.

[97] J. Gafni, L.M. Ellerby, Calpain activation in Huntington's disease, J. Neurosci. 22 (2002) 4842–4849.

[98] J. Gafni, et al. Inhibition of calpain cleavage of huntingtin reduces toxicity: accumulation of calpain/caspase fragments in the nucleus, J. Biol. Chem. 279 (2004) 20211–20220.

[99] C.E. Wang, et al. Accumulation of N-terminal mutant huntingtin in mouse and monkey models implicated as a pathogenic mechanism in Huntington's disease, Hum. Mol. Genet. 17 (2008) 2738–2751.

[100] C. Landles, et al. Proteolysis of mutant huntingtin produces an exon 1 fragment that accumulates as an aggregated protein in neuronal nuclei in Huntington disease, J. Biol. Chem. 285 (2010) 8808–8823.

[101] M.T. El-Daher, et al. Huntingtin proteolysis releases non-polyQ fragments that cause toxicity through dynamin 1 dysregulation, EMBO J. 34 (2015) 2255–2271.

[102] C.A. Gutekunst, et al. Identification and localization of huntingtin in brain and human lymphoblastoid cell lines with anti-fusion protein antibodies, Proc. Natl. Acad. Sci. USA 92 (1995) 8710–8714.

[103] G. Hoffner, P. Kahlem, P. Djian, Perinuclear localization of huntingtin as a consequence of its binding to microtubules through an interaction with beta-tubulin: relevance to Huntington's disease, J. Cell Sci. 115 (2002) 941–948.

[104] M. Damiano, L. Galvan, N. Déglon, E. Brouillet, Mitochondria in Huntington's disease, Biochim. Biophys. Acta. 1802 (2010) 52–61.

[105] C. Tourette, et al. A large scale Huntingtin protein interaction network implicates Rho GTPase signaling pathways in Huntington disease, J. Biol. Chem. 289 (2014) 6709–6726.

[106] L.S. Kaltenbach, et al. Huntingtin interacting proteins are genetic modifiers of neurodegeneration, PLoS Genet. 3 (2007) e82.

[107] J. Velier, et al. Wild-type and mutant huntingtins function in vesicle trafficking in the secretory and endocytic pathways, Exp. Neurol. 152 (1998) 34–40.

[108] J. Modregger, N.A. DiProspero, V. Charles, D.A. Tagle, M. Plomann, PACSIN 1 interacts with huntingtin and is absent from synaptic varicosities in presymptomatic Huntington's disease brains, Hum. Mol. Genet. 11 (2002) 2547–2558.

[109] E. Trushina, et al. Mutant huntingtin impairs axonal trafficking in mammalian neurons in vivo and in vitro, Mol. Cell Biol. 24 (2004) 8195–8209.

[110] J.R. McGuire, J. Rong, S.H. Li, X.J. Li, Interaction of Huntingtin-associated protein-1 with kinesin light chain: implications in intracellular trafficking in neurons, J. Biol. Chem. 281 (2006) 3552–3559.

[111] D. Zala, et al. Vesicular glycolysis provides on-board energy for fast axonal transport, Cell 152 (2013) 479–491.

[112] S. Engelender, et al. Huntingtin-associated protein 1 (HAP1) interacts with the p150Glued subunit of dynactin, Hum. Mol. Genet. 6 (1997) 2205–2212.

[113] S. Gunawardena, et al. Disruption of axonal transport by loss of huntingtin or expression of pathogenic polyQ proteins in Drosophila, Neuron 40 (2003) 25–40.

[114] L.R. Gauthier, et al. Huntingtin controls neurotrophic support and survival of neurons by enhancing BDNF vesicular transport along microtubules, Cell 118 (2004) 127–1388.

[115] W.C. Lee, M. Yoshihara, J.T. Littleton, Cytoplasmic aggregates trap polyglutamine-containing proteins and block axonal transport in a Drosophila model of Huntington's disease, Proc. Natl. Acad. Sci. USA 101 (2004) 3224–3229.

[116] J.P. Caviston, J.L. Ross, S.M. Antony, M. Tokito, E.L. Holzbaur, Huntingtin facilitates dynein/dynactin-mediated vesicle transport, Proc. Natl. Acad. Sci. USA 104 (2007) 10045–10050.

[117] J.P. Caviston, E.L. Holzbaur, Huntingtin as an essential integrator of intracellular vesicular trafficking, Trends Cell Biol. 19 (2009) 147–155.

[118] J.P. Caviston, A.L. Zajac, M. Tokito, E.L. Holzbaur, Huntingtin coordinates the dynein-mediated dynamic positioning of endosomes and lysosomes, Mol. Biol. Cell 22 (2011) 478–4923.

[119] D. Zala, M.V. Hinckelmann, F. Saudou, Huntingtin's function in axonal transport is conserved in Drosophila melanogaster, PLoS One 8 (2013) e60162.

[120] J.A. White 2nd, et al. Huntingtin differentially regulates the axonal transport of a sub-set of Rab-containing vesicles in vivo, Hum. Mol. Genet. 24 (2015) 7182–7195.

[121] A.M. Cuervo, S. Zhang, Selective autophagy and Huntingtin: learning from disease, Cell Cycle 14 (2015) 1617–1618.

[122] A. Gelman, M. Rawet-Slobodkin, Z. Elazar, Huntingtin facilitates selective autophagy, Nat. Cell Biol. 17 (2015) 214–2155.

[123] F. Saudou, S. Humbert, The biology of Huntingtin, Neuron 89 (2016) 910–926.

[124] R.A. De Souza, B.R. Leavitt, Neurobiology of Huntington's disease, Curr. Top. Behav. Neurosci. 22 (2015) 81–100.

[125] G.P. Bates, et al. Huntington disease, Nat. Rev. Dis. Primers 1 (2015) 15005.

[126] R.J. Ferrante, et al. Heterogeneous topographic and cellular distribution of huntingtin expression in the normal human neostriatum, J. Neurosci. 17 (1997) 3052–3063.

[127] A. Reiner, et al. Striatal parvalbuminergic neurons are lost in Huntington's disease: implications for dystonia, Mov. Disord. 28 (2013) 1691–1699.

[128] R.J. Ferrante, et al. Selective sparing of a class of striatal neurons in Huntington's disease, Science 230 (1985) 561–563.

[129] R.J. Ferrante, N.W. Kowall, E.P. Richardson Jr., E.D. Bird, J.B. Martin, Topography of enkephalin, substance P and acetylcholinesterase staining in Huntington's disease striatum, Neurosci. Lett. 71 (1986) 283–288.

[130] R.J. Ferrante, et al. Morphologic and histochemical characteristics of a spared subset of striatal neurons in Huntington's disease, J. Neuropathol. Exp. Neurol. 46 (1987) 12–27.

[131] R.J. Ferrante, M.F. Beal, N.W. Kowall, E.P. Richardson Jr., J.B. Martin, Sparing of acetylcholinesterase-containing striatal neurons in Huntington's disease, Brain Res. 411 (1987) 162–166.

[132] F.R. Fusco, et al. Huntingtin distribution among striatal output neurons of normal rat brain, Neurosci. Lett. 339 (2003) 53–56.

[133] A. Reiner, et al. Differential loss of striatal projection neurons in Huntington disease, Proc. Natl. Acad. Sci. USA 85 (1988) 5733–5737.

[134] R.L. Albin, A. Reiner, K.D. Anderson, J.B. Penney, A.B. Young, Striatal and nigral neuron subpopulations in rigid Huntington's disease: implications for the functional anatomy of chorea and rigidity-akinesia, Ann. Neurol. 27 (1990) 357–365.

[135] R.L. Albin, et al. Abnormalities of striatal projection neurons and N-methyl-D-aspartate receptors in presymptomatic Huntington's disease, N. Engl. J. Med. 322 (1990) 1293–1298.

[136] R.L. Albin, et al. Preferential loss of striato-external pallidal projection neurons in presymptomatic Huntington's disease, Ann. Neurol. 31 (1992) 425–430.

[137] H. Kiyama, A. Seto-Ohshima, P.C. Emson, Calbindin D28K as a marker for the degeneration of the striatonigral pathway in Huntington's disease, Brain Res. 525 (1990) 209–214.

[138] E.K. Richfield, K.A. Maguire-Zeiss, C. Cox, J. Gilmore, P. Voorn, Reduced expression of preproenkephalin in striatal neurons from Huntington's disease patients, Ann. Neurol. 37 (1995) 335–343.

[139] J.P. Vonsattel, Huntington disease models and human neuropathology: similarities and differences, Acta. Neuropathol. 115 (2008) 55–69.

[140] M. Cudkowicz, N.W. Kowall, Degeneration of pyramidal projection neurons in Huntington's disease cortex, Ann. Neurol. 27 (1990) 200–204.

[141] J.C. Hedreen, C.E. Peyser, S.E. Folstein, C.A. Ross, Neuronal loss in layers V and VI of cerebral cortex in Huntington's disease, Neurosci. Lett. 133 (1991) 257–261.

[142] E. Cattaneo, C. Zuccato, M. Tartari, Normal huntingtin function: an alternative approach to Huntington's disease, Nat. Rev. Neurosci. 6 (2005) 919–9306.

[143] M. Borrell-Pages, D. Zala, S. Humbert, F. Saudou, Huntington's disease: from huntingtin function and dysfunction to therapeutic strategies, Cell Mol. Life Sci. 63 (2006) 2642–2660.

[144] S. Imarisio, et al. Huntington's disease: from pathology and genetics to potential therapies, Biochem. J. 412 (2008) 191–209.

[145] N.S. Wexler, et al. Homozygotes for Huntington's disease, Nature 326 (1987) 194–197.

[146] R.H. Myers, et al. Homozygote for Huntington disease, Am. J. Hum. Genet. 45 (1989) 615–618.

[147] A. Durr, et al. Homozygosity in Huntington's disease, J. Med. Genet. 36 (1999) 172–173.

[148] F. Squitieri, et al. Homozygosity for CAG mutation in Huntington disease is associated with a more severe clinical course, Brain 126 (2003) 946–955.

[149] C.M. Ambrose, et al. Structure and expression of the Huntinton's disease gene: evidence against simple inactivation due to an expanded CAG repeat, Somat. Cell Mol. Genet. 20 (1994) 27–38.

[150] J. Nasir, et al. Targeted disruption of the Huntington's disease gene results in embryonic lethality and behavioral and morphological changes in heterozygotes, Cell 81 (1995) 811–823.

[151] J.R. O'Kusky, J. Nasir, F. Cicchetti, A. Parent, M.R. Hayden, Neuronal degeneration in the basal ganglia and loss of pallido-subthalamic synapses in mice with targeted disruption of the Huntington's disease gene, Brain Res. 818 (1999) 468–479.

[152] K.T. Dixon, J.A. Cearley, J.M. Hunter, P.J. Detloff, Mouse Huntington's disease homolog mRNA levels: variation and allele effects, Gene Expr. 11 (2004) 221–231.

[153] M.P. Duyao, et al. Inactivation of the mouse Huntington's disease gene homolog Hdh, Science 269 (1995) 407–410.

[154] S. Zeitlin, J.P. Liu, D.L. Chapman, V.E. Papaioannou, A. Efstratiadis, Increased apoptosis and early embryonic lethality in mice nullizygous for the Huntington's disease gene homologue, Nat. Genet. 11 (1995) 155–163.

[155] I. Dragatsis, A. Efstratiadis, S. Zeitlin, Mouse mutant embryos lacking huntingtin are rescued from lethality by wild-type extraembryonic tissues, Development 125 (1998) 1529–1539.

[156] J.K. White, et al. Huntingtin is required for neurogenesis and is not impaired by the Huntington's disease CAG expansion, Nat. Genet. 17 (1997) 404–410.

[157] I. Dragatsis, M.S. Levine, S. Zeitlin, Inactivation of Hdh in the brain and testis results in progressive neurodegeneration and sterility in mice, Nat. Genet. 26 (2000) 300–306.

[158] J.G. Hodgson, et al. Human huntingtin derived from YAC transgenes compensates for loss of murine huntingtin by rescue of the embryonic lethal phenotype, Hum. Mol. Genet. 5 (1996) 1875–1885.

[159] G. Wang, X. Liu, M.A. Gaertig, S. Li, X.J. Li, Ablation of huntingtin in adult neurons is nondeleterious but its depletion in young mice causes acute pancreatitis, Proc. Natl. Acad. Sci. USA 113 (2016) 3359–3364.

[160] X. Liu, et al. N-terminal Huntingtin knock-in mice: implications of removing the N-terminal region of Huntingtin for therapy, PLoS Genet. 12 (2016) e1006083.

[161] E. Cattaneo, Dysfunction of wild-type huntingtin in Huntington disease, News Physiol. Sci. 18 (2003) 34–37.

[162] J.C. Jacobsen, et al. HD CAG-correlated gene expression changes support a simple dominant gain of function, Hum. Mol. Genet. 20 (2011) 2846–2860.

[163] M. Banez-Coronel, et al. RAN translation in Huntington disease, Neuron 88 (2015) 667–677.

[164] J.D. Cleary, L.P. Ranum, Repeat-associated non-ATG (RAN) translation in neurological disease, Hum. Mol. Genet. 22 (2013) R45–R51.

[165] M.G. Kearse, P.K. Todd, Repeat-associated non-AUG translation and its impact in neurodegenerative disease, Neurotherapeutics 11 (2014) 721–731.

[166] H. Girstmair, et al. Depletion of cognate charged transfer RNA causes translational frameshifting within the expanded CAG stretch in huntingtin, Cell Rep. 3 (2013) 148–159.

[167] A. Fiszer, W.J. Krzyzosiak, RNA toxicity in polyglutamine disorders: concepts, models, and progress of research, J. Mol. Med. (Berl.) 91 (2013) 683–691.

[168] A. Mykowska, K. Sobczak, M. Wojciechowska, P. Kozlowski, W.J. Krzyzosiak, CAG repeats mimic CUG repeats in the misregulation of alternative splicing, Nucleic Acids Res. (2011) 39.

[169] J. Krol, et al. Ribonuclease dicer cleaves triplet repeat hairpins into shorter repeats that silence specific targets, Mol. Cell 25 (2007) 575–586.

[170] H. Tsoi, C.K. Lau, K.F. Lau, H.Y. Chan, Perturbation of U2AF65/NXF1-mediated RNA nuclear export enhances RNA toxicity in polyQ diseases, Hum. Mol. Genet. 20 (2011) 3787–3797.

[171] A.C. Woerner, et al. Cytoplasmic protein aggregates interfere with nucleocytoplasmic transport of protein and RNA, Science 351 (2016) 173–176.

[172] M.E. Ehrlich, Huntington's disease and the striatal medium spiny neuron: cell-autonomous and non-cell-autonomous mechanisms of disease, Neurotherapeutics 9 (2012) 270–284.

[173] C.S. Lobsiger, D.W. Cleveland, Glial cells as intrinsic components of non-cell-autonomous neurodegenerative disease, Nat. Neurosci. 10 (2007) 1355–1360.

[174] J. Bradford, et al. Mutant huntingtin in glial cells exacerbates neurological symptoms of Huntington disease mice, J. Biol. Chem. 285 (2010) 10653–10661.

[175] M. Faideau, et al. In vivo expression of polyglutamine-expanded huntingtin by mouse striatal astrocytes impairs glutamate transport: a correlation with Huntington's disease subjects, Hum. Mol. Genet. 19 (2010) 3053–3067.

[176] A. Crotti, et al. Mutant Huntingtin promotes autonomous microglia activation via myeloid lineage-determining factors, Nat. Neurosci. 17 (2014) 513–521.

[177] E.A. Thomas, et al. In vivo cell-autonomous transcriptional abnormalities revealed in mice expressing mutant huntingtin in striatal but not cortical neurons, Hum. Mol. Genet. 20 (2011) 1049–1060.

[178] T.B. Brown, A.I. Bogush, M.E. Ehrlich, Neocortical expression of mutant huntingtin is not required for alterations in striatal gene expression or motor dysfunction in a transgenic mouse, Hum. Mol. Genet. 17 (2008) 3095–3104.

[179] E. Pecho-Vrieseling, et al. Transneuronal propagation of mutant huntingtin contributes to non-cell autonomous pathology in neurons, Nat. Neurosci. 17 (2014) 1064–1072.

[180] F. Cicchetti, et al. Mutant huntingtin is present in neuronal grafts in Huntington disease patients, Ann. Neurol. 76 (2014) 31–42.

[181] M.A. Nance, Therapy in Huntington's disease: where are we?, Curr. Neurol. Neurosci. Rep. 12 (2012) 359–366.

[182] C.A. Ross, S.J. Tabrizi, Huntington's disease: from molecular pathogenesis to clinical treatment, Lancet Neurol. 10 (2011) 83–98.

[183] I. Munoz-Sanjuan, G.P. Bates, The importance of integrating basic and clinical research toward the development of new therapies for Huntington disease, J. Clin. Invest. 121 (2011) 476–483.

[184] I. Magen, E. Hornstein, Oligonucleotide-based therapy for neurodegenerative diseases, Brain Res. 1584 (2014) 116–128.

[185] P.S. Ramachandran, M.S. Keiser, B.L. Davidson, Recent advances in RNA interference therapeutics for CNS diseases, Neurotherapeutics 10 (2013) 473–485.

[186] B.M. Godinho, M. Malhotra, C.M. O'Driscoll, J.F. Cryan, Delivering a disease-modifying treatment for Huntington's disease, Drug Discov. Today 20 (2015) 50–64.

[187] N. Aronin, M. DiFiglia, Huntingtin-lowering strategies in Huntington's disease: antisense oligonucleotides, small RNAs, and gene editing, Mov. Disord. 29 (2014) 1455–1461.

[188] M.S. Keiser, H. Kordasiewicz, J. McBride, Gene Suppression strategies for dominantly inherited neurodegenerative diseases: lessons from Huntington's disease and spinocerebellar ataxia, Hum. Mol. Genet. 25 (2016) R53–64.

[189] M.L. Stephenson, P.C. Zamecnik, Inhibition of Rous sarcoma viral RNA translation by a specific oligodeoxyribonucleotide, Proc. Natl. Acad. Sci. USA 75 (1978) 285–288.

[190] A. Fire, et al. Potent and specific genetic interference by double-stranded RNA in Caenorhabditis elegans, Nature 391 (1998) 806–811.

[191] M. Jinek, et al. RNA-programmed genome editing in human cells, eLife 2 (2013) e00471.

[192] J.H. Chan, S. Lim, W.S. Wong, Antisense oligonucleotides: from design to therapeutic application, Clin. Exp. Pharmacol. Physiol. 33 (2006) 533–540.

[193] R.A. Smith, et al. Antisense oligonucleotide therapy for neurodegenerative disease, J. Clin. Invest. 116 (2006) 2290–2296.

[194] O.V. Matveeva, et al. Identification of sequence motifs in oligonucleotides whose presence is correlated with antisense activity, Nucleic Acids Res. 28 (2000) 2862–2865.

[195] C.F. Bennett, E.E. Swayze, RNA targeting therapeutics: molecular mechanisms of antisense oligonucleotides as a therapeutic platform, Annu. Rev. Pharmacol. Toxicol. 50 (2010) 259–293.

[196] H. Wu, et al. Determination of the role of the human RNase H1 in the pharmacology of DNA-like antisense drugs, J. Biol. Chem. 279 (2004) 17181–17189.

[197] R.J. Boado, A. Kazantsev, B.L. Apostol, L.M. Thompson, W.M. Pardridge, Antisense-mediated downregulation of the human huntingtin gene, J. Pharmacol. Exp. Ther. 295 (2000) 239–243.

[198] C. Nellemann, et al. Inhibition of Huntington synthesis by antisense oligodeoxynucleotides, Mol. Cell Neurosci. 16 (2000) 313–323.

[199] L. Hasholt, et al. Antisense downregulation of mutant huntingtin in a cell model, J. Gene Med. 5 (2003) 528–538.

[200] L.M. Stanek, et al. Antisense oligonucleotide-mediated correction of transcriptional dysregulation is correlated with behavioral benefits in the YAC128 mouse model of Huntington's disease, J. Huntingtons Dis. 2 (2013) 217–228.

[201] H.B. Kordasiewicz, et al. Sustained therapeutic reversal of Huntington's disease by transient repression of huntingtin synthesis, Neuron 74 (2012) 1031–1044.

[202] P.H. van Bilsen, et al. Identification and allele-specific silencing of the mutant huntingtin allele in Huntington's disease patient-derived fibroblasts, Hum. Gene Ther. 19 (2008) 710–719.

[203] E.L. Pfister, et al. Five siRNAs targeting three SNPs may provide therapy for three-quarters of Huntington's disease patients, Curr. Biol. 19 (2009) 774–778.

[204] M.S. Lombardi, et al. A majority of Huntington's disease patients may be treatable by individualized allele-specific RNA interference, Exp. Neurol. 217 (2009) 312–319.

[205] J.B. Carroll, et al. Potent and selective antisense oligonucleotides targeting single-nucleotide polymorphisms in the Huntington disease gene/allele-specific silencing of mutant huntingtin, Mol. Ther. 19 (2011) 2178–2185.

[206] J. Hu, M. Matsui, D.R. Corey, Allele-selective inhibition of mutant huntingtin by peptide nucleic acid-peptide conjugates, locked nucleic acid, and small interfering RNA, Ann. NY Acad. Sci. 1175 (2009) 24–31.

[207] M.E. Ostergaard, et al. Rational design of antisense oligonucleotides targeting single nucleotide polymorphisms for potent and allele selective suppression of mutant Huntingtin in the CNS, Nucleic Acids Res. 41 (2013) 9634–9650.

[208] N.H. Skotte, et al. Allele-specific suppression of mutant huntingtin using antisense oligonucleotides: providing a therapeutic option for all Huntington disease patients, PLoS One 9 (2014) e107434.

[209] K.A. Whitehead, R. Langer, D.G. Anderson, Knocking down barriers: advances in siRNA delivery, Nat. Rev. Drug Discov. 8 (2009) 129–138.

[210] T.R. Brummelkamp, R. Bernards, R. Agami, A system for stable expression of short interfering RNAs in mammalian cells, Science 296 (2002) 550–553.

[211] Y. Zeng, E.J. Wagner, B.R. Cullen, Both natural and designed micro RNAs can inhibit the expression of cognate mRNAs when expressed in human cells, Mol. Cell 9 (2002) 1327–1333.

[212] K. Terasawa, K. Shimizu, G. Tsujimoto, Synthetic pre-miRNA-based shRNA as potent RNAi triggers, J. Nucleic Acids 2011 (2011) 131579.

[213] C. Fellmann, et al. An optimized microRNA backbone for effective single-copy RNAi, Cell Rep. 5 (2013) 1704–1713.

[214] R. Calloni, D. Bonatto, Scaffolds for artificial miRNA expression in animal cells, Hum. Gene Ther. Methods 26 (2015) 162–174.

[215] G.J. Hannon, RNA interference, Nature 418 (2002) 244–251.

[216] J.C. Burnett, J.J. Rossi, K. Tiemann, Current progress of siRNA/shRNA therapeutics in clinical trials, Biotechnol. J. 6 (2011) 1130–1146.

[217] J.C. Burnett, J.J. Rossi, RNA-based therapeutics: current progress and future prospects, Chem. Biol. 19 (2012) 60–71.

[218] P. Kubowicz, D. Zelaszczyk, E. Pekala, RNAi in clinical studies, Curr. Med. Chem. 20 (2013) 1801–1816.

[219] N.J. Caplen, et al. Rescue of polyglutamine-mediated cytotoxicity by double-stranded RNA-mediated RNA interference, Hum. Mol. Genet. 11 (2002) 175–184.

[220] V.M. Miller, et al. Allele-specific silencing of dominant disease genes, Proc. Natl. Acad. Sci. USA 100 (2003) 7195–7200.

[221] H. Xia, et al. RNAi suppresses polyglutamine-induced neurodegeneration in a model of spinocerebellar ataxia, Nat. Med. 10 (2004) 816–820.

[222] Z.J. Chen, B.T. Kren, P.Y. Wong, W.C. Low, C.J. Steer, Sleeping Beauty-mediated down-regulation of huntingtin expression by RNA interference, Biochem. Biophys. Res. Commun. 329 (2005) 646–652.

[223] S.Q. Harper, et al. RNA interference improves motor and neuropathological abnormalities in a Huntington's disease mouse model, Proc. Natl. Acad. Sci. USA 102 (2005) 5820–5825.

[224] B. Huang, S. Kochanek, Adenovirus-mediated silencing of huntingtin expression by shRNA, Hum. Gene Ther. 16 (2005) 618–626.

[225] E. Rodriguez-Lebron, E.M. Denovan-Wright, K. Nash, A.S. Lewin, R.J. Mandel, Intrastriatal rAAV-mediated delivery of anti-huntingtin shRNAs induces partial reversal of disease progression in R6/1 Huntington's disease transgenic mice, Mol. Ther. 12 (2005) 618–633.

[226] Y.L. Wang, et al. Clinico-pathological rescue of a model mouse of Huntington's disease by siRNA, Neurosci. Res. 53 (2005) 241–249.

[227] Y. Machida, et al. rAAV-mediated shRNA ameliorated neuropathology in Huntington disease model mouse, Biochem. Biophys. Res. Commun. 343 (2006) 190–197.

[228] M. DiFiglia, et al. Therapeutic silencing of mutant huntingtin with siRNA attenuates striatal and cortical neuropathology and behavioral deficits, Proc. Natl. Acad. Sci. USA 104 (2007) 17204–17209.

[229] D.S. Gary, A. Davidson, O. Milhavet, H. Slunt, D.R. Borchelt, Investigation of RNA interference to suppress expression of full-length and fragment human huntingtin, Neuromol. Med. 9 (2007) 145–155.

[230] B. Huang, et al. High-capacity adenoviral vector-mediated reduction of huntingtin aggregate load in vitro and in vivo, Hum. Gene Ther. 18 (2007) 303–311.

[231] J.L. McBride, et al. Artificial miRNAs mitigate shRNA-mediated toxicity in the brain: implications for the therapeutic development of RNAi, Proc. Natl. Acad. Sci. USA 105 (2008) 5868–5873.

[232] N.R. Franich, et al. AAV vector-mediated RNAi of mutant huntingtin expression is neuroprotective in a novel genetic rat model of Huntington's disease, Mol. Ther. 16 (2008) 947–956.

[233] R.L. Boudreau, et al. Nonallele-specific silencing of mutant and wild-type huntingtin demonstrates therapeutic efficacy in Huntington's disease mice, Mol. Ther. 17 (2009) 1053–1063.

[234] R.L. Boudreau, I. Martins, B.L. Davidson, Artificial microRNAs as siRNA shuttles: improved safety as compared to shRNAs in vitro and in vivo, Mol. Ther. 17 (2009) 169–175.

[235] V. Drouet, et al. Sustained effects of nonallele-specific Huntingtin silencing, Ann. Neurol. 65 (2009) 276–285.

[236] V. Drouet, et al. Allele-specific silencing of mutant huntingtin in rodent brain and human stem cells, PLoS One 9 (2014) e99341.

[237] A. Burgess, Y. Huang, W. Querbes, D.W. Sah, K. Hynynen, Focused ultrasound for targeted delivery of siRNA and efficient knockdown of Htt expression, J. Control Release 163 (2012) 125–129.

[238] B.M. Godinho, J.R. Ogier, R. Darcy, C.M. O'Driscoll, J.F. Cryan, Self-assembling modified beta-cyclodextrin nanoparticles as neuronal siRNA delivery vectors: focus on Huntington's disease, Mol. Pharm. 10 (2013) 640–649.

[239] P. Kumar, et al. Transvascular delivery of small interfering RNA to the central nervous system, Nature 448 (2007) 39–43.

[240] A.C. Bonoiu, et al. Nanotechnology approach for drug addiction therapy: gene silencing using delivery of gold nanorod-siRNA nanoplex in dopaminergic neurons, Proc. Natl. Acad. Sci. USA 106 (2009) 5546–5550.

[241] A. Yamamoto, J.J. Lucas, R. Hen, Reversal of neuropathology and motor dysfunction in a conditional model of Huntington's disease, Cell 101 (2000) 57–66.

[242] J.L. McBride, et al. Preclinical safety of RNAi-mediated HTT suppression in the rhesus macaque as a potential therapy for Huntington's disease, Mol. Ther. 19 (2011) 2152–2162.

[243] R. Grondin, et al. Onset time and durability of Huntingtin suppression in rhesus putamen after direct infusion of antihuntingtin siRNA, Mol. Ther. Nucleic Acids 4 (2015) e245.

[244] D.K. Stiles, et al. Widespread suppression of huntingtin with convection-enhanced delivery of siRNA, Exp. Neurol. 233 (2012) 463–471.

[245] Y. Zhang, J. Engelman, R.M. Friedlander, Allele-specific silencing of mutant Huntington's disease gene, J. Neurochem. 108 (2009) 82–90.

[246] A.M. Monteys, M.J. Wilson, R.L. Boudreau, R.M. Spengler, B.L. Davidson, Artificial miRNAs targeting mutant Huntingtin show preferential silencing in vitro and in vivo, Mol. Ther. Nucleic Acids 4 (2015) e234.

[247] I. Lokody, Genetic therapies: correcting genetic defects with CRISPR-Cas9, Nat. Rev. Genet. 15 (2014) 63.

[248] C. Long, et al. Prevention of muscular dystrophy in mice by CRISPR/Cas9-mediated editing of germline DNA, Science 345 (2014) 1184–1188.

[249] X. Huang, et al. Production of gene-corrected adult beta globin protein in human erythrocytes differentiated from patient iPSCs after genome editing of the sickle point mutation, Stem Cells 33 (2015) 1470–1479.

[250] H. Yin, et al. Genome editing with Cas9 in adult mice corrects a disease mutation and phenotype, Nat. Biotechnol. 32 (2014) 551–553.

[251] Y. Li, et al. Excision of expanded GAA repeats alleviates the molecular phenotype of Friedreich's ataxia, Mol. Ther. 23 (2015) 1055–1065.

[252] G. Schwank, et al. Functional repair of CFTR by CRISPR/Cas9 in intestinal stem cell organoids of cystic fibrosis patients, Cell Stem Cell 13 (2013) 653–658.

[253] J.K. Joung, J.D. Sander, TALENs: a widely applicable technology for targeted genome editing, Nat. Rev. Mol. Cell Biol. 14 (2013) 49–55.

[254] M. Garriga-Canut, et al. Synthetic zinc finger repressors reduce mutant huntingtin expression in the brain of R6/2 mice, Proc. Natl. Acad. Sci. USA 109 (2012) E3136–3145.

[255] M. Heidenreich, F. Zhang, Applications of CRISPR-Cas systems in neuroscience, Nat. Rev. Neurosci. 17 (2016) 36–44.

[256] H. Wang, et al. One-step generation of mice carrying mutations in multiple genes by CRISPR/Cas-mediated genome engineering, Cell 153 (2013) 910–918.

[257] O. Shalem, et al. Genome-scale CRISPR-Cas9 knockout screening in human cells, Science 343 (2014) 84–87.

[258] T. Wang, J.J. Wei, D.M. Sabatini, E.S. Lander, Genetic screens in human cells using the CRISPR-Cas9 system, Science 343 (2014) 80–84.

[259] N.E. Sanjana, O. Shalem, F. Zhang, Improved vectors and genome-wide libraries for CRISPR screening, Nat. Methods 11 (2014) 783–784.

[260] K. Yoshimi, T. Kaneko, B. Voigt, T. Mashimo, Allele-specific genome editing and correction of disease-associated phenotypes in rats using the CRISPR-Cas platform, Nat. Commun. 5 (2014) 4240.

[261] W.Y. Hwang, et al. Efficient genome editing in zebrafish using a CRISPR-Cas system, Nat. Biotechnol. 31 (2013) 227–229.

[262] L.E. Jao, S.R. Wente, W. Chen, Efficient multiplex biallelic zebrafish genome editing using a CRISPR nuclease system, Proc. Natl. Acad. Sci. USA 110 (2013) 13904–13909.

[263] L. Swiech, et al. In vivo interrogation of gene function in the mammalian brain using CRISPR-Cas9, Nat. Biotechnol. 33 (2015) 102–106.

[264] H. Armitage, Gene-editing method halts production of brain-destroying proteins, Science (2015), doi: 10.1126/science.aad4739.

[265] N. Merienne, et al. Annual Meeting of the Society for Neuroscience Vol. Poster 303.06, Chicago, USA, 2015.

[266] E. Dorsey, Huntington Study Group, C.I.Characterization of a large group of individuals with Huntington disease and their relatives enrolled in the COHORT study, PLoS One 7 (2012) e29522.

[267] M. Orth, et al. Observing Huntington's disease: the European Huntington's Disease Network's REGISTRY, J. Neurol. Neurosurg. Psychiatry 82 (2011) 1409–1412.

[268] R. Andre, R.I. Scahill, S. Haider, S.J. Tabrizi, Biomarker development for Huntington's disease, Drug Discov. Today 19 (2014) 972–979.

[269] R.I. Scahill, E.J. Wild, S.J. Tabrizi, Biomarkers for Huntington's disease: an update, Expert Opin. Med. Diagn. 6 (2012) 371–375.

[270] D.W. Weir, A. Sturrock, B.R. Leavitt, Development of biomarkers for Huntington's disease, Lancet Neurol. 10 (2011) 573–590.

[271] H. Runne, et al. Analysis of potential transcriptomic biomarkers for Huntington's disease in peripheral blood, Proc. Natl. Acad. Sci. USA 104 (2007) 14424–14429.

[272] H.D. Rosas, et al. PRECREST: A phase II prevention and biomarker trial of creatine in at-risk Huntington disease, Neurology 82 (2014) 850–857.

[273] S.N. Wassef, et al. T1rho imaging in premanifest Huntington disease reveals changes associated with disease progression, Mov. Disord. 30 (2015) 1107–1114.

[274] M.J. Novak, et al. Basal ganglia-cortical structural connectivity in Huntington's disease, Hum. Brain Mapp. 36 (2015) 1728–1740.

[275] P. McColgan, et al. Selective vulnerability of Rich Club brain regions is an organizational principle of structural connectivity loss in Huntington's disease, Brain 138 (2015) 3327–3344.

[276] J. Goveas, et al. Diffusion-MRI in neurodegenerative disorders, Magn. Reson. Imaging 33 (2015) 853–876.

[277] E.J. Wild, et al. Quantification of mutant huntingtin protein in cerebrospinal fluid from Huntington's disease patients, J. Clin. Invest. 125 (2015) 1979–1986.

[278] T.M. Miller, et al. An antisense oligonucleotide against SOD1 delivered intrathecally for patients with SOD1 familial amyotrophic lateral sclerosis: a phase 1, randomised, first-in-man study, Lancet Neurol. 12 (2013) 435–442.

[279] S. Sivanesan, M.D. Howell, C.J. Didonato, R.N. Singh, Antisense oligonucleotide mediated therapy of spinal muscular atrophy, Transl. Neurosci. 4 (2013), doi: 10.2478/s13380-013-0109-2.

[280] C. Janson, et al. Clinical protocol. Gene therapy of Canavan disease: AAV-2 vector for neurosurgical delivery of aspartoacylase gene (ASPA) to the human brain, Hum. Gene Ther. 13 (2002) 1391–1412.

[281] M.H. Tuszynski, et al. A phase 1 clinical trial of nerve growth factor gene therapy for Alzheimer disease, Nat. Med. 11 (2005) 551–555.

[282] J.L. Eberling, et al. Results from a phase I safety trial of hAADC gene therapy for Parkinson disease, Neurology 70 (2008) 1980–1983.

[283] M.G. Kaplitt, et al. Safety and tolerability of gene therapy with an adeno-associated virus (AAV) borne GAD gene for Parkinson's disease: an open label, phase I trial, Lancet 369 (2007) 2097–2105.

[284] S. Worgall, et al. Treatment of late infantile neuronal ceroid lipofuscinosis by CNS administration of a serotype 2 adeno-associated virus expressing CLN2 cDNA, Hum. Gene Ther. 19 (2008) 463–474.

[285] W.J. Marks Jr., et al. Gene delivery of AAV2-neurturin for Parkinson's disease: a double-blind, randomised, controlled trial, Lancet Neurol. 9 (2010) 1164–1172.

[286] E. Montini, et al. Hematopoietic stem cell gene transfer in a tumor-prone mouse model uncovers low genotoxicity of lentiviral vector integration, Nat. Biotechnol. 24 (2006) 687–696.

[287] E. Montini, et al. The genotoxic potential of retroviral vectors is strongly modulated by vector design and integration site selection in a mouse model of HSC gene therapy, J. Clin. Invest. 119 (2009) 964–975.

[288] C.C. Bartholomae, et al. Lentiviral vector integration profiles differ in rodent postmitotic tissues, Mol. Ther. 19 (2011) 703–710.

[289] D. Cesana, et al. Uncovering and dissecting the genotoxicity of self-inactivating lentiviral vectors in vivo, Mol. Ther. 22 (2014) 774–785.

[290] L. Naldini, et al. In vivo gene delivery and stable transduction of nondividing cells by a lentiviral vector, Science 272 (1996) 263–267.

[291] A. Biffi, et al. Lentiviral hematopoietic stem cell gene therapy benefits metachromatic leukodystrophy, Science 341 (2013) 1233158.

[292] A. Aiuti, et al. Lentiviral hematopoietic stem cell gene therapy in patients with Wiskott-Aldrich syndrome, Science 341 (2013) 1233151.

[293] N. Cartier, et al. Hematopoietic stem cell gene therapy with a lentiviral vector in X-linked adrenoleukodystrophy, Science 326 (2009) 818–823.

[294] H.J. Stewart, et al. A stable producer cell line for the manufacture of a lentiviral vector for gene therapy of Parkinson's disease, Hum. Gene Ther. 22 (2011) 357–369.

[295] B. Jarraya, et al. Dopamine gene therapy for Parkinson's disease in a nonhuman primate without associated dyskinesia, Sci. Transl. Med. 1 (2009) ra4.

AMYOTROPHIC LATERAL SCLEROSIS: MECHANISMS AND THERAPEUTIC STRATEGIES

Ludo Van Den Bosch*,**

*Leuven Research Institute for Neuroscience
and Disease (LIND), KU Leuven, Leuven, Belgium*
***VIB, Center for Brain & Disease Research, Leuven, Belgium*

CHAPTER OUTLINE

Introduction 277
Excitotoxicity 280
Hyperexcitability 281
Pathogenic Role of Non-neuronal Cells 281
   Astrocyte Dysfunction 281
   Oligodendrocyte Dysfunction 282
   Microglial Dysfunction and T Cells 282
Shortage of Neurotrophic Factors 283
Mitochondrial Dysfunction 284
Axonal Defects 284
Altered Proteostasis and Autophagy 285
Altered RNA Metabolism and Stress Granule Formation 286
Hexanucleotide Repeats in C9ORF72 and Disturbances in Nucleocytoplasmic Transport 286
Conclusions 287
Acknowledgments 288
References 288

INTRODUCTION

Amyotrophic lateral sclerosis (ALS) is an adult-onset neurodegenerative disorder, which usually starts between the ages of 50 and 65. This relatively rare disease has an incidence between 1.5 and 2.7 per 100,000 in Europe [1], which is comparable to that of multiple sclerosis. ALS is characterized by the selective death of upper motor neuron in the motor cortex and of lower motor neuron in the brainstem and in the ventral horn of the spinal cord. This results in a progressive muscle phenotype characterized by spasticity, hyperreflexia, fasciculations, muscle atrophy, and paralysis. ALS is a dramatic disease and is usually fatal within 3–5 years after the diagnosis, which is mainly due to respiratory failure. About 20% of ALS patients have bulbar-onset ALS and these patients have a worse prognosis [2].

In 90% of patients, there is no family history of ALS and these patients suffer from sporadic ALS (SALS). In approximately 10% of ALS patients, a clear family history is present, and these patients are classified as familial ALS (FALS). In almost all cases, the disease is inherited in an autosomal dominant way. However, also autosomal recessive and X-linked forms of ALS exist. Mutations in more than 10 different genes are known to cause FALS. The most common genetic causes of FALS are mutations in the genes that encode superoxide dismutase 1 (SOD1; ~20%), fused in sarcoma (FUS; 1%–5%), TAR DNA-binding protein 43 (TDP-43 encoded by the TARDBP gene; 1%–5%) and hexanucleotide repeats in the C9ORF72 gene (~40%) (for a review: [3]).

The more than 170 mutations in SOD1 are distributed over the whole length of the protein and seem to impair the conformational stability of the protein leading to its aggregation. In FUS, the majority of ALS-causing mutations are clustered in the C-terminus of the protein, which contains the nuclear localization signal, while pathogenic mutations in the TARDBP gene cluster in the C-terminal glycine-rich region of TDP-43. "Loss-of-function" of FUS and TDP-43 in the nucleus as well as toxic "gain-of-functions" of cytoplasmic aggregating FUS or TDP-43 have been suggested to contribute to the disease pathogenesis. For FUS, it was recently claimed that it is the cytoplasmic mislocalization that is responsible for a gain-of-function leading to motor neuron loss [4]. The C9ORF72 gene contains a hexanucleotide repeat expansion $(GGGGCC)_n$ in the 5′ noncoding region and this repeat was not only identified as the most frequent cause of FALS (~40%) in the Western population, it is also an important cause of frontotemporal lobar degeneration (FTLD) [5,6]. Normal individuals most often have 2–8 repeats, while patients usually have hundreds to thousands of these hexanucleotide repeats.

ALS is clinically a very heterogeneous disease (for a review: [2]). The age and site of onset, as well as the rate of progression are extremely variable. Variation in disease progression is even observed in families with the same causal gene mutation [7]. These differences strongly suggest that there are environmental and/or genetic factors that modify the clinical phenotype of the disease. Genetic modifiers of ALS are genes or gene variants that can influence the onset age for developing ALS, as well as the presentation and severity of the disease. These genetic modifiers are very important as they could also play an important role in the susceptibility for SALS. So far, there are no clear environmental factors identified as inducers of ALS. However, a significantly higher incidence of ALS was found in a cohort of professional Italian soccer players [8]. In addition, an excitotoxic amino acid, β-methylamino-L-alanine (BMAA), present in the seeds of a cycad and produced by symbiotic cyanobacteria in the roots of this cycad was linked to the higher incidence of ALS in Guam [9,10]. It is generally accepted that SALS is caused by a combination of environmental factors, the genetic susceptibility of an individual and aging.

The presence and degree of cognitive dysfunction are also variable in ALS. In some patients, neurons in the prefrontal and temporal cortex are affected. This leads to cognitive and/or behavioral problems. These are usually very subtle. However, about 15% of ALS patients show signs of FTLD, a neurodegenerative disorder characterized by neuronal degeneration in the prefrontal and/or temporal lobe which leads to behavioral, personality, and/or language dysfunction (see also Chapter 8). The patients with a combination of motor and behavioral signs are diagnosed as suffering from ALS-FTLD [11,12]. On the other hand, 15% of FTLD patients show signs of motor neuron degeneration [13,14]. Based on this clinical overlap, it was suggested that ALS and FTLD are at the ends of one disease spectrum [15]. This is further confirmed by gene mutations that are common to both diseases (e.g., C9ORF72,

VCP, UBQLN2, and TBK1) and by a similar neuropathology (e.g., TDP-43 and FUS aggregates) (for reviews: [16,17]).

ALS cannot be cured. However, symptomatic and supportive care is available not only improving the quality of life, but also increasing the life expectancy of ALS patients. The clinical interventions include gastrostomy and noninvasive ventilation. In addition, riluzole was approved by the US Food and Drug Administration (FDA) for the treatment of ALS [18]. Until today, riluzole is the only approved therapeutic treatment for ALS. Riluzole increases the life span of the patients several months [18], a significant effect that was reproduced in different clinical trials [19–21].

The hope of the ALS community is to find additive treatments that have cumulative effects on top of riluzole and that can stop the disease process in an early phase. There are very good rodent models available that could be very helpful to reach this goal (for a review: [22]). A number of animal models based on the first discovered ALS-causing mutations in the SOD1 gene were created and these transgenic mice faithfully reproduce the human disease. The first transgenic mouse model overexpresses human SOD1 containing a substitution of glycine to alanine at position 93 (SOD1^{G93A}) under the control of the human SOD1 promoter [23]. This ubiquitous overexpression of human mutant SOD1 causes a progressive motor phenotype that closely resembles the human disease. Extensive efforts were made to generate other rodent models that recapitulate ALS based on other mutations in the genes encoding SOD1, TDP-43, FUS, and C9ORF72. While mouse and rat models overexpressing different missense mutations in the SOD1 gene show an adult-onset and progressive neurodegeneration, most of the TDP-43, FUS, and C9ORF72 models generated so far don't show extensive motor neuron death [22,24,25]. As a consequence, rodent mutant SOD1 models have been the most frequently used models in translational research as these are the only ones that resemble both clinically and pathologically the human situation. However, one should always keep in mind that SOD1 mutations are only found in approximately 2% of ALS patients and that the pathological hallmarks observed in *postmortem* tissue of these patients (SOD1 aggregates) are different from those found in almost all other ALS patients (TDP-43 or FUS aggregates). In addition, the mutant SOD1 mouse models are frequently blamed for the fact that all therapeutic strategies that were effective in these ALS mouse models subsequently failed in clinical trials. At least partially, these failures could be due to the fact that most therapeutic strategies were started presymptomatically, although inherent differences between mice and humans should also be taken into account.

Our expanding knowledge on the biology of ALS resulted in a number of potential new therapeutic strategies. Based on their target, these strategies can be subdivided into two main categories: causal or modifying treatments (for a review: [26]). While the first strategy targets disease-causing genes in order to avoid their expression and pathogenic effects, the second approach targets factors or mechanisms that could avoid or restore pathological processes involved in ALS. Causal treatments will have a large effect in a limited numbers of patients, while modifying therapies could result in a relatively small therapeutic effect in a larger number of patients.

In this chapter, I will discuss a number of biological processes that are involved in ALS and I will explain how insights into these mechanisms can eventually be translated into new therapeutic approaches.I will focus on the efforts to obtain disease-modifying therapies that could be effective in all ALS patients. The aim of these modifying treatment strategies is to interfere with pathological mechanisms that play a crucial role in the disease process. Many different mechanisms have already been suggested. Among others, these include excitotoxicity, hyperexcitability, astrocytosis, neuroinflammation, lack of neurotrophic support, mitochondrial dysfunction, dysregulated autophagy, axonal dysfunctions, abnormal RNA metabolism, problems with stress granules, and nucleocytoplasmic transport defects. I will introduce these pathogenic processes, illustrate how these can be modified and discuss whether these different strategies have been or could be translated into new therapies.

EXCITOTOXICITY

The only pathogenic mechanism for which there is clear evidence that it is indeed important in ALS patients is excitotoxicity. Excitotoxicity is the process of neuronal degeneration caused by overstimulation of glutamate receptors. One of the important arguments for an involvement of excitotoxicity in the ALS disease process is that the only FDA-approved drug proven to slow the disease process, riluzole, has antiexcitotoxic properties [18]. In addition, consumption of certain excitotoxins can result in motor neuron death, indicating that motor neurons are extremely sensitive to excessive stimulation of glutamate receptors (for reviews: [27,28]). Moreover, higher glutamate levels have been detected in cerebrospinal fluid of ALS patients [29]. This could be due to the lower expression of GLT-1, the major glutamate transporter in astrocytes responsible for the removal of glutamate from the synaptic cleft, implicating an important role for these non-neuronal cells as discussed further [30].

There are also a number of intrinsic properties of motor neurons that could render these cells particularly sensitive to excitotoxicity, which could explain the selective vulnerability of motor neurons in ALS. Most of these characteristics correlate to the way motor neurons handle Ca^{2+}. Motor neurons combine two exceptional characteristics: a high number of Ca^{2+}-permeable α-amino-3-hydroxy-5-methyl-4-isoxazolepropionic acid (AMPA) receptors [31,32] and a low Ca^{2+} buffering capacity [33]. These two properties seem to be essential to perform their normal motor neuron function, but under pathological conditions they could result in the selective death of motor neurons due to the fact that excessive amounts of Ca^{2+} enter the cells.

The high Ca^{2+} permeability of the AMPA receptors in motor neurons is due to a low expression of the GluR2 subunit [34]. AMPA receptors are tetramers composed of a variable association of four subunits (GluR1-4) and the Ca^{2+} permeability of the receptor is determined by the GluR2 subunit. Receptors containing at least one GluR2 subunit have a very low Ca^{2+} permeability compared to GluR2-lacking receptors. The Ca^{2+} impermeability of GluR2-containing AMPA receptors is due to the presence of a positively charged arginine instead of the genetically encoded neutral glutamine. This arginine residue is introduced by the editing of GluR2 pre-mRNA, a process that is virtually complete under normal conditions. However, this editing efficiency of the GluR2 mRNA is much lower in the ventral spinal gray matter of ALS patients compared to controls [35]. This deficiency in GluR2 editing was confirmed in single motor neurons isolated with laser microdissection from ALS patients [36]. The consequence of this editing deficiency is that AMPA receptors that contain GluR2 translated from unedited mRNA will have a neutral glutamine instead of the positively charged arginine and will be Ca^{2+} permeable. Transgenic mice lacking ADAR2, the enzyme responsible for GluR2 editing, develop a slow motor phenotype confirming that edited GluR2 is crucial for motor neuron survival [37].

In the absence of Ca^{2+}-binding proteins, intracellular Ca^{2+} handling in motor neurons relies on the endoplasmic reticulum (ER)-mitochondria Ca^{2+} cycle of which a disturbance could lead to the increased ER stress observed in ALS (for reviews: [28,38]). We showed that overexpression of one of the Ca^{2+} release channels in the ER (IP_3R2) of neurons has a negative effect on the disease process in the mutant SOD1 mouse model [39], while genetic removal of one of the enzymes (PLCδ1) responsible for the formation of the agonist (=IP$_3$) prolongs survival [40]. In addition, mitochondrial defects have been associated with ALS as we will discuss subsequently and this could also have a negative effect on the Ca^{2+} handling in motor neurons.

Therapeutically, the use of specific AMPA receptor antagonists is hindered by the severe side effects caused by these drugs. Other drugs with antiexcitotoxic properties gave positive results in the mutant SOD1 mouse, but failed in the clinic.

HYPEREXCITABILITY

More recently, at least part of the therapeutic effect of riluzole was linked to a (transient) effect on axonal and cortical hyperexcitability [41]. Peripheral and central hyperexcitability mirrors clinical features (fasciculations) and neurophysiological findings (increased intracortical excitability). Moreover, SALS patients with motor axonal or intracortical hyperexcitability have a worse prognosis [42,43]. Hyperexcitability could also link the negative effects of astrocytes (see the subsequent section) on motor neuron survival as mutant SOD1 expressing astrocytes increase hyperexcitability of motor neurons [44]. Hyperexcitability with increased spontaneous action potentials also seems to be a cell autonomous phenomenon as it is recapitulated in induced pluripotent stem cell (iPSC)-derived motor neurons from ALS patients harboring SOD1 and FUS mutations, as well as hexanucleotide repeats in C9ORF72 [45]. The K$^+$ channel activator retigabine both blocks this hyperexcitability and improves motor neuron survival in vitro when tested in mutant SOD1 cases [45]. These findings have led to the initiation of a clinical trial as retigabine is an FDA-approved drug with antiepileptic properties.

PATHOGENIC ROLE OF NON-NEURONAL CELLS

Based on a large number of experiments with the mutant SOD1 mouse model, ALS is considered as a non-cell autonomous disorder. Exclusive expression of mutant SOD1 in motor neurons, astrocytes, or microglia does not seem to be sufficient to cause the disease [46–49]. The opposite, deleting mutant SOD1 selectively from motor neurons, delays disease onset. In addition, deletion of mutant SOD1 only in astrocytes or in microglia delays disease progression [50,51]. These studies suggest that expression of mutant SOD1 in neurons is crucial for the disease to occur, while the surrounding cells are important for the progression of the disease. A study using chimeric mice with mixtures of normal and mutant SOD1 expressing cells provided additional evidence for a role of non-neuronal cells [52]. Mutant SOD1containing motor neurons surrounded by normal glial cells survive better [52]. These results suggest a significant role for astrocytes and microglia in the pathogenesis of mutant SOD1-induced neurodegeneration. More recently, it was shown that also expression of mutant SOD1 in oligodendrocytes plays an important role in the disease process [53]. Whether ALS is also a non-cell autonomous disease when it is induced by other genetic causes or in SALS is less clear. Expression of mutant TDP-43 in astrocytes doesn't show the same toxic effect on motor neurons as compared to the expression of mutant SOD1 [54].

ASTROCYTE DYSFUNCTION

Astrocytes are glial cells that stain positive for glial fibrillary acid protein (GFAP). These cells are very abundant in the nervous system and play an important role in supporting neurons. Astrocytes provide neurons with nutrients, metabolic precursors, and trophic factors, and are responsible for the uptake of glutamate released by neurons (for a review: [55] and see also Chapter 3).

It was already reported some time ago that the number of activated astrocytes in the mutant SOD1 mouse model correlates with disease progression [56]. Moreover and as already mentioned, selective deletion of mutant SOD1 in astrocytes significantly extends survival of mutant SOD1 mice [51]. In addition, coculturing mouse embryonic stem cell-derived motor neurons with astrocytes derived from *postmortem* spinal cord neural progenitor cells from SALS or FALS patients induces motor neuron death [57].

Astrocytes seem to have a dual function. On one hand, activated astrocytes release toxic factors. The identity of these factors is largely unknown. They seem to be soluble in nature as mutant astrocyte-conditioned medium is toxic for primary motor neurons of embryonic stem cell-derived motor neurons [58]. On the other hand, mutant SOD1 expressing astrocytes fail to protect neurons from excitotoxicity. Astrocytes can increase the excitability and Ca^{2+} influx in motor neurons by increasing the Na^+ current or by decreasing the neuronal GluR2 expression leading to a higher Ca^{2+} permeability of the AMPA receptor, ultimately leading to neuronal excitotoxicity [44,59].

In addition, ALS astrocytes show a lower reuptake of glutamate as the level of GLT-1, an excitatory amino acid transporter, is decreased which could lead to excitotoxicity. As a consequence, restoring the levels of the glial glutamate transporters in astrocytes was suggested as treatment strategy [60]. Presymptomatic administration of ceftriaxone, a compound increasing the expression of glial glutamate transporters in astrocytes, prolonged survival in the mutant SOD1 mouse model [60]. However, a phase III clinical trial investigating the clinical efficacy of ceftriaxone in ALS patients was negative [61].

OLIGODENDROCYTE DYSFUNCTION

Oligodendrocytes are another type of glial cells and these cells are responsible for the myelination of axons in the central nervous system (CNS). There is recent evidence that oligodendrocytes could also play an important role in ALS (for a review: [62]). Apart from their myelinating function, oligodendrocytes also provide trophic support to neuronal cells by releasing lactate. This release is mediated by the monocarboxylate transporter-1 (MCT1) [63]. While the number of oligodendrocytes remains the same [53,64], MCT1 levels significantly decrease in ALS patients and in the mutant SOD1 mouse model [63,64]. Oligodendrocytes start to degenerate before disease onset and are replaced by new oligodendrocytes. These new oligodendrocytes arise from precursor cells, also known as NG2+ cells, [53,64]. However, these new oligodendrocytes have morphologic abnormalities and do not fully differentiate as indicated by the reduced expression of MCT1 and myelin basic protein (MBP) [53,64]. As a consequence, these oligodendrocytes are not fully mature and are not able to keep the motor neurons alive [63]. Strikingly, specific deletion of mutant SOD1 in the oligodendrocytes of mutant SOD1 mice significantly delays disease onset and survival [53]. All together, these results indicate that therapeutic strategies that preserve oligodendrocyte function and/or that induce the differentiation of NG2+ oligodendrocyte precursor cells into fully differentiated oligodendrocytes could have therapeutic potential in ALS.

MICROGLIAL DYSFUNCTION AND T CELLS

Microglia are cells that stain positive for cluster of differentiation molecule 11b (CD11b). Microglia are the local macrophages of the nervous system and are responsible for the first and main form of active immune defense in the CNS. Microglia constantly screen the extracellular environment for potentially dangerous components using their processes. These cells are in close contact with neurons and astrocytes. As already indicated before, specific deletion of mutant SOD1 from the microglia of mutant SOD1 expressing mice significantly prolongs survival [50].

Similar to astrocytes, microglia also acquire an activated state during the disease [65]. The cell body enlarges and the processes get thicker. Microglia get properties of antigen-presenting cells and start to interact with T cells, which infiltrate in the spinal cord and cortex [66,67]. This microglial activation starts before disease onset and the number of activated microglia and infiltrated T cells increases with disease progression [56,67–70].

Reactive microglia have two different appearances: a deleterious one (M1 microglia) and a benign one (M2 microglia) [71]. While M2 microglia mainly secrete trophic and antiinflammatory factors, M1 microglia secrete proinflammatory molecules and reactive oxygen species (ROS). During disease progression, microglial cells change from M2 microglia into toxic M1 microglia. This was shown in *postmortem* patient material as well as in the mutant SOD1 mouse model [70,72,73].

The appearance of either M1 or M2 microglia is dependent on the T-cell subtype that is present in the spinal cord after infiltration. Secretion of different interleukins and cytokines induces either a shift to the neuroprotective (M2) or to the neurotoxic (M1) state of microglia. During an early ALS phase, regulatory T cells (T_{reg}) are abundantly present in the spinal cord, stimulating the appearance of antiinflammatory M2 microglia [74]. Later during the disease, the number of T_{reg} cells decreases, shifting the microglia to the neurotoxic state (M1) [74]. In both disease stages, signals between microglia and T cells regulate their appearance and phenotype [75]. These T-cell alterations are present in the mutant SOD1 mouse models as well as in blood samples of ALS patients [74,76]. All together, these data indicate that T cells could play an important role in the disease process.

Different therapeutic strategies targeting inflammation were already investigated in ALS mouse models and in patients. Administration of minocycline, a compound that reduces microglial activation, delays onset and prolongs survival in mutant SOD1 mouse models [77–79]. However, minocycline failed in a phase III clinical trial [80], although concerns were raised on the trial design [81]. Celecoxib, a cyclooxygenase 2 (cox2) inhibitor, is effective in the mutant SOD1 mouse model, but not in patients [82,83]. Attempts to neutralize the released ROS with vitamin E, creatine, or coenzyme Q10 (CoQ10) were unsuccessful in patients [84–86]. Also modulation of microglial function with glatiramer acetate failed in clinical trials [87].

Immunotherapies that suppress the neurotoxic state and retain or promote the protective functions of microglia could form an interesting strategy to treat ALS. One possibility is enhancing the levels of T_{reg}. A passive transfer of early phase mutant SOD1 T_{reg} cells prolongs the early disease phase in SOD1 mice by increasing M2 markers and suppressing M1 markers [74].

Our recent data indicate that deleting a Ca^{2+} release channel (IP$_3$R2) from the ER, a strategy that was initially developed to counteract excitotoxicity (see previous sections), shortens the life span of the mutant SOD1 mouse model [88]. This effect correlates with an increase of the innate immunity in the absence of IP$_3$R2 and illustrates how an expected positive effect in relation to one pathogenic mechanism (excitotoxicity) is outweighed by a negative effect on another mechanism (inflammation).

SHORTAGE OF NEUROTROPHIC FACTORS

Neurotrophic factors are essential for neuronal survival and differentiation. They influence dendrite and synapse formation, proliferation, migration, and differentiation of stem cells. Neurotrophic factors also play an important role in the repair and maintenance of neurons and their axons during disease or injury of the nervous system (for a review: [89]). Abnormal levels of neurotrophic factors and/or their receptors are observed in spinal cord and brain tissues of ALS patients [90].

The administration of neurotrophic factors is considered as an important therapeutic strategy to maintain neurons and their connections. One example of such a neurotrophic factor currently tested in a phase I clinical ALS trial is vascular endothelial growth factor (VEGF) (for a review: [91]). VEGF plays a crucial role in the formation of new blood vessels [92]. Hence, the discovery of VEGF as a modifier of ALS was unexpected. A transgenic mouse model with the "hypoxia response element" deleted in the promoter region of the VEGF gene (VEGF$^{delta/delta}$ mouse) shows reduced VEGF expression and develops motor neuron loss, muscle weakness, and a decline in motor function [93]. The loss of the VEGF neurotrophic effect can explain this motor neuron degeneration [93,94] and VEGF has a positive effect on the activated levels of the anti-apoptotic protein, Akt [95].

The enhancement of disease severity after crossing the VEGF$^{delta/delta}$ mouse with the mutant SOD1 mouse [96], the lower VEGF levels in the spinal cords of mutant SOD1 mice even before disease onset [97], and the improvement of the phenotype when VEGF is genetically overexpressed [98] also point toward an important modifying role of VEGF in ALS. Moreover, VEGF reduces astrogliosis in the spinal cord [99], stabilizes neuromuscular junctions in ALS mice [99], and upregulates the expression of the GluR2 subunit of the AMPA receptor counteracting excitotoxicity [100]. The therapeutic potential of VEGF was extensively explored in mutant SOD1 rodent models. Delivery of viral vectors containing the human VEGF transgene into the muscle or the ventricular system increases the lifespan of the mutant SOD1 mice [101,102]. Moreover, intracerebroventricular administration of VEGF ameliorates the phenotype and increases the life span of ALS rats [103].

MITOCHONDRIAL DYSFUNCTION

Both in animal models and in ALS patients, dysfunctional mitochondria are found. Mitochondria are involved in ATP synthesis, intracellular Ca^{2+} homeostasis, and induction of apoptosis. Hence, abnormal functioning of mitochondria can cause cell death. In the mutant SOD1 mouse model, swollen and vacuolated mitochondria are observed in motor neurons and in the skeletal muscle before the first signs of motor neuron degeneration appear [104–106]. Similar abnormalities are found in SALS patients [107,108]. In general, dysfunctions in mitochondrial respiration and ATP synthesis, axonal transport of mitochondria, mitochondrial dynamics, Ca^{2+} buffering (which could lead to excitotoxicity), and induction of apoptosis are all seen in SOD1 mouse models (for a review: [88]). Moreover, these abnormalities are not restricted to the mutant SOD1 pathogenesis as also mice overexpressing human TDP-43 have abnormal distribution of mitochondria and changes in mitochondrial dynamics [109,110], while neuronal cultures expressing ALS-linked FUS mutants contain smaller mitochondria [111].

A number of drugs that target mitochondrial dysfunction were already tested. Dexpramipexole, olesoxime, and creatine, three compounds modulating mitochondrial function, showed promising results in preclinical studies, but failed in different clinical studies [85,112–116].

AXONAL DEFECTS

Studies in rodent models and *postmortem* patient tissue showed that axons from motor neurons retract from the neuromuscular junction long before the loss of cell bodies [117,118]. After axon retraction, nerve sprouting can occur in an attempt to form new compensatory connections. Synapses with low sprouting capacities are more vulnerable to degeneration, while axons with a high sprouting capacity are more resistant [118]. From a therapeutic point of view, axonal retraction could be prevented by inducing the sprouting capacity of axons or by preserving the remaining neuromuscular junctions. This could lead to improved motor neuron function and survival.

Neurons in general and motor neurons in particular are highly dependent on axonal transport mechanisms along their very long axons to bring proteins, organelles, and other cargoes to their sites of function. Microtubules form tracks along axons on which different cargoes are transported with the help of motor proteins. Since axonal swellings containing neurofilaments were observed in sporadic and FALS patients [119,120], axonal transport defects have been intensively studied in ALS (for a review: [121]). In the mutant SOD1 mouse model, defects in axonal transport are an early event [122–126]. Moreover, point mutations in the DCTN1 gene encoding the p150 subunit of dynactin were found in sporadic and familial ALS patients [127,128], and a missense mutation in DCTN1 also gives rise to a distal hereditary motor neuropathy with vocal cord involvement [129]. In addition, a defect in axonal transport in larval motor neurons was a common denominator in *Drosophila* models for different genetic forms of ALS [130]. An active contribution of axonal transport deficits in motor neuron degeneration and more specifically in ALS could be envisaged as both the accumulation and site-specific loss of proteins, organelles, and vesicles can lead to an aggravation of the phenotype. As a consequence, attempts to improve axonal transport could be possible therapeutic interventions [121].

One possible strategy to improve axonal transport is the use of microtubule-modulating agents [131,132]. In mutant SOD1 mice, microtubules are extremely unstable due to enhanced microtubule dynamics [131]. Pharmacological modulation of microtubules in mutant SOD1 mice can restore axonal transport, is able to increase the survival of motor neurons, and has a positive effect on the lifespan of the mutant SOD1 mouse model [131,132].

ALTERED PROTEOSTASIS AND AUTOPHAGY

Neurodegenerative diseases are characterized by the appearance of intracellular protein aggregates that can become toxic for the cells. ALS is also characterized by the accumulation of ubiquitinated proteins. In the majority of ALS cases, these accumulations contain ubiquitinated TDP-43 even when there are no mutations in the TARDBP gene [133]. Increasing the clearance of these aggregates could be a possible therapeutic strategy for neurodegenerative disorders in general and for ALS in particular (for a review: [134]). Autophagy and the ubiquitin–proteasome system (UPS) are the major clearance pathways in the cell. Both systems are responsible for the proteostasis in eukaryotes and the prevention of protein aggregation. While the UPS is indispensable for the clearance of short-living proteins, the process of autophagy is important for the degradation of long-living proteins, abnormally folded proteins, and for organelle turnover (for a review: [135]). Autophagy subtypes differ in the way that the substrate to be degraded is delivered to the lysosomes. Macroautophagy, further referred to as autophagy, is the most studied subtype and is characterized by the formation of a double membrane enclosing a part of the cytoplasm that contains the substrate that needs to be degraded. This results in the formation of the autophagosome. In a later step of this process, autophagosomes fuse with lysosomes to degrade their cargo. When two different essential genes for autophagy are deleted (Atg5 or Atg7), each of these transgenic mice develops a neurodegenerative phenotype, which suggests an essential role for autophagy in neurons [136,137].

An increased number of autophagosomes is present in motor neurons of mutant SOD1 mice and SALS patients [138–141]. As autophagy is a multistep process, it can get disrupted at many different levels [142]. Whether this is a causal event, the result of an upregulation of autophagy for the clearance of toxic proteins or due to a disrupted autophagosome–lysosome fusion is not yet clear. Different pharmacological approaches to stimulate autophagy were already tested. Two independent studies showed a negative effect on disease progression by treating mouse models of ALS with rapamycin, which is an FDA-approved drug that induces autophagy [143,144]. This negative effect was probably due to the immunosuppressive function of rapamycin, as the disease slowed down in immunodeficient ALS mice treated with rapamycin [145]. This indicates that autophagy should be targeted with more specific compounds that do not have (negative) side effects. Other compounds, such as progesterone or trehalose have been administered to mutant SOD1 mice and showed promising results, such as delayed disease onset and longer survival which were correlated with an induction of the autophagic flux [146–148]. All together, these data indicate that the modulation of autophagy can be a potential new therapeutic strategy for ALS.

ALTERED RNA METABOLISM AND STRESS GRANULE FORMATION

Disturbances in RNA metabolism were linked to ALS after the discovery of mutations in TARDBP and FUS, encoding two different RNA-binding proteins (TDP-43 and FUS). Under normal conditions, both TDP-43 and FUS are localized in the nucleus. In brain and spinal cord of ALS patients, TDP-43 and FUS are in the cytoplasm of the neurons and sometimes also of the glial cells. The nuclear localization of TDP-43 and FUS relates to the physiological function of these DNA/RNA-binding proteins and they seem to be involved in almost every aspect of DNA and RNA processing (for a review: [149]). TDP-43 binds to more than 6000 RNA targets in the brain, with a preference for binding to transcripts with long introns [150]. Downregulation of TDP-43 has a major influence on RNA processing [150] and also FUS regulates RNA processing, especially of genes that regulate dendritic growth and synaptic functions [151].

In response to stress, both TDP-43 and FUS localize to the stress granules present in the cytoplasm. In these stress granules, mRNA is translationally inactive. Both TDP-43 and FUS contain low-complexity domains that are essential for their localization in these stress granules. After the stress period, the stress granules resolve and mRNA becomes available again to be translated. Recent data indicate that FUS mutations accelerate the transition from the liquid to the aggregated state in the stress granules [152]. At present, it is not yet clear whether and how these new insights into the defective RNA processing and stress granule formation can be translated into new therapeutic strategies.

HEXANUCLEOTIDE REPEATS IN C9ORF72 AND DISTURBANCES IN NUCLEOCYTOPLASMIC TRANSPORT

The exact function of the protein encoded by the C9ORF72 gene containing the hexanucleotide repeats is unknown. At this moment, three not mutually exclusive disease mechanisms are possible: (1) loss of expression of C9ORF72 transcripts (haploinsufficiency), (2) direct toxicity from sense or antisense RNA containing the expanded repeat which could be due to the sequestration of RNA-binding proteins to the hexanucleotide repeats present in the RNA foci, and (3) non-ATG dependent translation (RAN translation) of the repeat present in sense and/or antisense transcripts into aggregating dipeptide repeat proteins (DPRs). There is more and more evidence that haploinsufficiency is not the major driving force of the disease process in the presence of the hexanucleotide repeats in C9ORF72. As a consequence, the focus is shifted to the RNA foci and DPRs that are clearly present in brain and spinal cord tissue from FTLD and ALS patients. However, the correlation between these RNA foci and DPR aggregates and the pathology is not always straightforward as in many cases the RNA foci and DPRs are found in regions without pathology and vice versa. In addition, it is unclear how RNA foci and DPRs relate to each other.

A number of recent papers link the toxicity of the hexanucleotide repeats to an interference with the nucleocytoplasmic transport process (for a review: [153]). However, it is not yet clear whether this is due to the RNA, the DPRs, or a combination of both. The paper from the Rothstein group uses a *Drosophila* screen, in vitro binding assays, and studies in iPSCs-derived neurons from ALS patients harboring C9ORF72 mutations. They identify nucleocytoplasmic transport defects and find that genes functioning in this pathway modify the phenotype in flies [154]. They conclude that the sense RNA foci but not the DPRs drive these impairments. A second paper uses essentially the same approach as the previous one and they find nearly identical results and hits from the fly screens [155]. However, they conclude that the defects could be caused by RNA, DPRs, or a combination of both. We used a yeast model that specifically expressed the DPRs and did not have repeat RNAs [156]. Therefore, we only focused on the protein aspect of C9ORF72 toxicity. We conclude that the arginine-rich DPRs (e.g., GR and PR) disrupt nucleocytoplasmic transport. This interpretation is in line with the recent paper that compares flies expressing C9ORF72 GGGGCC repeats [157]. Flies that express the repeat from within an intron have high levels of sense RNA foci in the nucleus, low levels of RAN translation, and no neurodegeneration. On the other hand, flies that express the GGGGCC expansion in the context of an mRNA with a 3'UTR, which allows efficient export to the cytoplasm, produce high levels of DPRs and this causes neurodegeneration. From these results it can be concluded that accumulation of sense RNA foci in the nucleus is not sufficient to drive neurodegeneration in this fly model [157]. Whatever the cause of the neurodegeneration is, interference with the nucleocytoplasmic transport seems to be a potential therapeutic target and inhibition of nuclear export as well as antisense oligonucleotides targeting the hexanucleotide repeats can rescue the deficits [154].

CONCLUSIONS

Since the discovery of riluzole, no other therapeutic options for ALS patients were obtained and this already for more than 20 years. However, large research investments undoubtedly resulted in a better understanding of the disease. Both additional causal genes and new pathological processes underlying the selective motor neuron death were identified. The discovery of new and important causal genes could lead to the development of a causal therapy targeting these specific disease-causing genes. This could have a huge impact by preventing the pathological events responsible for the genetic forms of the disease. However, only a limited number of ALS patients will benefit from this therapy as in 90% of ALS patients there is no clear genetic cause. As a consequence, the alternative approach is to target common pathogenic processes that are involved in all forms of ALS. The effect of such a modifying treatment strategy could be smaller in comparison to that of a causal treatment, but the target population will be much larger as theoretically all ALS patients could benefit from this treatment. In order for such a strategy to work, one has to accept the premise that all different genetic and/or environmental factors causing ALS result in similar disease mechanisms leading to ALS. The major argument for this supposition is that the different forms of ALS present similarly in the hospital and that it is impossible to discriminate the different forms based on a clinical examination. Most likely, different disease-modifying therapies will have to be combined to "cure" ALS. In that case, one hopes that this combination will result in an additive and hopefully even a synergistic therapeutic effect.

A large number of preclinical studies were already performed using the different mutant SOD1 rodent models that prevent, reverse, or modulate the disease process. So far, none of these therapeutic strategies was replicated successfully in subsequent clinical trials. This raises serious doubts about the validity of the results obtained using this transgenic mouse model. As a consequence, alternatives are urgently needed. It is crucial to improve the translational efficiency of preclinical observations into new therapeutic strategies for ALS. There is an urgent need for new ALS models, for an increased rigorousness of the preclinical studies and for the identification of new biomarkers to facilitate clinical studies. The use of motor neurons differentiated from iPSCs derived from ALS patients seems to be a promising new strategy, not only to get better insight(s) into the pathogenic mechanism(s) leading to motor neuron death, but also to test new therapeutic strategies. In addition to the fact that the mutated genes are not overexpressed, these cells can also be derived from SALS patients. At this moment, besides riluzole administration, only symptomatic and supportive treatments based on multidisciplinary care can be offered to improve the quality of life of ALS patients.

ACKNOWLEDGMENTS

The author receives funding from the Fund for Scientific Research Flanders (FWO), Government agency for Innovation by Science and Technology (IWT), University of Leuven, Interuniversity Attraction Poles program of the Belgian Federal Science Policy Office, the European Community's Health Seventh Framework Programme, the Association Belge contre les Maladies neuro-Musculaires (ABMM), the Association Française contre les Myopathies (AFM), the ALS Liga (Belgium), the ALS Therapy Alliance, the ALS Association (ALSA), and the Muscular Dystrophy Association (MDA).

REFERENCES

[1] G. Logroscino, B.J. Traynor, O. Hardiman, A. Chio, D. Mitchell, R.J. Swingler, et al. Incidence of amyotrophic lateral sclerosis in Europe, J. Neurol. Neurosurg. Psychiatry 81 (4) (2010) 385–390.

[2] B. Swinnen, W. Robberecht, The phenotypic variability of amyotrophic lateral sclerosis, Nat. Rev. Neurol. 10 (11) (2014) 661–670.

[3] A.E. Renton, A. Chio, B.J. Traynor, State of play in amyotrophic lateral sclerosis genetics, Nat. Neurosci. 17 (1) (2014) 17–23.

[4] G. Shiihashi, D. Ito, T. Yagi, Y. Nihei, T. Ebine, N. Suzuki, Mislocated FUS is sufficient for gain-of-toxic-function amyotrophic lateral sclerosis phenotypes in mice, Brain 139 (2016) 2380–2394.

[5] M. DeJesus-Hernandez, I.R. Mackenzie, B.F. Boeve, A.L. Boxer, M. Baker, N.J. Rutherford, et al. Expanded GGGGCC hexanucleotide repeat in noncoding region of C9ORF72 causes chromosome 9p-linked FTD and ALS, Neuron 72 (2) (2011) 245–256.

[6] A.E. Renton, E. Majounie, A. Waite, J. Simon-Sanchez, S. Rollinson, J.R. Gibbs, et al. A hexanucleotide repeat expansion in C9ORF72 is the cause of chromosome 9p21-linked ALS-FTD, Neuron 72 (2) (2011) 257–268.

[7] L. Regal, L. Vanopdenbosch, P. Tilkin, L. Van den Bosch, V. Thijs, R. Sciot, et al. The G93C mutation in superoxide dismutase 1: clinicopathologic phenotype and prognosis, Arch. Neurol. 63 (2) (2006) 262–267.

[8] A. Chio, G. Benzi, M. Dossena, R. Mutani, G. Mora, Severely increased risk of amyotrophic lateral sclerosis among Italian professional football players, Brain 128 (Pt 3) (2005) 472–476.

[9] S.J. Murch, P.A. Cox, S.A. Banack, A mechanism for slow release of biomagnified cyanobacterial neurotoxins and neurodegenerative disease in Guam, Proc. Natl. Acad. Sci. USA 101 (33) (2004) 12228–12231.

[10] P.A. Cox, S.A. Banack, S.J. Murch, Biomagnification of cyanobacterial neurotoxins and neurodegenerative disease among the Chamorro people of Guam, Proc. Natl. Acad. Sci. USA 100 (23) (2003) 13380–13383.

[11] G.M. Ringholz, S.H. Appel, M. Bradshaw, N.A. Cooke, D.M. Mosnik, P.E. Schulz, Prevalence and patterns of cognitive impairment in sporadic ALS, Neurology 65 (4) (2005) 586–590.

[12] P. Lillo, J.R. Hodges, Frontotemporal dementia and motor neurone disease: overlapping clinic-pathological disorders, J. Clin. Neurosci. 16 (9) (2009) 1131–1135.

[13] J.R. Burrell, M.C. Kiernan, S. Vucic, J.R. Hodges, Motor neuron dysfunction in frontotemporal dementia, Brain 134 (Pt 9) (2011) 2582–2594.

[14] C. Lomen-Hoerth, T. Anderson, B. Miller, The overlap of amyotrophic lateral sclerosis and frontotemporal dementia, Neurology 59 (7) (2002) 1077–1079.

[15] W. Robberecht, T. Philips, The changing scene of amyotrophic lateral sclerosis, Nat. Rev. Neurosci. 14 (4) (2013) 248–264.

[16] S. Lattante, S. Ciura, G.A. Rouleau, E. Kabashi, Defining the genetic connection linking amyotrophic lateral sclerosis (ALS) with frontotemporal dementia (FTD), Trends Genet. 31 (5) (2015) 263–273.

[17] A.S. Ng, R. Rademakers, B.L. Miller, Frontotemporal dementia: a bridge between dementia and neuromuscular disease, Ann. NY Acad. Sci. 1338 (2015) 71–93.

[18] R.G. Miller, J.D. Mitchell, D.H. Moore, Riluzole for amyotrophic lateral sclerosis (ALS)/motor neuron disease (MND), Cochrane Database Syst. Rev. 3 (2012) CD001447.

[19] G. Bensimon, L. Lacomblez, V. Meininger, A controlled trial of riluzole in amyotrophic lateral sclerosis. ALS/Riluzole Study Group, N. Engl. J. Med. 330 (9) (1994) 585–591.

[20] L. Lacomblez, G. Bensimon, P.N. Leigh, P. Guillet, V. Meininger, Dose-ranging study of riluzole in amyotrophic lateral sclerosis. Amyotrophic Lateral Sclerosis/Riluzole Study Group II, Lancet 347 (9013) (1996) 1425–1431.

[21] G. Bensimon, L. Lacomblez, J.C. Delumeau, R. Bejuit, P. Truffinet, V. Meininger, et al. A study of riluzole in the treatment of advanced stage or elderly patients with amyotrophic lateral sclerosis, J. Neurol. 249 (5) (2002) 609–615.

[22] P. McGoldrick, P.I. Joyce, E.M.C. Fisher, L. Greensmith, Rodent models of amyotrophic lateral sclerosis, BBA Mol. Basis Dis. 1832 (9) (2013) 1421–1436.

[23] M.E. Gurney, H.F. Pu, A.Y. Chiu, M.C. Dalcanto, C.Y. Polchow, D.D. Alexander, et al. Motor-neuron degeneration in mice that express a human Cu,Zn superoxide-dismutase mutation, Science 264 (5166) (1994) 1772–1775.

[24] O.M. Peters, G.T. Cabrera, H. Tran, T.F. Gendron, J.E. McKeon, J. Metterville, et al. Human C9ORF72 hexanucleotide expansion reproduces RNA foci and dipeptide repeat proteins but not neurodegeneration in BAC transgenic mice, Neuron 88 (5) (2015) 902–909.

[25] J.G. O'Rourke, L. Bogdanik, A.K. Muhammad, T.F. Gendron, K.J. Kim, A. Austin, et al. C9orf72 BAC transgenic mice display typical pathologic features of ALS/FTD, Neuron 88 (5) (2015) 892–901.

[26] L. Poppe, L. Rue, W. Robberecht, L. Van Den Bosch, Translating biological findings into new treatment strategies for amyotrophic lateral sclerosis (ALS), Exp. Neurol. 262 (2014) 138–151.

[27] L. Van Den Bosch, P. Van Damme, E. Bogaert, W. Robberecht, The role of excitotoxicity in the pathogenesis of amyotrophic lateral sclerosis, Biochim. Biophys. Acta. 1762 (11–12) (2006) 1068–1082.

[28] J. Grosskreutz, L. Van Den Bosch, B.U. Keller, Calcium dysregulation in amyotrophic lateral sclerosis, Cell Calcium 47 (2) (2010) 165–174.

[29] M.L. Fiszman, K.C. Ricart, A. Latini, G. Rodriguez, R.E. Sica, In vitro neurotoxic properties and excitatory amino acids concentration in the cerebrospinal fluid of amyotrophic lateral sclerosis patients. Relationship with the degree of certainty of disease diagnoses, Acta Neurol. Scand. 121 (2) (2010) 120–126.

[30] J.D. Rothstein, M. Van Kammen, A.I. Levey, L.J. Martin, R.W. Kuncl, Selective loss of glial glutamate transporter GLT-1 in amyotrophic lateral sclerosis, Ann. Neurol. 38 (1) (1995) 73–84.

[31] S.G. Carriedo, H.Z. Yin, J.H. Weiss, Motor neurons are selectively vulnerable to AMPA/kainate receptor-mediated injury in vitro, J. Neurosci. 16 (13) (1996) 4069–4079.

[32] L. Van Den Bosch, W. Vandenberghe, H. Klaassen, E. Van Houtte, W. Robberecht, Ca(2+)-permeable AMPA receptors and selective vulnerability of motor neurons, J. Neurol. Sci. 180 (1–2) (2000) 29–34.

[33] B.K. Vanselow, B.U. Keller, Calcium dynamics and buffering in oculomotor neurones from mouse that are particularly resistant during amyotrophic lateral sclerosis (ALS)-related motoneurone disease, J. Physiol. 525 (Pt 2) (2000) 433–445.

[34] P. Van Damme, L. Van Den Bosch, E. Van Houtte, G. Callewaert, W. Robberecht, GluR2-dependent properties of AMPA receptors determine the selective vulnerability of motor neurons to excitotoxicity, J. Neurophysiol. 88 (3) (2002) 1279–1287.

[35] H. Takuma, S. Kwak, T. Yoshizawa, I. Kanazawa, Reduction of GluR2 RNA editing, a molecular change that increases calcium influx through AMPA receptors, selective in the spinal ventral gray of patients with amyotrophic lateral sclerosis, Ann. Neurol. 46 (6) (1999) 806–815.

[36] Y. Kawahara, K. Ito, H. Sun, H. Aizawa, I. Kanazawa, S. Kwak, Glutamate receptors: RNA editing and death of motor neurons, Nature 427 (6977) (2004) 801.

[37] T. Yamashita, S. Kwak, The molecular link between inefficient GluA2 Q/R site-RNA editing and TDP-43 pathology in motor neurons of sporadic amyotrophic lateral sclerosis patients, Brain Res. 1584 (2014) 28–38.

[38] V. Tadic, T. Prell, J. Lautenschlaeger, J. Grosskreutz, The ER mitochondria calcium cycle and ER stress response as therapeutic targets in amyotrophic lateral sclerosis, Front. Cell Neurosci. 8 (2014) 147.

[39] K.A. Staats, E. Bogaert, N. Hersmus, T. Jaspers, T. Luyten, G. Bultynck, et al. Neuronal overexpression of IP(3) receptor 2 is detrimental in mutant SOD1 mice, Biochem. Biophys. Res. Commun. 429 (3–4) (2012) 210–213.

[40] K.A. Staats, L. Van Helleputte, A.R. Jones, A. Bento-Abreu, A. Van Hoecke, A. Shatunov, et al. Genetic ablation of phospholipase C delta 1 increases survival in SOD1(G93A) mice, Neurobiol. Dis. 60 (2013) 11–17.

[41] S. Vucic, C.S. Lin, B.C. Cheah, J. Murray, P. Menon, A.V. Krishnan, et al. Riluzole exerts central and peripheral modulating effects in amyotrophic lateral sclerosis, Brain 136 (Pt 5) (2013) 1361–1370.

[42] K. Shibuya, S.B. Park, N. Geevasinga, P. Menon, J. Howells, N.G. Simon, et al. Motor cortical function determines prognosis in sporadic ALS, Neurology 87 (2016) 513–520.

[43] K. Kanai, K. Shibuya, Y. Sato, S. Misawa, S. Nasu, Y. Sekiguchi, et al. Motor axonal excitability properties are strong predictors for survival in amyotrophic lateral sclerosis, J. Neurol. Neurosurg. Psychiatry 83 (7) (2012) 734–738.

[44] E. Fritz, P. Izaurieta, A. Weiss, F.R. Mir, P. Rojas, D. Gonzalez, et al. Mutant SOD1-expressing astrocytes release toxic factors that trigger motoneuron death by inducing hyperexcitability, J. Neurophysiol. 109 (11) (2013) 2803–2814.

[45] B.J. Wainger, E. Kiskinis, C. Mellin, O. Wiskow, S.S. Han, J. Sandoe, et al. Intrinsic membrane hyperexcitability of amyotrophic lateral sclerosis patient-derived motor neurons, Cell Rep. 7 (1) (2014) 1–11.

[46] A. Pramatarova, J. Laganiere, J. Roussel, K. Brisebois, G.A. Rouleau, Neuron-specific expression of mutant superoxide dismutase 1 in transgenic mice does not lead to motor impairment, J. Neurosci. 21 (10) (2001) 3369–3374.

[47] M.M. Lino, C. Schneider, P. Caroni, Accumulation of SOD1 mutants in postnatal motoneurons does not cause motoneuron pathology or motoneuron disease, J. Neurosci. 22 (12) (2002) 4825–4832.

[48] D.R. Beers, J.S. Henkel, Q. Xiao, W. Zhao, J. Wang, A.A. Yen, et al. Wild-type microglia extend survival in PU.1 knockout mice with familial amyotrophic lateral sclerosis, Proc. Natl. Acad. Sc. USA 103 (43) (2006) 16021–16026.

[49] Y.H. Gong, A.S. Parsadanian, A. Andreeva, W.D. Snider, J.L. Elliott, Restricted expression of G86R Cu/Zn superoxide dismutase in astrocytes results in astrocytosis but does not cause motoneuron degeneration, J. Neurosci. 20 (2) (2000) 660–665.

[50] S. Boillee, K. Yamanaka, C.S. Lobsiger, N.G. Copeland, N.A. Jenkins, G. Kassiotis, et al. Onset and progression in inherited ALS determined by motor neurons and microglia, Science 312 (5778) (2006) 1389–1392.

[51] K. Yamanaka, S.J. Chun, S. Boillee, N. Fujimori-Tonou, H. Yamashita, D.H. Gutmann, et al. Astrocytes as determinants of disease progression in inherited amyotrophic lateral sclerosis, Nat. Neurosci. 11 (3) (2008) 251–253.

[52] A.M. Clement, M.D. Nguyen, E.A. Roberts, M.L. Garcia, S. Boillee, M. Rule, et al. Wild-type non-neuronal cells extend survival of SOD1 mutant motor neurons in ALS mice, Science 302 (5642) (2003) 113–117.

[53] S.H. Kang, Y. Li, M. Fukaya, I. Lorenzini, D.W. Cleveland, L.W. Ostrow, et al. Degeneration and impaired regeneration of gray matter oligodendrocytes in amyotrophic lateral sclerosis, Nat. Neurosci. 16 (5) (2013) 571–579.

[54] A.M. Haidet-Phillips, S.K. Gross, T. Williams, A. Tuteja, A. Sherman, M. Ko, et al. Altered astrocytic expression of TDP-43 does not influence motor neuron survival, Exp. Neurol. 250 (2013) 250–259.

[55] C.S. Lobsiger, D.W. Cleveland, Glial cells as intrinsic components of non-cell-autonomous neurodegenerative disease, Nat. Neurosci. 10 (11) (2007) 1355–1360.

[56] E.D. Hall, J.A. Oostveen, M.E. Gurney, Relationship of microglial and astrocytic activation to disease onset and progression in a transgenic model of familial ALS, Glia 23 (3) (1998) 249–256.

[57] A.M. Haidet-Phillips, M.E. Hester, C.J. Miranda, K. Meyer, L. Braun, A. Frakes, et al. Astrocytes from familial and sporadic ALS patients are toxic to motor neurons, Nat. Biotechnol. 29 (9) (2011) 824–828.

[58] M. Nagai, D.B. Re, T. Nagata, A. Chalazonitis, T.M. Jessell, H. Wichterle, et al. Astrocytes expressing ALS-linked mutated SOD1 release factors selectively toxic to motor neurons, Nat. Neurosci. 10 (5) (2007) 615–622.

[59] P. Van Damme, E. Bogaert, M. Dewil, N. Hersmus, D. Kiraly, W. Scheveneels, et al. Astrocytes regulate GluR2 expression in motor neurons and their vulnerability to excitotoxicity, Proc. Natl. Acad. Sci. USA 104 (37) (2007) 14825–14830.

[60] J.D. Rothstein, S. Patel, M.R. Regan, C. Haenggeli, Y.H. Huang, D.E. Bergles, et al. Beta-lactam antibiotics offer neuroprotection by increasing glutamate transporter expression, Nature 433 (7021) (2005) 73–77.

[61] M.E. Cudkowicz, S. Titus, M. Kearney, H. Yu, A. Sherman, D. Schoenfeld, et al. Safety and efficacy of ceftriaxone for amyotrophic lateral sclerosis: a multi-stage, randomised, double-blind, placebo-controlled trial, Lancet Neurol. 13 (11) (2014) 1083–1091.

[62] A. Nonneman, W. Robberecht, L. Van Den Bosch, The role of oligodendroglial dysfunction in amyotrophic lateral sclerosis, Neurodegener. Dis. Manag. 4 (3) (2014) 223–239.

[63] Y. Lee, B.M. Morrison, Y. Li, S. Lengacher, M.H. Farah, P.N. Hoffman, et al. Oligodendroglia metabolically support axons and contribute to neurodegeneration, Nature 487 (7408) (2012) 443–448.

[64] T. Philips, A. Bento-Abreu, A. Nonneman, W. Haeck, K. Staats, V. Geelen, et al. Oligodendrocyte dysfunction in the pathogenesis of amyotrophic lateral sclerosis, Brain 136 (Pt 2) (2013) 471–482.

[65] M.R. Turner, A. Cagnin, F.E. Turkheimer, C.C. Miller, C.E. Shaw, D.J. Brooks, et al. Evidence of widespread cerebral microglial activation in amyotrophic lateral sclerosis: an [11C](R)-PK11195 positron emission tomography study, Neurobiol. Dis. 15 (3) (2004) 601–609.

[66] J.I. Engelhardt, J. Tajti, S.H. Appel, Lymphocytic infiltrates in the spinal cord in amyotrophic lateral sclerosis, Arch. Neurol. 50 (1) (1993) 30–36.

[67] M.E. Alexianu, M. Kozovska, S.H. Appel, Immune reactivity in a mouse model of familial ALS correlates with disease progression, Neurology 57 (7) (2001) 1282–1289.

[68] G. Gowing, T. Philips, B. Van Wijmeersch, J.N. Audet, M. Dewil, L. Van Den Bosch, et al. Ablation of proliferating microglia does not affect motor neuron degeneration in amyotrophic lateral sclerosis caused by mutant superoxide dismutase, J. Neurosci. 28 (41) (2008) 10234–10244.

[69] I.M. Chiu, A. Chen, Y. Zheng, B. Kosaras, S.A. Tsiftsoglou, T.K. Vartanian, et al. T lymphocytes potentiate endogenous neuroprotective inflammation in a mouse model of ALS, Proc. Natl. Acad. Sci. USA 105 (46) (2008) 17913–17918.

[70] D.R. Beers, J.S. Henkel, W. Zhao, J. Wang, S.H. Appel, CD4+ T cells support glial neuroprotection, slow disease progression, and modify glial morphology in an animal model of inherited ALS, Proc. Natl. Acad. Sci. USA 105 (40) (2008) 15558–15563.

[71] K.A. Kigerl, J.C. Gensel, D.P. Ankeny, J.K. Alexander, D.J. Donnelly, P.G. Popovich, Identification of two distinct macrophage subsets with divergent effects causing either neurotoxicity or regeneration in the injured mouse spinal cord, J. Neurosci. 29 (43) (2009) 13435–13444.

[72] D.R. Beers, W. Zhao, B. Liao, O. Kano, J. Wang, A. Huang, et al. Neuroinflammation modulates distinct regional and temporal clinical responses in ALS mice, Brain Behav. Immun. 25 (5) (2011) 1025–1035.

[73] B. Liao, W. Zhao, D.R. Beers, J.S. Henkel, S.H. Appel, Transformation from a neuroprotective to a neurotoxic microglial phenotype in a mouse model of ALS, Exp. Neurol. 237 (1) (2012) 147–152.

[74] D.R. Beers, J.S. Henkel, W. Zhao, J. Wang, A. Huang, S. Wen, et al. Endogenous regulatory T lymphocytes ameliorate amyotrophic lateral sclerosis in mice and correlate with disease progression in patients with amyotrophic lateral sclerosis, Brain 134 (Pt 5) (2011) 1293–1314.

[75] Z. Gao, S.E. Tsirka, Animal models of MS reveal multiple roles of microglia in disease pathogenesis, Neurol. Res. Int. 2011 (2011) 383087.

[76] J.S. Henkel, D.R. Beers, S. Wen, A.L. Rivera, K.M. Toennis, J.E. Appel, et al. Regulatory T-lymphocytes mediate amyotrophic lateral sclerosis progression and survival, EMBO Mol. Med. 5 (1) (2013) 64–79.

[77] L. Van Den Bosch, P. Tilkin, G. Lemmens, W. Robberecht, Minocycline delays disease onset and mortality in a transgenic model of ALS, Neuroreport 13 (8) (2002) 1067–1070.

[78] J. Kriz, M.D. Nguyen, J.P. Julien, Minocycline slows disease progression in a mouse model of amyotrophic lateral sclerosis, Neurobiol. Dis. 10 (3) (2002) 268–278.

[79] S. Zhu, I.G. Stavrovskaya, M. Drozda, B.Y. Kim, V. Ona, M. Li, et al. Minocycline inhibits cytochrome c release and delays progression of amyotrophic lateral sclerosis in mice, Nature 417 (6884) (2002) 74–78.

[80] P.H. Gordon, D.H. Moore, R.G. Miller, J.M. Florence, J.L. Verheijde, C. Doorish, et al. Efficacy of minocycline in patients with amyotrophic lateral sclerosis: a phase III randomised trial, Lancet Neurol. 6 (12) (2007) 1045–1053.

[81] P.N. Leigh, V. Meininger, G. Bensimon, M. Cudkowicz, W. Robberecht, Minocycline for patients with ALS, Lancet Neurol. 7 (2) (2008) 119–120.

[82] D.B. Drachman, K. Frank, M. Dykes-Hoberg, P. Teismann, G. Almer, S. Przedborski, et al. Cyclooxygenase 2 inhibition protects motor neurons and prolongs survival in a transgenic mouse model of ALS, Ann. Neurol. 52 (6) (2002) 771–778.

[83] M.E. Cudkowicz, J.M. Shefner, D.A. Schoenfeld, H. Zhang, K.I. Andreasson, J.D. Rothstein, et al. Trial of celecoxib in amyotrophic lateral sclerosis, Ann. Neurol. 60 (1) (2006) 22–31.

[84] A. Galbussera, L. Tremolizzo, L. Brighina, D. Testa, R. Lovati, C. Ferrarese, et al. Vitamin E intake and quality of life in amyotrophic lateral sclerosis patients: a follow-up case series study, Neurol. Sci. 27 (3) (2006) 190–193.

[85] J.M. Shefner, M.E. Cudkowicz, D. Schoenfeld, T. Conrad, J. Taft, M. Chilton, et al. A clinical trial of creatine in ALS, Neurology 63 (9) (2004) 1656–1661.

[86] P. Kaufmann, J.L. Thompson, G. Levy, R. Buchsbaum, J. Shefner, L.S. Krivickas, et al. Phase II trial of CoQ10 for ALS finds insufficient evidence to justify phase III, Ann. Neurol. 66 (2) (2009) 235–244.

[87] V. Meininger, V.E. Drory, P.N. Leigh, A. Ludolph, W. Robberecht, V. Silani, Glatiramer acetate has no impact on disease progression in ALS at 40 mg/day: a double-blind, randomized, multicentre, placebo-controlled trial, Amyotrop. Lateral Scler. 10 (5–6) (2009) 378–383.

[88] H. Kawamata, G. Manfredi, Mitochondrial dysfunction and intracellular calcium dysregulation in ALS, Mech. Ageing Dev. 131 (7–8) (2010) 517–526.

[89] A. Henriques, C. Pitzer, A. Schneider, Neurotrophic growth factors for the treatment of amyotrophic lateral sclerosis: where do we stand?, Front. Neurosci. 4 (2010) 32.

[90] E. Ekestern, Neurotrophic factors and amyotrophic lateral sclerosis, Neurodegener. Dis. 1 (2–3) (2004) 88–100.

[91] A.C. Pronto-Laborinho, S. Pinto, M. de Carvalho, Roles of vascular endothelial growth factor in amyotrophic lateral sclerosis, Biomed. Res Int. 2014 (2014) 947513.

[92] P. Carmeliet, E. Storkebaum, Vascular and neuronal effects of VEGF in the nervous system: implications for neurological disorders, Semin. Cell Dev. Biol. 13 (1) (2002) 39–53.

[93] B. Oosthuyse, L. Moons, E. Storkebaum, H. Beck, D. Nuyens, K. Brusselmans, et al. Deletion of the hypoxia-response element in the vascular endothelial growth factor promoter causes motor neuron degeneration, Nat. Genet. 28 (2) (2001) 131–138.

[94] L. Van Den Bosch, E. Storkebaum, V. Vleminckx, L. Moons, L. Vanopdenbosch, W. Scheveneels, et al. Effects of vascular endothelial growth factor (VEGF) on motor neuron degeneration, Neurobiol. Dis. 17 (1) (2004) 21–28.

[95] M. Dewil, D. Lambrechts, R. Sciot, P.J. Shaw, P.G. Ince, W. Robberecht, et al. Vascular endothelial growth factor counteracts the loss of phospho-Akt preceding motor neurone degeneration in amyotrophic lateral sclerosis, Neuropathol. Appl. Neurobiol. 33 (5) (2007) 499–509.

[96] D. Lambrechts, E. Storkebaum, M. Morimoto, J. Del-Favero, F. Desmet, S.L. Marklund, et al. VEGF is a modifier of amyotrophic lateral sclerosis in mice and humans and protects motoneurons against ischemic death, Nat. Genet. 34 (4) (2003) 383–394.

[97] L. Lu, L. Zheng, L. Viera, E. Suswam, Y. Li, X. Li, et al. Mutant Cu/Zn-superoxide dismutase associated with amyotrophic lateral sclerosis destabilizes vascular endothelial growth factor mRNA and downregulates its expression, J. Neurosci. 27 (30) (2007) 7929–7938.

[98] Y. Wang, X.O. Mao, L. Xie, S. Banwait, H.H. Marti, D.A. Greenberg, et al. Vascular endothelial growth factor overexpression delays neurodegeneration and prolongs survival in amyotrophic lateral sclerosis mice, J. Neurosci. 27 (2) (2007) 304–307.

[99] C. Zheng, M.K. Skold, J. Li, I. Nennesmo, B. Fadeel, J.I. Henter, VEGF reduces astrogliosis and preserves neuromuscular junctions in ALS transgenic mice, Biochem. Biophys. Res. Commun. 363 (4) (2007) 989–993.

[100] E. Bogaert, P. Van Damme, K. Poesen, J. Dhondt, N. Hersmus, D. Kiraly, et al. VEGF protects motor neurons against excitotoxicity by upregulation of GluR2, Neurobiol. Aging 31 (12) (2010) 2185–2191.

[101] M. Azzouz, G.S. Ralph, E. Storkebaum, L.E. Walmsley, K.A. Mitrophanous, S.M. Kingsman, et al. VEGF delivery with retrogradely transported lentivector prolongs survival in a mouse ALS model, Nature 429 (6990) (2004) 413–417.

[102] J.C. Dodge, C.M. Treleaven, J.A. Fidler, M. Hester, A. Haidet, C. Handy, et al. AAV4-mediated expression of IGF-1 and VEGF within cellular components of the ventricular system improves survival outcome in familial ALS mice, Mol. Ther. 18 (12) (2010) 2075–2084.

[103] E. Storkebaum, D. Lambrechts, M. Dewerchin, M.P. Moreno-Murciano, S. Appelmans, H. Oh, et al. Treatment of motoneuron degeneration by intracerebroventricular delivery of VEGF in a rat model of ALS, Nat. Neurosci. 8 (1) (2005) 85–92.

[104] P.C. Wong, C.A. Pardo, D.R. Borchelt, M.K. Lee, N.G. Copeland, N.A. Jenkins, et al. An adverse property of a familial ALS-linked SOD1 mutation causes motor neuron disease characterized by vacuolar degeneration of mitochondria, Neuron 14 (6) (1995) 1105–1116.

[105] J. Kong, Z. Xu, Massive mitochondrial degeneration in motor neurons triggers the onset of amyotrophic lateral sclerosis in mice expressing a mutant SOD1, J. Neurosci. 18 (9) (1998) 3241–3250.

[106] C.M. Higgins, C. Jung, Z. Xu, ALS-associated mutant SOD1^{G93A} causes mitochondrial vacuolation by expansion of the intermembrane space and by involvement of SOD1 aggregation and peroxisomes, BMC Neurosci. 4 (2003) 16.

[107] A.K. Afifi, F.P. Aleu, J. Goodgold, B. MacKay, Ultrastructure of atrophic muscle in amyotrophic lateral sclerosis, Neurology 16 (5) (1966) 475–481.

[108] S. Sasaki, M. Iwata, Mitochondrial alterations in the spinal cord of patients with sporadic amyotrophic lateral sclerosis, J. Neuropathol. Exp. Neurol. 66 (1) (2007) 10–16.

[109] W. Wang, L. Li, W.L. Lin, D.W. Dickson, L. Petrucelli, T. Zhang, et al. The ALS disease-associated mutant TDP-43 impairs mitochondrial dynamics and function in motor neurons, Hum. Mol. Genet. 22 (23) (2013) 4706–4719.

[110] Y.F. Xu, T.F. Gendron, Y.J. Zhang, W.L. Lin, S. D'Alton, H. Sheng, et al. Wild-type human TDP-43 expression causes TDP-43 phosphorylation, mitochondrial aggregation, motor deficits, and early mortality in transgenic mice, J. Neurosci. 30 (32) (2010) 10851–10859.

[111] M.L. Tradewell, Z. Yu, M. Tibshirani, M.C. Boulanger, H.D. Durham, S. Richard, Arginine methylation by PRMT1 regulates nuclear-cytoplasmic localization and toxicity of FUS/TLS harbouring ALS-linked mutations, Hum. Mol. Genet. 21 (1) (2012) 136–149.

[112] C. Sunyach, M. Michaud, T. Arnoux, N. Bernard-Marissal, J. Aebischer, V. Latyszenok, et al. Olesoxime delays muscle denervation, astrogliosis, microglial activation and motoneuron death in an ALS mouse model, Neuropharmacology 62 (7) (2012) 2346–2352.

[113] L.J. Martin, Olesoxime, a cholesterol-like neuroprotectant for the potential treatment of amyotrophic lateral sclerosis, IDrugs 13 (8) (2010) 568–580.

[114] M.E. Cudkowicz, L.H. van den Berg, J.M. Shefner, H. Mitsumoto, J.S. Mora, A. Ludolph, et al. Dexpramipexole versus placebo for patients with amyotrophic lateral sclerosis (EMPOWER): a randomised, double-blind, phase 3 trial, Lancet Neurol. 12 (11) (2013) 1059–1067.

[115] T. Bordet, B. Buisson, M. Michaud, C. Drouot, P. Galea, P. Delaage, et al. Identification and characterization of cholest-4-en-3-one, oxime (TRO19622), a novel drug candidate for amyotrophic lateral sclerosis, Journal Pharmacol. Exp. Ther. 322 (2) (2007) 709–720.

[116] D.M. Pastula, D.H. Moore, R.S. Bedlack, Creatine for amyotrophic lateral sclerosis/motor neuron disease, Cochrane Database Syst. Rev. 12 (2012) CD005225.

[117] L.R. Fischer, D.G. Culver, P. Tennant, A.A. Davis, M. Wang, A. Castellano-Sanchez, et al. Amyotrophic lateral sclerosis is a distal axonopathy: evidence in mice and man, Exp. Neurol. 185 (2) (2004) 232–240.

[118] D. Frey, C. Schneider, L. Xu, J. Borg, W. Spooren, P. Caroni, Early and selective loss of neuromuscular synapse subtypes with low sprouting competence in motoneuron diseases, J. Neurosci. 20 (7) (2000) 2534–2542.

[119] K. Okamoto, S. Hirai, M. Shoji, Y. Senoh, T. Yamazaki, Axonal swellings in the corticospinal tracts in amyotrophic lateral sclerosis, Acta Neuropathol. 80 (2) (1990) 222–226.

[120] S. Sasaki, S. Maruyama, Increase in diameter of the axonal initial segment is an early change in amyotrophic lateral sclerosis, J. Neurol. Sci. 110 (1–2) (1992) 114–120.

[121] K.J. De Vos, A.J. Grierson, S. Ackerley, C.C.J. Miller, Role of axonal transport in neurodegenerative diseases, Annu. Rev. Neurosci. 31 (2008) 151–173.

[122] K.J. De Vos, A.L. Chapman, M.E. Tennant, C. Manser, E.L. Tudor, K.F. Lau, et al. Familial amyotrophic lateral sclerosis-linked SOD1 mutants perturb fast axonal transport to reduce axonal mitochondria content, Hum. Mol. Genet. 16 (22) (2007) 2720–2728.

[123] D. Kieran, M. Hafezparast, S. Bohnert, J.R. Dick, J. Martin, G. Schiavo, et al. A mutation in dynein rescues axonal transport defects and extends the life span of ALS mic, J. Cell Biol. 169 (4) (2005) 561–567.

[124] T.L. Williamson, D.W. Cleveland, Slowing of axonal transport is a very early event in the toxicity of ALS-linked SOD1 mutants to motor neurons, Nat. Neurosci. 2 (1) (1999) 50–56.

[125] B. Zhang, P. Tu, F. Abtahian, J.Q. Trojanowski, V.M. Lee, Neurofilaments and orthograde transport are reduced in ventral root axons of transgenic mice that express human SOD1 with a G93A mutation, J. Cell Biol. 139 (5) (1997) 1307–1315.

[126] J.F. Collard, F. Cote, J.P. Julien, Defective axonal-transport in a transgenic mouse model of amyotrophic-lateral-sclerosis, Nature 375 (6526) (1995) 61–64.

[127] C. Munch, R. Sedlmeier, T. Meyer, V. Homberg, A.D. Sperfeld, A. Kurt, et al. Point mutations of the p150 subunit of dynactin (DCTN1) gene in ALS, Neurology 63 (4) (2004) 724–726.

[128] M. Stockmann, M. Meyer-Ohlendorf, K. Achberger, S. Putz, M. Demestre, H. Yin, et al. The dynactin p150 subunit: cell biology studies of sequence changes found in ALS/MND and Parkinsonian syndromes, J. Neural Transm. 120 (5) (2013) 785–798.

[129] I. Puls, C. Jonnakuty, B.H. LaMonte, E.L. Holzbaur, M. Tokito, E. Mann, et al. Mutant dynactin in motor neuron disease, Nat. Genet. 33 (4) (2003) 455–456.

[130] K.R. Baldwin, V.K. Godena, V.L. Hewitt, A.J. Whitworth, Axonal transport defects are a common phenotype in Drosophila models of ALS, Hum. Mol. Genet. 25 (12) (2016) 2378–2392.

[131] P. Fanara, J. Banerjee, R.V. Hueck, M.R. Harper, M. Awada, H. Turner, et al. Stabilization of hyperdynamic microtubules is neuroprotective in amyotrophic lateral sclerosis, J. Biol. Chem. 282 (32) (2007) 23465–23472.

[132] Y. Jouroukhin, R. Ostritsky, Y. Assaf, G. Pelled, E. Giladi, I. Gozes, NAP (davunetide) modifies disease progression in a mouse model of severe neurodegeneration: protection against impairments in axonal transport, Neurobiol. Dis. 56 (2013) 79–94.

[133] M. Neumann, D.M. Sampathu, L.K. Kwong, A.C. Truax, M.C. Micsenyi, T.T. Chou, et al. Ubiquitinated TDP-43 in frontotemporal lobar degeneration and amyotrophic lateral sclerosis, Science 314 (5796) (2006) 130–133.

[134] R. Banerjee, M.F. Beal, B. Thomas, Autophagy in neurodegenerative disorders: pathogenic roles and therapeutic implications, Trends Neurosci. 33 (12) (2010) 541–549.

[135] R.A. Nixon, The role of autophagy in neurodegenerative disease, Nat. Med. 19 (8) (2013) 983–997.

[136] T. Hara, K. Nakamura, M. Matsui, A. Yamamoto, Y. Nakahara, R. Suzuki-Migishima, et al. Suppression of basal autophagy in neural cells causes neurodegenerative disease in mice, Nature 441 (7095) (2006) 885–889.

[137] M. Komatsu, S. Waguri, T. Chiba, S. Murata, J. Iwata, I. Tanida, et al. Loss of autophagy in the central nervous system causes neurodegeneration in mice, Nature 441 (7095) (2006) 880–884.

[138] L. Li, X. Zhang, W. Le, Altered macroautophagy in the spinal cord of SOD1 mutant mice, Autophagy 4 (3) (2008) 290–293.

[139] N. Morimoto, M. Nagai, Y. Ohta, K. Miyazaki, T. Kurata, M. Morimoto, et al. Increased autophagy in transgenic mice with a G93A mutant SOD1 gene, Brain Res. 1167 (2007) 112–117.

[140] F. Tian, N. Morimoto, W. Liu, Y. Ohta, K. Deguchi, K. Miyazaki, et al. In vivo optical imaging of motor neuron autophagy in a mouse model of amyotrophic lateral sclerosis, Autophagy 7 (9) (2011) 985–992.

[141] C. Hetz, P. Thielen, S. Matus, M. Nassif, F. Court, R. Kiffin, et al. XBP-1 deficiency in the nervous system protects against amyotrophic lateral sclerosis by increasing autophagy, Genes Dev. 23 (19) (2009) 2294–2306.

[142] E. Wong, A.M. Cuervo, Autophagy gone awry in neurodegenerative diseases, Nat. Neurosci. 13 (7) (2010) 805–811.

[143] X. Zhang, L. Li, S. Chen, D. Yang, Y. Wang, X. Zhang, et al. Rapamycin treatment augments motor neuron degeneration in SOD1(G93A) mouse model of amyotrophic lateral sclerosis, Autophagy 7 (4) (2011) 412–425.

[144] A. Bhattacharya, A. Bokov, F.L. Muller, A.L. Jernigan, K. Maslin, V. Diaz, et al. Dietary restriction but not rapamycin extends disease onset and survival of the H46R/H48Q mouse model of ALS, Neurobiol. Aging 33 (8) (2012) 1829–1832.

[145] K.A. Staats, S. Hernandez, S. Schonefeldt, A. Bento-Abreu, J. Dooley, P. Van Damme, et al. Rapamycin increases survival in ALS mice lacking mature lymphocytes, Mol. Neurodegener. 8 (2013) 31.

[146] J. Kim, T.Y. Kim, K.S. Cho, H.N. Kim, J.Y. Koh, Autophagy activation and neuroprotection by progesterone in the G93A-SOD1 transgenic mouse model of amyotrophic lateral sclerosis, Neurobiol. Dis. 59 (2013) 80–85.

[147] K. Castillo, M. Nassif, V. Valenzuela, F. Rojas, S. Matus, G. Mercado, et al. Trehalose delays the progression of amyotrophic lateral sclerosis by enhancing autophagy in motoneurons, Autophagy 9 (9) (2013) 1308–1320.

[148] X. Zhang, S. Chen, L. Song, Y. Tang, Y. Shen, L. Jia, et al. MTOR-independent, autophagic enhancer trehalose prolongs motor neuron survival and ameliorates the autophagic flux defect in a mouse model of amyotrophic lateral sclerosis, Autophagy 10 (4) (2014) 588–602.

[149] S.C. Ling, M. Polymenidou, D.W. Cleveland, Converging mechanisms in ALS and FTD: disrupted RNA and protein homeostasis, Neuron 79 (3) (2013) 416–438.

[150] M. Polymenidou, C. Lagier-Tourenne, K.R. Hutt, S.C. Huelga, J. Moran, T.Y. Liang, et al. Long pre-mRNA depletion and RNA missplicing contribute to neuronal vulnerability from loss of TDP-43, Nat. Neurosci. 14 (4) (2011) 459–468.

[151] H. Qiu, S. Lee, Y. Shang, W.Y. Wang, K.F. Au, S. Kamiya, et al. ALS-associated mutation FUS-R521C causes DNA damage and RNA splicing defects, J. Clin. Invest. 124 (3) (2014) 981–999.

[152] A. Patel, H.O. Lee, L. Jawerth, S. Maharana, M. Jahnel, M.Y. Hein, et al. A liquid-to-solid phase transition of the ALS protein FUS accelerated by disease mutation, Cell 162 (5) (2015) 1066–1077.

[153] S. Boeynaems, E. Bogaert, P. Van Damme, L. Van Den Bosch, Inside out: the role of nucleocytoplasmic transport in ALS and FTLD, Acta Neuropathol. 132 (2) (2016) 159–173.

[154] K. Zhang, C.J. Donnelly, A.R. Haeusler, J.C. Grima, J.B. Machamer, P. Steinwald, et al. The C9orf72 repeat expansion disrupts nucleocytoplasmic transport, Nature 525 (7567) (2015) 56–61.

[155] B.D. Freibaum, Y. Lu, R. Lopez-Gonzalez, N.C. Kim, S. Almeida, K.H. Lee, et al. GGGGCC repeat expansion in C9orf72 compromises nucleocytoplasmic transport, Nature 525 (7567) (2015) 129–133.

[156] A. Jovicic, J. Mertens, S. Boeynaems, E. Bogaert, N. Chai, S.B. Yamada, et al. Modifiers of C9orf72 dipeptide repeat toxicity connect nucleocytoplasmic transport defects to FTD/ALS, Nat. Neurosci. 18 (9) (2015) 1226–1229.

[157] H. Tran, S. Almeida, J. Moore, T.F. Gendron, U. Chalasani, Y. Lu, et al. Differential toxicity of nuclear RNA foci versus dipeptide repeat proteins in a drosophila model of C9ORF72 FTD/ALS, Neuron 87 (6) (2015) 1207–1214.

Index

A

Acetylcholinesterase inhibitors
  donepezil, 116
  galantamine, 116
  rivastigmine, 116
Acidic phospholipids, 25
Actin cytoskeleton, 9
AD. *See* Alzheimer's disease (AD)
Adeno-associated virus (AAV), 262
Adult brain regeneration
  cell transplantation, role of, 109
  endogenous stem cells pool, role of, 108–109
    environmental enrichment, 108
    pharmaco- and gene therapy, 108–109
  as model system, 109
Adult neurogenesis (AN), 101
  Alzheimer's disease (AD), 106
    human patients, 106
    transgenic animal models, 106
  Huntington's disease
    animal models, 107
    human patients, 107
  mechanism, 101–103
    dentate gyrus (DG), role of, 102
    stem cell niche modulation, role of, 103
    subventricular zone (SVZ), role of, 101–102
  neurodegenerative environment, role of, 104–107
  Parkinson's disease (PD), 104–105
    acute lesion models, 104–105
    human patients, 105
    transgenic animal models, 105
Adult stem cells (ASCs), 83
Advanced glycation end products (AGE), 32
AGD. *See* Argyrophilic grain disease (AGD)
Aggregation-prone proteins, 7
Agonist, 280
Allogeneic cells, 89
ALP. *See* Autophagy-lysosome pathway (ALP)
Alpha-synuclein gene *(SNCA)*, 177
ALS. *See* Amyotrophic lateral sclerosis (ALS)
Altered proteostasis, 285
Alzheimer's disease (AD), 50, 58–60, 106, 116
  amyloid cascade hypothesis, 118
  clinical features, 116
  early-onset familial (EOFAD), 117
  genetics, 117
  genome-wide association study (GWAS) analyses of sporadic, 59
  late-onset (LOAD), 117

neuropathology, 116
NFT as diagnostic hallmark of, 124
pathological hallmark of, 58
risk factor for, 58
symptomatic treatment, 116
  acetylcholinesterase inhibitors, 116
  NMDA antagonist, 116
symptoms
  cognitive impairment, 116
  episodic deficits in short-term memory, 116
  impairment of declarative and nondeclarative memory, 116
  memory loss, 116
  neuropsychiatric, 116
γ−Aminobutyric acid (GABA), 53
α-Amino-3-hydroxy-5-methyl-4-isoxazolepropionic acid (AMPA), 280
AMPA. *See* α-Amino-3-hydroxy-5- methyl-4-isoxazolepropionic acid (AMPA)
Amyloid β (Aβ) peptide, 2, 118. *See also* Amyloid precursor protein (APP)
  catabolism, 121
  degradation, 121
    by enzymes
      angiotensin-converting enzyme, 121
      insulin-degrading enzyme (IDE), 121
      neprilysin, 121
  immunotherapy, 121
  induced tau-pathology, 116, 134, 135
  plaques, 106. *See also* Amyloid plaque
Amyloid cascade hypothesis, 86, 118
  Aβ role, for pathogenesis of AD, 118
Amyloid plaque, 4, 22, 116
  deposition, 135
  pathology, recapitulation of, 118
Amyloid precursor protein (APP), 85, 106, 116
  cleavage of, 116
    by β-secretase, 116
    by γ-secretase, 116, 119
Amyloid toxicity, 87
Amyotrophic lateral sclerosis (ALS), 50, 64, 91, 200, 277
  affected areas, 64
  clinical features of, 64
  respiratory complications, 64
  riluzole for treatment of, 208
AN. *See* Adult neurogenesis (AN)
Animal models, 138. *See also* Transgenic models
  for identification and preclinical validation of therapeutic strategies, 116

Anti-aβ therapies, 118
Anti-α syn immunohistochemistry, 178
Antiparkisonian drugs, 182
Antisense oligonucleotide (ASO), 7, 93, 213, 218, 260, 287
Aphasia, 208. *See also* Primary progressive aphasia (PPA)
apoE4. *See* Apolipoprotein E4 (apoE4)
Apolipoprotein E4 (apoE4), 86
APP. *See* Amyloid precursor protein (APP)
Argyrophilic grain disease (AGD), 200
ASCs. *See* Adult stem cells (ASCs)
ASO. *See* Antisense oligonucleotides (ASO)
Astrocytes, 6, 53
    derived from, 53
    dysfunction, 281
    morphology, 53
    in propagation of action potentials, 53
Autophagosome, 285
Autophagy, 8, 33, 285
Autophagy-lysosome pathway (ALP), 178, 183
Autosomal dominant mutations, 118
Axonal defects, 284
Axonal swellings, 285
Axon retraction, 284

B

BBB. *See* Blood-brain barrier (BBB)
BDNF. *See* Brain-derived neurotrophic factor (BDNF)
Biological membranes, 5
Blood-brain barrier (BBB), 9, 182, 226
Braak and Thal staging, 116
Bradykinesia, 85, 106, 251
Brain-derived neurotrophic factor (BDNF), 54
Brain homogenates, 4
Bromocriptine, 208

C

CAG-Age Product (CAP) score, 253
Calcium signaling, 6
Calpains, 8
Cardiolipin, 25
Casein kinase 2 inhibitors, 184
Cas9 nickase, 83
Catechol-O-methyl transferase (COMT), 207
    inhibitor, 207
Causal mutations, 202
    GRN gene, 202
CBS. *See* Corticobasal syndrome (CBS)
CDK5. *See* Cyclin-dependent kinase 5 (CDK5)
Cell therapy trials, 81
Central nervous system (CNS), 282
    activation of, 52
    major glial cell types involved in, 52

Cerebrospinal fluid (CSF), 54, 181, 260
Chaperone-mediated autophagy (CMA), 186
Chaperones, 33
Cholinesterase inhibitors, 86, 183
Chromatin maintenance, 91
Chronic neuroinflammation, 54
CLR01 molecular tweezer, 36
Clustered regularly interspaced short palindromic repeats (CRISPR)-Cas9 technologies, 83
CMA. *See* Chaperone-mediated autophagy (CMA)
CNS. *See* Central nervous system (CNS)
Cochaperone HDJ-1, 181
Cortical neurons generation
    control lines, role of, 92
    frontotemporal dementia (FTD)-iPSC, role of, 92
Corticobasal syndrome (CBS), 176, 200
CRISPR. *See* Clustered regularly interspaced short palindromic repeats (CRISPR)
CRISPR RNAs (crRNAs), 83
crRNAs. *See* CRISPR RNAs (crRNAs)
αB-Crystallin, 181
CSF. *See* Cerebrospinal fluid (CSF)
Cyclin-dependent kinase 5 (CDK5), 126
Cytosolic folding states, 24
Cytosolic proteases, 8

D

Danger-associated molecular patterns (DAMPs), 53
Deep brain stimulation (DBS), 160, 187
Deletion mutations, 164
Delirium, 177
Dementia, 50
Dementia with Lewy bodies (DLB), 175
Dentate gyrus (DG), 103
DG. *See* Dentate gyrus (DG)
Dipeptide repeat proteins (DPRs), 286
Disease-modifying drug development, 209–224
    cellular waste clearance systems, 220–222
    C9orf72, 202, 216–218, 219
    granulin, 214
    heterogeneity, 222–224
    tau, 209–214
    TDP-43, 219–220
Disease-modifying therapy development, 224–226
    toward early and targeted intervention, 226
    understanding of FTD, 224–226
DJ-1 genes, 166
DLB. *See* Dementia with Lewy bodies (DLB)
DNA repair, 91
Dopamine, 56
    agonist, 85
    precursor, 85

replacement drug, 182
transporter, 86
Dopaminergic (DA) neurons, 4, 85
Double-strand DNA breaks, 83
DPRs. *See* Dipeptide repeat proteins (DPRs)

E

Early-onset familial Alzheimer's disease (EOFAD), 117
EGCG. *See* Epigallocatachin-3-gallate (EGCG)
Embryonic cells grafting, 11
Embryonic stem cell (ESC), 82
Endocytic vesicle membranes, 7
Endogenous stem cells, 101–104
    human relevance, 103–104
        dentate gyrus (DG), 104
        subventricular zone (SVZ), 103
    as therapeutic target, 101–104
EOFAD. *See* Early-onset familial Alzheimer's disease (EOFAD)
Eosin staining, 178
Ependymal cells, 54, 108
    layer, 103
Epigallocatachin-3-gallate (EGCG), 7, 36
ESC. *See* Embryonic stem cell (ESC)
Etanercept, 68
Excitotoxicity, 280
Exosomes, 6
    associated exocytosis, 185
Extracellular serine protease, 8
Extracellular vesicles
    interneuronal communication route, role as, 6

F

FACS. *See* Fluorescence-activated cell sorting (FACS)
FALS. *See* Familial ALS (FALS)
Familial ALS (FALS), 278
FDA. *See* US Food and Drug Administration (FDA)
FGF-2. *See* Fibroblast growth factor 2 (FGF-2)
Fibrillar polyQ assemblies, 5
Fibroblast growth factor 2 (FGF-2), 108
Fluorescence-activated cell sorting (FACS), 85
Frontotemporal dementia (FTD), 66, 91, 117
    behavioral variant (bvFTD), 200
    causes
        SQSTM1 mutation, 204
        TREM2 mutation, 204
    clinical presentations, 200–201
    disease management, 206–209
        nonpharmacologic management, 209
            behavioral interventions, 209
            education of caregivers, 209
            psychosocial support, 209

pharmacologic symptomatic treatment, 206–209
    aphasia, 208
    behavioral and cognitive symptoms, 206–208
    motor symptom management, 208–209
    genetics of, 202–204
    causal genes, 202–204
    disease risk, 204
    pathological and clinical correlations in, 205
  heterogeneous disorder, 205
  overview, 199
  pathological heterogeneity, 201–202
  risk factors, 204
  selective serotonin reuptake inhibitors, effect on, 206
  treatment, 206
    cholinesterase inhibitors, efficacy in, 207
    NMDA receptor antagonists, efficacy in, 207
Frontotemporal lobar degeneration (FTLD), 199, 278
    FTLD-fused in sarcoma (FTLD-FUS), 201
    FTLD-TAR DNA-binding protein 43 (FTLD-TDP), 201
    FTLD-ubiquitin proteasome system (FTLD-UPS), 201
FTD. *See* Frontotemporal dementia (FTD)
FTLD. *See* Frontotemporal lobar degeneration (FTLD)
FTLD-FUS. *See* FTLD-fused in sarcoma (FTLD-FUS)
FTLD-fused in sarcoma (FTLD-FUS), 201
FTLD-TAR DNA-binding protein 43 (FTLD-TDP), 201
    pathology, 220
FTLD-tau pathology, 201
FTLD-TDP. *See* FTLD-TAR DNA-binding protein 43 (FTLD-TDP)
FTLD-ubiquitin proteasome system (FTLD-UPS), 201
FTLD-UPS. *See* FTLD-ubiquitin proteasome system (FTLD-UPS)
FUS. *See* Fused-in-sarcoma (FUS)
Fused-in-sarcoma (FUS), 93

G

GA. *See* Geldanamycin (GA)
GABAergic neurons, 86, 109
GCase enzyme, 38
Geldanamycin (GA), 186
Genome editing, 262–263
Genome-wide association (GWA), 251
GFAP. *See* Glial fibrillary acid protein (GFAP)
Glial cells, 6, 51
  in CNS development, 51
  homeostasis, and pathology, 51
Glial fibrillary acid protein (GFAP), 281
Glial fibrillary astrocytic protein *(GFAP)* promoter, 65
Glial glutamate transporters, 282
β−Glucocerebrosidase gene *(GBA)*, 177
Glucocorticoids, 10
Glycogen synthase kinase-3 (GSK-3), 126

Glycosphingolipid (GM1), 25
GM1. *See* Glycosphingolipid (GM1)
Goldilocks principle, 164
GRN expression, 214
GSK-3. *See* Glycogen synthase kinase-3 (GSK-3)

H

Heat shock factor 1 (HSF1), 186
Heat shock proteins (HSPs), 181
Heparan sulfate proteoglycans (HSPG), 131
HLA. *See* Human leukocyte antigen (HLA)
Holdases, 7
hPSC-derived cells, 88
HSF1. *See* Heat shock factor 1 (HSF1)
HSPG. *See* Heparan sulfate proteoglycans (HSPG)
HSPs. *See* Heat shock proteins (HSPs)
HTTExon1. *See* Huntingtin exon 1 (HTTExon1)
HTT gene, 253
    5′ end of human HTT gene, 256
    open reading frame (ORF) encoding, 253
    representation and corresponding transcription
        start sites, 255
    transcription start sites, 253
HTT protein, 253
Human fibroblasts, 87
Human genetic disorders
    genotype-phenotype causal relationships, 83
Human immortalized cells, 92
Human leukocyte antigen (HLA), 204
Human matrigel, 95
Huntingtin aggregates, 108
Huntingtin exon 1 (HTTExon1), 2
Huntingtin protein, 257
Huntington's disease (HD), 7, 50, 65, 89, 184,
        251, 252
    animal models, 107
    caused by, 65
    gene and transcripts, 253–256
    human patients, 107
    molecular strategies for, 259–260
    neurological disorder, 251
    pathogenic mechanisms, 258–259
    prevalence rate in western populations, 253
    primary neuropathology of, 65
6-Hydroxy-dopamine (6-OHDA), 104
Hyperexcitability, 281

I

IDE. *See* Insulin-degrading enzyme (IDE)
Immunization, 9
Immunosuppressive therapy, 89
Inclusion body myopathy, 200

Induced pluripotent stem cell (iPSC), 281
    technology, 81
Insulin-degrading enzyme (IDE), 121
Intercellular adhesion molecule 1 (ICAM1), 55
Interferon-β signaling, 181
International Society for Stem Cell Research (ISSCR), 88
In vitro and in vivo cells manipulation
    possible ways, 83
iPSC. *See* Induced pluripotent stem cell (iPSC)
IPSC characteristics
    neurodegenerative disease
        development, role of, 91–92
iPSC-derived teratomas, 89
IPSC technology
    disease modeling, role in, 92–95
        alzheimer's disease, 95
        amyotrophic lateral sclerosis, 93–95
        parkinson's disease (PD), 95
    drug screening, role in, 92–95
iPSC technology. *See* Induced pluripotent stem cell (iPSC)
    technology

K

Kinase inhibitors
    lithium, 211
    valproate, 211
KTKEGV sequence, 24
Kv7 channel activator, 93

L

Late-onset Alzheimer's disease (LOAD), 117
    risk factors for, 117
LB. *See* Lewy bodies (LB)
LBD. *See* Lewy body dementias (LBD)
L-3,4-Dihydroxyphenylalanine (L-DOPA), 158, 182
Levodopa, 104, 182
Lewy bodies (LB), 4, 21, 157
Lewy body dementias (LBD), 58, 66, 175–188
    alpha-synuclein, 178
    amyloid, 181
    chaperones proteins, 181
    clinical symptoms, 177
    genetic association, 177–178
    management of, 182–188
        current symptomatic treatment, 182–183
        disease-modifying therapy, 183–188
            deep brain stimulation (DBS), 187–188
            αsyn aggregation, modulation of, 186
            αsyn protein targeting, 183–185
    neuroinflammation, 181
    pathophysiology, 178–181
    tau, 181

Lewy body disorders
    molecular mechanisms, 180
Lewy body precursor-like inclusions, 92
Lewy neurites (LN), 21, 157, 178
LFA1. *See* Lymphocyte function-associated antigen
    1 (LFA1)
Lipid bilayer curving, 6
LN. *See* Lewy neurites (LN)
LOAD. *See* Late-onset Alzheimer's disease (LOAD)
Long-term potentiation (LTP), 118
LRRK2 gene, 95, 165–166
LRRK2 mutations, 93, 105
LTP. *See* Long-term potentiation (LTP)
Lymphocyte function-associated antigen 1 (LFA1), 55

M

Macroautophagy, 285
MACS. *See* Magnetic-activated cell sorting (MACS)
Magnetic-activated cell sorting (MACS), 85
Maillard reaction, 32
MAPT. *See* Microtubule-associated protein tau (MAPT)
MARK4, Microtubule affinity–regulating kinase
    4 (MARK4)
MBP. *See* Myelin basic protein (MBP)
MCT1. *See* Monocarboxylate transporter-1 (MCT1)
Memantine, 116
Membrane-bound folding states, 24–25
Membrane-bound naked misfolded pathogenic protein
    assemblies, 6
Mesenchymal stem cell (MSC), 83
2′-O-Methoxyethylribose (MOE) gapmer
    oligonucleotides, 260
N-Methyl-D-aspartic acid (NMDA), 207
    antagonist, 116
1-Methyl-4-phenyl-1,2,3,6-tetrahydropyridine
    (MPTP), 104
Microglia, 51
    dysfunction, 282
    functional role in CNS development and homeostasis, 52
    numbers and densities, 51
    origins of, 51
    resting state, 51
Microtubule affinity-regulating kinase 4 (MARK4), 126
Microtubule-associated protein tau (MAPT), 116
Microtubule-associated protein tau (*MAPT*) gene, 116, 177,
    178, 202
    mutation, 202, 226
Microtubule-binding (MTB), 201
miR-Embedded siRNA, 261
Misfolded pathologic protein aggregates
    therapeutic strategies targeting accumulation, 8
    therapeutic strategies targeting cell-to-cell propagation, 9

Misfolded proteins, 21
    aggregates
        diseases, 2
        spread and transmission of disease, role in, 4
    assemblies
        prion-like propagation
            therapeutic strategies, expected effect of, 12
            therapeutic strategies, targets of, 11
    neurodegenerative disorders, hallmark of, 21
Mitochondrial dysfunction, 178, 284
Mitral cells, 101
MND. *See* Motor neuron disease (MND)
Molecular chaperones, 1
Molecular crowding, 6
Molten globules, 26
Monocarboxylate transporter-1 (MCT1), 282
Monocyte chemoattractant protein (MCP1), 55
Morphogens, 103
Motor neuron disease (MND), 200
Motor neurons, 280
Mouse somatic cells, 82
MPTP. *See* 1-Methyl-4-phenyl-1,2,3,6-tetrahydropyridine
    (MPTP)
MSC. *See* Mesenchymal stem cell (MSC)
Multiple sclerosis, 89
Mutant huntingtin synthesis, 7
M337V TDP-43 mutation, 95
Myelin basic protein (MBP), 282
Myeloid cells 2, 87

N

NAC. *See* Nonamyloid component of Aβ plaques (NAC)
NBM. *See* Nucleus basalis of Meynert (NBM)
NDDs. *See* Neurodegenerative diseases (NDDs)
Nedd4 as ubiquitin ligase, 36
Nerve growth factor (NGF), 54, 87
Neural stem cells (NSCs), 84
Neuregulin-dependent myelination, 119
Neuroblasts, 102
NeuroD1, 84
Neurodegeneration, 4. *See also* Neurodegenerative diseases
    (NDDs)
    aggregation-prone proteins, role of
        therapeutic strategies targeting misfolding, 7
        therapeutic strategies targeting physiological
            levels, 7
    mechanisms of neuroinflammation-mediated, 56
    microglial activation, 55
    prevention
        adult brain regeneration, role of, 108–109
    reactive microgliosis during, 55
Neurodegenerative dementia, 199

Neurodegenerative diseases (NDDs), 50, 81
  challenge, of developing effective therapeutics, 50
  chronic neuroinflammation
    pathophysiological mediator, 54
  danger associated molecular patterns (DAMPs), 57
  feature, 57
  global burden, 50
  multiple hit hypotheses, 50
  pattern recognition receptors (PRRs), 57
  reactive microgliosis, during progressive
    neurodegeneration, 55
  therapeutic strategies targeting neuroinflammation
    in, 66–68
Neurodegenerative protein misfolding diseases, 21
Neurofibrillary tangles (NFTs), 116, 181
Neurogenin-2, 84
Neuroinflammation, 57
  mechanisms of, 56
  misfolded pathologic protein aggregate-mediated
    therapeutic strategies, 10
  selected clinical trials for therapeutics targeting, 69
Neuronal membrane proteins, 6
Neuropsychiatric inventory (NPI), 206
Neurosin, 8
Neurotrophic factors, 283
  shortage of, 283
Neurotrophin receptors, 103
NFTs. See Neurofibrillary tangles (NFTs)
NGF. See Nerve growth factor (NGF)
Nicotine, 161
Nilotinib, 184
NMDA. See N-Methyl-D-aspartic acid (NMDA)
Nonamyloid component of Aβ plaques
    (NAC), 22
Non-ATG translation, 93
Noncoding region, 278
Nonneuronal cells, 281
  pathogenic role, 281
Nonsteroidal antiinflammatory drugs, 181
NPI. See Neuropsychiatric inventory (NPI)
NSCs. See Neural stem cells (NSCs)
Nuclear localization signal (NLS), 257
Nuclear membrane receptors, 10
Nucleation-dependent polymerization reaction, 28
Nucleocytoplasmic transport, 287
Nucleotide hydrolysis, 2
Nucleus basalis of Meynert (NBM), 187

O

OB. See Olfactory bulb (OB)
6-OHDA. See 6-Hydroxy-dopamine (6-OHDA)
Olfactory bulb (OB), 101

Oligodendrocyte, 6, 54
  dysfunction, 282
  progenitor cells, 84
Oligomerization, 26
Oral PD medication, 85

P

PA. See Phosphatidic acid (PA)
Paget disease of bone (PDB), 200
Parkin genes, 166
Parkinsonism dementia complex
  affected brain areas, 179
  neural pathways, 179
Parkinson's disease (PD), 1, 50, 60–63, 157
  aging, 158
  antineurodegenerative approach for, 164
  cause of, 158
  characterization by extrapyramidal motor deficits, 60
  current treatment approaches in, 158
    deep brain stimulation and cell replacement, 160
    nonpharmacological approaches, 160
    pharmacological approaches, 158
    replacing dopamine, 158
  development of disease-modifying therapy for, 161
  dopamine, 158
  drugs, 160–161
  IL6 promoter, 61
  L-DOPA therapy, 159
  microglial activation, 60
  motor symptoms, 85
  nonmotor symptoms, 85
  novel targets for disease-modifying therapies, 161
  postmortem studies of PD patients, 60
  potential targets, for disease modification, 162
  role for caffeine consumption, 61
  substantia nigra, 61
  symptomatic treatment of, 159
  symptoms, 157
Parkinson's disease dementia (PDD), 175. See also
    Frontotemporal dementia (FTD)
  clinical signs, 176
PARK2 iPSC-derived neurons, 95
Pathogenic protein aggregates
  propagation
    limiting steps, 6–7
    propagation routes, 5–6
  therapeutic strategies to restore damage, 9–10
Pattern recognition receptors (PRRs), 53
PCA. See Protein-fragment complementation assay (PCA)
PC12 cells, 184
PD. See Parkinson's disease (PD)
PDB. See Paget disease of bone (PDB)

PDD. *See* Parkinson's disease dementia (PDD)
Periglomerular neurons, 101
PGRN. *See* Progranulin (PGRN)
Phosphatidic acid (PA), 25
Phosphatidylserines (PS), 25
PiB. *See* Pittsburgh compound B (PiB)
PINK1 genes, 166
    mutation, 93
Pioglitazone, 160
Pittsburgh compound B (PiB), 181
PKA. *See* Protein kinase A (PKA)
Plasmin, 8
Plasminogen, 8
Pluripotency, 81
p38 mitogen-activated protein kinase, 184
Posttranslational modifications (PTM), 29, 32
PPA. *See* Primary progressive aphasia (PPA)
p62 protein, 204
Presenilin 1 (PSEN1), 106
Primary progressive aphasia (PPA), 200
    logopenic variant, 200
    nonfluent variant, 200
    semantic variant, 200
Prion-like spreading, 116
Prion protein (PrP), 2
Prion protein receptor (PrP receptor), 133
Progerin-induced aging, 92
Progranulin (PGRN), 91
Progressive supranuclear palsy (PSP), 176, 200
Protein aggregation, 1–2
    fibrillar assemblies, into model, 3
Protein assemblies
    intercellular propagation, 5
    mechanism, 2
    prion-like propagation, 4
Protein catabolism, 91
Protein-fragment complementation assay
        (PCA), 184
Protein homeostasis mechanism, 25
Protein kinase A (PKA), 221
    phosphorylation, 221
Protein misfolding, 1–2
    model, 3
Protein translocation, 5
Proteostasis, 7, 27
Protofibrils, 30
PrP. *See* Prion protein (PrP)
PrP receptor. *See* Prion protein receptor
        (PrP receptor)
PS. *See* Phosphatidylserines (PS)
PSEN1. *See* Presenilin 1 (PSEN1)
PSP. *See* Progressive supranuclear palsy (PSP)
PTM. *See* Posttranslational modifications (PTM)

R
rAAV. *See* Recombinant adeno-associated viral vector
        (rAAV)
Radial glia, 53
Radial glia-like cells, 102
RAN. *See* Repeat-associated non-ATG-initiated (RAN)
RBANS. *See* Repeatable battery for the assessment of
        neuropsychological status (RBANS)
RBP. *See* RNA-binding proteins (RBP)
Reactive oxygen species (ROS), 283
Recessive genes, 166–167
Recombinant adeno-associated viral vector (rAAV), 123, 186
Repeatable battery for the assessment of neuropsychological
        status (RBANS), 207
Repeat-associated non-ATG-initiated (RAN), 217
    translation, 217
Retinal pigment epithelium cells (RPECs), 86
Riluzole, 279
Rivastigmine, 183
RMS. *See* Rostral migratory stream (RMS)
RNA-binding proteins (RBP), 213
RNA-binding ribonucleoproteins, 203
    FUS, 203
    TDP-43, 203
RNAi. *See* RNA interference (RNAi)
RNA-induced silencing complex (RISC), 261
RNA interference (RNAi), 184, 261–262
Rodent subventricular zone, 102
ROS. *See* Reactive oxygen species (ROS)
Rostral migratory stream (RMS), 101
RPECs. *See* Retinal pigment epithelium cells (RPECs)

S
SALS. *See* Sporadic ALS (SALS)
SCI. *See* Spinal cord injury (SCI)
Secondary gliosis, 88
β-Secretase-cleaving enzyme, 119
γ−Secretase inhibitor, 95
Selective serotonin reuptake inhibitors (SSRI), 206
    as antidepressants, 206
sgRNA. *See* Single-guide RNA (sgRNA)
β-Sheet blockers, 35
Short-hairpin RNA (shRNA), 7, 261
Single-guide RNA (sgRNA), 83
Single nucleotide polymorphisms (SNPs), 177, 253
SNCA. *See* α-SYN encoding gene *(SNCA)*
SNCA DA neurons, 95
SNP. *See* Single nucleotide polymorphism (SNP)
SOD1. *See* Superoxide dismutase (SOD1)
SOD1D90V iPSC-derived motor neurons, 94
Sodium potassium ATPase, 9
Solanezumab, 86

SOX2 lentiviral overexpression, 84
Specific cell types generation, 84–85
Spinal cord injury (SCI), 89
Spinal cord motor neurons, 84
Sporadic ALS (SALS), 278
SSRI. *See* Selective serotonin reuptake inhibitors (SSRI)
Stem cells, 82–85
    genome editing, 83
    neurodegenerative diseases modeling, role in, 91–95
    regenerative therapy, role in, 85–90
        cell transplants, track of, 90
        clinical translation, use in, 88–90
        immunology, use in, 88–90
Steric zippers, 29
Substantia nigra, 85
Subventricular zone (SVZ), 101–103
Superoxide dismutase (SOD1), 2, 93
SVZ. *See* Subventricular zone (SVZ)
α-SYN. *See* α-Synuclein (α-SYN)
Synaptogenesis, 53, 91
α-SYN encoding gene *(SNCA)*, 22
αSyn protein targeting, 183–185
    clearance, 183–184
    extracellular propagation, 185
    oligomerization modulation, 184
α-Synuclein (α-SYN), 22–23, 105
    accumulation, 86
    aggregation, 28, 30
        Gaucher's disease, influence of, 34
        lipid-protein ratio, role in, 31
        membrane-induced, 30
    amphipathic character of, 23
    amyloid formation, 27, 28, 31
        oligomeric assemblies, 27
    β-sheet assemblies, 27, 28
    chaperone-mediated delivery of, 36
    conformational flexibility of, 23
    C-terminal region, 25
        truncation of, 26
    cytosolic states of, 23
    disease pathogenesis, role in, 26
    etiopathogenesis of heterogeneous diseases, role in, 22
    fibrillization of, 28
    folding stochastic process of, 27
    glycation of, age-associated posttranslational modification, 32–33
    helical curvature of, 24
    hydrophobicity of, 23
    membrane-bound, 23
    misfolded variants, 25–33
        fibrillar variants, 29–30
        folding intermediates, 25–27
        misfolding in complex environments, 30–31

        oligomeric variants, 27–28
            two-step nucleation model, 27
        synuclein strains, 32
    misfolding of
        membrane interaction, relationship between, 30
        targeting aggregation
            direct inhibition, aggregation, 36–37
            lowering expression levels, 35–36
        as therapeutic target, 33–38
            preserving native state, 33
            preventing formation of folding intermediates, 33–35
            stabilizing membrane interactions, 38
            targeting aggregation, 35–37
    native folding of, genetic and environmental factors affecting, 22
    native state, 22–23
    net negative charge of, 23
    nucleation
        primary, 31
        secondary, 29
    Parkinson's disease (PD), role in, 22
    partial folding of, 26
    physicochemical properties of, 23
    protein, 161–162
        amount of, 164
        biophysical characteristic of, 163
        homeostasis mechanism, folding monitoring, 25
        strain-like properties, 163
    protein targeting. *See* αSyn protein targeting
    N-terminal region, 25
    therapeutic interventions for reducing pathogenicity, 34
Synucleinopathies, 1, 22, 32, 35

T

TALEN. *See* Transcription activator-like effector nucleases (TALEN)
Tau aggregation, 4
Tau-hyperphosphorylation, 118
Tau kinases
    cyclin-dependent kinase-5 (Cdk5), 211
    glycogen synthase kinase-3 beta (GSK3β), 211
Tauopathies, 117, 128, 133
    therapeutic strategies under investigation for, 210
Tau phosphorylation, 202
    inhibition, 211
        lithium, role in, 212
        valproate, role in, 212
Tau-phosphorylation, 126
Tau splicing, 213
T cells, 282

Thal stages, 116
Therapeutic strategies, limitations, 11
TNTs. *See* Tunneling nanotubes (TNTs)
Tolcapone, 207
Transcription-activator like effector nucleases (TALEN),
    83, 262
Transcription factors *Oct4, Sox2, Klf4,* and *c-Myc*
    (OSKM), 82
Transcriptome, 91
Transforming growth factor beta 1 (TGFβ1), 52
Transgenic mice, 280
Transgenic models, 118
    amyloid pathology as preclinical models to identify targets
        for anti-Aβ therapies, 118–120, 123
        APP and APP/PS1 preclinical models, limitations
          of, 123
        APP and APP/PS1 transgenic models, 118, 121
        clearance of Aβ by immunotherapy, 121
        combination of anti-Aβ strategies and other
          approaches, 123
        secretases and their validation as therapeutic
          targets, 119
    mouse models. *See* Transgenic mouse models
    tau-pathology as preclinical models to identify targets for
        anti-tau-directed therapies, 124–125, 127
        microtubule-stabilizing agents, 126
        targeting intracellular tau-degradation, 127
        targeting tau
          by immunization therapy, 127
          phosphorylation and other posttranslational
            modifications of tau, 126
        tau-transgenic models recapitulating tau-pathology,
          124, 130
Transgenic mouse models, 128
    Aβ-to-tau axis, venues to identify therapeutic targets
        aiming at molecular and cellular mechanisms linking
          Aβ and tau, 132–136
        animal models of AD robustly recapitulate
          Aβ-induced tau-pathology, 132
        Aβ-induced inflammation as a potential mechanism
          of Aβ-induced tau-pathology, 135
        heterotypic seeding of tau-aggregation by
          preaggregated Aβ as a potential mechanism
          of Aβ-induced tau-pathology, 136
        molecular and cellular mechanisms of Aβ-induced
          tau-pathology, 133
        neuronal signaling pathways downstream of Aβ as
          a potential mechanism of Aβ-induced tau-
          pathology, 133
    prion-like spreading and propagation of tau-pathology,
        venues for therapeutic strategies, 128
        immunotherapy targeting extracellular misfolded forms
          of tau to inhibit prion-like propagation of tau-
          pathology, 131
        prion-like spreading of tau-pathology in preclinical
          tauopathy models in vivo, demonstration of,
          129
        targeting extracellular release and uptake of misfolded
          tau, acting as tau-seeds for prion-like
          propagation of tau-pathology, 129–131
        targeting tau-misfolding and aggregation to inhibit
          prion-like propagation of tau-pathology, 132
Translocons, 5
Transmembrane proteins
    type I, 116
Transmission electron microscopy, 5
Transthyretin tetramers, 33
Tremor bradykinesia, 157
Trojan horses of neurodegeneration, 185
Truncating mutations, 164
Tumor cells, immunological surveillance, 88
Tumor necrosis factor alpha (TNFα), 53
Tunneling nanotubes (TNTs), 6

U

Ubiquitin proteasome system (UPS), 7, 33, 127, 285
Umbilical cord blood, 83
Unfoldases, 7
Unified Huntington's disease rating scale (UHDRS), 251
Unified Parkinson's disease rating scale (UPDRS), 208
UPDRS. *See* Unified Parkinson's disease rating scale
    (UPDRS)
UPS. *See* Ubiquitin-proteasome system (UPS)
Urate, 161
US Food and Drug Administration (FDA), 279

V

Valosin-containing protein *(VCP)*, 202
Valproate, 211
Vascular cell adhesion molecule (VCAM1), 55
Vascular endothelial growth factor (VEGF), 284
VCP. *See* Valosin-containing protein *(VCP)*
VEGF. *See* Vascular endothelial growth factor (VEGF)

W

Watson-Crick base pairing, 260

Z

ZFN. *See* Zinc finger nuclease (ZFN)
Zinc finger nucleases (ZFN), 83, 262

Printed in the United States
By Bookmasters